SOLAR AND STELLAR CORONAL MASS EJECTIONS

IAU SYMPOSIUM 388

COVER ILLUSTRATION:

Hubble Space Telescope image of NGC 6822 (credit: NASA, ESA, and The Hubble Heritage Team (STScI/AURA); Acknowledgment: C. R. O'Dell (Vanderbilt University) and L. Bianchi (Johns Hopkins University and Osservatorio Astronomico, Torinese, Italy). The superposed artwork of a star is from science illustration by Chuck Carter (chuckcarterart@icloud.com)

IAU SYMPOSIUM PROCEEDINGS SERIES

INTERNATIONAL ASTRONOMICAL UNION

UNION ASTRONOMIQUE INTERNATIONALE

International Astronomical Union

IAU

SOLAR AND STELLAR CORONAL MASS EJECTIONS

PROCEEDINGS OF THE 388th SYMPOSIUM OF THE INTERNATIONAL ASTRONOMICAL UNION KRAKOW, POLAND 5–10 MAY 2024

Edited by

NAT GOPALSWAMY

NASA Goddard Space Flight Center, USA

OLGA MALANDRAKI

National Observatory of Athens IAASARS, Greece

ALINE VIDOTTO

Leiden Observatory, Netherlands

and

WARD MANCHESTER IV

University of Michigan, USA

CAMBRIDGE
UNIVERSITY PRESS

CAMBRIDGE UNIVERSITY PRESS
University Printing House, Cambridge CB2 8BS, United Kingdom
1 Liberty Plaza, Floor 20, New York, NY 10006, USA
10 Stamford Road, Oakleigh, Melbourne 3166, Australia

First published 2026

Printed in Great Britain by Henry Ling Limited, The Dorset Press, Dorchester, DT1 1HD

Typeset in System LaTeX 2ε

A catalogue record for this book is available from the British Library of Congress Cataloguing in Publication data

This journal issue has been printed on FSC™-certified paper and cover board. FSC is an independent, non-governmental, not-for-profit organization established to promote the responsible management of the world's forests. Please see www.fsc.org for information.

ISBN 9781009545884 hardback
ISSN 1743-9213

For EU product safety concerns, contact us at Calle de José Abascal, 56, 1°, 28003 Madrid, Spain, or email eugpsr@cambridge.org

Table of Contents

Preface

The IAU symposium 388 on *Solar and Stellar Coronal Mass Ejections* brought together solar and stellar researchers on coronal mass ejections (CMEs) under one roof during 5-10 May 2024 to discuss the commonalities and differences in understanding this energetic phenomenon. Solar CME researchers were challenged to think like stellar researchers and vice versa. About 140 people from 23 different countries attended the symposium, which was hosted by the Astronomical Observatory of the Jagiellonian University, Krakow, Poland. Thanks to the superb organization by Professor Grzegorz Michalek and his resourceful team, the symposium proceeded smoothly. We are extremely thankful for the outstanding technical and financial support provided by the host institute.

A one-day, pre-symposium school on the same topic to provide refresher to the students preparing them to better participation in the main symposium. The main program featured 83 oral talks and 57 posters grouped under 8 sessions followed by a panel discussion on future directions in instrumentation, observations, modeling, and solar-stellar collaborations. Almost all invited participants promptly accepted to give invited reviews and travelled to Poland. The contributed papers were of high quality, carefully selected by the scientific organizing committee. Solar and stellar topics were purposely interspersed to encourage cross-fertilization of ideas. Each session started with a review talk followed by one or two invited talks and remaining contributed papers. Extensive discussions took place following the talks and continued during dedicated poster sessions that helped solar and stellar CME researchers to tend to the same page.

CMEs are of great importance not only because of the intriguing physics involved in their origin but also how they impact planets around the Sun and stars. The scientific program consisted of these two aspects discussed in various sessions. The first day of the program started with the solar sources of CMEs, viz., closed magnetic regions followed by talks on CME initiation from such regions. The second day was devoted to the inter relationship among observed phenomena such as CMEs, flares, and eruptive prominences. Also discussed was the detection and modeling of stellar CMEs. Acceleration of energetic particles by CME-driven shocks and flares that have far-reaching consequences were discussed on the third day. As the CMEs start propagating into the solar and stellar environment, they interact with the ambient medium, modifying their physical properties. This was the subject of the fourth day including the generation of interplanetary radio bursts and gamma-rays. The fifth day was focused on the impact of CMEs on the planets surrounding the parent star. This included a detailed discussion on extreme events both modern and historical. The panel discussion had experts from various subfields with the panel members initiating discussion after brief presentations. In-depth discussions on various topics emphasized the commonality of CMEs representing energetic expulsions that a wide variety of stars produce. The symposium made it clear that a synergistic approach to solar and stellar CMEs is inevitable to understand how the magnetic energy is stored and released from spatially confined magnetic regions on the stars and their consequences for star-planet relations over various timescales.

This proceedings is a collection of papers presented at the symposium. The papers were all reviewed by competent colleagues, whose comments helped improve the presentation of many papers.

By all accounts, this symposium was a great success, which can be attributed to the hard work and dedication of the local organizing committee and the scientific organizing committee and quality presentations and discussions provided by the attendees.

July 2025
Nat Gopalswamy
Olga Malandraki
Aline Vidotto
Ward Manchester IV

Committees

Scientific Organizing Committee

Nat Gopalswamy, NASA/GSFC, USA (co-chair)
Olga Malandraki, National Observatory of Athens, Greece (co-chair)
Aline Vidotto, Leiden Observatory, Netherlands
Ward Manchester, U. Michigan, USA
Iver Cairns, U. Sydney, Australia
Piyali Chatterjee, Indian Institute of Astrophysics, India
Hebe Cremades, Universidad de Mendoza, Argentina
Mateja Dumbovic, U. Zagreb, Croatia
Christina Kay, Catholic University of America, USA
Kyoko Watanabe, National Defense Academy, Japan
Martin Leitzinger, U. Graz, Austria
Hiroyuki Maehara, NAOJ, Japan
Pertti Mäkelä, Catholic University of America, USA
Grzegorz Michalek, Jagiellonian Observatory, Poland
Dhani Herdiwijaya, Institut Teknologi Bandung, Indonesia
Alexander Nindos, U. Ioannina, Greece (ex-officio)
Susan Samwel, NRIAG, Egypt
Fang Shen, Chinese Academy of Sciences, China
Alphonse Sterling, NASA/MSFC, USA

Local Organizing Committee

Grzegorz Michalek, Astronomical Observatory, Jagiellonian U., Poland (chair)
Seiji Yashiro, Catholic University of America, USA
Pertti Mäkelä, NASA, Catholic University of America, USA
Vasanth Velusamy, Astronomical Observatory, Jagiellonian U., Poland
Ashutosh Pattnaik, Astronomical Observatory, Jagiellonian U., Poland

List of Participants

Airapetian Vladimir
Alterman Benjamin
Alvarado-Gomez Julian
Amerstorfer Tanja
Amerstorfer Ute
Awasthi Arun Kumar
Barnes Graham
Berlicki Arkadiusz
Bicz Kamil
Bilokon Oleksandr
Bucik Radoslav
Callingham Joseph
Carcaboso Fernando
Chen Hechao
Chen Yuehong
Chen Jun
Cremades Hebe
Dbrowski Bartosz
De Leo Yara
Desai Mihir
Devi Pooja
Dima Gabriel
Dissauer Karin
Dudnik Oleksiy
Effenberger Frederic
Evensberget Dag
Gandhi Harshita
Gopalswamy Nat
Green Lucie
Gunaseelan Sindhuja
Hayakawa Hisashi
Heber Bernd
Heinzel Petr
Herbst Konstantin
Hoerloeck Malte
Hu Huidong
Iglesias Francisco
Inoue Shun
Jardine Moira
Jin Meng
Joshi Bhuwan
Kansabanik Devojyoti
Kay Christina
Khuntia Soumyaranjan
Kilpua Emilia
Kliem Bernhard
Koleva Kostadinka
Konijn David
Koya Shafina
Kramar Maxim
Krupar Vratislav
Kumari Anshu
Kusano Kanya
Kuźma Błażej
Kyeremateng Kwabena
Lamy Philippe
Leitzinger Martin
Linton Mark
Litwicka Michalina
Lu Hong-Peng
MacCormack Cecilia
Maehara Hiroyuki
Magdalenic Jasmina
Maharana Anwesha
Maity Samriddhi Sankar
Makela Pertti
Malandraki Olga
Manchester Ward
Michalek Grzegorz
Mishev Alexander
Miyake Fusa
Mohamed Heba
Mohan Atul
Mrozek Tomasz
Mugatwala Ronish
Namekata Kosuke
Nandi Dibyendu
Nitta Nariaki
Notsu Yuta
Odert Petra
Osten Rachel
Pajdosz-Śmierciak Urszula
Pattnaik Ashutosh
Perri Barbara
Petrov Vasily
Poedts Stefaan
Raeside Shauna Rose
Reyes Paula
Rigney Jeremy
Rodgers-Lee Donna
Rukdee Surangkhana
Sangalli Mattia
Sarkar Ranadeep
Schmieder Brigitte
Shen Chenglong
Shen Fang
Sheoran Jyoti
Sieyra Maria Valeria
Sivabalasundar Aswin Amirtha Raj
Srivasrava Nnandita

Sterling Alphonse
Sun Zheng
Sun Xudong
Sylwester Janusz
Sylwester Barbara
Talpeanu Dana-Camelia
Tishler Noam
Toriumi Shin
Torok Tibor
Tristan Isaiah
Usoskin Ilya
Valentino Angelos
Vashishtha Nitin
Veluchamy Vasanth
Vemareddy P
Verbeke Christine
Veronig Astrid
Vida Krisztian
Vidotto Aline
Wang Rui
Waszewski Angelica
Welsch Brian
Wimmer-Schweingruber Robert F
Wollmann Jiri
Wyleżoł Patryk
Xie Hong
Xu Yu
Yang Shangbin
Yang Zihao
Yang Yi
Yashiro Seiji
Yuan Ding
Zhang Jie
Zhao Zhanhao
Zhukov Andrei
Łaguz Tomasz

UNIWERSYTET
JAGIELLOŃSKI
AUDITORIUM
MAXIMUM

Solar and Stellar Coronal Mass Ejections
Proceedings IAU Symposium No. 388, 2026
N. Gopalswamy, O. Malandraki, A. Vidotto & W. Manchester, eds.
doi:10.1017/S1743921324001674

Solar Sources of Flares and CMEs

Shin Toriumi

Institute of Space and Astronautical Science, Japan Aerospace Exploration Agency,
Sagamihara, Japan

Abstract. Strong solar flares and coronal mass ejections (CMEs) are prone to originate within and near active regions (ARs) with a high magnetic complexity. Therefore, to better understand the generation mechanism of flares and the resultant CME eruption and to gain insight into their stellar counterparts, it is crucial to reveal how solar flare-productive ARs are generated and developed. In this review, first, we summarize some general aspects of solar flares and key observational characteristics of such ARs. Then, we discuss a series of flux emergence simulations that were performed to elucidate the subsurface origins of their complexity and introduce state-of-the-art models that consider the effect of turbulent thermal convection. Future flare observations using SOLAR-C, a next-generation high-throughput extreme ultraviolet spectroscopy mission, are also discussed.

Keywords. Sun, Solar flare, Coronal mass ejection, Active region

1. Introduction

Solar flares, especially strong ones, which may develop into coronal mass ejections (CMEs), are known to emanate from active regions (ARs) (see Figure 1) (Toriumi & Wang 2019). Recent stellar observations have shown that even more massive events occur in late-type dwarf stars (for example, Maehara et al. 2012). However, with the current telescope capabilities, it is difficult to observe flare-hosting starspots in spatially resolved images. Therefore, in order to understand the solar and stellar magnetic activities common to these stars, it is important to deepen our knowledge of the solar ARs that produce flares and CMEs. Accordingly, this review summarizes the key observational features of flare-productive ARs and numerical modeling approaches to understand their formation mechanisms.

The remainder of this paper is organized as follows. Section 2 presents key aspects and observational features of flare-productive ARs. Sections 3 and 4 present the numerical modeling. In Section 5, we discuss future flare observations with the introduction of the next-generation mission SOLAR-C. Finally, Section 5 summarizes this review and provides perspectives for future studies.

2. Observations

In astronomy, the term *flare* describes the sudden amplification of electromagnetic waves over various wavelength ranges. Solar flares may be observed as brightening in the X-ray, ultraviolet (UV), visible, infrared, and radio bands. The energy released in a single event ranges from 10^{29} erg to 10^{32} erg, with time scales ranging from tens of minutes to hours and size scales ranging in the order of 10 Mm to 100 Mm.

From an energetics perspective, solar flares can be viewed as a phenomenon that releases excess magnetic energy, or free energy, which accumulates in the coronal magnetic field. Figure 2 shows the energy diagram for the pre-flare magnetic field $\mathbf{B}$ and the

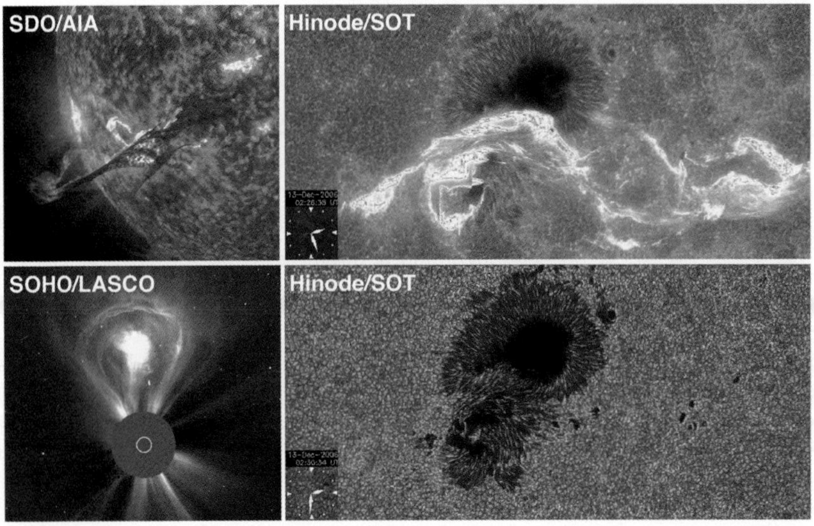

Figure 1. Collection of solar flare and CME observations. Clockwise from top left: Filament eruption captured by SDO/AIA (credit: NASA), flare ribbons and sunspots captured by Hinode/SOT (NAOJ/JAXA), and a light-bulb-shaped CME captured by SOHO/LASCO (NASA/ESA).

potential magnetic field $\mathbf{B}_{\rm pot}$, which is the minimum energy state. The free energy can be described as

$$\Delta E_{\rm mag} = E_{\rm mag} - E_{\rm pot} = \int \frac{B^2}{8\pi}\, dV - \int \frac{B_{\rm pot}^2}{8\pi}\, dV. \tag{1}$$

It is released through magnetic reconnection and plasma instability during a flare event, although only a (small) fraction of the stored free energy is released during a single event. During the flare, free energy is converted into thermal energy (hot plasma above 1 MK), kinetic energy (i.e., CME), and particle acceleration (solar energetic particles). This suggests that the magnetic field must be non-potential for a flare to occur.

A key observational feature of flare-productive ARs is the sheared polarity inversion line (PIL). Such PILs have been shown to have a strong horizontal field (>4000 G), a strong B_z gradient (~ 100 G Mm^{-1}), and a strong magnetic shear (80°–90°) (e.g., Severnyj 1958; Hagyard *et al.* 1984; Schrijver 2007), and are sometimes accompanied by sunspot rotation (Min & Chae 2009; Brown *et al.* 2003). Helical magnetic structures or flux ropes can often be observed above the PILs before an eruption. They are seen as filaments in Hα and other chromospheric lines and as sigmoids in soft X-rays and UV lines (Rust & Kumar 1996; Canfield *et al.* 1999; Pevtsov 2002). When it erupts, the flux rope evolves into the magnetic skeleton of the CME. All the above-mentioned structures indicate that the magnetic field is non-potential.

A representative statistical study on flaring ARs is Sammis *et al.* (2000), where it was shown that there is a positive correlation between the spot area and maximum flare magnitude in terms of soft X-ray flux.† Thus, one of the key conditions is the AR area, and this tendency may indicate that larger ARs can hold larger amounts of magnetic energy.

† It is pointed out by Fisher *et al.* (1998), Pevtsov *et al.* (2003), Toriumi & Airapetian (2022), and other studies that the quiescent X-ray flux also correlates with the AR area (or the total magnetic flux).

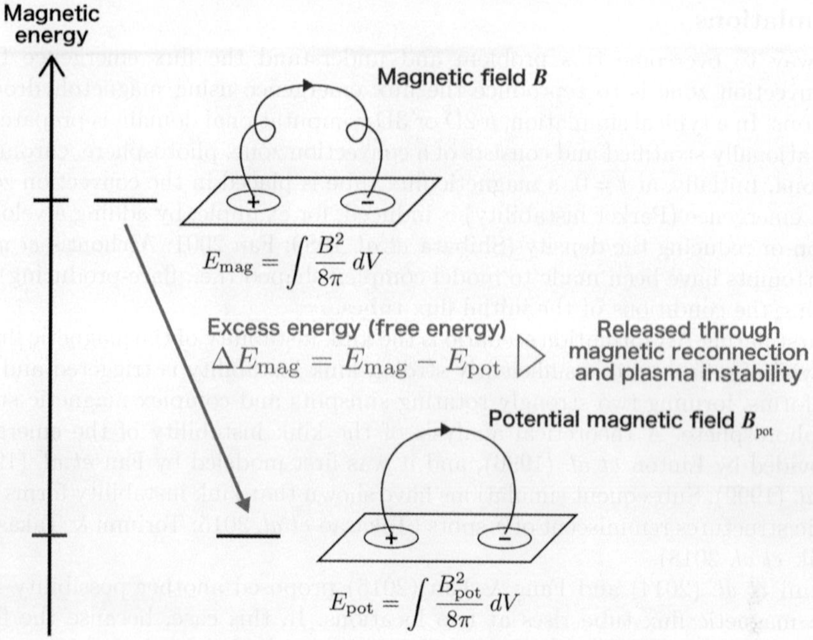

Figure 2. Energetics of solar flares. The differential energy between the two levels, the free energy ΔE_{mag}, is released through the flare. Note that only a fraction of the free energy is actually released in each event.

Sammis *et al.* (2000) also showed that the difference in sunspot configuration is also an important factor. Compared to the β-spots, which have a simple bipolar structure, δ-spots, in which umbrae of positive and negative polarities reside in a common penumbra, are prone to produce massive flares (Künzel 1960). A number of statistical studies (e.g., Mayfield & Lawrence 1985; Tian *et al.* 2002; Toriumi *et al.* 2017) support this finding. The importance of δ-spots as flare-prolific ARs is most evident in NOAA AR 5395, which was a typical δ-spot with numerous positive and negative umbrae within a massive penumbra (Wang *et al.* 1991). This AR produced more than 200 flares, including the X15-class event, and was responsible for the magnetic storm that caused a blackout in Quebec and damaged a transformer in New Jersey in March 1989. Therefore, the magnetic and morphological complexity of ARs is another indication, which is natural because the free energy accumulates in a non-potential magnetic field.

In addition, the rapid evolution of an AR is also an important factor, as magnetic energy must accumulate sufficiently faster than it dissipates (Sun & Norton 2017; Norton *et al.* 2022). The key elements of flare-producing ARs are summarized as follows:

- Area: The AR has a large amount of magnetic energy.
- Complexity: The AR has a large non-potentiality, i.e., large free energy.
- Rapid evolution: The accumulation of magnetic energy is sufficiently faster than its dissipation.

How do these complexly shaped ARs form? ARs are formed by the emergence of magnetic flux from the deep convection zone; then, what happens down there? It is impossible to observe the convection zone via direct optical observation, and helioseismology may not be sensitive enough to detect emerging fluxes with high spatial and temporal resolutions (for example, see Kosovichev 2009).

3. Simulations

One way to overcome this problem and understand the flux emergence from the deep convection zone is to reproduce the flux emergence using magnetohydrodynamic simulations. In a typical simulation, a 2D or 3D computational domain is prepared, which is gravitationally stratified and consists of a convection zone, photosphere, chromosphere, and corona. Initially, at $t = 0$, a magnetic flux tube is placed in the convection zone, and buoyant emergence (Parker instability) is induced, for example, by adding a velocity perturbation or reducing the density (Shibata *et al.* 1989; Fan 2001; Archontis *et al.* 2004). Many attempts have been made to model complex-shaped (i.e., flare-producing) ARs by controlling the conditions of the initial flux tubes.

The first flaring AR formation scenario is the kink instability of the magnetic flux tubes. If the twist of a flux tube is sufficiently strong, kink instability is triggered and the flux tube deforms, forming two strongly rotating sunspots and complex magnetic structures in the photosphere. A theoretical analysis of the kink instability of the emerging flux was provided by Linton *et al.* (1996), and it was first modeled by Fan *et al.* (1998) and Fan *et al.* (1999). Subsequent simulations have shown that kink instability forms complex magnetic structures reminiscent of δ-spots (Takasao *et al.* 2015; Toriumi & Takasao 2017; Knizhnik *et al.* 2018).

Toriumi *et al.* (2014) and Fang & Fan (2015) proposed another possibility in which a single magnetic flux tube rises at two locations. In this case, because the flux tube produces a double-arched (i.e., M-shaped) loop instead of a classical Ω-loop, two pairs of bipoles are formed in the photosphere. The positive and negative spots collide at the center of the domain, forming a densely packed δ-spot.

The third concept is the emergence of multiple magnetic flux tubes intertwined with each other. From the sunspot record over two solar cycles, Jaeggli & Norton (2016) found that the fraction of γ- and/or δ-spots increases during the solar maximum, and suggested the possibility that the pileup of emerging magnetic fluxes produces complex-shaped ARs. Simulation results by Jouve *et al.* (2018) showed that when magnetic flux tubes collide below the photosphere, depending on their mutual angle, the structure of the photospheric magnetic fields becomes complex.

However, these numerical simulations were highly idealized. In some models, the subsurface layer was just an adiabatically-stratified atmosphere; thus, no convection occurred. Even in the convective-flux-emergence simulations, flux tubes were kinematically inserted from the shallow bottom boundary for computational constraints (Cheung *et al.* 2010; Rempel & Cheung 2014). Thus, it remains unclear how deep large-scale thermal convection affects flux emergence.

4. Recent Attempts: R2D2 Simulations

To overcome this difficulty and reproduce realistic thermal convection in a 3D computational box that spans the entire convection zone, the Radiation and Reduced speed of sound technique (RSST) for Deep Dynamics, or R2D2, simulation code was developed by Hotta *et al.* (2019). R2D2 solves the magnetohydrodynamic equations with realistic radiation transfer and the equation of state, and implements RSST, enabling the reproduction of an inserted flux tube forming an AR in a self-consistent manner as a result of interaction with convective upflows/downflows in a deep domain (down to -200 Mm). Using R2D2, Toriumi & Hotta (2019) and Hotta & Toriumi (2020) simulated δ-spot formation via the collision of positive and negative polarities, as a single twisted flux tube emerges at two locations. The δ-spot is formed by two co-rotating sunspots, creating a sheared magnetic arcade or a flux rope-like structure above the PIL. This supports the concept of idealized simulations proposed by Toriumi *et al.* (2014) and Fang & Fan

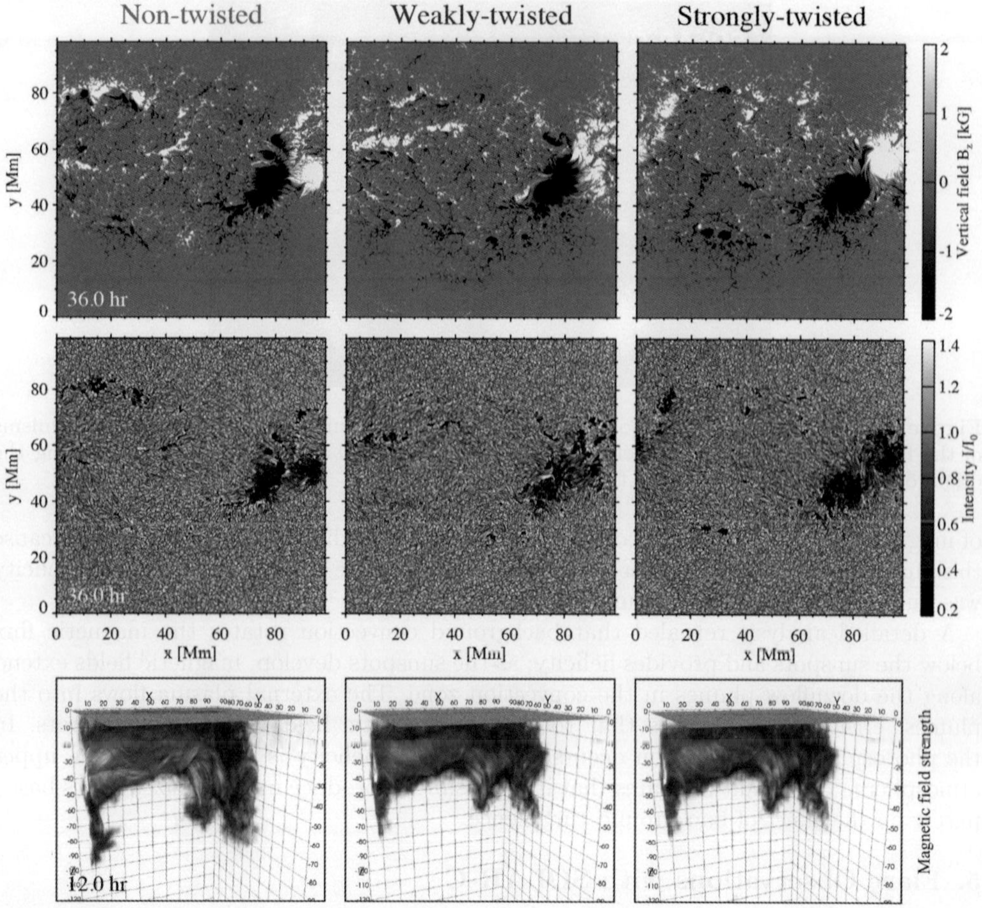

Figure 3. (Top) Vertical magnetic field strength (B_z) in the photosphere, (Middle) emergent intensity, and (Bottom) 3D volume rendering of the total magnetic field ($|\mathbf{B}|$) for the non-twisted ($q/q_{\rm cr} = 0$), weakly-twisted ($q/q_{\rm cr} = 1/4$), and strongly-twisted ($q/q_{\rm cr} = 1/2$) flux tube cases. Image reproduced with permission from Toriumi *et al.* (2023), copyright by the authors.

(2015) (Section 3) and indicates that such situations could occur in the actual Sun. In addition, Kaneko *et al.* (2022) found that the success or failure of emergence depends significantly on the initial position of the magnetic flux tube relative to the background convection.

Recently, Toriumi *et al.* (2023) investigated the difference in flux emergence by varying only the initial twist of the flux tubes while keeping the background convection identical for the three cases (Figure 3). The twist strength $q/q_{\rm cr}$ was varied in three cases: $q/q_{\rm cr} = [0, 1/4, 1/2]$, where $q_{\rm cr}$ is the critical twist strength for kink instability (Linton *et al.* 1996). Consequently, in all cases, the flux tube reached the photosphere and formed a bipolar AR. After the positive spot crossed the horizontal boundary (the $x = 0$ boundary in the top panels of Figure 3), the positive and negative spots collided head-on around $x = 90$ Mm owing to the periodic boundary condition, and eventually formed a δ-spot. There was little difference in the amount of magnetic flux appearing in the photosphere for the three cases. However, the magnetic helicity measured in the photosphere was greater for $q/q_{\rm cr} = 1/2$ than that for $q/q_{\rm cr} = 1/4$. The amount of injected relative magnetic helicity was also estimated by measuring the relative helicity flux through the photosphere. Interestingly, even in the un-twisted case (i.e., $q/q_{\rm cr} = 0$), a finite amount

Figure 4. SOLAR-C with its onboard telescope EUVST will unravel the formation mechanisms of the hot plasma and the Sun's effects on Earth and the solar system ultimately seeking the origin of the solar system and life (Credit: JAXA/NAOJ).

of magnetic helicity was detected, which explains the medium-sized solar flares. Because this simulation did not consider solar rotation, the finite injection of magnetic helicity was purely an effect of background convection.

A detailed analysis revealed that background convection rotates the magnetic flux below the sunspots and provides helicity; as the sunspots develop, magnetic fields extend along the downflow plumes in the convection zone. The external plasma flows into the plumes, creating local eddies that rotate the magnetic fields below the sunspots. In the photosphere, spot rotation occurs and magnetic helicity is injected into the upper atmosphere. This result indicates that it is possible to produce medium-sized flares based purely on the effect of background turbulence.

5. Flare Observations with SOLAR-C

This section focuses on the introduction to the solar flare observations to be performed by JAXA's next-generation solar-observing satellite, SOLAR-C (Figure 4), which is scheduled to be launched in 2028 (Shimizu *et al.* 2020). SOLAR-C will aim at two science objectives, namely, it will aim to

I. Understand how fundamental processes lead to the formation of the solar atmosphere and the solar wind; and
II. Understand how the solar atmosphere becomes unstable, releasing the energy that drives solar flares and eruptions.

SOLAR-C is equipped with the EUV High-throughput Spectroscopic Telescope (EUVST) onboard, whose three unique aspects are as follows:

(1) Seamless and broad temperature coverage: from $\log T = 4$ to 7;
(2) High spatial/temporal resolution: 0.4 arcsec, 0.5-s exposure; and
(3) High dispersion spectroscopy: velocity resolution of 2 km s^{-1}.

Magnetic reconnection is thought to be an important process in solar flares. However, it has been difficult to reveal these processes using the capabilities of previous telescopes. For example, the magnetic reconnection zone is much darker than the surrounding area, and the resolution of the telescopes is insufficient. By taking advantage of spectroscopy with high spatial and temporal resolutions, SOLAR-C will reveal how solar flares occur.

In particular, it answers the questions of how magnetic reconnection occurs fast enough to explain the observed flares and how the magnetic energy is stored and suddenly released. Repeated scanning of the entire AR over days also reveals how energy accumulates in the AR and how local changes destabilize the entire magnetic system.

The spectral coverage of EUVST is 17–21.5 nm (SW) and 46–122 nm (LW), which includes a variety of EUV lines that are sensitive to the chromospheric, transition-region, coronal, and flare temperature plasmas (0.02–15 MK). The field of view is designed as 280×280 arcsec^2, which is sufficient to cover regular-sized ARs.

SOLAR-C is an international mission with the participation of the USA (NASA) and European countries (ESA and the space agencies of Germany, France, Italy, Switzerland, and Belgium). These institutions will build the detectors, fabricate the grating, and lead the integration and verification of the EUVST components. The SOLAR-C project also plans the coordinated observation campaigns with NASA's MUSE mission (De Pontieu *et al.* 2020, target launch in 2027) and other ground-based facilities to follow the energy transport from the photosphere to the corona, which allows for detailed investigations of the occurrence mechanisms of solar flares and CMEs.

6. Summary and Perspectives

A series of flare observations have revealed that the characteristics of flare-prolific ARs can be summarized as area, complexity, and rapid evolution, which correspond to the amount of magnetic energy to be stored, the non-potentiality of the magnetic field, and fast energy storage, respectively. One future direction for exploration is to determine whether these features are also observed in starspots. Statistical studies have been conducted on starspot areas (for example, Notsu 2013). However, is it possible to obtain information on complexity?

Flux emergence simulations revealed that the strong twist of the emerging flux and the interaction between the fluxes contributed to AR complexity. However, these simulations are highly ideal, and realistic modeling that includes convection has become available in recent years, as represented by the R2D2 code. It was found that background convection alone can supply sufficient magnetic helicity to explain medium-sized solar flares. Future starspot modeling would also be of interest.

In addition, for future observational studies, the SOLAR-C satellite with a new ultraviolet spectrometer, the EUVST, is scheduled to be launched in 2028. SOLAR-C will cooperate with other solar-observing satellites and ground-based telescopes to reveal the energy storage in ARs and the energy releasing during flares. Collaboration with stellar XUV studies should also be promoted.

Acknowledgments

The author wishes to thank the organizers of the IAU Symposium No. 388 Solar and Stellar Coronal Mass Ejections for inviting him to the symposium. This work was supported by JSPS KAKENHI Grant Nos. JP20KK0072 (PI: S. Toriumi), JP21H01124 (PI: T. Yokoyama), and JP21H04492 (PI: K. Kusano).

References

Archontis, V., Moreno-Insertis, F., Galsgaard, K., *et al.* 2004, *A&A*, 426, 1047
Brown, D. S., Nightingale, R. W., Alexander, D., *et al.* 2003, *Solar Phys.*, 216, 79
Canfield, R. C., Hudson, H. S., McKenzie, D. E. 1999, *Geophysical Research Letters*, 26, 627
Cheung, M. C. M., Rempel, M., Title, A. M., Schüssler, M. 2010, *ApJ*, 720, 233
De Pontieu, B., Martínez-Sykora, J., Testa, P., *et al.* 2020, *ApJ*, 888, 3
Fan, Y. 2001, *ApJ* (Letters), 554, L111
Fan, Y., Zweibel, E. G., Linton, M. G., Fisher, G. H. 1998, *ApJ* (Letters), 505, L59

Fan, Y., Zweibel, E. G., Linton, M. G., Fisher, G. H. 1999, *ApJ*, 521, 460
Fang, F., Fan, Y. 2015, *ApJ*, 806, 79
Fisher, G. H., Longcope, D. W., Metcalf, T. R., Pevtsov, A. A. 1998, *ApJ*, 508, 885
Hagyard, M. J., Smith, J. B., Jr., Teuber, D., West, E. A. 1984, *Solar Phys.*, 91, 115
Hotta, H., Iijima, H., Kusano, K. 2019, *Science Advances*, 5, 2307
Hotta, H., Toriumi, S. 2020, *MNRAS*, 498, 2925
Jaeggli, S. A., Norton, A. A. 2016, *ApJ* (Letters), 820, L11
Jouve, L., Brun, A. S., Aulanier, G. 2018, *ApJ*, 857, 83
Kaneko, T., Hotta, H., Toriumi, S., Kusano, K. 2022, *MNRAS*, 517, 2775
Knizhnik, K. J., Linton, M. G., DeVore, C. R. 2018, *ApJ*, 864, 89
Kosovichev, A. G. 2009, *Space Sci. Revs*, 144, 175
Künzel, H. 1960, *AN*, 285, 271
Linton, M. G., Longcope, D. W., Fisher, G. H. 1996, *ApJ*, 469, 954
Maehara, H. Shibayama, T., Notsu, S., *et al.* 2012, *Nature*, 485, 478
Mayfield, E. B., Lawrence, J. K. 1985, *Solar Phys.*, 96, 293
Min, S., Chae, J. 2009, *Solar Phys.*, 258, 203
Norton, A. A., Levens, P. J., Knizhnik, K. J., *et al.* 2022, *ApJ*, 938, 117
Notsu, Y., Shibayama, T., Maehara, H. 2013, *ApJ*, 771, 127
Pevtsov, A. A. 2002, *Solar Phys.*, 207, 111
Pevtsov, A. A., Fisher, G. H., Acton, L. W., *et al.* 2003, *ApJ*, 598, 1387
Rempel, M., & Cheung, M. C. M. 2014, *ApJ*, 785, 90
Rust, D. M., Kumar, A. 1996, *ApJ* (Letters), 464, L199
Sammis, I., Tang, F., Zirin, H. 2000, *ApJ*, 540, 583
Schrijver, C. J. 2007, *ApJ* (Letters), 655, L117
Severnyj, A. B. 1958, *Izvestiya Ordena Trudovogo Krasnogo Znameni Krymskoj Astrofizicheskoj Observatorii (Bulletin of the Crimean Astrophysical Observatory)*, 20, 22
Shibata, K., Tajima, T., Steinolfson, R. S., Matsumoto, R. 1989, *ApJ*, 345, 584
Shimizu, T., Imada, S., Kawate, T., *et al.* 2020, *Proceedings of the SPIE*, 11444, 114440N
Sun, X. & Norton, A. A. 2017, *Research Notes of the American Astronomical Society*, 1, 24
Takasao, S., Fan, Y., Cheung, M. C. M., Shibata, K. 2015, *ApJ*, 813, 112
Tian, L., Liu, Y., Wang, J. 2002, *Solar Phys.*, 209, 361
Toriumi, S. & Airapetian, V. S. 2022, *ApJ*, 927, 179
Toriumi, S., Hotta, H. 2019, *ApJ* (Letters), 886, L21
Toriumi, S., Hotta, H., Kusano, K. 2023, *Scientific Reports*, 13, 8994
Toriumi, S., Iida, Y., Kusano, K. *et al.* 2014, *Solar Phys.*, 289, 3351
Toriumi, S., Schrijver, C. J., Harra, L. K., Hudson, H., Nagashima, K. 2017, *ApJ*, 834, 56
Toriumi, S., Takasao, S. 2017, *ApJ*, 850, 39
Toriumi, S., Wang, H. 2019, *Living Reviews in Solar Physics*, 16, 3
Wang, H., Tang, F., Zirin, H., Ai, G. 1991, *ApJ*, 380, 282

Discussion

Jie Zhang: Thanks for a very nice review talk. In your opinion or based on your experience, what is the dominant mechanism that produces twists in the corona leading to solar eruption? To be more specific, is the twist mainly from the bodily emergence of twisted flux tubes from the sub-photosphere, or is it mainly from the cancellation/shearing motion on the photospheric surface?

Shin Toriumi: Thank you for your positive comments and interesting question. I think that both mechanisms are possible. However, my simulation results show that while the twist of the flux tube itself is important, the external flow field of thermal convection, which probably causes cancellation/shearing in the photosphere, makes a non-negligible, or perhaps significant, contribution.

Andrei Zhukov: You rightly placed emphasis on δ-configurations, as we know from observations that they are very flare-prolific. However, I would like to understand why is it the δ that is so important? Why is it important that the two strong magnetic polarities are engulfed by the same penumbra? Why does the penumbra matter in comparison with the situation of two strong magnetic polarities adjacent to each other but without a common penumbra?

Shin Toriumi: It is just empirically known that δ-spots produce strong events. However, there may be several reasons for this. From my perspective, if an AR has δ-spots, the chances that the positive and negative polarities are very close or that the penumbra is large (i.e., the AR harbors a large amount of free energy) increase.

Xudong Sun: You mentioned the possibility to try realistic flux emergence models for starspots. What would be the targets?

Shin Toriumi: The easiest task would be to make the magnetic flux tubes thicker in order to create starspots with larger amounts of magnetic flux. It would also be interesting to attempt flux emergence simulations for non-G-type stars by changing the background stratification profile.

Solar and Stellar Coronal Mass Ejections
Proceedings IAU Symposium No. 388, 2026
N. Gopalswamy, O. Malandraki, A. Vidotto & W. Manchester, eds.
doi:10.1017/S1743921324001881

Characterization of the source regions of CMEs with polar position angles as seen from Earth

Hebe Cremades[1], Francisco A. Iglesias[1], Luciano A. Merenda[2], Fernando M. López[1] and Diego G. Lloveras[1]

[1]Grupo de Estudios en Heliofísica de Mendoza, CONICET, Universidad de Mendoza, Mendoza, Argentina
email: hebe.cremades@um.edu.ar

[2]Facultad de Ciencias Exactas y Naturales, Universidad Nacional de Cuyo, Mendoza, Argentina

Abstract. Despite the proliferation of solar and heliospheric observations, many questions regarding the nature and evolution of coronal mass ejections (CMEs) persist. We focus on exploring relationships between the three-dimensional characteristics of CMEs and their source region properties. We analyze the sources of a set of CMEs having polar position angles as seen from Earth, for which their morphological characteristics can be estimated with less uncertainties than typically possible. We rely on observations of different solar atmospheric regimes, provided by the STEREO, SOHO, and SDO missions. In the low corona and chromosphere, we characterize associated filaments and eruptive phenomena; while in the photosphere we investigate polarity inversion lines. The joint analysis of source regions and CMEs enables studying how the attributes of source regions impact the morphological and dynamical characteristics of CMEs, with implications for space weather forecasting and for the understanding of processes involved in the initiation and early evolution of CMEs.

Keywords. Sun: coronal mass ejections, Sun: atmosphere, Sun: activity

1. Introduction

Coronal mass ejections (CMEs) are undoubtedly the most spectacular but also threatening solar events. Their potential to disrupt technology and harm humans in space (e.g. Schwenn 2006; Bothmer and Daglis 2007) calls for a comprehensive understanding not only of CMEs themselves, but also their solar sources, related phenomena, and processes involved in their initiation and evolution. Once a CME erupts, it is of utmost importance to predict whether it will impact Earth (or any other object of interest in the heliosphere), when will that happen, and how magnetic fields will be organized within them. A proper assessment of three-dimensional (3D) CME morphology is crucial with regard to those three aspects (Cremades 2018). Multi-viewpoint coronagraphic imaging has played a vital role in characterizing CME morphology, although with limitations imposed by the nature of this type of observations (see, e.g. Verbeke et al. 2023).

To identify CME events for which uncertainties in their 3D reconstruction can be minimized, Cremades et al. (2020) proposed a specific selection criterion (see next section). This enabled a careful analysis of the evolution of CME morphology in the first $\approx 8\ R_\odot$ for 12 events. Among other findings, they measured an average ratio of angular width in the broadside direction to that in the axial direction of ≈ 1.8. The value nicely compares with the one found by previous studies on CME morphology based on single-viewpoint coronagraph observations: ≈ 1.6 (St. Cyr et al. 2004; Cremades 2005). The latter also found the same ratio for incipient CMEs in the low corona.

In the present study, we characterize the source regions of the 12 CME events carefully studied by Cremades et al. (2020), with the ultimate goal of understanding how their 3D attributes impact the morphological and dynamical characteristics of CMEs. To this end, we rely on observations of different regimes of the solar atmosphere, provided by the STEREO, SOHO, and SDO missions. In the photosphere we investigate polarity inversion line (PIL) lengths and tilts, while in the chromosphere we characterize the associated filaments. In the low corona, we measure attributes of post-eruptive arcades related to the eruptions. The connection of these source region characteristics with the properties of the ensuing CMEs are presented and discussed.

2. The set of twelve analyzed CMEs

The set of twelve CMEs that inspires this study is special in that 3D reconstructions can be achieved with less uncertainties than typically possible. As demonstrated by Cabello et al. (2016), a proper combination of multiple spacecraft location and CME propagation direction is key to better constrain the 3D properties of CMEs. For the past almost two decades, all space-borne coronagraphs providing multiple vantage points have lied on the ecliptic plane, which is not enough to achieve a reliable 3D reconstruction of a CME propagating close to the same plane. Therefore, in the case of such a CME, a well-constrained 3D reconstruction would require at least one viewpoint out of the ecliptic (which will be soon the case with Solar Orbiter's increasing latitude). However, under recent and current conditions, well-constrained 3D reconstructions can be achieved for CMEs propagating at relatively high latitudes during spacecraft quadrature periods. More details on the criteria used to identify the twelve CMEs can be found in Cremades et al. (2020).

One of the important properties of the set of 12 CMEs is that their 3D reconstructions have been performed for various time stamps, including the first instances of evolution in the low corona. The method used for the 3D reconstruction is the graduated cylindrical shell (GCS) forward model (Thernisien et al. 2006, 2009), which enables the fitting of an ad-hoc geometrical shape that resembles a croissant, to the outer shell of a CME as simultaneously viewed from multiple perspectives. The success of the GCS technique in concurrently matching the multiple views of the analyzed events, from birth and until $\approx 8\ R_{\odot}$, implies that CMEs can be significantly wider in one direction (that of their main symmetry axis, lateral, broadside, or face-on view) than in the other (axial or edge-on view). Moreover, Cremades et al. (2020) found that the height evolution of the angular width in orthogonal directions follows the same behavior, though at a different pace. They also found that the angular width in the lateral direction (AW_L) grows faster than that in the axial direction (AW_D), and the difference between the growth rates is much larger in the first $\approx 3\ R_{\odot}$, surpassing the $10^{\circ}/R_{\odot}$. For most events, after $\approx 4\ R_{\odot}$ both expansion rates and their difference drop to become negligible. The ratio between angular widths in orthogonal directions, AW_L/AW_D, shows steep variations for individual events in the first two solar radii, stabilizing at different heights ranging $1.3-2.7\ R_{\odot}$with an average AW_L/AW_D of ≈ 1.8.

3. Relationship between properties of source regions and CMEs

In this section, we examine the properties of the source regions where the set of 12 CMEs originate, and how these are related to the 3D characteristics of the CMEs. As mentioned above, source region length is determined for the associated PIL in the photosphere, filaments in the chromosphere, and post-eruptive arcades (PEAs) in the low corona. Source region tilts are also determined from photospheric, chromospheric, and low coronal data. Finally, PEA separation and propagation speeds are measured in low coronal data. Relationships of these quantities with CME characteristics arising from the 3D reconstructions are presented for each case in the following subsections.

3.1. *Source region length*

The length of source regions was characterized in three different regimes: PILs in the photosphere, filaments in the chromosphere, and PEAs in the low corona. The lengths of PILs were determined as part of a procedure used to characterize magnetic tongues (see Poisson et al. 2015), which is also used to determine the tilt of PILs (Section 3.2). Filament lengths were measured on 304 Å images provided by the EUV telescopes SDO/AIA and/or STEREO/SECCHI EUVI (depending on source region location), as the distance between the farthest footpoints of the erupting prominence (see Figure 1a). An erupting prominence could be found for all twelve events, regardless of whether they originated in active or quiescent regions. PEA length is measured in 193/195 Å EUV images (from SDO/AIA or STEREO/SECCHI EUVI depending on PEA location), as the average extent of the arcade footpoints at both sides (see Figure 1b). Footpoints were measured for all events but one, which did not exhibit a PEA in connection with the eruption.

Since a flux rope would have its main symmetry axis aligned with its source region neutral line, it is expected that there is some relationship between source region length and AW_L. This is however not the case, but longer PEAs do tend to be associated with CMEs with smaller AW_L/AW_D ratios, *i.e.* less asymmetric CMEs (Figure 2). Given that long PEAs are related to long PILs, typically the case of high-latitude decaying bipolar regions with long, quiescent filaments, this result can be interpreted in terms of the energy budget available for the eruption, which for that type of regions is presumably scarce. CMEs originating in such regions would not expand much in the lateral direction, thus yielding less asymmetric CMEs.

3.2. *Source region tilt*

As for the case of lengths, source region tilts are characterized in the photospheric, chromospheric, and low coronal regimes. PILs were measured in SDO/HMI data only for front-sided active regions, by means of the technique developed by (Poisson et al. 2015). Tilts of the associated filaments are obtained as part of the measurements performed to measure filament length in EUV 304 Å data. PEA tilts also arise from the characterization in EUV 193/195 Å data addressed in Section 3.1. The resulting tilt values of the three types of source region features are plotted in Figure 3, as a scatter plot against the CME tilt yielded by the GCS reconstruction. Discrepancies are evident not only between CME and source region tilts, but also between tilts measured for the different source region features. The first type of discrepancy can be mainly attributed to rotations during the process of eruption. Discrepancies among the different source region features measured for each event can differ for various reasons. For instance, filament tilts were measured for the portion of filament that erupted, whose end points may differ from those of the full PIL; or PEAs can extend beyond the former location of the erupted filament and follow a bent path, because of series of reconnections that might take place after the eruption. It is worth noting that groups of points to the top-left and bottom-right of the plot represent high tilts (with 90° being vertical), and even if the x- and y-axes have

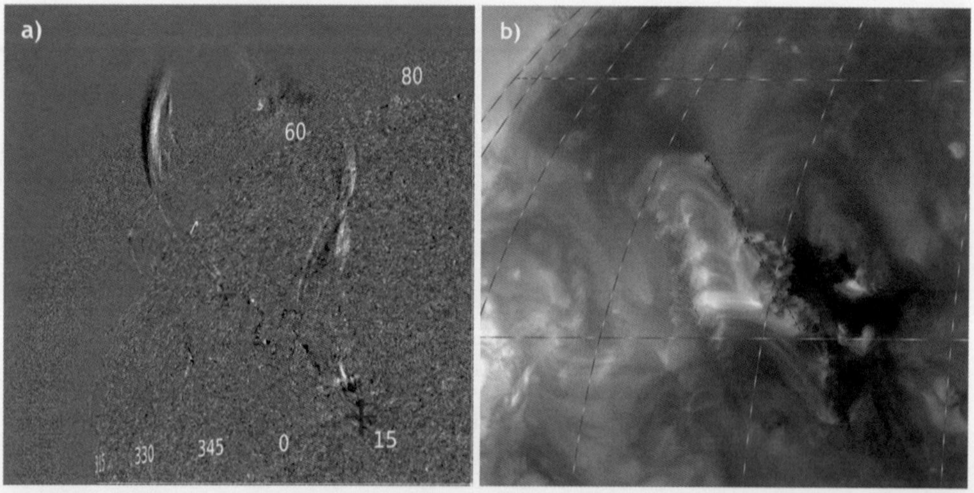

Figure 1. Example of coordinate measurements of source region phenomena associated with the eruption on 23 January 2013. a) Farthest footpoints of the erupting prominence, indicated by red and blue crosses in a running-difference image of SDO/AIA 304 Å at 14:00 UT. b) PEA selected footpoints (red and blue) overlaid on an SDO/AIA 193 Å image of the PEA at 14:49 UT. The red and blue dashed lines are linear fits to the footpoints at either side of the arcade, and their average is considered as the extent of the arcade.

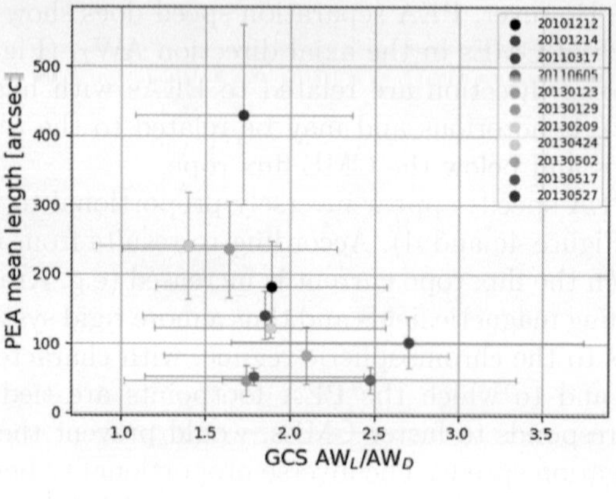

Figure 2. PEA mean length *vs.* AW_L/AW_D ratio arising from the GCS reconstructed parameters.

opposite sign values for those two groups of points (e.g. -75° and 75°), those tilts do not differ greatly from each other.

3.3. *Coronal properties*

The dynamics of PEAs can be characterized by repeating the procedure introduced in Section 3.1 (also see Figure 1b) for various times during PEA evolution after the eruptions. Inspired by the work of Tripathi et al. (2006); Li and Zhang (2009), we determined the propagation and separation speeds of PEAs, namely the speed at which PEAs propagate along their main axis, and the speed at which footpoints separate orthogonally from the main axis. Both speeds have been found to range between 0 − 15 km s^{-1} for all analyzed events. These values are in agreement with the mentioned analyses by Tripathi et al. (2006); Li and Zhang (2009). In analogy with what was mentioned in

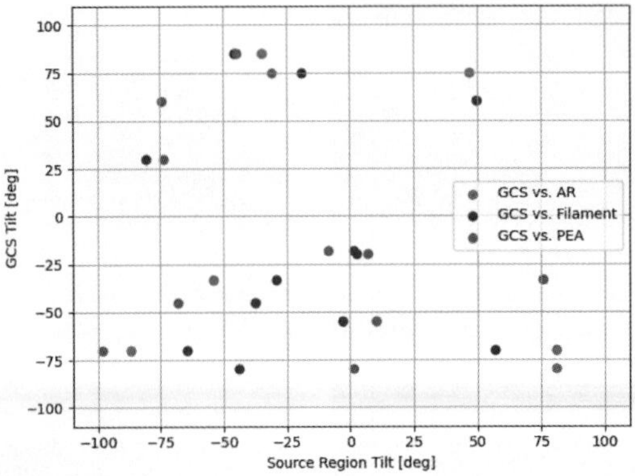

Figure 3. CME tilt from GCS reconstructions *vs.* tilts determined from photospheric magnetic fields (red), chromospheric filaments (blue), and PEAs in the low corona (green).

Section 3.1, since a flux rope would have its main symmetry axis aligned with the central spine of the associated PEA, a consistency between PEA propagation speed and CME lateral expansion speed is expected, as well as between PEA separation speed and CME axial expansion speed. However, no direct relationships were found between PEA speeds and CME expansion speeds. Following a similar reasoning, the PEA propagation speed is expected to hold a correlation with the final value of AW_L, but this is not clear from Figure 4a). However, PEA separation speed does show some correlation with the final angular width of CMEs in the axial direction AW_D (Figure 4b). The fact that wider CMEs in the axial direction are related to PEAs with higher separation speeds (except for an outlier), is notorious and may be related to the reconnection rate above the post-eruptive loops and below the CME flux rope.

In addition, both PEA speeds appear inversely proportional to the radial propagation speed of the CMEs (Figure 4c and d). According to results from numerical simulations, faster CMEs arise when the flux rope current is increased (e.g. Krause et al. 2018), which yields higher surrounding magnetic fields and thus a more rigid system. We argue that this constrain also extends to the chromospheric regime, with characteristic speeds governed by the sound speed, and to which the PEA footpoints are tied. Therefore, the more rigid system that corresponds to faster CMEs, would prevent the PEAs to reach faster separation and propagation speeds. The inverse proportionality between PEA speeds and CME radial propagation speed is clearer at low coronal heights (*e.g.* at $\sim 2\ R_{\odot}$, as in Figure 4c and d), while at larger distances it dilutes, possibly due to the influence of the background coronal structure on CME kinematics.

4. Summary and Outlook

In this work we have examined various properties of the source regions of a particular set of CMEs that have been reconstructed in 3D by Cremades et al. (2020) with less uncertainties than typically possible. Correlations between source regions and CME properties are not systematically observed; however some interesting relationships were found, that are promising toward the goal of early characterization of CMEs on the basis of their source region properties. The observed relationships are as follows: (i) source regions exhibiting longer PEAs tend to be related with CMEs having lower values of AW_L/AW_D (i.e. less asymmetric CMEs), (ii) PEA separation speed shows a correlation with the angular width in the axial direction AW_D, and (iii) PEA separation and propagation speeds appear inversely proportional to the radial propagation speed of the CMEs arising from the 3D reconstruction.

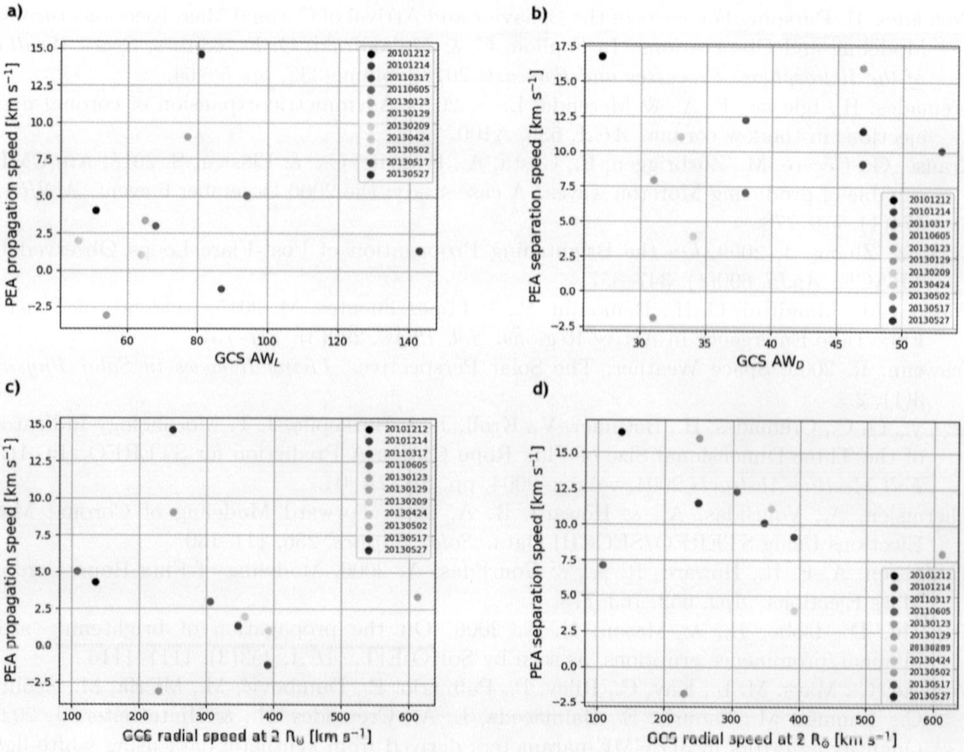

Figure 4. (a) PEA propagation speed *vs.* CME angular width in the axial direction (b) PEA separation speed *vs.* CME angular width in the axial direction. (c) PEA propagation and (d) separation speeds *vs.* CME radial speed at 2 R$_\odot$.

The very distinct nature of the events here analyzed calls for an analysis on a case-by-case basis. This is required to understand the key factors that play a role in determining a more (or less) asymmetric expansion in CMEs, also considering multi-causal relationships. Moreover, the limited number of analyzed events may mask existing correlations, due to the differing characteristics of these events. We are currently working on extending the sample to include around 20 more events, which will significantly enrich this analysis. Finally, we are expectant to analyze low-latitude events observed in the near future by the Solar Orbiter mission, which is to enable better-constrained 3D reconstructions for low-latitude propagating events.

Acknowledgements. HC, FAI & FML are researchers of "Carrera del Investigador Científico" of CONICET. HC and FAI are thankful to the Organizing Committees of IAUS388 for the provided support that ensured participation in the Symposium. The authors thank Dr. M. Poisson for providing his code for measurement of magnetic tongues, an anonymous reviewer for valuable comments and M. Cécere for fruitful discussions. This work is supported by PICT-2021-GRF-TI-00803 (Agencia I+D+i), PIP 11220200102710CO (CONICET), and co-funded by the European Union through the Dynasun project.

References

Bothmer, V. & Daglis, I. A. 2007,. *Space Weather – Physics and Effects.*

Cabello, I., Cremades, H., Balmaceda, L., & Dohmen, I. 2016, First Simultaneous Views of the Axial and Lateral Perspectives of a Coronal Mass Ejection. *Solar Physics*, 291, 1799–1817.

Cremades, H. 2005,. *Three-Dimensional Configuration and Evolution of Coronal Mass Ejections.* PhD thesis, Technische Universität Carolo-Wilhelmina zu Braunschweig, Brauschweig, Germany. ISBN-13: 978-3936586404.

Cremades, H. Pursuing Forecasts of the Behavior and Arrival of Coronal Mass Ejections through Modeling and Observations. In Foullon, C. & Malandraki, O. E., editors, *Space Weather of the Heliosphere: Processes and Forecasts* 2018, volume 335, pp. 58–64.

Cremades, H., Iglesias, F. A., & Merenda, L. A. 2020, Asymmetric expansion of coronal mass ejections in the low corona. *A&A*, 635, A100.

Krause, G., Cécere, M., Zurbriggen, E., Costa, A., Francile, C., & Elaskar, S. 2018, Are CMEs capable of producing Moreton waves? A case study: the 2006 December 6 event. *MNRAS*, 474(1), 770–778.

Li, L. & Zhang, J. 2009, On the Brightening Propagation of Post-Flare Loops Observed by TRACE. *ApJS*, 690(1), 347–357.

Poisson, M., Mandrini, C. H., Démoulin, P., & López Fuentes, M. 2015, Evidence of Twisted Flux-Tube Emergence in Active Regions. *Sol. Phys.*, 290(3), 727–751.

Schwenn, R. 2006, Space Weather: The Solar Perspective. *Living Reviews in Solar Physics*, 3(1), 2.

St. Cyr, O. C., Cremades, H., Bothmer, V., Krall, J., & Burkepile, J. T. Morphology Indicators of the Three-Dimensional Size of Flux Rope CMEs: A Prediction for STEREO. In *AGU Fall Meeting Abstracts* 2004, volume 2004, pp. SH22A–04.

Thernisien, A., Vourlidas, A., & Howard, R. A. 2009, Forward Modeling of Coronal Mass Ejections Using STEREO/SECCHI Data. *Solar Physics*, 256, 111–130.

Thernisien, A. F. R., Howard, R. A., & Vourlidas, A. 2006, Modeling of Flux Rope Coronal Mass Ejections. *ApJ*, 652, 763–773.

Tripathi, D., Isobe, H., & Mason, H. E. 2006, On the propagation of brightening after filament/prominence eruptions, as seen by SoHO-EIT. *A&A*, 453(3), 1111–1116.

Verbeke, C., Mays, M. L., Kay, C., Riley, P., Palmerio, E., Dumbović, M., Mierla, M., Scolini, C., Temmer, M., Paouris, E., Balmaceda, L. A., Cremades, H., & Hinterreiter, J. 2023, Quantifying errors in 3D CME parameters derived from synthetic data using white-light reconstruction techniques. *Advances in Space Research*, 72(12), 5243–5262.

Solar and Stellar Coronal Mass Ejections
Proceedings IAU Symposium No. 388, 2026
N. Gopalswamy, O. Malandraki, A. Vidotto & W. Manchester, eds.
doi:10.1017/S1743921324001807

Magnetic Decay Index Profile and Coronal Mass Ejection Speed

Bernhard Kliem[1,2], Georgios Chintzoglou[3], Tibor Török[4] and Jie Zhang[2]

[1]Institute of Physics & Astronomy, University of Potsdam, 14476 Potsdam, Germany
[2]George Mason University, Fairfax, VA, USA
[3]Lockheed Martin Solar & Astrophysics Laboratory, Palo Alto, CA, USA
[4]Predictive Science, Inc., San Diego, CA, USA

Abstract. We study the relationship between the speed of coronal mass ejections (CMEs) and the height profile of the ambient magnetic field, quantified by its decay index, $n(h)$. Our sample is composed of 15 very fast CMEs ($V_{\rm CME} \geq 1500$ km s^{-1}; all halo CMEs) and 22 halo CMEs below this speed limit from Solar Cycle 23. The very fast CMEs yield a high correlation of 0.81 between $V_{\rm CME}$ and the slope of $n(h)$ in a height range above the onset height of the torus instability if one extremely fast outlier, which closely followed another very fast CME, is excluded. This is consistent with the hypothesis that the torus instability plays a decisive role in CME acceleration. The whole sample yields a weaker correlation, which is still significant if events with a broad torus-stable dip in $n(h)$ are excluded. A parametric simulation study of flux-rope eruptions from quadrupolar and two-scale bipolar source regions confirms the decelerating effect of such dips. Very fast, moderate-velocity, and confined eruptions are found.

Keywords. Solar coronal mass ejections, Solar magnetic fields, Magnetohydrodynamics

1. Introduction

The speed of coronal mass ejections (CMEs), $V_{\rm CME}$, is a major factor in controlling their geoeffectiveness. It is decisive for the formation of CME-driven shocks, which may produce large solar energetic particle events, strongly influences (joint with the frontside field direction and ram pressure) the strength of the geomagnetic perturbation in cases of direct interaction with the magnetosphere, and is correlated with the magnitude of the usually associated flare (Srivastava and Venkatakrishnan 2004; Gopalswamy *et al.* 2008; Zhang *et al.* 2021). Most fast CMEs ($V_{\rm CME} > 700$km s^{-1}) are fully accelerated in the inner corona, within $h < 1R_\odot$ above the photosphere (MacQueen and Fisher 1983; Zhang and Dere 2006). Therefore, the capability to predict the coronal velocity of CMEs is of great practical interest.

Source-region parameters that were found to correlate with CME speed mostly refer to the photospheric magnetic field and can be divided into *magnitude parameters* such as magnetic flux, area, average field strength, *structural parameters* such as the number of polarity inversion lines (PILs), a measure of complexity (Wang and Zhang 2008), the effective distance of the main polarities (Guo *et al.* 2006), and the change of the shear angle at the PIL during the eruption (Su *et al.* 2007), and *combined parameters* that scale with both the magnitude of the source region and its structural properties, such as the free magnetic energy and helicity of the source region (Liu 2007b; Sung *et al.* 2009), PIL length (Wang and Zhang 2008), and non-neutralized current (Kontogiannis *et al.* 2019;

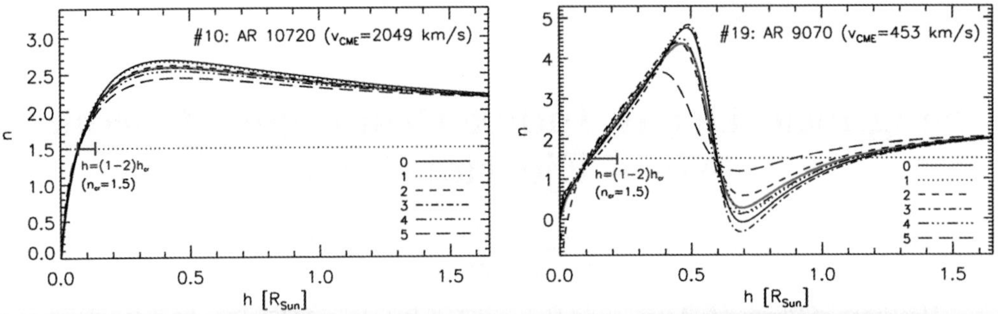

Figure 1. Decay index height profiles for the CMEs on 2005-01-15, 06:30 UT (left) and 2000-07-07, 10:26 UT (right). Each case has 6 sampling lines, and the average $n(h)$ is plotted in red. The height range Δh used to compute $\langle n'(h)\rangle_{\Delta h}$ is marked in blue.

Liu *et al.* 2024). Parameters of the coronal field have received relatively little attention so far, with CMEs originating under pseudostreamers (more open field) being faster on average than those originating under regular streamers (Liu 2007a) and (probably related to this) higher decay index of the coronal field being associated with faster CMEs (Xu *et al.* 2012; Deng and Welsch 2017). CME speeds are strongly correlated with the amount of flux reconnected in the associated flare (Qiu and Yurchyshyn 2005; Deng and Welsch 2017; Zhu *et al.* 2020). However, the cause-effect relationship—whether stronger reconnection causes a faster rise of the CME flux rope or vice versa—is not yet known.

All associations with source region parameters show scatter, especially when the sample is not restricted to fast CMEs, indicating that $V_{\rm CME}$ is controlled by a combination of source-region parameters and that non-magnetic forces (aerodynamic drag) may play a significant role for the slower CMEs within the coronagraph's field of view. The highest correlation coefficients, $c \gtrsim 0.8$ for fast CMEs, were found between $V_{\rm CME}$ and PIL length, total non-neutralized current, and reconnected flux (Kontogiannis *et al.* 2019; Qiu and Yurchyshyn 2005).

Here, we study the dependence of $V_{\rm CME}$ on the height profile of the coronal field. Assuming a major role for the torus instability (TI) in the acceleration of CMEs (Kliem and Török 2006), one can expect that the height profile of the decay index, $n(h) = -{\rm d}\log B_{\rm ep}(h)/{\rm d}\log h$, above the critical height for TI onset, $h > h_{\rm cr}$ (determined by $n(h) > n_{\rm cr}$), is related to CME acceleration. For a steeper height profile, the external poloidal ("strapping") field, $B_{\rm ep}$, decreases faster with height, so that the mismatch between the upward Lorentz self-force of the erupting flux rope and the downward tension force of the strapping field rapidly increases, equivalent to strong CME acceleration and suggesting a correlation between the slope $n'(h{\geq}h_{\rm cr})$ and $V_{\rm CME}$. The TI threshold depends on parameters and is usually found to lie in the range $n_{\rm cr} \approx 1$–2, with a canonical value of $3/2$. Because $n(h)$ is a structural parameter of the source region, its slope can only control the resulting velocity of a torus-unstable flux rope relative to the Alfvén velocity, $V_{\rm A}$, in the source volume, which limits the correlation with $V_{\rm CME}$. Nevertheless, $n'(h{\geq}h_{\rm cr})$ deserves attention as a potentially important parameter in the acceleration of CMEs.

2. Data analysis

We extract a sample of 15 very fast front-side CMEs ($V_{\rm CME} \geq 1500$ km s^{-1}), all halo CMEs, from the complete sample of 57 such CMEs in 1996/06–2007/01 in Wang and Zhang (2008) by restricting the source location to $\leq 30°$ from disk center (for reliable magnetogram data) and by requiring that flare ribbon observations are available. We

Table 1. Event sample.

Event	Date	Time [UT]	NOAA AR	$V_{\rm CME}$ [km s^{-1}]	$h_{\rm cr}$ [Mm]	$\langle n'(h)\rangle_{\Delta h}$ [$R_\odot^{-1}$]
0	2000/07/14	10:54:07	9077	1674	51	5.2
1	2000/09/12	11:54:05	9163	1550	177	2.4
2	2001/09/24	10:30:59	9632	2402	51	13.7
3	2002/08/16	12:30:05	10069	1585	135	3.5
4	2002/09/05	16:54:06	10102	1748	147	4.0
5	2002/11/09	13:31:45	10180	1838	42	9.4
6	2003/10/28	11:30:05	10486	2459	59	10.8
7	2003/10/29	20:54:05	10486	2029	55	12.4
8	2003/11/18	08:50:05	10501	1660	97	4.1
9	2004/11/07	16:54:05	10696	1759	59	8.1
10	2005/01/15	06:30:05	10720	2049	46	8.4
11	2005/01/15	23:06:50	10720	2861	51	6.0
12	2005/01/17	09:30:05	10720	2094	55	5.5
13	2005/05/13	17:12:05	10759	1689	76	5.7
14	2006/12/13	02:54:04	10930	1774	59	5.5
15	1997/05/12	05:30:05	8038	464	46	8.5
16	1997/08/30	01:30:35	8076	371	164	1.7
17	1997/09/28	01:08:33	–	359	160	2.0
18	1999/06/12	21:26:08	8569	465	139	2.8
19	2000/07/07	10:26:05	9070	453	80	9.1
20	2000/09/15	21:50:07	9165	257	29	29.3
21	2001/03/19	05:26:05	9380	389	206	0.6
22	2000/02/17	21:30:08	8872	728	93	4.4
23	2001/03/24	20:50:05	9390	906	72	4.8
24	2001/09/28	08:54:34	9636	846	118	0.0
25	2001/11/21	14:06:05	9704	518	63	1.2
26	2002/04/15	03:50:05	9906	720	67	2.6
27	2005/05/26	15:06:05	10767	586	46	4.9
28	2006/04/30	09:54:04	10876	544	114	4.5
29	2000/02/08	09:30:05	8858	1079	244	2.7
30	2000/02/12	04:31:20	8858	1107	164	1.0
31	2000/06/06	15:54:05	9026	1119	59	1.4
32	2000/11/24	05:30:05	9236	1289	55	15.2
33	2001/04/09	15:54:02	9415	1192	223	2.9
34	2001/10/22	15:06:05	9672	1336	51	-0.9
35	2002/12/19	22:06:05	10229	1092	76	14.0
36	2005/05/06	17:28:31	10758	1128	122	2.3

Notes: The sample consists of 15 very fast ($V_{\rm CME} \geq 1500$ km s^{-1}) and 22 slower halo CMEs from Solar Cycle 23. Time of first appearance in the *SOHO*/LASCO/C2 coronagraph (Brueckner *et al.* 1995), source region (NOAA active region [AR] number), and linear CME speed in the combined LASCO/C2 & C3 field of view are given. Parameter values maximizing the correlation between $V_{\rm CME}$ and $\langle n'(h)\rangle_{\Delta h}$ for the very fast CMEs, $n_{\rm cr} = 1.5$ and $\delta = 1$, are used in the computation of critical height, $h_{\rm cr}$, and average slope of the decay index, $\langle n'(h)\rangle_{\Delta h}$.

add slower CMEs from the CDAW Halo CME catalog (Gopalswamy *et al.* 2010) using the same selection criteria, in a first step limited to 22 additional events with good data of the flare ribbons, to allow careful individual analysis of each event in this sub-sample. The events are listed in Table 1.

From the extent of the flare ribbons, we determine the erupting section of the PIL and place 3–7 nearly equally distributed points along that section. The coronal potential field, $B_{\rm pot}$, is computed along vertical "sampling" lines rooted in these points using a spherical-harmonic expansion (with 248 harmonics for high accuracy) and a source surface, where the field is prescribed to be radial, at $r = 2.66\ R_\odot$ (Schatten *et al.* 1969; Altschuler and Newkirk 1969). For each CME, the synoptic map from the Michelson Doppler Imager (MDI, Scherrer *et al.* 1995) aboard *SOHO* is patched with the final pre-eruption MDI magnetogram of the source region. The line-of-sight field component is converted to

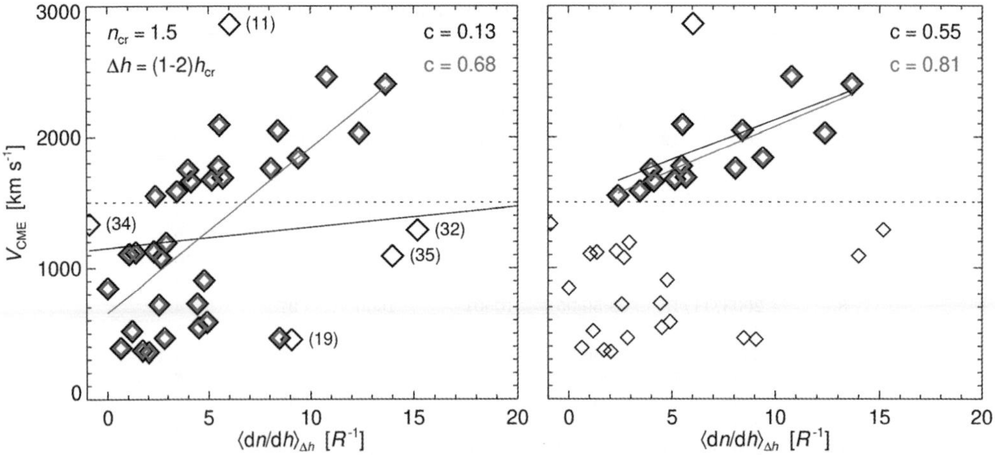

Figure 2. $V_{\rm CME}$ vs. $\langle n'(h)\rangle_{\Delta h}$ for the parameters $n_{\rm cr} = 1.5$ and $\delta = 1$ that maximize the correlation coefficients for the very fast CMEs. The whole sample yields a correlation coefficient $c = 0.13$ (left panel, black line is the linear fit). Excluding Event #11 and all 5 events with a strong dip in $n(h)$ (#19, 20, 32, 34, 35; open diamonds; see text), raises the correlation to $c = 0.68$ (red fit line). Event #20 lies outside the range shown (see Table 1). Results for the sub-sample of very fast CMEs are shown in the right panel.

the radial one, assuming that the photospheric field is approximately radial (Wang and Sheeley 1992).

The decay index along each sampling line is computed using the horizontal component of $B_{\rm pot}$ as an approximation of $B_{\rm ep}$. Averaging over all sampling lines for each source region yields one function $n(h)$ for each CME. Different from many earlier studies, which used a fixed height range in the analysis of $n(h)$, often $h = 42$–105 Mm, we use an event-specific height range, Δh, by determining $h_{\rm cr}$ for each event individually (Table 1) and setting $\Delta h \in [h_0, h_{\rm max}] = [1, 1+\delta]h_{\rm cr}$. The free parameter δ is set uniformly across the sample, but a range of values, $0.1 \le \delta \le 2$, is considered in the analysis, to check the robustness of the results. For the same purpose, TI threshold values in the range $1.3 \le n_{\rm cr} \le 1.7$ are considered. The upper edge of the height range is limited by the location of the source surface in the $B_{\rm pot}$ model. The height range scales with $h_{\rm cr}$, hence, with the related main spatial scale of the source region; this improves the correlations found. Finally, the average slope of the decay index profile in the range Δh, $\langle n'(h)\rangle_{\Delta h}$, is computed. Figure 1 illustrates $n(h)$ and Δh for two representative events and our standard values of $n_{\rm cr} = 1.5$ and $\delta = 1$.

3. Observational results

Figure 2 shows that $\langle n'(h)\rangle_{\Delta h}$ and $V_{\rm CME}$ are significantly correlated for the sub-sample of very fast CMEs ($c = 0.55$), consistent with the TI playing a significant role in their acceleration. Similar to all other parameters studied in the literature, the correlation degrades when the whole range of CME velocities is included. However, the correlations for both samples are considerably improved when two categories of events are excluded. The first of these is a CME (#11) which is particularly fast, the likely reason being that it followed another fast CME from the same source region (#10 [Fig. 1 left]) within 16.5 hours; the resulting higher acceleration would then not primarily be related to the $n(h)$ profile, but rather to a coronal density depletion. The second are cases with a dip in $n(h)$ deep and broad enough to introduce a significant torus-stable height range above the unstable height range between $h_{\rm cr}$ and the dip (#19 [Fig. 1 right], 20, 32, 34,

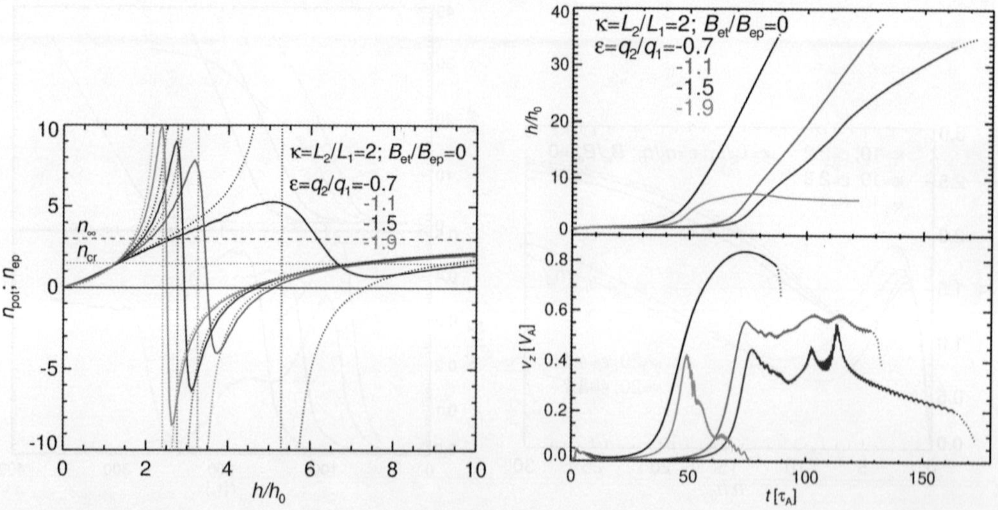

Figure 3. Decay index profiles $n(h)$ (left) and rise profiles of TD flux ropes, $h(t)/h_0$ and $v_z(t)/V_A$, (right) in a line quadrupole with $B_{et}=0$ for a range of source strength ratios $\epsilon=q_2/q_1$. Dotted profiles $n(h)$ are the exact profiles computed from B_{ep}, which have a pole at the position of the null (X-) line, and the solid profiles are computed from the corresponding potential field. The rise profiles are dotted during the period when the flux rope apex closely approaches the closed upper boundary.

35). Flux ropes erupting from the unstable height range just above h_{cr} are significantly decelerated in such overlying torus-stable height ranges. If they still enter the upper torus-unstable range, a CME results, otherwise the eruption remains confined. As Fig. 1 illustrates, the corresponding $n(h)$ profiles tend to be steep between h_{cr} and the peak value of $n(h)$ below the dip (as explained in Sect. 4), producing the high-slope lower-speed data points in Fig. 2, which, consequently, deviate strongly from the expected correlation between $\langle n'(h)\rangle_{\Delta h}$ and V_{CME}. The resulting correlation for the very fast CMEs, $c=0.81$ with a confidence interval of [0.69,0.89] at the 1σ (68%) confidence level, belongs to the highest obtained so far for any source-region parameter in spite of the fact that $\langle n'(h)\rangle_{\Delta h}$ is only a structural parameter not covering the spread in source-region Alfvén speed. The correlation coefficient changes (decreases) by no more than 10% in the range $n_{cr}=1.3$–1.7 if $\delta=1$ is used and by 25% in the whole range of the free parameters considered. The correlation for the whole sample is also significantly improved to $c=0.68$ (with a confidence interval of [0.57,0.77]) by excluding the events with dipped $n(h)$ and Event #11.

4. Simulation study

In order to quantify the effect of dipped $n(h)$ profiles on V_{CME} and to establish criteria for their exclusion from our correlation analysis, we have started a parametric simulation study of flux rope eruptions. Dipped $n(h)$ profiles occur in multipolar regions with a magnetic null line or point, in particular in quadrupolar regions (Török and Kliem 2007), and in two-scale bipolar regions if the scales of the photospheric flux distribution are sufficiently different (Kliem *et al.* 2021). Both configurations can be modeled through a simple extension of the Titov-Démoulin force-free flux rope equilibrium (Titov and Démoulin 1999, henceforth TD99) by replacing the sub-photospheric bipole, which is the source of B_{ep}, with a pair of bipoles, both placed symmetrically to the plane of the toroidal flux rope and on the axis of rotational symmetry. Denoting the magnetic charges as q_1 and q_2 and the half-spacings of the bipoles as L_1 and L_2, a line quadrupole results

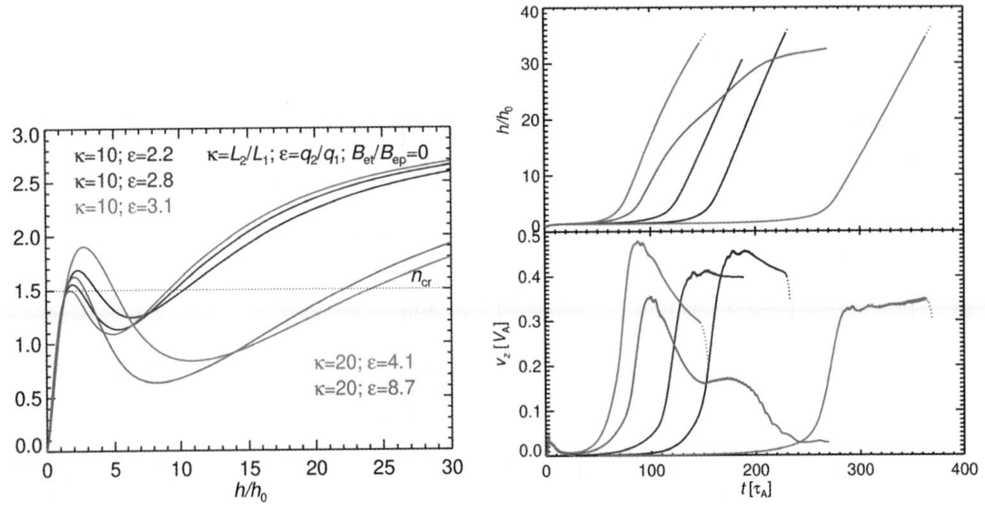

Figure 4. Decay index and rise profiles for the two-scale bipole with $B_{\rm et}=0$ for a range of source strength ratios $\epsilon=q_2/q_1$ and two values of the scale ratio $\kappa=L_2/L_1$, displayed as in Fig. 3. The very different evolution times until considerable acceleration commences result from small differences of the initial equilibria from the exact marginal stability configurations.

for $q_1q_2<0$ ($L_2>L_1$; see Fig. 5 in Kliem *et al.* 2014 for an illustration) and a two-scale bipolar region results for $q_1q_2>0$, $L_2\gg L_1$. These configurations are used as the initial condition of zero-beta ideal MHD simulations in large boxes of size 40^3 constructed with a stretched grid of highest resolution 0.04 in the area of the initial flux rope, whose apex height is normalized to $h_0=1$ (see, e.g., Hassanin *et al.* 2022 for the numerical details). The ratios $\epsilon=q_2/q_1$ and $\kappa=L_2/L_1$ are used as independent parameters and q_1, q_2 are determined by the equilibrium condition (TD99) for given geometry (L_1, L_2), which is chosen such that the equilibrium is very close to marginal stability. A short, transient phase of initial numerical relaxation of the approximate analytical equilibrium causes the flux rope to expand slightly, whereby it enters the unstable domain, so that the TI develops spontaneously.

Figure 3 shows the decay index and rise profiles for a TD flux rope embedded in a line quadrupole with zero external toroidal (shear) field, $B_{\rm et}=0$. The X-line above the flux rope thus turns into a null line, where $n(h)$ has a pole, which introduces a dip above the pole and leads to a high slope $n'(h)$ below the pole. When $n(h)$ is approximated by using the horizontal components of $B_{\rm pot}$ instead of $B_{\rm ep}$, as in our observational study, the pole turns into an inflection point with a dip similar in shape to our dipped observational $n(h)$ profiles (Fig. 1). $B_{\rm pot}$ is computed in the box from the initial TD field values at the boundaries. The flux in the footprints of the flux rope acts as the source of a guide field component in the potential field along the null line of $B_{\rm ep}$, turning the pole into an inflection point. As ϵ increases, corresponding to increasing exterior flux, the dip in $n(h)$ turns deeper and wider, and shifts to smaller heights. The corresponding rise profiles show an increasing deceleration of the erupting flux rope until a confined eruption results for the highest ϵ chosen; they depend only weakly upon κ.

Figure 4 shows the results for the two-scale bipole. Again, a broader and deeper dip in $n(h)$ increasingly decelerates the torus-unstable flux rope until a confined eruption results for the maximum values of ϵ and κ shown. The minimum scale ratio for the effect to occur is found to be $\kappa\sim 10$. Both simulation series confirm our rationale for excluding cases with deeply dipped $n(h)$ from the correlation analysis in Sect. 3.

5. Discussion

By restricting the study to halo CMEs with good data of flare ribbons, the sub-sample of slower CMEs ($V_{\rm CME} < 1500$ km s^{-1}) is strongly reduced in size compared to the whole population of slower CMEs. However, to provide reliable support to the forecast of very fast CMEs, our ultimate goal, the present study will be extended to include all CMEs with usable flare ribbon data from a representative period. With such a larger sample, it remains to be seen whether the exclusion of high-slope, lower-speed events based on the properties of a dip in $n(h)$ can be nearly as complete as with the sample used here.

The parametric simulation study of eruptions from source regions with dipped $n(h)$ profiles will be extended to quantify the relationship between the properties of $n(h)$ (torus-stable dip and underlying torus-unstable peak) and the resulting velocity of erupting flux ropes (normalized by $V_{\rm A}$). This will be related to the ratio of core-region flux forming the equilibrium and overlying exterior flux, and ultimately to the flux distribution in the photosphere.

For applicability in forecast schemes, a method to estimate the potentially erupting section of the PIL must be included. The computation of the coronal currents and their neutralization degree might indicate the section most clearly, but depends on a nonlinear force-free field extrapolation, which is computationally demanding and does not yield unique solutions. Selecting the strong-gradient section of the PIL (Falconer *et al.* 2003; Schrijver 2007; Dhakal and Zhang 2024)) is a promising option.

The correlation between $\langle n'(h)\rangle_{\Delta h}$ and $V_{\rm CME}$ might improve further if oblique propagation directions, which occur for many CMEs, are considered. The expected propagation direction can be estimated from the asymmetry of flux in the environment of the erupting PIL section.

6. Conclusions

From the presented stage of our ongoing study, we conclude the following.

(1) $\langle n'(h)\rangle_{\Delta h}$ shows a very high correlation with $V_{\rm CME}$ for very fast CMEs ($V_{\rm CME} \geq$ 1500 km s^{-1}) with $c = 0.81$ in Cycle 23, among the highest correlations found so far, although $\langle n'(h)\rangle_{\Delta h}$ is only a structural parameter not covering the spread in $V_{\rm A}$.

(2) This is consistent with the hypothesis that the TI plays a decisive role in the coronal acceleration of CMEs.

(3) A significant torus-stable dip in $n(h)$, reflecting a significant amount of exterior flux passing over the current-carrying core flux, indicates a deceleration of flux ropes erupting from the core flux, including the possibility of confinement.

Competing interests. The authors declare none.

Acknowledgments. We acknowledge very helpful comments by an anonymous referee and support from the DFG through grants No. 392171665 and 520867050 and from NASA through grants No. 80NSSC20K1274 and 80NSSC24K1108.

References

Altschuler, M. D. & Newkirk, G. 1969, *Solar Phys.*, 9, 131
Brueckner, G. E., Howard, R. A., Koomen, M. J., *et al.* 1995, *Solar Phys.*, 162, 357
Dhakal, S. K. & Zhang, J. 2024, *ApJ*, 960, 36
Deng, M. & Welsch, B. T. 2017, *Solar Phys.*, 292, 17
Falconer, D. A., Moore, R. L., & Gary, G. A. 2003, *J. Geophys. Res.*, 108, 1380
Gopalswamy, N., Akiyama, S., Yashiro, S., *et al.* 2008, *JASTP*, 70, 245
Gopalswamy, N., Yashiro, S., Michalek, G., *et al.* 2010, *Sun and Geosphere*, 5, 7
Guo, J., Zhang, H., Chumak, O. V., & Liu, Y. 2006, *Solar Phys.*, 237, 25

Hassanin, A., Kliem, B., Seehafer, N., & Török, T. 2022, *ApJ Lett.*, 929, L23
Kliem, B., Lee, J., Liu, R., *et al.* 2021, *ApJ*, 909, 91
Kliem, B., Lin, J., Forbes, T. G., *et al.* 2014, *ApJ*, 789, 46
Kliem, B. & Török, T. 2006, *Phys. Rev. Lett.*, 96, 255002
Kontogiannis, I., Georgoulis, M. K., Guerra, J. A., *et al.* 2019, *Solar Phys.*, 294, 130
Liu, Y. 2007a, *ApJ Lett.*, 654, L171
Liu, Y. 2007b, *Adv. Sp. Res.*, 39, 1767
Liu, Y., Török, T., Titov, V. S., *et al.* 2024, *ApJ*, 961, 148
MacQueen, R. M. & Fisher, R. R. 1983, *Solar Phys.*, 89, 89
Qiu, J. & Yurchyshyn, V. B. 2005, *ApJ Lett.*, 634, L121
Schatten, K. H., Wilcox, J. M., & Ness, N. F. 1969, *Solar Phys.*, 6, 442
Schrijver, C. J., 2007, *ApJ Lett.*, 655, L117
Scherrer, P. H., Bogart, R. S., Bush, R. I., *et al.* 1995, *Solar Phys.*, 162, 129
Srivastava, N. & Venkatakrishnan, P. 2004, *JGR*, 109, 10103
Su, Y., Van Ballegooijen, A., McCaughey, J., *et al.* 2007, *ApJ*, 665, 1448
Sung, S.-K., Marubashi, K., Cho, K.-S., *et al.* 2009, *ApJ*, 699, 298
Titov, V. S. & Démoulin, P. 1999, *A&A*, 351, 707
Török, T. & Kliem, B. 2007, *AN*, 328, 743
Wang, Y. & Zhang, J. 2008, *ApJ*, 680, 1516
Wang, Y.-M. & Sheeley, Jr., N. R. 1992, *ApJ*, 392, 310
Xu, Y., Liu, C., Jing, J., & Wang, H. 2012, *ApJ*, 761, 52
Zhang, J. & Dere, K. P. 2006, *ApJ*, 649, 1100
Zhang, J., Temmer, M., Gopalswamy, N., *et al.* 2021, *Progr. Earth Planetary Sci.*, 8, 56
Zhu, C., Qiu, J., Liewer, P., *et al.* 2020, *ApJ*, 893, 141

Solar and Stellar Coronal Mass Ejections
Proceedings IAU Symposium No. 388, 2026
N. Gopalswamy, O. Malandraki, A. Vidotto & W. Manchester, eds.
doi:10.1017/S1743921324001716

Exploring the Dynamics of CME-Driven Shocks by Comparing Numerical Modeling and Observations

Meng Jin[1], Gang Li[2], Nariaki Nitta[1], Wei Liu[1,3], Vahé Petrosian[4], Ward Manchester[5], Christina Cohen[6], Frederic Effenberger[3,7], Zheyi Ding[8], Melissa Pesce-Rollins[9], Nicola Omodei[4] and Nat Gopalswamy[10]

[1]Lockheed Martin Solar and Astrophysics Lab (LMSAL), Palo Alto, CA, USA
[2]General Linear Space Plasma Lab LLC, Foster City, CA, USA
[3]Bay Area Environmental Research Institute (BAERI), Mountain View, CA, USA
[4]Stanford University, Stanford, CA, USA
[5]University of Michigan Ann Arbor, Ann Arbor, MI, USA
[6]California Institute of Technology, Pasadena, CA, USA
[7]Ruhr University Bochum (RUB), Bochum, Germany
[8]KU Leuven, Leuven, Belgium
[9]Istituto Nazionale di Fisica Nucleare (INFN), Sezione di Pisa, Pisa, Italy
[10]NASA Goddard Space Flight Center, Greenbelt, MD, USA

Abstract. Shocks driven by coronal mass ejections (CMEs) are primary drivers of gradual solar energetic particle (SEP) events, posing significant risks to space technology and astronauts. Concurrently, particles accelerated at these shocks may also propagate back to the Sun, potentially generating gamma-ray emissions through pion decay. We incorporated advanced modeling and multi-messenger observations to explore the role of CME-driven shocks in gamma-ray emissions and SEPs. Motivated by Fermi-LAT long-duration solar flares, we used the AWSoM MHD model to investigate the connection between the shocks and the properties of observed gamma-ray emissions. By coupling the AWSoM with iPATH model, we evaluate the impact of shock evolution complexity near the Sun on SEP intensity and spectra. Our result points to the importance of accurate background coronal and solar wind modeling, as well as detailed observations of CME source regions, in advancing our understanding of CME-driven shocks and the dynamics of associated energetic particles.

Keywords. Solar Gamma Ray Flares, Coronal Mass Ejections, Energetic Particles, Particle Transport

1. Introduction

Shocks driven by Coronal Mass Ejections (CMEs) play an important role in accelerating particles. Particles cross the shock front back and forth due to interaction with magnetic turbulence and gain energy through the diffusive shock acceleration mechanism (e.g. Axford et al. 1977). Accelerated particles that escape upstream can propagate along the interplanetary magnetic field lines and are detected in situ as Solar Energetic Particles (SEPs), posing significant risks to both technology and astronauts in space.

Conversely, particles accelerated at the shock front can also escape downstream, traveling back to the Sun where they may generate γ-ray emissions through pion decay with the ambient plasma in the photosphere. The sustained/long-duration γ-ray emission (SGRE) events observed by the Fermi Large Area Telescope (LAT: Atwood et al. 2009) provide evidence supporting this "shock scenario", which potentially opens a new window for observing shock-accelerated particles in γ-rays. Gopalswamy et al. (2018) found a strong correlation between the SGRE events and Type II radio bursts, with a linear relationship between their durations, suggesting a common shock accelerating both protons and electrons that are respectively responsible for such emissions. In a recent study, Gopalswamy et al. (2021) found that the number of 500 MeV protons from the SGRE events is correlated with the protons propagating into space as SEPs.

In both cases, the CME-driven shock and its dynamical evolution through the inhomogeneous solar wind are crucial for understanding the particle acceleration and transport processes involved. Employing data-driven global magnetohydrodynamics (MHD) models enables the quantitative reconstruction of the global corona and the dynamic evolution of CME-driven shocks, offering unique insights into the role of CMEs in these phenomena (e.g. Plotnikov et al. 2017; Jin et al. 2018). In addition, to properly model SEP events, the MHD models need to be coupled with particle acceleration and transport models (e.g. Li et al. 2021). This study aims to investigate both γ-ray and SEP events to delve into the dynamics of CME-driven shocks by comparing advanced modeling and observations results.

The paper is structured as follows: Section 2 provides an overview of the models used; Section 3 presents the main results; discussions are in Section 4, followed by the summary and conclusions in Section 5.

2. Model

To reconstruct the background solar corona and wind, we use the Alfvén Wave Solar Model (AWSoM; van der Holst et al. 2014), a data-driven global MHD model within the Space Weather Modeling Framework (Tóth et al. 2012). By incorporating a physically consistent treatment of wave reflection, dissipation, and heat partitioning between electrons and protons, AWSoM has been demonstrated to reproduce high-fidelity global solar corona conditions (e.g., Sokolov et al. 2013; Jin et al. 2017b; Sachdeva et al. 2019). Specifically, the inner boundary condition of the magnetic field is specified with full-surface magnetic maps. Aflvén waves are driven at the inner boundary with a Poynting flux that scales with the surface magnetic field. The solar wind is heated by Alfvén wave dissipation and accelerated by thermal and Aflvén wave pressure. While electrons and protons are collisionally coupled, their temperatures are computed separately. This treatment is critical for reproducing physically correct CME-driven shocks (Jin et al. 2013).

To generate a CME eruption, we use a simulation tool called the Eruptive Event Generator Gibson-Low (EEGGL, now available at NASA CCMC, Jin et al. 2017a; Borovikov et al. 2017) that inserts a Gibson-Low (GL) flux rope (Gibson and Low 1998) into the corona. It automatically determines the flux rope parameters with a handful of quantities given by observations (i.e., magnetogram and CME speeds) so that the modeled CMEs can travel at the desired speeds near the Sun.

To model SEP events, we further couple the AWSoM MHD model with the improved Particle Acceleration and Transport in the Heliosphere model (iPATH: Hu et al. 2017). iPATH numerically solves the particle acceleration at the shock and transport in the solar wind with both along- and cross-field diffusion. One unique feature of iPATH is the adoption of a 2D onion shell module which tracks the acceleration of particles and their subsequent convection and diffusion downstream of the shock. Li et al. (2021) extended

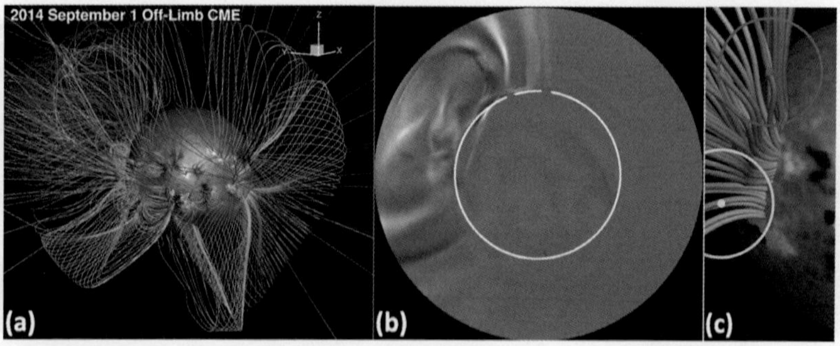

Figure 1. MHD modeling result of the SOL2014-09-01 Fermi BTL event. (a) Magnetic field configuration at $t = 16$ minutes after the flux rope eruption. The red and white field lines represent the flux rope and large-scale helmet streamers, respectively. The green field lines are selected surrounding active regions as well as open field lines. (b) Concurrent synthesized tri-color composite, running-ratio image for AIA 211 Å(red), 193 Å(green), and 171 Å(blue) channels showing the simulated EUV waves. (c) Selected field lines connecting to the CME shock (green) and source flare region (red). The red and yellow dots with 95% uncertainty circles represent >100 MeV emission centroids identified previously by Ackermann et al. (2017) and updated recently in Ajello et al. (2021), respectively.

the 2D shell structures to 3D using the AWSoM MHD shock and plasma inputs. In addition, an ensemble approach was developed to account for the model uncertainties in magnetic connectivity and shock profiles, i.e., multiple field lines were traced that were evenly distributed within 10° of a heliospheric observer.

3. Results

3.1. *CME-driven Shock in Relation to Fermi SGRE Solar Flares*

In this section, we show the modeling results for three Fermi SGRE events. Two of them (SOL2014-09-01 and SOL2021-07-17) are behind-the-limb (BTL) flares (Pesce-Rollins et al. 2015; Ackermann et al. 2017), in which the γ-ray emission region on the visible side of the Sun was located *away* from the flare site behind the limb by tens of degrees in longitude. The other, equally intriguing, event is SOL2012-03-07, in which the Fermi γ-ray emission centroid *migrated with time* over the solar disk for 10 hours, long past the impulsive phase of the flare (Ajello et al. 2021).

In the SOL2014-09-01 event, the solar flare region was located 43° behind the east limb. However, Fermi-LAT detected γ-ray emissions on the front side of the Sun for ~2 hours. This event was associated with a fast CME with a speed >1900 km s^{-1}. Figure 1 shows the MHD modeling result of this event. Specifically, Figure 1a depicts the magnetic field configuration after 16 minutes of the CME onset, while Figure 1b illustrates the simulated global EUV waves associated with the CME, where two different wave fronts are evident (marked in red arrows). The first wave front (to the right) is the fast-mode wave/shock driven by the eruption. The second one (to the left) is caused by the expanding CME flux rope shell (e.g., Jin et al. 2016). This double wave front feature is consistent with SDO/AIA observations of the event (e.g., Grechnev et al. 2018). Note that EUV waves are thought to be the low coronal counterparts of the CME-driven shock waves that propagate into the heliosphere. A recent study by Pesce-Rollins et al. (2022) demonstrated a strong correlation between the time derivative of the EUV wave intensity profile and Fermi/LAT γ-ray flux.

By modeling the evolution of the CME and the CME-driven shock over one hour, we derived the detailed history of 3D shock parameters including compression ratio, shock speed, shock Alfvén Mach number, and shock angle. We found that the shock compression

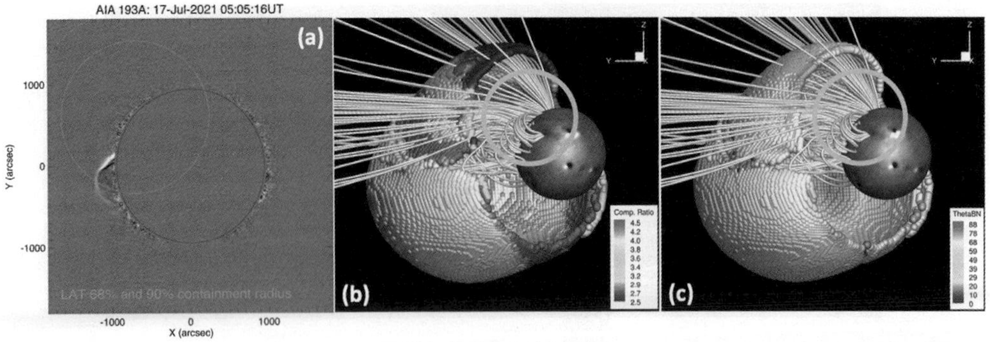

Figure 2. (a) Flare averaged Fermi-LAT emission centroid, shown as 90% (solid) and 68% (dashed) uncertainty circles, overlaid on the SDO/AIA 193 Å running difference image (adpated from Pesce-Rollins et al. 2022). (b)-(c) Modeled CME-driven shock compression ratio and θ_{Bn} at t = 20 minutes after eruption. The orange circle shows the same 90% uncertainty radius of the Fermi-LAT emission centroid as in (a).

ratio increases rapidly from ~1.8 at 10 minutes to ~4.6 at 20 minutes and then gradually decreases to ~3.7 at 60 minutes. This evolution trend is similar to the Fermi/LAT γ-ray intensity profile. Figure 1c shows the field lines connected to the CME shock (in green) and the source region (in red). Notably, in the recently updated Fermi solar flare catalog, the >100 MeV emission centroid has been revised further south (marked by a yellow circle), thus aligning more closely with the shock-connected field lines. See Jin et al. (2018) for detailed modeling results of this event.

SOL2021-07-17 is the most recent Fermi BTL event observed (Figure 2), and it exhibits the largest separation (>50°) between the source region and the emission centroid of all the Fermi BTL events (Pesce-Rollins et al. 2022). Preliminary modeling results of the event are shown in Figure 2b-c, which include the 3D shock compression ratio and shock angle from the simulation. The shock surface is notably non-spherical due to the inhomogeneous background solar wind, leading to significant variations in the shock parameters across different parts of the shock surface. Similar to the SOL2014-09-01 event, this event features shock-connecting open field lines that trace back to the Fermi emission centroid area. Additionally, the area where the shock intersects these field lines exhibits a quasi-perpendicular nature.

Lastly, we discuss the SOL2012-03-07 event. In this event Fermi observed the γ-ray emission centroid migration over the solar disk in 10 hours after flare impulsive phase (Figure 3a). This event is more challenging to model since there are two CMEs involved, separated by 1 hour. Although the two CMEs are from the same active region, they erupted from different parts of the polarity inversion line (PIL), with the first one started from the north, and the second one from the south (e.g., Patsourakos et al. 2016).

We model the SOL2012-03-07 event by initiating two separate GL flux ropes at different parts of PIL to match the AIA observation. To estimate strong shock location in the simulation, we define a new parameter that combines compression ratio and Alfvén Mach number as shock strength measure. This approach enabled us to identify areas of strong shock on the shock surface and trace the field lines from those regions back to the Sun's surface (Figure 3). At 30 minutes, these footpoints are situated south of the active region, aligning with the initial Fermi centroid marked in red. By 99 minutes into the simulation, the footpoints connected to the strong shock location begin moving northward, as indicated by the yellow circle, a movement consistent with observations.

Furthermore, we observed another strong shock region that developed at 99 minutes, with its footpoints situated near the Fermi centroid at 07:02, marked by a green circle.

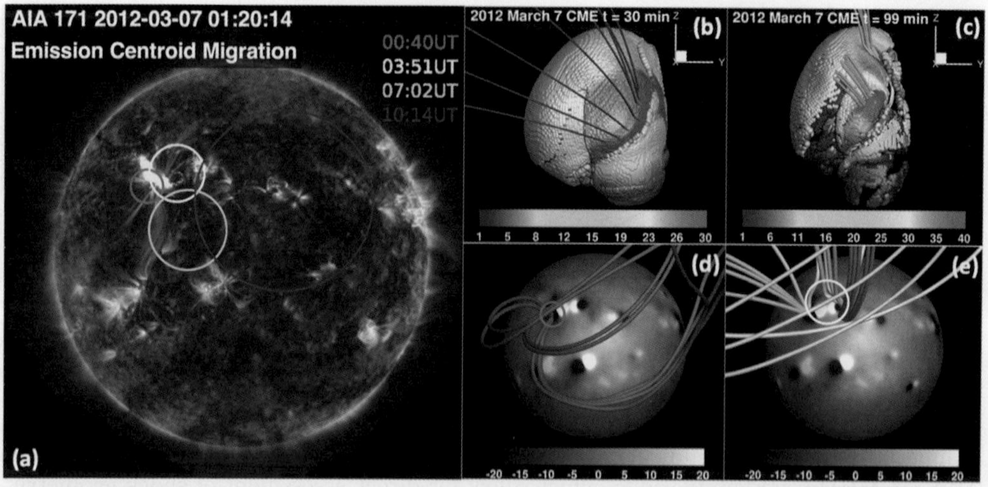

Figure 3. (a) Migration of the γ-ray emission centroid of the SOL2012-03-07 flare over 10 hours overlaid on an AIA 171 Å image at 00:40 UT. Circles show the 95% uncertainty. (b) & (c) Shock surface with the shock strength ($CR \cdot M_A$) in color derived from the simulation. (d) & (e) Magnetic field lines connecting to the strongest portion of the shock. The red and yellow circles are reproduced from (a), indicating good agreement of the LAT centroids with the roots of the field lines in corresponding colors.

While we continue to work on the simulation beyond 100 minutes, these preliminary results indicate that the temporal evolution of footpoints, those which are connected to the strongest shock locations, aligns with the migration of the Fermi centroid during the first 100 minutes.

3.2. *The Impact of CME-driven Shock Evolution on the in-situ SEP Intensity and Spectra*

In this section, we discuss the impact of the CME-driven shock on the properties of SEPs measured in situ. As demonstrated in the previous section, the parameters of the CME-driven shock can vary significantly across different parts of the shock surface. Given that heliospheric observers at various locations are connected to different parts of the shock, the characteristics of the SEPs observed can be influenced by the properties of the specific shock segment to which they are connected. Understanding this relationship is crucial for better interpreting SEP observations.

One example is the 2021 October 28 SEP event, which was observed by Parker Solar Probe (PSP), STEREO-A, and SOHO/ACE with almost identical Fe/O ratio, even there is ~60° separation in longitude (Cohen et al. 2022). To better understand this distinctive characteristic, we modeled the associated CME-driven shock of this SEP event and calculated the field line connectivity to different observer locations including the SOHO/ACE, STA/B, SolO, and PSP (Figure 4). Our preliminary results indicate that although the three observers (i.e., PSP, STEREO-A, and ACE) are connected to different parts of the shock, these areas nevertheless exhibit very similar properties, e.g., compression ratio, shock angle (see Figure 4a & b), as well as the upstream condition (see Figure 4c for upstream plasma density). This consistency might explain the similar Fe/O ratios observed at these locations.

Fe-rich events in gradual SEP events are thought to originate either directly from flare contributions (e.g., Cane et al. 2003, 2006) or from suprathermal flare particles that are preferentially accelerated by quasi-perpendicular shocks (Tylka et al. 2005). To

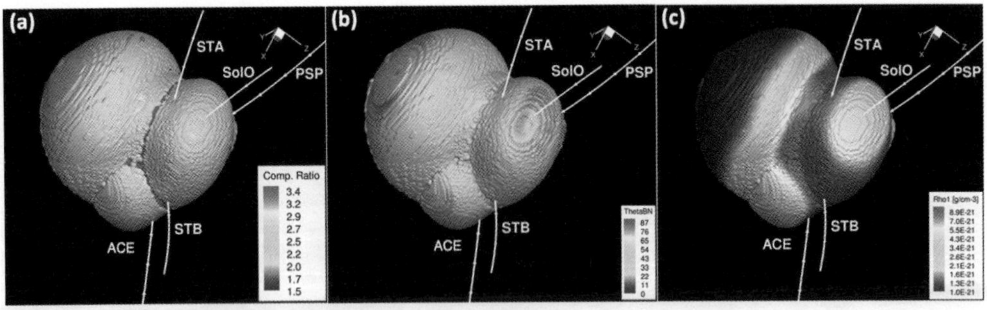

Figure 4. (a) The shock compression ratio, (b) shock angle, and (c) upstream plasma density along the shock surface at t = 1 hour in the 2021 October 28 SEP event. The white lines represent the field connectivity to different heliospheric observers.

evaluate the "quasi-perpendicular shock scenario", eight SEP events were modeled using AWSoM MHD CME simulations, which allow for direct assessment of shock geometry – a feature difficult to discern from near-Sun observations (Nitta et al. 2022). However, the preliminary results do not indicate a strong correlation between Fe-rich SEP events and quasi-perpendicular shocks. Suprathermal particles, which can be affected significantly by the inhomogeneity of the solar wind (Wijsen et al. 2023), or due to preceding impulsive SEP events, may play a significant role and warrant further investigation.

To quantitatively investigate the SEP events and their relationship to the CME-driven shocks, one needs both the global MHD model for tracking the CME-driven shock evolution through the background solar wind and the particle acceleration and transport model for calculating the transport of particles along the magnetic field lines. Most SEP models rely on simplified background solar wind and start at a larger distance from the Sun (e.g., 10 Rs). However, CME-driven shock could form at a height of just a few solar radii, with information on timing and formation height revealed through radio observations (e.g., Kozarev et al. 2011; Gopalswamy et al. 2013). There are previous attempts to include early phase shock evolution (e.g., Kozarev et al. 2013; Borovikov et al. 2018; Linker et al. 2019). By coupling two state-of-the-art models AWSoM and iPATH, Li et al. (2021) modeled the 2012 May 17 SEP event, including the very early phase. The new model successfully reproduced the SEP intensity profiles at both the Earth and STEREO-A locations as shown in Figure 5a-b. The new model also reproduces the high-energy end of SEP spectra up to 1 GeV observed by PAMELA (Bruno et al. 2018). The roll-over feature at high energy can be seen in both the observation and the simulation (Figure 5c). This study emphasizes the importance of early phase shock evolution on the particle acceleration and transport, especially on the high-energy end of SEP spectra. A new SEP forecasting model (SEPCaster) is under development based on this new capability (Whitman et al. 2023).

4. Discussion

Regarding the Fermi-LAT SGRE solar flares, our MHD simulations have shown some promising results supporting the "shock scenario". However, recent observational studies challenge the notion that the shock is the origin of the γ-ray producing particles (e.g., de Nolfo et al. 2019; Bruno et al. 2023). de Nolfo et al. (2019) analyzed the particles interacting with the Sun as observed by Fermi/LAT and the SEPs observed at 1 AU from PAMELA in 14 SGRE solar flare events. Their findings suggest that the two particle populations are not correlated. Gopalswamy et al. (2021) offered a possible explanation by addressing two observational effects: the γ-ray flux might be underestimated in limb events, and the latitudinal widths of SEP distributions are energy-dependent.

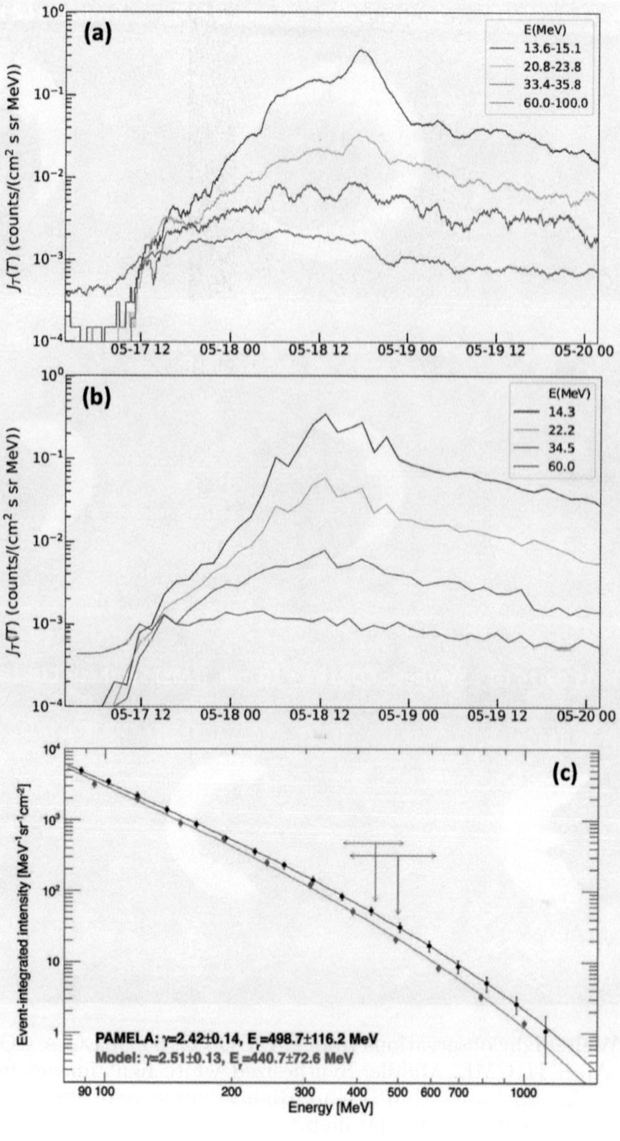

Figure 5. (a) Proton time-intensity profiles for 4 energy bins observed at STEREO-A during the 2012 May 17 SEP event. (b) Modeled proton time-intensity profiles from AWSoM+iPATH. (c) Spectrum comparison between PAMELA observation and model simulation. The black and red arrows indicated the location of the roll-over energies. Figures adapted from Li et al. (2021).

There are alternative scenarios coexisting with the "shock scenario". For example, flare-accelerated particles that are trapped locally within extended coronal loops after an eruption could be the source of the γ-ray producing particles (e.g., Ryan and Lee 1991; Hudson 2018). In the simulation of the SOL2014-09-01 event, we found some large-scale closed loops connected to the source region due to the interaction between the flux rope magnetic field and the global solar corona. These field lines could be significant for examining these trapped particles. In summary, further information remains to be explored in the simulations, and more advanced modeling is necessary to further differentiate between the scenarios mentioned above.

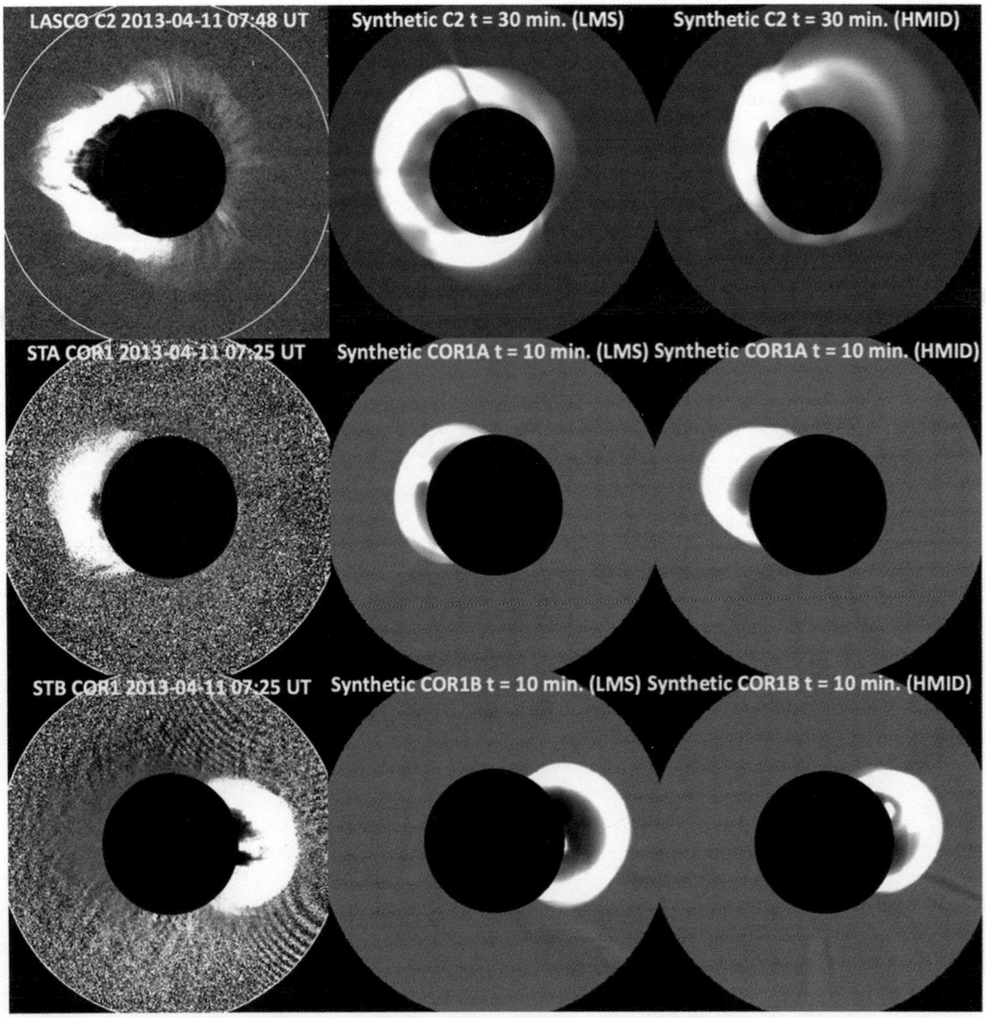

Figure 6. Left: White-light observations from LASCO C2, STEREO-A COR1 and STEREO-B COR1 of 2013 April 11 CME. Middle: Synthesized white light images from the simulation using Lockheed Martin Synchronic (LMS) map. Right: Synthesized white light images from the simulation using HMI Diachronic (HMID) map.

Another challenge to the "shock scenario" is the effect of magnetic mirroring. The strong magnetic mirroring, resulting from a high degree of convergence of magnetic field lines towards the Sun and extremely small loss cones, could potentially prevent particles from reaching the photosphere to produce γ-rays. To assess the impact of magnetic mirroring that inhibits particle transport back to the Sun, we calculated the evolution of the mirror ratio in the SOL2014-09-01 simulation. Using an approximate relationship between the escape time and mirror ratio (Malyshkin and Kulsrud 2001; Effenberger and Petrosian 2018), we determined that the particle escape timescale is shorter than the γ-ray emission period, which means a significant portion of the downstream protons can reach the photosphere and produce γ-rays in this event. Future studies should self-consistently calculate the transport of back-propagating protons downstream of the CME-driven shock by combining theory and MHD modeling results (e.g., Petrosian et al. 2023).

Finally, we would like to discuss one of the major challenges in CME-driven shock and SEP modeling: the influence of the background solar wind on the properties of the CME-driven shock (Jin et al. 2022). There are previous studies (e.g., Hinterreiter et al. 2019) show the uncertainties of the background solar wind modeling. Figure 6 illustrates how the ambient solar corona and wind can significantly affect the parameters of the CME-driven shock. The figure shows the synthesized CME white light images driven by an identical flux rope, but under slightly different background solar wind conditions caused by different types of input magnetic maps (one is a synchronic map, and the other is a diachronic/synoptic map). These variations in the ambient solar wind lead to distinctly different CME-driven shocks. The morphology of the shock surface shows greater latitudinal expansion in the Lockheed Martin Synchronic (LMS) case than in the HMI Diachronic (HMID) case. The resulting shock parameters are also noticeably different. For detailed comparison of the two cases, we refer to Jin et al. (2022). This result underscores the importance of obtaining better constraints on the plasma environment of the CME source region and its path, especially for accurately capturing the early phase of shock evolution.

5. Conclusions

The CME-driven shock plays a crucial role in accelerating particles in the heliosphere. By integrating advanced modeling with multi-messenger observations, we have demonstrated the significant impact CME-driven shocks have on both γ-ray emissions and in-situ SEP observations. Our results provide promising evidence that particles accelerated by CME shocks could travel back to the Sun, producing γ-rays observed by Fermi and potentially opening a new avenue for observing shock-accelerated particles. However, alternative scenarios also exist, and more detailed modeling and observational efforts are required to enhance our understanding of the physical mechanisms behind Fermi SGRE events. By coupling AWSoM with iPATH and involving the early-stage (<10 Rs) shock evolution, our new model shows promising potential to reproduce SEP intensity profiles and spectra at various locations in the heliosphere. Lastly, our findings underscore the importance of accurate background coronal and solar wind modeling, as well as detailed observations of CME source regions, in advancing our understanding of CME-driven shocks and the dynamics of associated energetic particles.

Acknowledgements. M.J., N.N., W.L. acknowledge support from NASA's SDO/AIA contract (NNG04EA00C) to LMSAL. AIA is an instrument onboard the Solar Dynamics Observatory, a mission for NASA's Living With a Star program. The authors also acknowledge support from multiple NASA grants 80NSSC24K0136, 80NSSC21K1782, 80NSSC21K1327, 80NSSC18K1126, 80NSSC24K0175, and 80NSSC22K0527.

References

Ackermann, M. *et al.* 2017, Fermi-LAT Observations of High-energy Behind-the-limb Solar Flares. ApJ, 835(2), 219.

Ajello, M. *et al.* 2021, First Fermi-LAT Solar Flare Catalog. ApJS, 252(2), 13.

Atwood, W. B. *et al.* 2009, The Large Area Telescope on the Fermi Gamma-Ray Space Telescope Mission. ApJ, 697(2), 1071–1102.

Axford, W. I., Leer, E., & Skadron, G. The Acceleration of Cosmic Rays by Shock Waves. In *International Cosmic Ray Conference* 1977, volume 11 of *International Cosmic Ray Conference*, 132.

Borovikov, D., Sokolov, I. V., Manchester, W. B., Jin, M., & Gombosi, T. I. 2017, Eruptive event generator based on the Gibson-Low magnetic configuration. *Journal of Geophysical Research (Space Physics)*, 122(8), 7979–7984.

Borovikov, D., Sokolov, I. V., Roussev, I. I., Taktakishvili, A., & Gombosi, T. I. 2018, Toward a Quantitative Model for Simulation and Forecast of Solar Energetic Particle Production during Gradual Events. I. Magnetohydrodynamic Background Coupled to the SEP Model. ApJ, 864(1), 88.

Bruno, A., de Nolfo, G. A., Ryan, J. M., Richardson, I. G., & Dalla, S. 2023, Statistical Relationship between Long-duration High-energy Gamma-Ray Emission and Solar Energetic Particles. ApJ, 953(2), 187.

Bruno, A. *et al.* 2018, Solar Energetic Particle Events Observed by the PAMELA Mission. ApJ, 862(2), 97.

Cane, H. V., Mewaldt, R. A., Cohen, C. M. S., & von Rosenvinge, T. T. 2006, Role of flares and shocks in determining solar energetic particle abundances. *Journal of Geophysical Research (Space Physics)*, 111(A6), A06S90.

Cane, H. V., von Rosenvinge, T. T., Cohen, C. M. S., & Mewaldt, R. A. 2003, Two components in major solar particle events. Geophys. Res. Lett., 30(12), 8017.

Cohen, C., mason, G., Jin, M., Nitta, N., Christian, E., Cummings, A., Desai, M., De Nolfo, G., Giacalone, J., Hill, M., Joyce, C., Labrador, A., Leske, R., Matthaeus, W., McComas, D., McNutt, R., Mewaldt, R., Mitchell, D., Mitchell, J., Rankin, J., Roelof, E., Schwadron, N., Stone, E., Szalay, J., Wiedenbeck, M., Ho, G., & Allen, R. Heavy Ion Characteristics Observed by Multiple Spacecraft Observations in the 28 October 2021 Solar Energetic Particle Event. In *Third Triennial Earth-Sun Summit (TESS)* 2022, volume 54, 2022n7i201p06.

de Nolfo, G. A., Bruno, A., Ryan, J. M., Dalla, S., Giacalone, J., Richardson, I. G., Christian, E. R., Stochaj, S. J., Bazilevskaya, G. A., Boezio, M., Martucci, M., Mikhailov, V. V., & Munini, R. 2019, Comparing Long-duration Gamma-Ray Flares and High-energy Solar Energetic Particles. ApJ, 879(2), 90.

Effenberger, F. & Petrosian, V. 2018, The Relation between Escape and Scattering Times of Energetic Particles in a Turbulent Magnetized Plasma: Application to Solar Flares. ApJ, 868(2), L28.

Gibson, S. E. & Low, B. C. 1998, A Time-Dependent Three-Dimensional Magnetohydrodynamic Model of the Coronal Mass Ejection. ApJ, 493, 460–473.

Gopalswamy, N., Mäkelä, P., Yashiro, S., Lara, A., Xie, H., Akiyama, S., & MacDowall, R. J. 2018, Interplanetary Type II Radio Bursts from Wind/WAVES and Sustained Gamma-Ray Emission from Fermi/LAT: Evidence for Shock Source. ApJ, 868(2), L19.

Gopalswamy, N., Xie, H., Akiyama, S., Yashiro, S., Usoskin, I. G., & Davila, J. M. 2013, The First Ground Level Enhancement Event of Solar Cycle 24: Direct Observation of Shock Formation and Particle Release Heights. ApJ, 765(2), L30.

Gopalswamy, N., Yashiro, S., Mäkelä, P., Xie, H., & Akiyama, S. 2021, The Common Origin of High-energy Protons in Solar Energetic Particle Events and Sustained Gamma-Ray Emission from the Sun. ApJ, 915(2), 82.

Grechnev, V. V., Kiselev, V. I., Kashapova, L. K., Kochanov, A. A., Zimovets, I. V., Uralov, A. M., Nizamov, B. A., Grigorieva, I. Y., Golovin, D. V., Litvak, M. L., Mitrofanov, I. G., & Sanin, A. B. 2018, Radio, Hard X-Ray, and Gamma-Ray Emissions Associated with a Far-Side Solar Event. Sol. Phys., 293(10), 133.

Hinterreiter, J., Magdalenic, J., Temmer, M., Verbeke, C., Jebaraj, I. C., Samara, E., Asvestari, E., Poedts, S., Pomoell, J., Kilpua, E., Rodriguez, L., Scolini, C., & Isavnin, A. 2019, Assessing the Performance of EUHFORIA Modeling the Background Solar Wind. Sol. Phys., 294(12), 170.

Hu, J., Li, G., Ao, X., Zank, G. P., & Verkhoglyadova, O. 2017, Modeling Particle Acceleration and Transport at a 2-D CME-Driven Shock. *Journal of Geophysical Research (Space Physics)*, 122(11), 10,938–10,963.

Hudson, H. S. The Relationship between Long-Duration Gamma-Ray Flares and Solar Cosmic Rays. In Foullon, C. & Malandraki, O. E., editors, *Space Weather of the Heliosphere: Processes and Forecasts* 2018, volume 335, pp. 49–53.

Jin, M., Manchester, W. B., van der Holst, B., Oran, R., Sokolov, I., Toth, G., Liu, Y., Sun, X. D., & Gombosi, T. I. 2013, Numerical Simulations of Coronal Mass Ejection on 2011 March 7: One-temperature and Two-temperature Model Comparison. ApJ, 773, 50.

Jin, M., Manchester, W. B., van der Holst, B., Sokolov, I., Tóth, G., Mullinix, R. E., Taktakishvili, A., Chulaki, A., & Gombosi, T. I. 2017,a Data-constrained Coronal Mass Ejections in a Global Magnetohydrodynamics Model. ApJ, 834a, 173.

Jin, M., Manchester, W. B., van der Holst, B., Sokolov, I., Tóth, G., Vourlidas, A., de Koning, C. A., & Gombosi, T. I. 2017,b Chromosphere to 1 AU Simulation of the 2011 March 7th Event: A Comprehensive Study of Coronal Mass Ejection Propagation. ApJ, 834b, 172.

Jin, M., Nitta, N. V., & Cohen, C. M. S. 2022, Assessing the Influence of Input Magnetic Maps on Global Modeling of the Solar Wind and CME-Driven Shock in the 2013 April 11 Event. *Space Weather*, 20(3), e2021SW002894.

Jin, M., Petrosian, V., Liu, W., Nitta, N. V., Omodei, N., Rubio da Costa, F., Effenberger, F., Li, G., Pesce-Rollins, M., Allafort, A., & Manchester, Ward, I. 2018, Probing the Puzzle of Behind-the-limb γ-Ray Flares: Data-driven Simulations of Magnetic Connectivity and CME-driven Shock Evolution. ApJ, 867(2), 122.

Jin, M., Schrijver, C. J., Cheung, M. C. M., DeRosa, M. L., Nitta, N. V., & Title, A. M. 2016, A Numerical Study of Long-range Magnetic Impacts during Coronal Mass Ejections. ApJ, 820(1), 16.

Kozarev, K. A., Evans, R. M., Schwadron, N. A., Dayeh, M. A., Opher, M., Korreck, K. E., & van der Holst, B. 2013, Global Numerical Modeling of Energetic Proton Acceleration in a Coronal Mass Ejection Traveling through the Solar Corona. ApJ, 778(1), 43.

Kozarev, K. A., Korreck, K. E., Lobzin, V. V., Weber, M. A., & Schwadron, N. A. 2011, Off-limb Solar Coronal Wavefronts from SDO/AIA Extreme-ultraviolet Observations—Implications for Particle Production. ApJ, 733(2), L25.

Li, G., Jin, M., Ding, Z., Bruno, A., de Nolfo, G. A., Randol, B. M., Mays, L., Ryan, J., & Larlo, D. 2021, Modeling the 2012 May 17 Solar Energetic Particle Event Using the AWSoM and iPATH Models. ApJ, 919(2), 146.

Linker, J. A., Caplan, R. M., Schwadron, N., Gorby, M., Downs, C., Torok, T., Lionello, R., & Wijaya, J. Coupled MHD-Focused Transport Simulations for Modeling Solar Particle Events. In *Journal of Physics Conference Series* 2019, volume 1225 of *Journal of Physics Conference Series*, 012007. IOP.

Malyshkin, L. & Kulsrud, R. 2001, Transport Phenomena in Stochastic Magnetic Mirrors. ApJ, 549(1), 402–415.

Nitta, N., Jin, M., & Cohen, C. Understanding the Origin of Iron-rich Gradual Solar Energetic Particle Events. In *44th COSPAR Scientific Assembly. Held 16-24 July* 2022, volume 44, 1168.

Patsourakos, S., Georgoulis, M. K., Vourlidas, A., Nindos, A., Sarris, T., Anagnostopoulos, G., Anastasiadis, A., Chintzoglou, G., Daglis, I. A., Gontikakis, C., Hatzigeorgiu, N., Iliopoulos, A. C., Katsavrias, C., Kouloumvakos, A., Moraitis, K., Nieves-Chinchilla, T., Pavlos, G., Sarafopoulos, D., Syntelis, P., Tsironis, C., Tziotziou, K., Vogiatzis, I. I., Balasis, G., Georgiou, M., Karakatsanis, L. P., Malandraki, O. E., Papadimitriou, C., Odstrčil, D., Pavlos, E. G., Podlachikova, O., Sandberg, I., Turner, D. L., Xenakis, M. N., Sarris, E., Tsinganos, K., & Vlahos, L. 2016, The Major Geoeffective Solar Eruptions of 2012 March 7: Comprehensive Sun-to-Earth Analysis. ApJ, 817(1), 14.

Pesce-Rollins, M., Omodei, N., Krucker, S., Di Lalla, N., Wang, W., Battaglia, A. F., Warmuth, A., Veronig, A. M., & Baldini, L. 2022, The Coupling of an EUV Coronal Wave and Ion Acceleration in a Fermi-LAT Behind-the-Limb Solar Flare. ApJ, 929(2), 172.

Pesce-Rollins, M., Omodei, N., Petrosian, V., Liu, W., Rubio da Costa, F., Allafort, A., & Chen, Q. 2015, First Detection of >100 MeV Gamma Rays Associated with a Behind-the-limb Solar Flare. ApJ, 805(2), L15.

Petrosian, V., Orlando, E., & Strong, A. 2023, Transport of Cosmic-Ray Electrons from 1 au to the Sun. ApJ, 943(1), 21.

Plotnikov, I., Rouillard, A. P., & Share, G. H. 2017, The magnetic connectivity of coronal shocks from behind-the-limb flares to the visible solar surface during γ-ray events. A&A, 608, A43.

Ryan, J. M. & Lee, M. A. 1991, On the Transport and Acceleration of Solar Flare Particles in a Coronal Loop. ApJ, 368, 316.

Sachdeva, N., van der Holst, B., Manchester, W. B., Tóth, G., Chen, Y., Lloveras, D. G., Vásquez, A. M., Lamy, P., Wojak, J., Jackson, B. V., Yu, H.-S., & Henney, C. J. 2019, Validation of the Alfvén Wave Solar Atmosphere Model (AWSoM) with Observations from the Low Corona to 1 au. ApJ, 887(1), 83.

Sokolov, I. V., van der Holst, B., Oran, R., Downs, C., Roussev, I. I., Jin, M., Manchester, Ward B., I., Evans, R. M., & Gombosi, T. I. 2013, Magnetohydrodynamic Waves and Coronal Heating: Unifying Empirical and MHD Turbulence Models. ApJ, 764(1), 23.

Tóth, G., van der Holst, B., Sokolov, I. V., De Zeeuw, D. L., Gombosi, T. I., Fang, F., Manchester, W. B., Meng, X., Najib, D., Powell, K. G., Stout, Q. F., Glocer, A., Ma, Y.-J., & Opher, M. 2012, Adaptive numerical algorithms in space weather modeling. *Journal of Computational Physics*, 231, 870–903.

Tylka, A. J., Cohen, C. M. S., Dietrich, W. F., Lee, M. A., Maclennan, C. G., Mewaldt, R. A., Ng, C. K., & Reames, D. V. 2005, Shock Geometry, Seed Populations, and the Origin of Variable Elemental Composition at High Energies in Large Gradual Solar Particle Events. ApJ, 625(1), 474–495.

van der Holst, B., Sokolov, I. V., Meng, X., Jin, M., Manchester, IV, W. B., Tóth, G., & Gombosi, T. I. 2014, Alfvén Wave Solar Model (AWSoM): Coronal Heating. ApJ, 782, 81.

Whitman, K. *et al.* 2023, Review of Solar Energetic Particle Prediction Models. *Advances in Space Research*, 72(12), 5161–5242.

Wijsen, N., Li, G., Ding, Z., Lario, D., Poedts, S., Filwett, R. J., Allen, R. C., & Dayeh, M. A. 2023, On the seed population of solar energetic particles in the inner heliosphere. *Journal of Geophysical Research: Space Physics*, 128(3), e2022JA031203.

Solar and Stellar Coronal Mass Ejections
Proceedings IAU Symposium No. 388, 2026
N. Gopalswamy, O. Malandraki, A. Vidotto & W. Manchester, eds.
doi:10.1017/S1743921324001856

The Relationship of Coronal Dimmings with CMEs and the Implication in Detecting Stellar CMEs

Nariaki V. Nitta[1], Meng Jin[1] and Karin Dissauer[2]

[1] Lockheed Martin Solar and Astrophysics Lab (LMSAL), Palo Alto, CA, USA

[2] NorthWest Research Associates, Boulder, CO, USA

Abstract. Coronal mass ejections (CMEs) play a significant role in space weather. In order to evaluate space weather effects, it is important to understand the origin of CMEs in the solar atmosphere. Among various proxies for CMEs in the low corona, the coronal dimming observed as intensity deficit in extreme ultraviolet (EUV) or soft X-ray wavelengths is thought to be the most reliable observable. We examine dimmings detected automatically by Solar Demon in images taken by the Atmospheric Imaging Assembly on board the Solar Dynamics Observatory (SDO), and associate strong dimmings with CMEs, flares and EUV waves. The association rate of dimmings with either CMEs or flares is about 75%, and that with EUV waves is about 35%. We also discuss the complementary use of EUV irradiance data, for example, from the Extreme Ultraviolet Experiment on board SDO to isolate dimmings. This may have implications in detecting CMEs from stars.

Keywords. Coronal Dimmings, Coronal Waves, Coronal Mass Ejections, Solar Flares

1. Introduction

Coronal mass ejections (CMEs) are the most energetic solar transient phenomenon, and can significantly disturb the heliosphere that includes the near-Earth space. CMEs are traditionally defined in white-light coronagraph images, in which an occulting disk is placed to block the Sun's light. This prevents us from observing the CME very close to the Sun. Therefore, we rely on its proxies in the low corona to understand how the CME is launched. Coronal dimmings, which represent areas at which the intensity decreases in extreme ultraviolet (EUV) or soft X-ray wavelengths, are considered the most reliable indicator of CMEs (e.g., Thompson et al. 2000; Reinard and Biesecker 2009; Nitta et al. 2014), especially since they often outline the spatial extent of CMEs (Thompson et al. 2000). It is possible that coronal dimmings may contain important information on the nascent and early evolutional stages of CMEs as shown in advanced numerical simulations (e.g., Jin et al. 2022) and in detailed case studies of dimmings (e.g., Cheng and Qiu 2016; Chikunova et al. 2023; Zhu et al. 2024; Jain et al. 2024). These studies suggest that dimmings may be used as a space weather forecasting tool.

Dimmings are also detected in spatially-integrated EUV irradiance data, even in the presence of strong flare emission (Mason et al. 2014, 2016, 2019). This is important for two reasons. Among several flare-related observables, only dimmings in EUV irradiance data can be used to distinguish X-class flares that are eruptive (associated with a CME) from those that are non-eruptive, according to Harra et al. (2016). Furthermore, the dimming in EUV irradiance data is one of the most promising ways of detecting stellar CMEs (Odert et al. 2017; Cranmer 2017; Moschou et al. 2019), as recently explored by

Veronig et al. (2021) and Loyd et al. (2022), in anticipation of forthcoming missions that will enhance the detection capabilities for EUV dimmings (e.g., France et al. 2022; Jin et al. 2020).

In order to properly evaluate the usefulness of dimmings for improving space weather forecasting and detecting CMEs from stars, it is important to conduct statistical or ensemble studies. Such studies of many dimming events have been carried out, for example, by Dissauer et al. (2018, 2019) and Chikunova et al. (2020). Most of the events used in these papers came from the list of large-scale coronal propagation fronts (LCPFs, or EUV waves as more commonly referred to) as published by Nitta et al. (2013) with only a few exceptions. Accordingly, the above-mentioned papers may have been subject to some selection bias, although the initial papers that reported on the two phenomena (Thompson et al. 1998, 1999) suggested that they are intimately related.

To further address the importance of coronal dimmings, it is meaningful to study those that are selected without referring to other phenomena. There is an extensive database of coronal dimmings automatically detected and characterized by the software package called Solar Demon (Kraaikamp and Verbeeck 2015). In this contribution, we show our on-going effort to study dimmings based on Solar Demon. In the next section, we briefly describe Solar Demon and show initial findings on strong dimming events. In Section 3, we discuss dimmings in irradiance data, including an example that was not detected by Solar Demon. We conclude in Section 4, briefly discussing the implications of what we have so far found. A full paper will be prepared in near future.

2. Solar Demon Dimmings

The Solar Demon software package (Kraaikamp and Verbeeck 2015) was developed to automatically detect and characterize solar flares, coronal dimmings, and EUV waves in EUV images taken by the Atmospheric Imaging Assembly (AIA: Lemen et al. 2012) on SDO. Solar Demon uses AIA images in two-minute cadence, rebinned to 1024×1024 pixels. The dimming detection makes Solar Demon unique. No other resources carry as extensive databases of dimmings as Solar Demon. A temporary catalog of dimmings is made in near real time using AIA quick look data, but the permanent database of dimmings, which this study accesses, is built on AIA science data that typically become available a few days later.

Solar Demon detects dimmings on running difference images of a six-minute temporal separation, and characterizes them on base difference images. The Solar Demon web page (`https://www.sidc.be/solardemon/science/dimmings.php`) provides basic information such as the dimming intensity, times (start, peak and end), maximum change of intensity drop, centroid of dimming regions in heliographic latitude and longitude. The pages specific to individual events show temporal variations of dimming parameters and images, still and animated, in intensity, base difference, dimming masks, and so on.

We concentrate on strong dimming events, expecting them to be more likely associated with CMEs. We first limited to events with the dimming intensity $<-300\times10^3$ DN/s and a known location of the centroid of the dimming areas on the solar disk. Then we eliminated those events that occurred while there were no LASCO data to isolate events intrinsically without a CME. A total of 380 events up to 2022 February were selected in this way. It appears that Solar Demon has acted differently, starting some time in 2022, and included many more events with large intensity deficits due possibly to changing instrument characteristics that may require modifying the parameters used in Solar Demon. This issue has been noted by the development team (E. Kraaikamp, 2024, private communication).

For each dimming event we separately search for flares, CMEs and EUV waves that are closest in time to the dimming. For flares, we use the NOAA event list that is

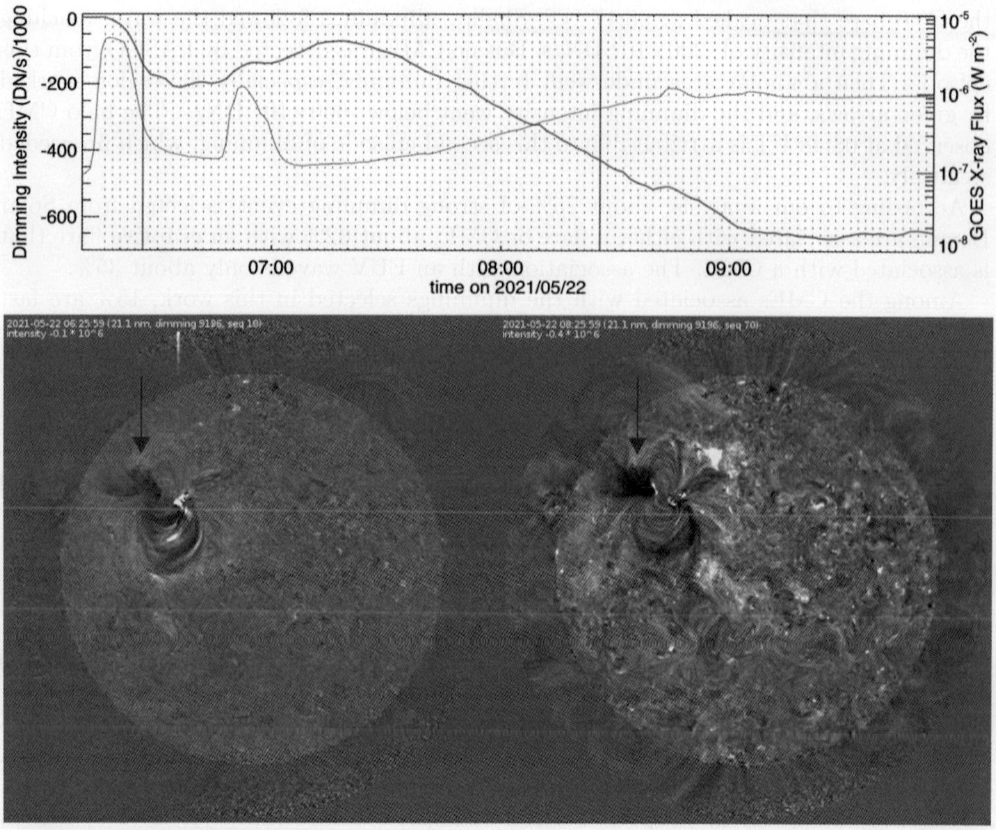

Figure 1. Example of Solar Demon outputs. The upper panel shows the time variations of the dimming intensity (red). We added the GOES X-ray (1 – 8 Å) light curves (blue), and colorized the dimming intensity. The lower part of the figure shows AIA 211 Å base difference (pre-event subtracted) images at two times as indicated by brown lines in the upper plot. Nearly the same area (pointed by green arrows) dimmed in the first and second episodes of events associated, respectively, with the C6.0 and more gradual C1.4 flares.

ingested to the Heliophysics Event Knoweledgebase (HEK: Hurlburt et al. 2012) with the added/updated positional information contained in the "SolarSoft latest events" archives linked on `https://www.lmsal.com/solarsoft/last_events/`. If the time difference is small and the location of the flare is at least in the same quadrant as the centroid of the dimming, we consider the flare to be associated.

The association of dimmings with CMEs is based on the CDAW LASCO CME catalog (`https://cdaw.gsfc.nasa.gov/CME_list/`), which provides not only the basic properties of CMEs, such as the time of the first detection, linear speed and acceleration, position angle, mass, etc., but also various plots and movies, including LASCO running difference images side by side with GOES X-ray light curves. Finally, the association of dimmings with EUV waves is based on the catalog of LCPFs (`https://aia.lmsal.com/AIA_Waves/index.html`), which has been extended from Nitta et al. (2013) and also contains movies of intensity and both running and base difference images.

In most (∼70%) of our strong dimmings, it is straightforward to determine whether they are associated with flares, CME and EUV waves, and which events if they are associated. However, there are cases we need a further examination of data as illustrated in Figure 1. As discussed by Nitta et al. (2023), an EUV wave was seen in association with the C6.0 flare that peaked at 06:17 UT. Shallow dimmings followed the wave, reaching the dimming intensity of -200$\times$10^3 DN/s. But no CME was detected by LASCO from this episode. Then a similar area underwent further dimming, especially the region marked by green arrows, and the dimming intensity went below -600$\times$10^3 DN/s. The halo CME observed at 08:48 UT is attributable to this second episode of dimming, which proceeded stealthily.

According to our analysis, about 75% of strong dimming events selected from Solar Demon are associated with either a flare or CME, about 62% with an eruptive flare that is associated with a CME. The association with an EUV wave is only about 35%.

Among the CMEs associated with the dimmings selected in this work, 15% are fast ($v \gtrsim 1000$ km s^{-1}), 44% are slow ($v \lesssim 500$ km s^{-1}) and 41% belong to the range of v between 500 and 1000 km s^{-1}. Full halo CMEs (angular width of 360°) account for 31% of the CMEs associated with our dimming events. About 74% of these CMEs are wider than 100°. For the flares associated with dimmings selected in this work, 6%, 34%, 50%, and 10% are X-class, M-class, C-class, and B-class, respectively. At this point, we have not investigated the associations in opposite directions, such as those of partial or full halo CMEs and X-class flares with dimmings.

3. Dimmings in Irradiance Data

Apparently, it is more challenging to observe dimmings in spatially integrated irradiance data than in EUV images such as those from SDO/AIA. However, the Extreme Ultraviolet Variability Experiment (EVE Woods et al. 2012) on board SDO, especially the MEGS-A instrument that covers the wavelength of 50 – 370 Å, have revealed dimmings in eruptive events (Mason et al. 2014, 2016, 2019). This is likely helped by the 1 Å spectral resolution that allows better temperature discrimination than AIA.

We study dimmings in the same set of 10 Fe spectral lines that were used by Mason et al. (2014, 2016). Here we just show in Figure 2 an interesting example of dimmings in MEGS-A data. It appears that the intensity declined after ≈22:00 UT on 2012 October 4 up to the highest temperature as shown in the bottom panel. This is highly unusual, since in most cases dimming is noted in EVE data only up to the Fe XII – Fe XIV lines. Note that this event was initially called a stealth CME as no clear low coronal signatures were found in AIA imags before taking difference of long (e.g., six hour) separation (Nitta and Mulligan 2017). In this case, dimming precedes the flare, which develops very slowly to the peak flux of B7.8.

Although not discussed here in detail, we have also isolated dimmings in heavily limb-occulted events, indicating that dimmings can extend to high altitudes. These events are free of "contamination" from the contemporaneous flare associated with the CME. We will report on the results in a separate paper.

4. Discussion

We present the first results of our investigation on dimmings selected automatically by Solar Demon. Limiting to strong (intensity deficit <-300$\times$10^3 DN/s) dimmings, the association rate of dimmings with CMEs and flares is found to be about 75%. This association rate of strong dimmings with CMEs can be taken as a lower limit of the CME visibility by modern coronagraphs such as SOHO/LASCO (e.g., Yashiro et al. 2005; Vourlidas et al. 2020). Vakhrusheva et al. (2024) also studied dimmings detected by Solar Demon, but without limits on the dimming strength, only 16% of their dimming events were associated with a CME.

We plan to conduct further analyses similar to those of Dissauer et al. (2018, 2019), using the dimming events from Solar Demon, primarily to see if there was a bias in these

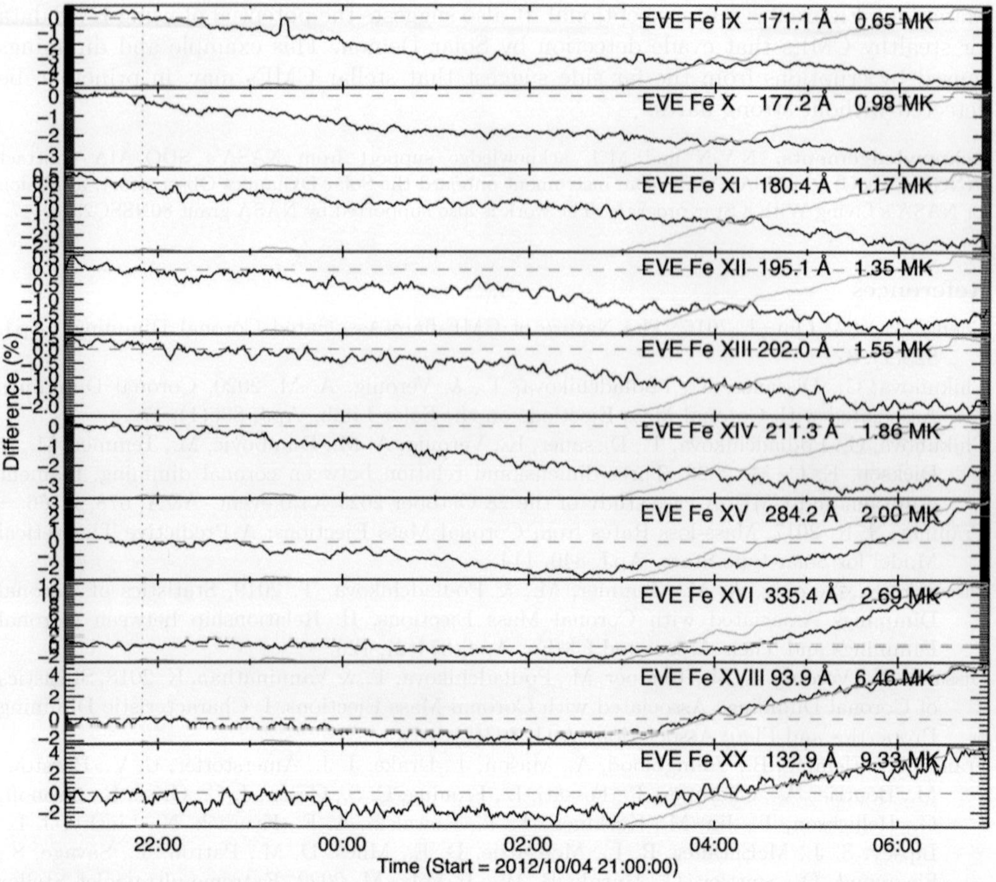

Figure 2. Eve MEGS-A light curves for a CME that was associated with a very slowly developing flare. As indicated, the temperature of the line goes up from top to bottom. GOES 1–8 Å X-ray light curves, scaled in each panel, are indicated in gray.

studies as a result of selecting events mostly from the catalog of EUV waves originally made public by Nitta et al. (2013). It is predicted, however, that the bias could be negligible. This is because the above-mentioned catalog picked far more EUV waves (362 vs 164, according to Long et al. (2017)) than those as detected by the automated software CorPITA (Long et al. 2014).

Although automated detection generally leads to databases less biased than those manually generated, Solar Demon has some restrictions in terms of detecting dimmings. It relies on running difference images and thus requires variations in intensity within the six-minute separation. Therefore, none of the stealthy CMEs, although geoeffective, as studied by Nitta et al. (2021), are included in the Solar Demon database. It would probably have failed to accommodate the CME-related dimming in Figure 1 without the first episode that was characterized by faster intensity drop. Another restriction may be the use of only 211 Å images, given that dimmings may be distributed differently at different temperatures.

This is where EVE data with good temperature discrimination become important and complementary to AIA images. As mentioned, dimming signals in EVE data are generally largest in the low-temperature Fe IX line, decreasing toward high-temperature lines. In most cases, dimming is seen only up to lines that represent ∼2 MK plasma. Therefore, the example in Figure 2 is quite exceptional. It also suggests the potential of using EVE data for stealthy CMEs that evade detection by Solar Demon. This example and dimmings caused by eruptions from the far side suggest that stellar CMEs may, in principle, be detected without strong flares.

Acknowledgements. N.V.N and M.J. acknowledge support from NASA's SDO/AIA contract (NNG04EA00C) to LMSAL. AIA is an instrument onboard the Solar Dynamics Observatory, a mission for NASA's Living With a Star program. The work is also supported by NASA grant 80NSSC23K0417.

References

Cheng, J. X. & Qiu, J. 2016, The Nature of CME-flare-Associated Coronal Dimming. ApJ, 825(1), 37.

Chikunova, G., Dissauer, K., Podladchikova, T., & Veronig, A. M. 2020, Coronal Dimmings Associated with Coronal Mass Ejections on the Solar Limb. ApJ, 896(1), 17.

Chikunova, G., Podladchikova, T., Dissauer, K., Veronig, A. M., Dumbović, M., Temmer, M., & Dickson, E. C. M. 2023, Three-dimensional relation between coronal dimming, filament eruption, and CME. A case study of the 28 October 2021 X1.0 event. A&A, 678, A166.

Cranmer, S. R. 2017, Mass-loss Rates from Coronal Mass Ejections: A Predictive Theoretical Model for Solar-type Stars. ApJ, 840, 114.

Dissauer, K., Veronig, A. M., Temmer, M., & Podladchikova, T. 2019, Statistics of Coronal Dimmings Associated with Coronal Mass Ejections. II. Relationship between Coronal Dimmings and Their Associated CMEs. ApJ, 874(2), 123.

Dissauer, K., Veronig, A. M., Temmer, M., Podladchikova, T., & Vanninathan, K. 2018, Statistics of Coronal Dimmings Associated with Coronal Mass Ejections. I. Characteristic Dimming Properties and Flare Association. ApJ, 863(2), 169.

France, K., Fleming, B., Youngblood, A., Mason, J., Drake, J. J., Amerstorfer, U. V., Barstow, M., Bourrier, V., Champey, P., Fossati, L., Froning, C. S., Green, J. C., Grisé, F., Gronoff, G., Hellickson, T., Jin, M., Koskinen, T. T., Kowalski, A. F., Kruczek, N., Linsky, J. L., Lipscy, S. J., McEntaffer, R. L., McKenzie, D. E., Miles, D. M., Patton, T., Savage, S., Siegmund, O., Spittler, C., Unruh, B. W., & Volz, M. 2022, Extreme-ultraviolet Stellar Characterization for Atmospheric Physics and Evolution mission: motivation and overview. *Journal of Astronomical Telescopes, Instruments, and Systems*, 8, 014006.

Harra, L. K., Schrijver, C. J., Janvier, M., Toriumi, S., Hudson, H., Matthews, S., Woods, M. M., Hara, H., Guedel, M., Kowalski, A., Osten, R., Kusano, K., & Lueftinger, T. 2016, The Characteristics of Solar X-Class Flares and CMEs: A Paradigm for Stellar Superflares and Eruptions? Sol. Phys., 291, 1761–1782.

Hurlburt, N., Cheung, M., Schrijver, C., Chang, L., Freeland, S., Green, S., Heck, C., Jaffey, A., Kobashi, A., Schiff, D., Serafin, J., Seguin, R., Slater, G., Somani, A., & Timmons, R. 2012, Heliophysics Event Knowledgebase for the Solar Dynamics Observatory (SDO) and Beyond. Sol. Phys., 275(1-2), 67–78.

Jain, S., Podladchikova, T., Chikunova, G., Dissauer, K., & Veronig, A. M. 2024, Coronal dimmings as indicators of the direction of early coronal mass ejection propagation. A&A, 683, A15.

Jin, M., Cheung, M. C. M., DeRosa, M. L., Nitta, N. V., & Schrijver, C. J. 2022, Coronal Mass Ejections and Dimmings: A Comparative Study Using MHD Simulations and SDO Observations. ApJ, 928(2), 154.

Jin, M., Cheung, M. C. M., DeRosa, M. L., Nitta, N. V., Schrijver, C. J., France, K., Kowalski, A., Mason, J. P., & Osten, R. 2020, Coronal Dimming as a Proxy for Stellar Coronal Mass Ejections. *arXiv e-prints*, arXiv:2002.06249.

Kraaikamp, E. & Verbeeck, C. 2015, Solar Demon - an approach to detecting flares, dimmings, and EUV waves on SDO/AIA images. *Journal of Space Weather and Space Climate*, 5(27), A18.

Lemen, J. R., Title, A. M., Akin, D. J., Boerner, P. F., Chou, C., Drake, J. F., Duncan, D. W., Edwards, C. G., Friedlaender, F. M., Heyman, G. F., Hurlburt, N. E., Katz, N. L., Kushner,

G. D., Levay, M., Lindgren, R. W., Mathur, D. P., McFeaters, E. L., Mitchell, S., Rehse, R. A., Schrijver, C. J., Springer, L. A., Stern, R. A., Tarbell, T. D., Wuelser, J.-P., Wolfson, C. J., Yanari, C., Bookbinder, J. A., Cheimets, P. N., Caldwell, D., Deluca, E. E., Gates, R., Golub, L., Park, S., Podgorski, W. A., Bush, R. I., Scherrer, P. H., Gummin, M. A., Smith, P., Auker, G., Jerram, P., Pool, P., Soufli, R., Windt, D. L., Beardsley, S., Clapp, M., Lang, J., & Waltham, N. 2012, The Atmospheric Imaging Assembly (AIA) on the Solar Dynamics Observatory (SDO). Sol. Phys., 275(1-2), 17–40.

Long, D. M., Bloomfield, D. S., Gallagher, P. T., & Pérez-Suárez, D. 2014, CorPITA: An Automated Algorithm for the Identification and Analysis of Coronal "EIT Waves". Sol. Phys., 289(9), 3279–3295.

Long, D. M., Murphy, P., Graham, G., Carley, E. P., & Pérez-Suárez, D. 2017, A Statistical Analysis of the Solar Phenomena Associated with Global EUV Waves. Sol. Phys., 292, 185.

Loyd, R. O. P., Mason, J. P., Jin, M., Shkolnik, E. L., France, K., Youngblood, A., Villadsen, J., Schneider, C., Schneider, A. C., Llama, J., Ramiaramanantsoa, T., & Richey-Yowell, T. 2022, Constraining the Physical Properties of Stellar Coronal Mass Ejections with Coronal Dimming: Application to Far-ultraviolet Data of ϵ Eridani. ApJ, 936(2), 170.

Mason, J. P., Attie, R., Arge, C. N., Thompson, B., & Woods, T. N. 2019, The SDO/EVE Solar Irradiance Coronal Dimming Index Catalog. I. Methods and Algorithms. ApJS, 244(1), 13.

Mason, J. P., Woods, T. N., Caspi, A., Thompson, B. J., & Hock, R. A. 2014, Mechanisms and Observations of Coronal Dimming for the 2010 August 7 Event. ApJ, 789, 61.

Mason, J. P., Woods, T. N., Webb, D. F., Thompson, B. J., Colaninno, R. C., & Vourlidas, A. 2016, Relationship of EUV Irradiance Coronal Dimming Slope and Depth to Coronal Mass Ejection Speed and Mass. ApJ, 830, 20.

Moschou, S.-P., Drake, J. J., Cohen, O., Alvarado-Gómez, J. D., Garraffo, C., & Fraschetti, F. 2019, The Stellar CME-Flare Relation: What Do Historic Observations Reveal? ApJ, 877(2), 105.

Nitta, N. V., Aschwanden, M. J., Freeland, S. L., Lemen, J. R., Wülser, J.-P., & Zarro, D. M. 2014, The Association of Solar Flares with Coronal Mass Ejections During the Extended Solar Minimum. Sol. Phys., 289, 1257–1277.

Nitta, N. V., Bučík, R., Mason, G. M., Ho, G. C., Cohen, C. M. S., Gómez-Herrero, R., Wang, L., & Balmaceda, L. A. 2023, Solar Activities Associated with 3He-rich Solar Energetic Particle Events Observed by Solar Orbiter. *Frontiers in Astronomy and Space Sciences*, 10, 50.

Nitta, N. V. & Mulligan, T. 2017, Earth-Affecting Coronal Mass Ejections Without Obvious Low Coronal Signatures. Sol. Phys., 292(9), 125.

Nitta, N. V., Mulligan, T., Kilpua, E. K. J., Lynch, B. J., Mierla, M., O'Kane, J., Pagano, P., Palmerio, E., Pomoell, J., Richardson, I. G., Rodriguez, L., Rouillard, A. P., Sinha, S., Srivastava, N., Talpeanu, D.-C., Yardley, S. L., & Zhukov, A. N. 2021, Understanding the Origins of Problem Geomagnetic Storms Associated with "Stealth" Coronal Mass Ejections. *Space Sci. Rev.*, 217(8), 82.

Nitta, N. V., Schrijver, C. J., Title, A. M., & Liu, W. 2013, Large-scale Coronal Propagating Fronts in Solar Eruptions as Observed by the Atmospheric Imaging Assembly on Board the Solar Dynamics Observatory – an Ensemble Study. ApJ, 776, 58.

Odert, P., Leitzinger, M., Hanslmeier, A., & Lammer, H. 2017, Stellar coronal mass ejections - I. Estimating occurrence frequencies and mass-loss rates. MNRAS, 472, 876–890.

Reinard, A. A. & Biesecker, D. A. 2009, The Relationship between Coronal Dimming and Coronal Mass Ejection Properties. ApJ, 705(1), 914–919.

Thompson, B. J., Cliver, E. W., Nitta, N., Delannée, C., & Delaboudinière, J.-P. 2000, Coronal dimmings and energetic CMEs in April-May 1998. Geophys. Res. Lett., 27, 1431–1434.

Thompson, B. J., Gurman, J. B., Neupert, W. M., Newmark, J. S., Delaboudinière, J.-P., Cyr, O. C. S., Stezelberger, S., Dere, K. P., Howard, R. A., & Michels, D. J. 1999, SOHO/EIT Observations of the 1997 April 7 Coronal Transient: Possible Evidence of Coronal Moreton Waves. ApJ, 517, L151–L154.

Thompson, B. J., Plunkett, S. P., Gurman, J. B., Newmark, J. S., St. Cyr, O. C., & Michels, D. J. 1998, SOHO/EIT observations of an Earth-directed coronal mass ejection on May 12, 1997. Geophys. Res. Lett., 25, 2465–2468.

Vakhrusheva, A. A., Shugai, Y. S., Kaportseva, K. B., Eremeev, V. E., & Kalegaev, V. V. 2024, Parameters of Coronal Dimmings and Their Variations during Solar Cycle 24. *Geomagnetism and Aeronomy*, 64(1), 1–10.

Veronig, A. M., Odert, P., Leitzinger, M., Dissauer, K., Fleck, N. C., & Hudson, H. S. 2021, Indications of stellar coronal mass ejections through coronal dimmings. *Nature Astronomy*,.

Vourlidas, A., Balmaceda, L. A., Xie, H., & St. Cyr, O. C. 2020, The Coronal Mass Ejection Visibility Function of Modern Coronagraphs. ApJ, 900(2), 161.

Woods, T. N., Eparvier, F. G., Hock, R., Jones, A. R., Woodraska, D., Judge, D., Didkovsky, L., Lean, J., Mariska, J., Warren, H., McMullin, D., Chamberlin, P., Berthiaume, G., Bailey, S., Fuller-Rowell, T., Sojka, J., Tobiska, W. K., & Viereck, R. 2012, Extreme Ultraviolet Variability Experiment (EVE) on the Solar Dynamics Observatory (SDO): Overview of Science Objectives, Instrument Design, Data Products, and Model Developments. Sol. Phys., 275(1-2), 115–143.

Yashiro, S., Gopalswamy, N., Akiyama, S., Michalek, G., & Howard, R. A. 2005, Visibility of coronal mass ejections as a function of flare location and intensity. *Journal of Geophysical Research (Space Physics)*, 110(A12), A12S05.

Zhu, C., DeVore, C. R., Dahlin, J. T., Qiu, J., Kazachenko, M. D., Uritsky, V. M., & Vandervelde, J. S. 2024, Large-scale Coronal Dimming Foreshadowing a Solar Eruption on 2011 October 1. ApJ, 961(2), 218.

Solar and Stellar Coronal Mass Ejections
Proceedings IAU Symposium No. 388, 2026
N. Gopalswamy, O. Malandraki, A. Vidotto & W. Manchester, eds.
doi:10.1017/S1743921324001492

Solar Cycle Variation of Axial Orientations and Favorable Locations of Eruptive MFRs

Hong Xie[1,2], Nat Gopalswamy[1], Sachiko Akiyama[1,2], Pertti Makela[1,2] and Seiji Yashiro[1,2]

[1]NASA Goddard Space Flight Center, 8800 Greenbelt Road Greenbelt, MD, 20771, USA
email: hong.xie@nasa.gov

[2]The Catholic University of America, 620 Michigan Avenue, N.E. Washington, DC 20064, USA

Abstract. Using multi-viewpoint observations from STEREO and SOHO during three solar cycles from 23 to 25, we study the magnetic flux rope (MFR) structures of coronal mass ejections (CMEs) near the Sun and magnetic clouds (MCs) at 1au. The study aims to investigate two phenomena: 1) the occurrence rate of CMEs near Hale sector boundaries (HBs) and 2) solar-cycle variation of MFR axial orientations in CMEs and MCs. Our preliminary results include: 1) the axes of MFRs in cycle 25 present a systematic northward orientation, which is the same as in cycle 23 but opposite to cycle 24; 2) the majority of the MFRs occurred near HBs (within 30 degrees) and some exceptional events occurred at non-HBs; 3) the axial fields in MCs present a similar north-south orientation, which changes from cycle to cycle. We discuss the implication of solar cycle variations of MFR axial orientations for space weather forecasts.

Keywords. Hale polarity law, Hale boundary, magnetic flux ropes, coronal mass ejections

1. Introduction

Since the start of the previous century, solar observations have revealed that active regions (ARs) spot polarities present a systematic east-west orientation that reverses in opposite hemispheres and across odd and even cycles (Hale's law, Hale & Nicholson 1925). They also present a north-south orientation, i.e., the leading spots appear closer to the equator than the following spots do and the tilt of the spots with opposite polarities grows with the latitude of AR to facilitate the polar field reversal in each cycle (Joy's law, Hale et al. 1919).

Solar eruptions, such as flares and coronal mass ejections (CMEs), are also influenced by large-scale neutral lines in the solar corona, i.e., the boundaries between opposite magnetic poles (Wilcox & Ness 1965). Svalgaard et al. (1975) showed that the sector boundary observed in magnetic fields at Earth can be mapped back to the large-scale neutral line in the photosphere and in the solar corona. Svalgaard & Wilcox (1976); Svalgaard et al. (2011) found that magnetic field fluxes tend to emerge near the Hale sector boundary (HB hereafter), and suggested a link between the HB effect to the deep interior of the Sun. A number of other studies have also found that flares and most complex CME-productive ARs tend to occur near the HB (e.g. Getachew et al. 2017; Gyenge et al. 2017; Loumou et al. 2018).

The magnetic flux rope (MFR) axial field in bipolar ARs and filament channels (FLs) can be approximated by the tilt of polarity inversion lines (PILs) between opposite polarities. Recent studies have shown that most CMEs and magnetic clouds (MCs) contain the MFR structures (e.g. Gopalswamy et al. 2013; Xie et al. 2013; Vourlidas 2014); and

Table 1. Numbers of MC-CMEs associated with three source types.

	Total	Stealth CMEs	Flare CMEs	Filament CMEs
SC 23	26	7	11	8
SC 24	27	7	9	11
SC 25	24	7	8	9

the axial tilts of CMEs and MCs are well correlated with the PILs of their source regions (e.g. Marubashi et al. 2015; Xie et al. 2021). Similar cycle variations have been found in MC axial tilts at 1au (e.g. Bothmer & Rust 1997; Gopalswamy et al. 2015; Lepping et al. 2011). In this paper, we investigate the solar-cycle variation of the MFR axial orientation and the relationship between HBs, PIL tilts and CME eruptions. The study aims to answer two questions: 1) how frequently or at what rate do CMEs erupt at HBs? 2) How does the solar-cycle variation of axial orientation of MFRs affect geomagnetic activity?

2. Data and Event Selection

Seventy-seven MC-CMEs during the first four years of three cycles from 23 to 25 have been used in this study. Events in cycles 23 and 24 are selected from Xie et al. (2021), where we have excluded complex events involving successive CMEs. The selected MC-CME pairs are divided into three types, i.e., stealth CMEs, CMEs associated with flares (flare CMEs), and CMEs associated with filaments (filament CMEs). In Table 1, we list the number of CMEs in each cycle and CME types.

To identify the MFR sources and PIL tilts, we used extreme ultraviolet (EUV) images from the Atmospheric Imaging Assembly (AIA; Lemen et al. 2012) and magnetograms from the Helioseismic and Magnetic Imager (HMI; Scherrer et al. 2012) aboard the Solar Dynamics observatory (SDO; Pesnell et al. 2012). For stealth CMEs, since the associated EUV and magnetogram signatures are too faint, we apply a flux-rope fit to determine their source locations, we then use the coronal neutral line (CNL) tilt combined with the hemispheric helicity rule as a proxy for the PIL orientation (see details in Xie et al. 2021).

3. Results and Discussion

Figure 1 shows distributions of the MFR tilts for each cycle: (top) Hale CMEs and (bottom) non-Hale CMEs, where Hale (non-Hale) CMEs refer to cases when associated source polarities are following (against) the Hale's polarity law. We can see that, for all three cycles, the majority of the CMEs (∼83%) are Hale-CMEs. It is shown that during odd (even) cycles, the Hale MFR axial field has a systematic northward (southward) component. In cycles 23 and 25, the Hale MFR axial directions in the northern hemisphere are northeast with a negative helicity (left-handed), and northwest in the southern hemisphere with a positive helicity (right-handed). In even cycle 24, the Hale MFR axial tilts are opposite to the odd cycle, i.e., change to southwest and southeast directions.

In Figure 2, we study the relative position of CME sources to the HB, where we overplot the CME source location (red circle) on the Wilcox Solar Observatory (WSO) source surface map, and HBs are marked by thick purple lines. Here HB refers to the portion of a sector boundary where the polarity change across its neutral line matches that in bipolar source regions (ARs of FLs).

Figure 2 shows (top) an example of a Hale MFR on 2021 Aug 23 and (bottom) a non-Hale MFR on 2023 Mar 19. In Figure 2 top panels, the associated CME erupted near a northern HB, resulting in a northeast left-handed MC at 1 au, while in Figure 2 bottom panels, a CME occurred near a southern non-HB, leading to a southward right-handed

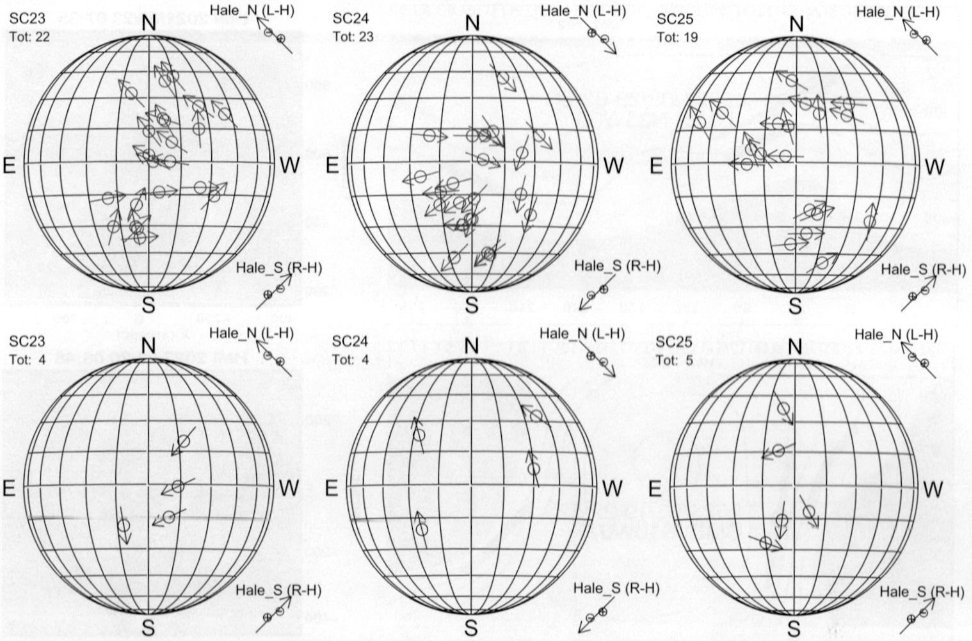

Figure 1. Comparison of PIL tilts for Hale and non-Hale source regions during odd cycles (23 and 25) and even cycle 24.

MC and well-known geomagnetic storm on 2023 Mar 24 during cycle 25 (e.g. Rajana et al. 2024; Paul et al. 2024).

By comparing the CME and HB locations, we found that ∼74% of the MC-CMEs in our dataset erupted near HBs (within ±30°); ∼26% of CME-MCs occurred at the non-HBs, suggesting that HB is one of the locations favorable to CME eruption. The non-HB CMEs tend to occur at inter-Hale ARs (FLs) source regions and/or complex ARs accompanied with new emerging fluxes.

To discuss the effect of solar cycle variation of MFR axial orientations on geomagnetic activity, in Figure 3a, we compare the correlations between solar wind electric field (VB_z) and Dst index for Hale MC-CMEs during cycle 24 (red diamond) and 25 (blue 'x'). It shows that the fitted slope in cycle 24 is ∼1.8 times larger than in cycle 25, i.e., 1.66 vs 0.93, suggesting that an even cycle with a prevailing southward axial field is more geoeffective than an odd cycle. There are 9 major geomagnetic storms during the first four years of cycle 24 and only 4 major storms in cycle 25 from 2020 to 2023. In Figure 3b, we overplotted all the major geomagnetic storms with Dst < -100 nT (marked by red and blue circles) for the two cycles. Table 2 lists the number of major geomagnetic storms, where the storms are grouped as ejecta, sheath, Hale-MC and non-Hale MC storms, respectively. It is shown that the difference of fit-slopes for the two cycles has been smoothed out after adding ejecta, sheath and non-Hale storms. During cycle 24, 4 out of 9 major storms are caused by the southward Hale MCs, 3 are associated with ejectas and 2 are sheath storms. During cycle 25, 2 storms are caused by shock sheaths, 1 by a low-inclination MC (with a tilt∼20°), 1 by a southward non-Hale MC.

Note that, the geoeffective activity is not only affected by the MC axial orientation but also the solar cycle strength, i.e., the solar wind electric field. As shown in Figure 3c, during an "active" cycle, the stronger cycle 23 has more number of major geomagnetic storms than both weaker cycles 24 and 25 due to the large VB_z. The geoeffectiveness during an acitve cycle may be enhanced due to interactions with other interplanetary

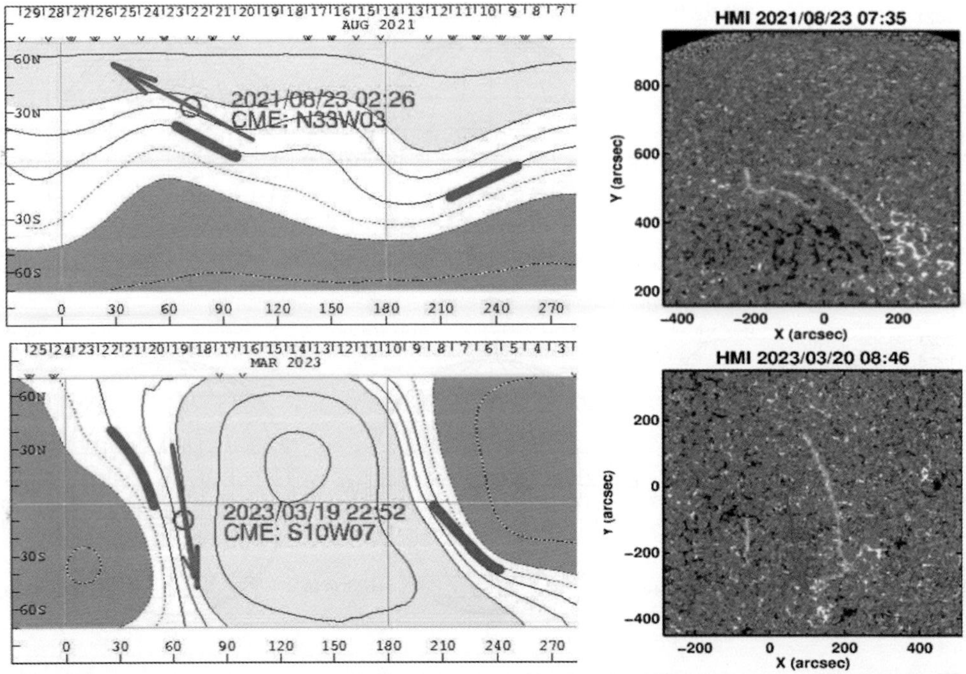

Figure 2. (Left) WSO source surface synoptic maps calculated for a height of 2.5 Rs. Dark gray contours represent negative (-) fields, light gray contours, positive (+) fields. The black solid line is the CNL. The CME source locations (red circles), PILs (red arrows) and HBs (thick purple lines) are overplotted on the maps. (Right) The CME PILs and post-eruption arcade footpoints (green lines) overplotted on the HMI magnetograms.

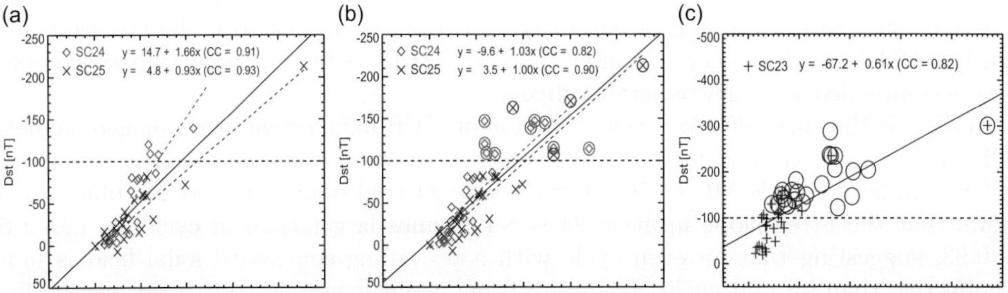

Figure 3. (a) Correlation between VB_z and Dst during cycles 24 (red diamond) and 25 (blue 'x'), (b) same as (a), with major geomagnetic storms added, marked by red and blue circles for the two cycles, and (c) correlation between VB_z and Dst (blue cross) with major geomagnetic storms (blue circles) during cycle 23.

magnetic structures, primarily due to the compression of the existing B_z magnetic field component (e.g. Farrugia et al. 2006; Lugaz et al. 2017).

4. Conclusions

1) The MFR PILs of cycle 25 present a systematic northward orientation as in cycle 23 but opposite to cycle 24 (southward orientation). This cycle variation may explain that, although cycle 25 is slightly stronger, it contains a relatively weaker geomagnetic

Table 2. Numbers of different major geomagnetic storms during two cycles.

	Total	MC storms		Ejecta storms	Sheath storms
		Hale	non-Hale		
SC 24	9	4	0	3	2
SC 25	4	2	1	0	2

activity than cycle 24 during its first four years. 2) ∼74% CME-MCs erupted near HBs (within 30°), suggesting HB is one of the locations favorable to CME eruption.

We thank SOHO, STEREO, and SDO teams for making their data available online. H.X., S.A., and S.Y. are partially supported by NSF grant AGS-2228967. P.M. is partially supported by NSF grant AGS-2043131.

References

Bothmer, V., & Rust, D. 1997, The Field Configuration of Magnetic Clouds and the Solar Cycle (American Geophysical Union (AGU)), 139–146, doi: https://doi.org/10.1029/GM099p0139

Farrugia, C. J., Jordanova, V. K., Thomsen, M. F., et al. 2006, Journal of Geophysical Research (Space Physics), 111, A11104, doi: 10.1029/2006JA011893

Getachew, T., Virtanen, I., & Mursula, K. 2017, *Solar Phys.*, 292, 174, doi: 10.1007/s11207-017-1198-9

Gopalswamy, N., Nieves-Chinchilla, T., Hidalgo, M., et al. 2013, *Solar Phys.*, 284, 1, doi: 10.1007/s11207-013-0280-1

Gopalswamy, N., Yashiro, S., Xie, H., Akiyama, S., & Mäkelä, P. 2015, Journal of Geophysical Research (Space Physics), 120, 9221, doi: 10.1002/2015JA021446

Gyenge, N., Singh, T., Kiss, T. S., Srivastava, A. K., & Erdélyi, R. 2017, *ApJ*, 838, 18, doi: 10.3847/1538-4357/aa62a8

Hale, G. E., Ellerman, F., Nicholson, S. B., & Joy, A. H. 1919, *ApJ*, 49, 153, doi: 10.1086/142452

Hale, G. E., & Nicholson, S. B. 1925, *ApJ*, 62, 270, doi: 10.1086/142933

Lemen, J. R., Title, A. M., Akin, D. J., et al. 2012, *Solar Phys.*, 275, 17, doi: 10.1007/s11207-011-9776-8

Lepping, R. P., Wu, C. C., Berdichevsky, D. B., & Szabo, A. 2011, *Solar Phys.*, 274, 345, doi: 10.1007/s11207-010-9646-9

Loumou, K., Hannah, I. G., & Hudson, H. S. 2018, *A&A*, 618, A9, doi: 10.1051/0004-6361/201731050

Lugaz, N., Temmer, M., Wang, Y., & Farrugia, C. J. 2017, *Solar Phys.*, 292, 64, doi: 10.1007/s11207-017-1091-6

Marubashi, K., Akiyama, S., Yashiro, S., et al. 2015, *Solar Phys.*, 290, 1371, doi: 10.1007/s11207-015-0681-4

Paul, K. S., Haralambous, H., & Oikonomou, C. 2024, Advances in Space Research, 73, 6029, doi: 10.1016/j.asr.2024.03.026

Pesnell, W. D., Thompson, B. J., & Chamberlin, P. C. 2012, *Solar Phys.*, 275, 3, doi: 10.1007/s11207-011-9841-3

Rajana, S. S. K., Panda, S. K., Jade, S., et al. 2024, *Ap&SS*, 369, 3, doi: 10.1007/s10509-024-04268-9

Scherrer, P. H., Schou, J., Bush, R. I., et al. 2012, *Solar Phys.*, 275, 207, doi: 10.1007/s11207-011-9834-2

Svalgaard, L., Hannah, I. G., & Hudson, H. S. 2011, *ApJ*, 733, 49, doi: 10.1088/0004-637X/733/1/49

Svalgaard, L., & Wilcox, J. M. 1976, *Solar Phys.*, 49, 177, doi: 10.1007/BF00221492

Svalgaard, L., Wilcox, J. M., Scherrer, P. H., & Howard, R. 1975, *Solar Phys.*, 45, 83, doi: 10.1007/BF00152219

Vourlidas, A. 2014, Plasma Physics and Controlled Fusion, 56, 064001, doi: 10.1088/0741-3335/56/6/064001

Wilcox, J. M., & Ness, N. F. 1965, *JGR*, 70, 5793, doi: 10.1029/JZ070i023p05793

Xie, H., Gopalswamy, N., & Akiyama, S. 2021, *ApJ*, 922, 64, doi: 10.3847/1538-4357/ac23cc

Xie, H., Gopalswamy, N., & St. Cyr, O. C. 2013, *Solar Phys.*, 284, 47, doi: 10.1007/s11207-012-0209-0

Solar and Stellar Coronal Mass Ejections
Proceedings IAU Symposium No. 388, 2026
N. Gopalswamy, O. Malandraki, A. Vidotto & W. Manchester, eds.
doi:10.1017/S1743921324001765

The Role of Magnetic Reconnection in the Structure & Dynamics of Fast Coronal Mass Ejections

Brian T. Welsch[1] **and C. R. DeVore**[2]

[1]Natural & Applied Sciences, University of Wisconsin - Green Bay,
2420 Nicolet Drive, Green Bay, WI 54311, USA

[2]Heliophysics Science Division, NASA Goddard Space Flight Center,
8800 Greenbelt Road, Greenbelt, MD 20771, USA

Abstract. Both observations and models of flare-associated coronal mass ejections (CMEs) suggest that magnetic reconnection in an ejection's wake substantially increases the net, outward Lorentz force accelerating the CME. A stronger outward force can cause a feedback loop, driving further magnetic reconnection in a "reconnective instability." The flux accretion model captures this by relating reconnected flux, $\Delta\Phi_{\rm rec}$, and magnetic field strength, $B_{\rm CME}$, to increased outward Lorentz force, ΔF_r. To better understand reconnection's role in CME dynamics, we analyze two snapshots from a 2.5D, MHD simulation of a breakout eruption. Outward Lorentz forces increase substantially as reconnection proceeds, caused primarily by "flank currents," which flow just inside the boundary of the rising ejection's wake and parallel to its axis. This model's reconnection jet also alters the ejection's internal structure, an effect that could be sought in observations. Analyzing reconnection-induced Lorentz forces in 3D simulations could provide additional insights into CME dynamics.

Keywords. Magnetic Reconnection, CMEs – dynamics, CMEs – structure, CMEs – acceleration

1. Introduction

Magnetic reconnection is a rich topic (e.g., Priest and Forbes 2000) that is relevant to magnetic energy release in plasmas of the Sun, other stars, and many other physical systems. In the case of solar flares, reconnection is widely believed to be the central process facilitating the release of stored magnetic energy (e.g., McKenzie 2002). For a flare-associated coronal mass ejection (CME), *flare reconnection* (Antiochos et al. 1999) is believed to occur in the wake of the rising ejection. Antiochos et al. (1999) contrast flare reconnection with *breakout reconnection*, which occurs in front of the erupting structure, and which they posit initiates the eruption by removing overlying, restraining flux. Lin and Forbes (2000) analyzed a 2.5D, Cartesian model of a CME, and describe a feedback loop between flare reconnection and the upward force accelerating the CME. Moore et al. (2001) proposed that the eruption is initiated by *tether cutting* reconnection under the pre-eruptive structure, which reduces the effect of downward magnetic tension from fields overlying that structure. Once an eruption is underway, tether cutting reconnection and flare reconnection are indistinguishable.

Observationally, significant correlations have been found between CMEs' speeds, $v_{\rm CME}$, and the amounts of magnetic flux that reconnected in their associated flares, $\Phi_{\rm rec}$, deduced from flare ribbons (Qiu and Yurchyshyn 2005) or post-eruption arcades (Gopalswamy et al. 2017). These correlations suggest, consistent with the theoretical models above, that reconnection plays an important role in altering the forces on ejections. From the temporal correlation between soft X-ray emission, interpreted as a proxy of magnetic reconnection, and CME acceleration, Zhang and Dere (2006) also suggested that reconnection and CME acceleration operate in a feedback loop early in the eruption process: more reconnection drives more acceleration, which in turn drives further reconnection, and further subsequent acceleration.

Based upon these and other observations and models, Welsch (2018) presented the *flux accretion* model of eruptions, in which flare reconnection below a rising ejection adds magnetic flux to the ejection. This (i) leads to momentum transfer from the outflow jet to the ejection and (ii) increases the net, outward component of the Lorentz force,$(\mathbf{J} \times \mathbf{B})/c$, acting on the ejection. In the model's treatment of Lorentz forces, the accreted underlying flux both reduces the net force due to downward magnetic tension from overlying fields (the mechanism of tether cutting) and increases the outward hoop force (e.g., Chen 1996) from these underlying fields exerting greater outward magnetic pressure on the ejection. For both components of the total force, the flux accretion model predicts that the change in outward force, ΔF_r, due to the reconnection of an amount of flux $\Delta\Phi_{\rm rec}$ scales as

$$\Delta F_r \propto \Delta\Phi_{\rm rec} B_{\rm CME} \; , \tag{1}$$

where $B_{\rm CME}$ is a characteristic field strength in the vicinity of the CME.

In this brief paper, we explore how reconnection affects the Lorentz force on models of CMEs. In Section 2, we consider a highly idealized model of magnetic fields and currents in a CME source region, and analyze the effect of reconnection upon these fields and currents. Then, in Section 3, we quantitatively investigate evolution of the Lorentz forces — and the magnetic fields and currents that contribute to it — in two snapshots from a numerical simulation of CME as flare reconnection proceeds.

2. Qualitative Analysis of Reconnection's Effect on Lorentz Forces

We first consider the overall structure of the magnetic field and currents in a generic eruptive configuration, and then consider how these are relevant for flare reconnection and evolving Lorentz forces during the eruption. To illustrate the large-scale morphology of the system, Figure 1 shows a snapshot from the axisymmetric, 2.5D MHD simulation that we will analyze in the next section. (Details of the simulation are described there.) The simulation employed spherical coordinates, and we interpolated its physical values onto a uniform, Cartesian, (1024 × 1024) grid, with $x \in [0, 6]R_\odot$ and $z \in [-3, 3]R_\odot$. The simulation's inner boundary, at $r = R_\odot$, corresponds to the base of the corona and can be seen near the image's left edge. Co-latitude θ increases clockwise from the Sun's north pole (out of the frame, at $z = +1$) toward its equator (at $z = 0$), and ϕ increases into the page. The grayscale shows B_ϕ, weighted by r^2 so its structure remains visible even as it weakens with larger r. The black lines show contours of the flux function, f, with the poloidal field, $\mathbf{B}_{\rm pol}$, obeying

$$\mathbf{B}_{\rm pol} = \nabla f \times \nabla\phi \; , \tag{2}$$

meaning contours of f correspond to field lines of $\mathbf{B}_{\rm pol}$. For $x \in [1.25, 2.5]R_\odot$, this flux system's radial field is outward below the $z = 0$ line and inward above it. An intense azimuthal current — a large-scale current sheet — flows between these oppositely directed radial fields, and is directed out of the page. In Figure 1, this current is concentrated between the innermost flux function contours straddling $z = 0$. The red and blue contours bracketing $(x, z) = (1.7, 0)$ correspond to 100 km/s inward and outward radial outflows, respectively, from the flare reconnection region.

An eruptive flare invariably occurs above a photospheric polarity inversion line (PIL), across which radial photospheric magnetic field changes sign. The horizontal photospheric

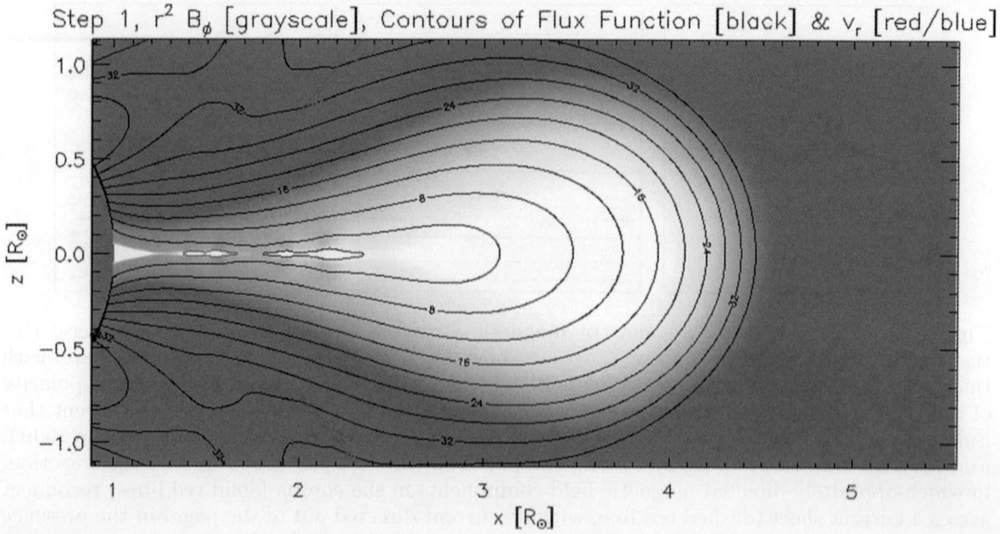

Figure 1. In this snapshot from the spherical, axisymmetric, 2.5D MHD simulation that we analyze, the simulation's spherical inner boundary ($r = R_\odot$) can be seen at frame left, x increases to the right, θ increases clockwise from the Sun's north pole (out of the frame, at $z = +1$) toward its equator (at $z = 0$), and ϕ increases into the page. The grayscale shows B_ϕ, weighted by r^2 to keep the structure of B_ϕ visible as it weakens with increasing r. The red and blue contours correspond to 100 km/s inward and outward radial velocities, respectively. The black lines show contours of the flux function, f, corresponding to poloidal projections of field lines.

field along a flaring PIL almost always exhibits a significant *shear* component, i.e., a component of **B** parallel to the PIL. Associations between magnetic shear and flare occurrence were noted long ago (e.g., Hagyard et al. 1984). In addition to being flare-prone, sheared fields along PILs have also been associated with increased likelihood of CME occurrence (e.g., Falconer 2001; Falconer et al. 2003). Typically, sheared fields are localized in the vicinity of PILs, with horizontal photospheric fields farther from PILs exhibiting less shear (e.g., Falconer 2001). This decrease in shear with distance from the PIL is also typically observed in overlying fields, inferred from the directions of chromospheric H-α fibrils and coronal loops' axes.

Because the coronal plasma in CME source regions is magnetically dominated (the plasma's β, its ratio of gas to magnetic pressures, is much less than unity), the $(\mathbf{J} \times \mathbf{B})/c$ (Lorentz) force is very small in slowly evolving, pre-eruptive fields, i.e., such fields are very nearly *force-free.* Thus, (i) $\mathbf{J} \neq 0$ implies that **J** must flow very nearly parallel to **B**, and (ii) the presence of a significant shear component of **B** implies a shear component of **J**. The localization of the shear component of **B** in the vicinity of the PIL implies that there must also be a current sheath surrounding the sheared field, to isolate or "switch off" the shear component of **B**. These elements are illustrated in the left panel Figure 2, which shows an overhead view of magnetic structure near a sheared photospheric PIL (red dashed line). The red lines with arrows show magnetic fields above the photosphere, with lines progressing from thicker to thinner with increasing height. The red $\odot$ and $\otimes$ show the polarity of the photospheric radial magnetic field. The blue $\odot$ and $\otimes$ symbols show the sheath current (similar to that in a solenoid) that isolates the magnetic field's shear component (which points to the right).

Flare reconnection alters the connectivity of both preflare coronal magnetic fields and current structures. We are primarily interested in how reconnection alters the radial

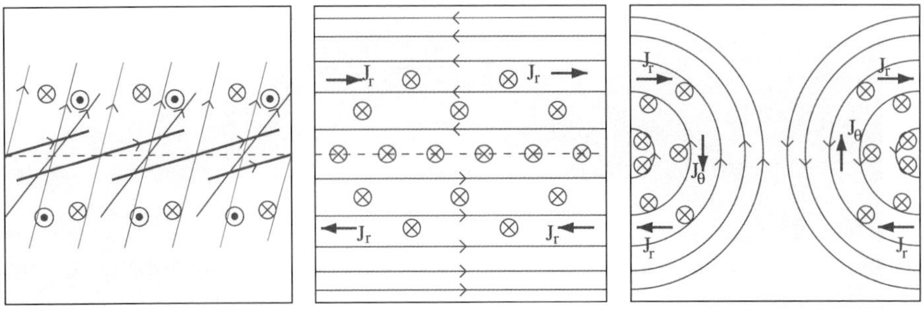

Figure 2. Left: An overhead view of magnetic structure near a sheared photospheric PIL (red dashed line). Solid red lines with arrows show magnetic fields above the photosphere, with thicker to thinner corresponding to fields at increasing altitudes. Red ⊙ and ⊗ show polarity of the photospheric radial magnetic field. Blue ⊙ and ⊗ symbols show the sheath current that isolates (or "switches off") the magnetic field's shear component (which points to the right). Middle: This side view, above and along the PIL from the left panel, shows flare reconnection, in which oppositely directed magnetic field components in the corona (solid red lines) reconnect across a current sheet (dashed red line, with its current directed out of the page) in the presence of guide field (red ⊗ symbols). Sheath currents associated with the localization of the guide field are also present (blue vectors). Right: Reconnection joins formerly oppositely directed fields into arched fields (solid red lines), and in the process reroutes the sheath currents (blue vectors), introducing oppositely directed currents between the post-reconnection flux systems. Each of these field and current components can, in principle, contribute to the radial Lorentz force that accelerates the ejection.

component of the Lorentz force, F_r,

$$cF_r = J_\theta B_\phi - J_\phi B_\theta \ . \tag{3}$$

Current in the reconnecting current sheet (which we assume is thin, but not infinitely so) also flows parallel and above to the PIL; in our geometry, this is J_ϕ. The shear component of **B** corresponds to a guide field, i.e., a nonzero field component along the reconnecting current sheet (see, e.g., Priest and Forbes 2000); in our geometry, this is B_ϕ. Reconnection in the presence of a guide field can introduce twist into originally untwisted fields (e.g., Wright and Berger 1989; Chae 1999). The resulting fields can be highly twisted (e.g., Longcope et al. 2007) and generally carry currents. Aspects of the reconfiguration of coronal fields and current components are illustrated schematically in the middle and right panels of Figure 2. These panels depict a view in the corona along the guide field (i.e., above and along the PIL in the Figure's left panel), showing the reversed field components that reconnect (solid red lines), the guide field (red ⊗ symbols), and the sheath currents (blue vectors) associated with the localization of the shear component of **B**. In the middle panel, a current sheet exists along the red dashed line, with current flowing out of the page. Reconnection joins oppositely directed fields (solid red lines in middle panel) into arched fields (solid red lines in right panel), which possess a B_θ component in our geometry. The process also reroutes the sheath currents (blue vectors), introducing oppositely directed currents between the post-reconnection flux systems, corresponding to a J_θ component. The repulsion of these antiparallel J_θ currents works against the attraction of the parallel J_ϕ currents in the post-reconnection fields (both out of the page here).

In the following section, we investigate the evolution of F_r as flare reconnection proceeds in the model CME, and compare the contributions from the $(J_\theta B_\phi)$ and $(J_\phi B_\theta)$ terms in Equation (3).

3. Quantitative Analysis of Reconnection's Effect on Lorentz Forces

In this section, we analyze two snapshots from a 2.5D, axisymmetric, breakout simulation using the Adaptively Refined Magnetohydrodynamic Solver (ARMS) that closely matches the simulations performed and analyzed by Karpen et al. (2012) (hereafter "KAD12"). The run analyzed here differs in only one potentially relevant detail: in the simulations that we analyze, six levels of grid refinement were permitted, whereas only five levels were enabled in KAD12. The first snapshot, our Step 1 (the code's step 241354), corresponded to the onset of flare reconnection, and at the second snapshot, our Step 2 (the code's step 254006), a substantial amount of flux has reconnected. (Data from additional time steps could be analyzed, but computation of the flux function for this simulation must be done *post facto*, which requires significant manual effort. Consequently, the analysis of additional data would require the availability of additional resources.)

Figure 3 shows the inward and outward (red and blue, respectively) components of the radial Lorentz force at each step, both heavily saturated and weighted by r^4, so their spatial structure can be discerned at large r, despite $\mathbf{J}$ and $\mathbf{B}$ being much weaker there than near $r = R_\odot$. The thin black contours show the flux function, which makes clear that a substantial amount of flux has reconnected between these snapshots. The thick contour outlines a *CME mask*, meant to capture the main body of the erupting structure. We have chosen to define this mask using a fixed-r inner boundary and an outer boundary set at the same flux-function contour in both snapshots. We also note the presence of several key structural features of the global magnetic field. Post-reconnection, poloidal fields in the flare arcade are prominent near $z = 0$ and $x \in [1, 1.5]R_\odot$ in the bottom panel. The breakout current sheet is near $x = 5R_\odot$ in the top panel. The separatrix current sheets, which flow between the erupting flux system and adjacent systems and are extensions of the breakout current sheet, run diagonally near $1.5R_\odot < x < 2.25R_\odot$ in both panels.

The distribution of the radial Lorentz force is richly structured, with features in the reconnection outflow region clearly making large contributions to the Lorentz force. Notably, forces present in the jet are strong enough to deform post-reconnection flux-function contours, making them concave inward — i.e., with a magnetic tension that points radially inward. It is unclear whether observed CMEs exhibit such structure. To quantify evolution of the radial Lorentz force as the reconnection proceeds, we can integrate the interpolated force density over the CME mask on our Cartesian grid, weighted by r (since $dV = dx\, dz\, r\, d\phi$, as in cylindrical coordinates). Applying the Step-1 mask to both steps, the outward Lorentz force increases by a factor of 2.3; applying the Step-2 mask to both steps, the outward Lorentz force increases by a factor of 1.4; and applying the Step-1 mask at Step 1 and the Step-2 mask at Step 2 (i.e., advecting the flux-function boundary), the outward Lorentz force increases by a factor of 1.5. Expansion of the eruption, combined with conservation of flux, weakens its magnetic field: we averaged each step's $|\mathbf{B}|$ over the CME mask at that step, and this average decreases by about 20% between Steps 1 and 2. (The area of Step 2's mask is 5/3 the size of Step 1's.) All else equal, a weaker $|\mathbf{B}|$ would decrease the Lorentz force. However, the average current density, $|\mathbf{J}|$, in Step 2's CME mask is about 4/3 that in Step 1's. This increase in Lorentz force can be explained both by reconnection and the development of an ideal MHD instability, e.g., the torus instability (Kliem and Török 2006).

In Figure 4, we investigate the radial profile of the radial component of the Lorentz force, $cF_r = (J_\theta B_\phi - J_\phi B_\theta)$, outward from the flare reconnection site in Steps 1 and 2. To compare contributions to F_r between small and large r, we weight all quantities plotted by r^2. The flare reconnection site was identified in each step by finding where the smoothed radial velocity along $z = 0$ changed sign. The values plotted in this figure are *strip-averaged*: to compute them, we averaged values of $[J_\theta B_\phi]$ and $-[J_\phi B_\theta]$ in a

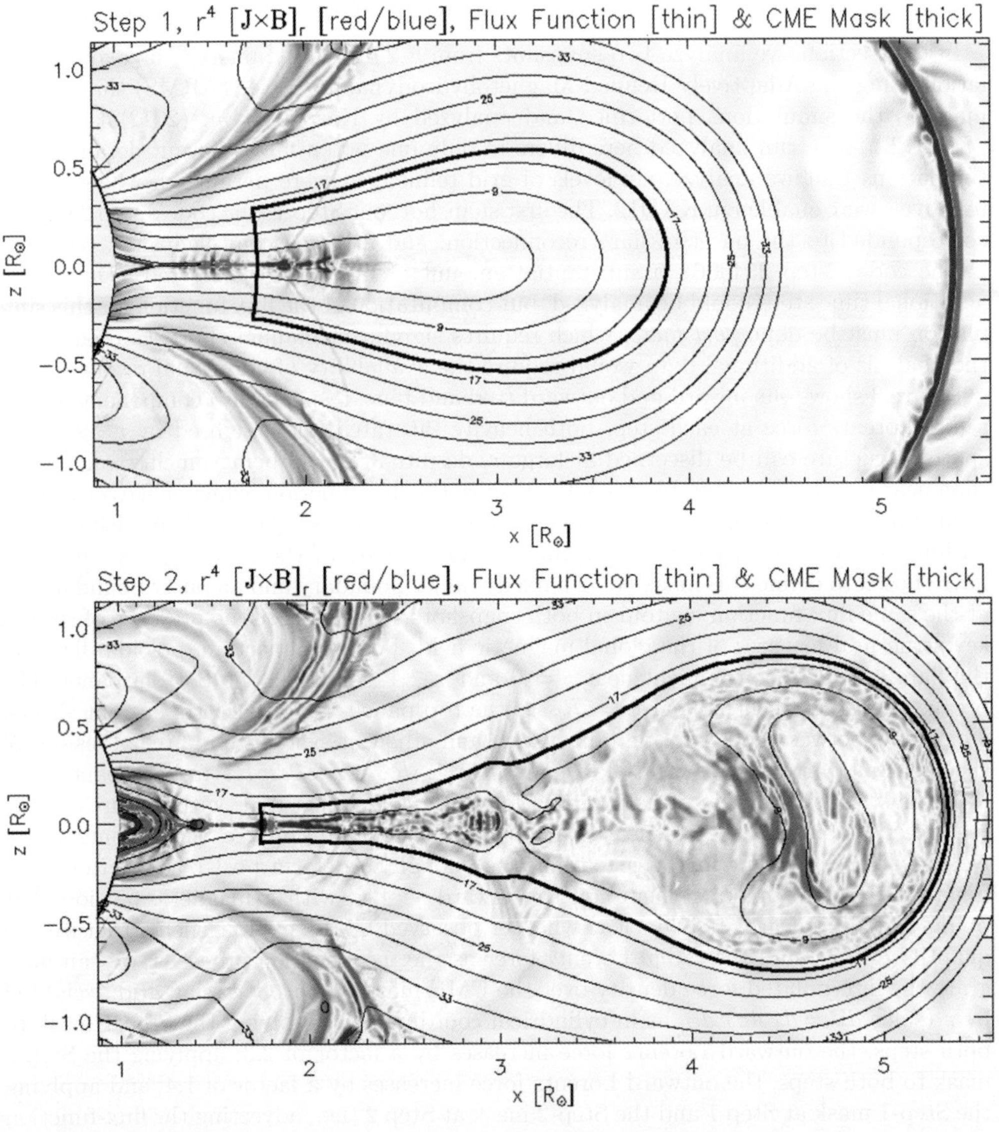

Figure 3. Red (blue) shows inward (outward) components of the radial Lorentz force at snapshots 1 and 2, weighted by r^4, so their spatial structure can be discerned at large r, where **J** and **B** are much weaker than near the model's inner boundary. The thin black contours show the flux function, from which evolution of the field, including some effects of reconnection, can be inferred. The thick contour shows a CME mask whose outer boundary is at the same flux-function contour in both snapshots (see text).

100-cell-wide strip, encompassing $z \in [-0.29, 0.29]R_\odot$. The highly oscillatory structure of $r^2(J_\theta B_\phi - J_\phi B_\theta)$ in the top panel of Figure 4 precludes clearly identifying how the primary contributions to the overall force arise. To clarify which areas contribute to the CME's acceleration, in the middle panel, at each r we plot the *cumulative* (summed) value of $r^2(J_\theta B_\phi - J_\phi B_\theta)$, from the base of the reconnection outflow to that r. In this cumulative plot, a positive (negative) slope over some interval in r indicates a net positive (negative) contribution to the CMEs outward acceleration in that interval. Clearly, Lorentz forces that develop in the reconnection jet begin making a much

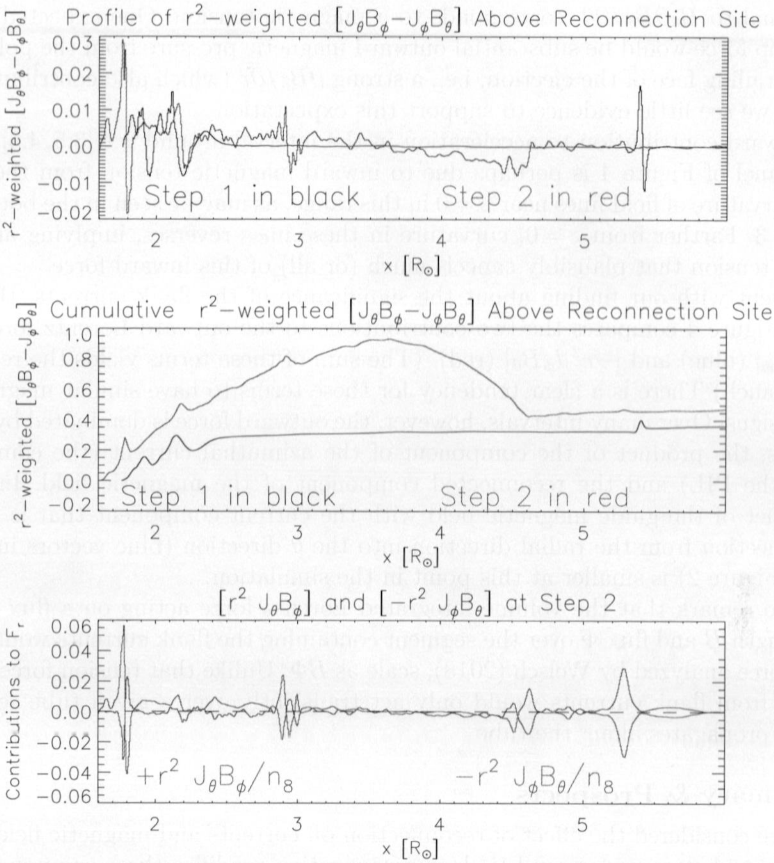

Figure 4. Top: Black (red) curves show r^2-weighted contributions to the radial Lorentz force at Step 1 (Step 2), averaged over a strip 100 cells (0.58 $R_\odot$) wide in z, outward in x from the reconnection site at that Step. Middle: Cumulative sum of the curves in the top panel from the reconnection site to each x. The excess outward force in Step 2 compared to Step 1 arises almost entirely from the diagonal, blue bands in Figure 3 just inside the CME mask around $x \in [2.5, 3.5] R_\odot$. Flank currents (see text) flow out of the page in these bands (parallel to the flare-reconnection current sheet). Bottom: For Step 2, the blue curve shows $[r^2 J_\theta B_\phi]$ and the red curve shows $[-r^2 J_\phi B_\theta]$. The latter generally makes a greater contribution to the outward Lorentz force.

larger contribution to the CME's acceleration between Steps 1 and 2. These findings support the idea that reconnection plays a key role in the ejection's outward acceleration, consistent with the expectations of Lin and Forbes (2000) and Zhang and Dere (2006), and the findings of KAD12.

Crucially, by decreasing the width of the strip over which the inverse-density-weighted radial force is averaged, we found that essentially *all* of the excess contribution to the outward force for Step 2 compared to Step 1 comes from *flank currents* — the tilted, blue bands just inside the CME mask (bottom panel of Figure 3) for $r \in [2.5, 3.5] R_\odot$. These structures are reminiscent of the standing waves (slow-mode shocks) in the Petschek reconnection model (Petschek 1964), but are macroscopic regions instead of the idealized discontinuities in that model. We found that these regions contain strong azimuthal currents parallel to that in the flare reconnection sheet, i.e., out of the page, resulting from a strong $\partial B_r / \partial \theta$ near the CME mask boundary, in which B_r exterior to the CME decreases relative to B_θ within the CME. The associated component of the Lorentz force, proportional to $B_\theta \partial B_r/\partial\theta$, corresponds to a magnetic tension. One expected signature of the hoop force would be substantial outward magnetic pressure from the poloidal field over the trailing face of the ejection, i.e., a strong $\partial B_\theta/\partial r$ (which also contributes to J_ϕ). However, we see little evidence to support this expectation.

The inward contribution to acceleration in the interval around $x \in [3.5, 4.5]R_\odot$ in the middle panel of Figure 4 is perhaps due to inward magnetic tension from the concave-inward curvature of field lines near $z = 0$ in this range, as may be seen in the bottom panel of Figure 3. Farther from $z = 0$, curvature in these lines reverses, implying an outward magnetic tension that plausibly cancels much (or all) of this inward force.

Consistent with our finding about the significance of the flank currents, the bottom panel of Figure 4 compares the two contributions to the outward Lorentz force at Step 2, $[r^2 J_\theta B_\phi]$ (blue) and $[-r^2 J_\phi B_\theta]$ (red). (The sum of these terms yields the red curve in the top panel.) There is a clear tendency for these terms to have similar magnitude but opposite signs. Over many intervals, however, the outward force is dominated by $(-J_\phi B_\theta)$ — that is, the product of the component of the azimuthal current (the component of **J** along the PIL) and the reconnected component of the magnetic field. In contrast, the product of the guide magnetic field with the current component that is redirected by reconnection from the radial direction into the θ direction (blue vectors in the right panel of Figure 2) is smaller at this point in the simulation.

We also remark that the volume-integrated Lorentz force acting on a flux tube with field strength B and flux Φ over the segment containing the flank currents would, like the tension force analyzed by Welsch (2018), scale as $B\Phi$. Unlike that tension force, however, the force from flank currents would only act transiently over a given tube segment, as the bend propagates along the tube.

4. Summary & Prospects

We have considered the effect of reconnection on currents and magnetic fields in eruptive flares, to better understand (i) how reconnection modifies these components of the Lorentz force and thus (ii) how flare-associated CMEs are accelerated.

We were not able to analyze enough snapshots from available ARMS simulations to test the flux accretion model's scaling in Equation (1). The necessary analysis would require computing radial Lorentz force changes between several closely spaced simulation steps, with each time interval involving reconnection of a known amount of flux. If sufficient resources — essentially, funding — becomes available, such an analysis will be pursued.

We found that, in this simulation, PIL-aligned "flank" current structures in the reconnection outflow make the primary contribution to the total outward Lorentz force. These findings accord with those reported by Jiang et al. (2021) in their analysis of another CME simulation (see bottom row of their Fig. 4). Contrary to the prediction of the flux accretion model, we see little evidence of additional hoop force from reconnected flux driving further acceleration. This conclusion, however, should not be hastily generalized to other configurations, for two reasons, both related to this simulation's axisymmetry: this model's reconnection jet is probably stronger than that in real, 3D events; and this model's hoop force is likely weaker than that in shorter-length-scale systems. Regarding the reconnection jet: the axial invariance here implies that flow and magnetic structures created in the reconnection process are coherent over much longer length scales than would be expected in a more realistic 3D geometry, within which reconnection would likely be patchy (Linton and Longcope 2006; Linton et al. 2009), producing a much more fragmented structure in the outflow. Regarding the erupting system's length scale: the axisymmetry here implies that the radius of curvature of the model ejection's axis, r_c, is on the order of the Sun's radius, $r_c \sim R_\odot$. This might be an apt model for the eruption of a polar-crown filament, but does not accurately represent the much smaller r_c expected for an ejection from an active region, for which $r_c \sim R_\odot/10$ would often be more appropriate. Because the strength of the hoop force increases steeply for fixed ejection width as r_c shrinks (e.g., Figure 6 in Welsch 2018), a much smaller r_c should, all else equal, yield a much larger hoop force.

We conclude with recommendations for future work. In the simulation that we analyzed, the forces from the reconnection jet have deformed the eruption's core, making its trailing edge concave. We are unaware of observational reports of such concave structure within CMEs; if found, this would indicate that forces in the reconnection outflow play a substantial role in the interior structure of CMEs. Given the considerations in the two preceding paragraphs, we believe that analysis of the evolution of Lorentz forces in fully 3D, active-region-scale simulations is warranted. We also note that, while the association between CME speeds $v_{\rm CME}$ and reconnection fluxes $\Phi_{\rm rec}$ for eruptive flares is relatively well established, the flux accretion model's prediction, per the scaling in Equation (1), that CMEs' accelerations (and probably their final speeds) should be correlated with coronal magnetic field strengths in their source regions, $B_{\rm CME}$, has not been tested. We suggest that potential-field extrapolation might be an easy approach to test this prediction, but note that radio observations might enable direct estimation of $B_{\rm CME}$.

Acknowledgements. We thank the U.S. taxpayer, whose taxes helped support the work performed here. We thank B. J. Lynch for supplying code to read this ARMS output into IDL, and for taking the time to write a detailed how-to for executing this conversion. BTW is also grateful to M. Linton and B. Kliem for their time spent in engaging discussions of forces acting on CMEs. BTW gratefully acknowledges support from NASA HSR-80NSSC23K0092 and NSF SHINE AGS-2302697. CRD was supported by a grant from NASA's H-ISFM program to Goddard Space Flight Center.

References

Antiochos, S. K., DeVore, C. R., & Klimchuk, J. A. 1999, A Model for Solar Coronal Mass Ejections. *The Astrophysical Journal*, 510, 485–493.

Chae, J. 1999, Non-Coplanar Magnetic Reconnection as a Magnetic Twist Origin. *Journal of Korean Astronomical Society*, 32(2), 137–147.

Chen, J. 1996, Theory of prominence eruption and propagation: Interplanetary consequences. *Journal of Geophysical Research*, 101(10), 27499–27520.

Falconer, D. A. 2001, A prospective method for predicting coronal mass ejections from vector magnetograms. *Journal of Geophysical Research*, 106, 25185–25190.

Falconer, D. A., Moore, R. L., & Gary, G. A. 2003, A measure from line-of-sight magnetograms for prediction of coronal mass ejections. *Journal of Geophysical Research (Space Physics)*, 108, 1380.

Gopalswamy, N., Yashiro, S., Akiyama, S., & Xie, H. 2017, Estimation of Reconnection Flux Using Post-eruption Arcades and Its Relevance to Magnetic Clouds at 1 AU. *Solar Physics*, 292, 65.

Hagyard, M. J., Teuber, D., West, E. A., & Smith, J. B. 1984, A quantitative study relating observed shear in photospheric magnetic fields to repeated flaring. *Solar Physics*, 91, 115–126.

Jiang, C., Feng, X., Liu, R., Yan, X., Hu, Q., Moore, R. L., Duan, A., Cui, J., Zuo, P., Wang, Y., & Wei, F. 2021, A fundamental mechanism of solar eruption initiation. *Nature Astronomy*, 5, 1126.

Karpen, J. T., Antiochos, S. K., & DeVore, C. R. 2012, The Mechanisms for the Onset and Explosive Eruption of Coronal Mass Ejections and Eruptive Flares. *The Astrophysical Journal*, 760, 81.

Kliem, B. & Török, T. 2006, Torus Instability. *Physical Review Letters*, 96(25), 255002–+.

Lin, J. & Forbes, T. G. 2000, Effects of reconnection on the coronal mass ejection process. *Journal of Geophysical Research*, 105(A2), 2375–2392.

Linton, M. G., DeVore, C. R., & Longcope, D. W. 2009, Patchy reconnection in a Y-type current sheet. *Earth, Planets, and Space*, 61, 573–576.

Linton, M. G. & Longcope, D. W. 2006, A Model for Patchy Reconnection in Three Dimensions. *The Astrophysical Journal*, 642, 1177–1192.

Longcope, D. W., Ravindra, B., & Barnes, G. 2007, Determining the Source of Coronal Helicity through Measurements of Braiding and Spin Helicity Fluxes in Active Regions. *The Astrophysical Journal*, 668, 571–585.

McKenzie, D. E. Signatures of Reconnection in Eruptive Flares. In Martens, P. C. H. & Cauffman, D., editors, *Multi-Wavelength Observations of Coronal Structure and Dynamics* 2002, volume 13 of *COSPAR Colloq. Series*, pp. 155–164.

Moore, R. L., Sterling, A. C., Hudson, H. S., & Lemen, J. R. 2001, Onset of the Magnetic Explosion in Solar Flares and Coronal Mass Ejections. *ApJ*, 552, 833–848.

Petschek, H. E. Magnetic Field Annihilation. In *NASA Special Publication* 1964, volume 50, 425.

Priest, E. R. & Forbes, T. G. 2000,. *Magnetic reconnection: MHD theory and applications.* Cambridge University Press, New York.

Qiu, J. & Yurchyshyn, V. B. 2005, Magnetic Reconnection Flux and Coronal Mass Ejection Velocity. *The Astrophysical Journal Letters*, 634, L121–L124.

Welsch, B. T. 2018, Flux Accretion and Coronal Mass Ejection Dynamics. *Solar Physics*, 293(7), 113.

Wright, A. N. & Berger, M. A. 1989, The effect of reconnection upon the linkage and interior structure of magnetic flux tubes. *Journal of Geophysical Research*, 94(A2), 1295–1302.

Zhang, J. & Dere, K. P. 2006, A Statistical Study of Main and Residual Accelerations of Coronal Mass Ejections. *The Astrophysical Journal*, 649, 1100–1109.

Discussion — Questions & Responses

Question, by Dr. Bernhard Kliem: It appears possible that the increase of the outward hoop force from the flux accretion to the erupting rope by flare reconnection underneath is weaker than estimated so far, because this flux carries little current. The currents driving the accretion of the flux flow in the slow-mode shocks attached to the diffusion region (or to the uppermost diffusion region if several such regions exist). The Lorentz force of these currents accelerates the upward reconnection outflow jet, hence, drives the accretion of flux to the erupting flux rope. The new flux itself carries little current; it "dipolarizes" while joining the rope flux. Therefore, it appears possible that the primary addition to the upward forces consists in the momentum transfer from the upward reconnection jet to the erupting flux rope (if the jet moves faster than the rope). There is no doubt on the reduction of the downward tension force of overlying flux by flare reconnection. This latter effect is included in the treatment of the torus instability by Kliem and Török (2006).

Response, by Dr. Welsch: Regarding the presence of currents in the outflow region, the flux accretion model does indeed posit that the post-reconnection flux *dipolarizes* (e.g., Priest and Forbes 2000) — i.e., evolves toward a roughly dipolar shape. But only an exactly dipolar field is current-free; we may sensibly speak of a compressed or stretched dipolar field, which contains currents, but is still essentially dipolar in character (versus unipolar or quadrupolar). Thus, I do not use the term "dipolarization" to imply that significant currents are not present. (This term appears to have originally been applied to retraction of highly stretched magnetotail fields in geospace toward the Earth after reconnecting; but even when fully dipolarized, these fields form a stretched dipole.)

Clearly, however, we did not find evidence of the outward hoop force acting in the simulation analyzed here. However, as noted in the concluding section of this article (above), compared to the solar case, this simulation's axisymmetry probably makes both (1) the reconnection jet stronger and (2) the hoop force weaker. Hence, determining whether the hoop force from post-reconnection magnetic fields is dynamically important or not requires analysis of other models with more realistic geometry.

Solar and Stellar Coronal Mass Ejections
Proceedings IAU Symposium No. 388, 2026
N. Gopalswamy, O. Malandraki, A. Vidotto & W. Manchester, eds.
doi:10.1017/S1743921324001844

The Role of Helicity Injection and Poloidal Magnetic Field in a Breakout CME

Nitin Vashishtha[1,2], Vaibhav Pant[1], Dana-Camelia Talpeanu[3], Dipankar Banerjee[1,4,5] and Shantanu Rastogi[2]

[1]Aryabhatta Research Institute of Observational Sciences, Beluwakhan, Nainital, India

[2]Department of Physics, Deen Dayal Upadhyaya Gorakhpur University, Gorakhpur 273009, India
email: nitin.vashishtha@aries.res.in

[3]Solar-Terrestrial center of Excellence - SIDC, Royal Observatory of Belgium, Avenue Circulaire 3, Brussels 1180, Belgium

[4]Indian Institute of Astrophysics, 2nd Block Koramangala, Bangalore, India

[5]Center of Excellence in Space Science, IISER Kolkata, Kolkata, India

Abstract. Coronal mass ejections (CMEs) profoundly impact space weather, necessitating a comprehensive understanding of their initiation and evolution in the solar corona. Solar Cycle 24 exhibited reduced sunspot numbers and heliospheric magnetic pressure compared to Solar Cycle 23, leading to increased weaker CME occurrences. Accurate forecasting relies on global magnetic parameters like those from SHARP datasets. We used MPI-AMRVAC to conduct numerical simulations employing the Breakout model in a multipolar magnetic field configuration embedded within a dipolar background. Shear-induced initiation of CMEs was studied to assess the influence of the global poloidal magnetic field on eruptions. Our research investigates the evolution of magnetic parameters under different scenarios (failed, single, multiple eruptions) by varying helicity injection. Findings suggest a stronger poloidal magnetic field inhibits CMEs, and the rate of absolute net current helicity may provide a clear picture of CME eruption.

Keywords. Coronal Mass Ejection (CME), Sun, Current Helicity, Magnetohydrodynamics (MHD)

1. Introduction

Coronal Mass Ejections (CMEs) are powerful and dynamic solar events that release massive amounts of plasma and magnetic fields from the Sun's corona into the heliosphere (Hundhausen et al. (1984); Gopalswamy (2004); Yashiro et al. (2004); Webb & Howard (2012)). These ejections are key drivers of space weather disturbances on Earth, capable of causing shock waves and geomagnetic storms that can disrupt technology (Hapgood (2017); Schrijver (2015); Schrijver & Higgins (2015)). CMEs can travel at speeds ranging from a few hundred to several thousand km s^{-1} (Webb & Howard (2012)), and their occurrence varies with the solar cycle, ranging from one per day during solar minimum to nearly five per day during solar maximum (St. Cyr et al. (1999); Gopalswamy et al. (2005, 2006); Webb & Howard (2012)). Understanding the initiation and propagation of CMEs is essential for improving space weather predictions. Numerical simulation provides a valuable tool for studying these processes, from their origins in the Sun's lower atmosphere to their impact on Earth.

During Solar Cycle 24, lower heliospheric pressure significantly influenced CME evolution, leading to a higher occurrence of low-energy CMEs compared to previous cycles (Gopalswamy et al. (2014, 2020)). This illustrates the influence of global-scale phenomena, such as the heliospheric magnetic field, on events like CMEs. Predicting CMEs remains challenging due to the absence of a singular physical pathway for their initiation. Furthermore, various other parameters also influence CME eruptions. Numerous characteristics of active regions (ARs) have been recognized as precursors to flares and CMEs (Toriumi & Wang (2019) and reference therein), with extensive studies focusing on global magnetic parameters and the injection of magnetic energy within ARs (Zhang (2001); Pariat et al. (2017); Zuccarello et al. (2018); Moraitis et al. (2021); Li et al. (2021); Liokati et al. (2022); Liu et al. (2023)). The Spaceweather HMI Active Region Patch (SHARP) data products, used for tracking magnetic flux on the solar disk, have been instrumental in improving predictions, especially with the advancement of machine learning methods (Bobra & Couvidat (2014); Sinha et al. (2022); Raju & Das (2023)). To understand the physical processes and identify indicators for space weather forecasting, we conducted numerical MHD simulations using the breakout CME model (Antiochos (1998); Antiochos et al. (1999); van der Holst et al. (2007); Zuccarello et al. (2009, 2012a); Hosteaux et al. (2018); Talpeanu, D.-C. et al. (2020); Talpeanu et al. (2022)). This allowed us to examine the progression of magnetic parameters under varying helicity injections, leading to both eruption and failed eruption, and investigate the global coronal field's effect on CME initiation.

2. Simulation Setup

We use the Message Passing Interface - Adaptive Mesh Refinement Versatile Advection Code (MPI-AMRVAC) to solve the MHD equations in a 2.5 D axisymmetric spherical domain to simulate a breakout CME. The domain ranges from the North to the South Pole angularly and from 1 $R_\odot$ to 10 $R_\odot$ radially. Initially, the domain is divided into 128×128 uniformly stretched cells. Furthermore, to achieve high resolution and observe detailed system dynamics, we implement Adaptive Mesh Refinement (AMR) with three additional levels beyond the initial resolution. Refinement criteria are selected to focus on regions exhibiting strong current density. We implement the TVDLF scheme with a minmod slope limiter and enforce magnetic field divergence-free conditions using the generalised Lagrange multiplier (GLM) method (Dedner et al., 2002). The CFL number is maintained at 0.3 during the solution of the MHD equations. Our units are scaled, with the solar radius as the length scale, 100 km s^{-1} for velocity, and 1.93 h for time.

Our simulation setup includes a background global dipole magnetic field, a multipolar arcade system similar to (van der Holst et al. (2007); Zuccarello et al. (2012b)), and a shearing velocity for injecting helicity once the system reaches equilibrium. The gravitational field is modelled with an additional source term in the momentum equation. We incorporate a heating term into the MHD equations to replicate the solar wind's fast and slow bimodal distribution (Jacobs et al. (2005); Chané et al. (2006, 2008); Hosteaux et al. (2019)). This empirical heating source varies with latitude and radial distance from the Sun's center. The vector potential for this quadrupole magnetic field system is given by (van der Holst et al. (2007)):

$$A_\phi = A_0 \frac{\cos^2(180^\circ \lambda / 2\Delta a)}{r^4 \sin\theta} \tag{1}$$

The solar latitude, λ, is defined as $\lambda = 90^\circ - \theta$, where A_0 = -0.73 G $R_\odot^5$ and $\Delta a = 28.6^\circ$. This additional term modifies the dipole field only when the absolute value of λ is less than 28.6° ($|\lambda| < 28.6^\circ$). Initially, we allow the system to achieve a steady state. Subsequently, once the steady state is established, we introduce time-dependent shearing

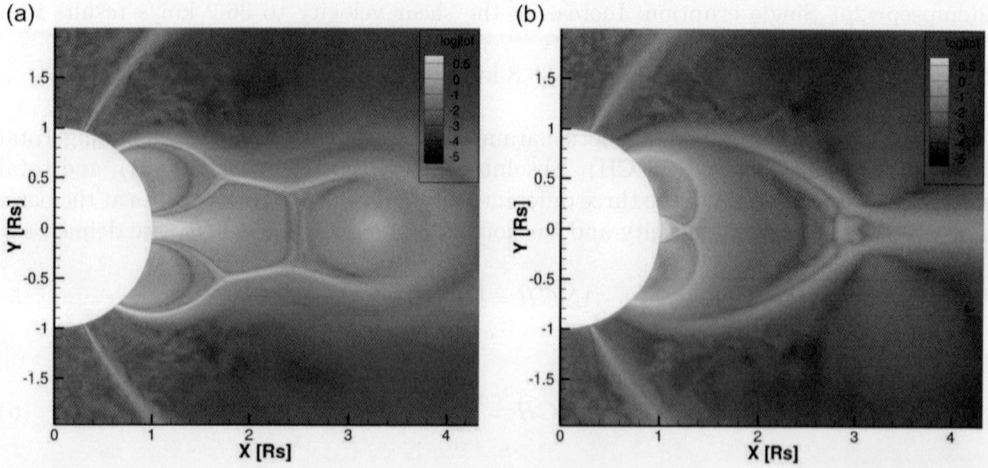

Figure 1. Simulated current density maps (A m^{-2}) at different times following the initiation of shearing, with varying polar magnetic field strengths: (a) t = 15.46 h; B*pol* = 2.2 G, (b) t = 15.46 h; B*pol* = 2.5 G

at the inner boundary, described as follows:

$$v_\phi = v_0(\lambda^2 - \Delta b^2)^2 \sin\lambda \, \sin[\pi(t - t_0)/\Delta t] \quad (2)$$

We applied the shearing velocity within the latitude range $|\lambda| < 8.59°$, which predominantly covers the central arcade. Outside this range, v_ϕ is set to zero. We maintained Δb = 8.59°. Shearing was applied over a period of 27 h.

2.1. *Effect of the poloidal magnetic field strength*

The dipole magnetic field in spherical coordinates is given by:

$$Bpol_r = \frac{2B\ Cos\theta}{r^3} \quad (3)$$

$$Bpol_\theta = \frac{B\ Sin\theta}{r^3} \quad (4)$$

We examine the effect of varying the poloidal magnetic field strength by adjusting the parameter B in equation 3, incrementing from 2.2 G to 2.5 G in steps of 0.1 G. For a poloidal magnetic field strength of 2.2 G (Figure 1a), we observe an elevation in the central arcade of the multipolar system, leading to a breakout CME eruption. Increasing the poloidal field to 2.3 G results in the ascent of the central arcade, but no apparent formation of a flux rope is observed. Further increasing the poloidal field to 2.5 G (Figure 1b) reduces the maximum height reached by the central arcade during shearing motion at the base. All other parameters, such as domain size and shear velocity, remained constant across all cases. This demonstrates the significant influence of background magnetic field strength on CME eruption.

2.2. *Effect of change in the helicity injection*

The shear velocity at the base of the magnetic arcade system determines the helicity injection into the system, leading to different evolutionary scenarios:

1tempsep=1pt Failed eruption: At a shear velocity of 35.8 km/s at the base.

2tempsep=2pt Single eruption: Increasing the shear velocity to 36.2 km/s results in a successful eruption.

3tempsep=3pt Multiple eruptions: At 38.8 km/s, multiple eruptions occur.

We investigate various global magnetic parameters across these scenarios, including Total Unsigned Current Helicity (TUCH), Absolute Net Current Helicity (ANCH), and total magnetic energy, for each of the three different maximum shear velocity values at the base. The absolute net current helicity and the total unsigned current helicity are defined as:

$$ANCH = |\sum_{i=0}^{n} H_{c_i}| \tag{5}$$

$$TUCH = \sum_{i=0}^{n} |H_{c_i}| \tag{6}$$

where n represents the total number of cells, and H_c is the current helicity, defined as:

$$H_c = \int_V \boldsymbol{B} \cdot \boldsymbol{J} \mathrm{d}\boldsymbol{V} \tag{7}$$

$\boldsymbol{J}$ is the current density and $\boldsymbol{B}$ is the magnetic field. The excess magnetic energy consistently increases over time, peaking before breakout reconnection, then decreases in failed and single eruption cases but rises and falls again in multiple eruptions. Similarly, the total unsigned current helicity continuously increases, with a gradual rate for lower shear and a persistent rise even after flux rope formation in all cases. Figure 2 illustrates the time evolution of absolute net current helicity in the arcade system after the initiation of shearing at the base. The three distinct curves are teal for the failed eruption scenario, orange for the single eruption scenario, and purple for the multiple eruption scenario. Vertical lines in each plot denote the times of flux rope formation for each scenario: solid teal line for the failed eruption, solid orange line for the single eruption, solid purple line for the first CME flux rope, and dotted purple line for the second CME flux rope. A black dotted vertical line indicates the end of the shearing velocity period.

The absolute net current helicity continuously increases until the first breakout reconnection, with a more gradual rise for lower shear. It decreases after peaking (points A, B, and C for multiple, single, and failed eruptions, respectively). During this decline, flux rope formation occurs in all cases. For failed and single eruptions, helicity continues to decrease. In multiple eruptions, it rises again between points D and E, but the second peak is lower than the first. The rate of increase between D and E is steeper than the initial phase in the failed eruption case, leading to another eruption. Thus, it is evident that not only the absolute magnitude but also the rate of increase of absolute net current helicity plays a crucial role in determining the occurrence of eruptions.

3. Summary

In this study, MHD simulations using the MPI-AMRVAC code were conducted to model breakout CMEs facilitated by shear-induced reconnection in a multipolar magnetic setup. Variations in background magnetic field strength were explored to understand their impact on CME occurrences, particularly during weaker solar cycles with lower heliospheric pressure similar to Solar Cycle 24. Contrary to significant changes observed in other studies with larger magnetic field adjustments, our simulations showed marginal (<5%) increases in background dipole strength, resulting in consistent magnetic configurations across different setups but yielding varied eruption outcomes. Additionally, adjustments in shear amplitude at the base influenced helicity injection, leading to

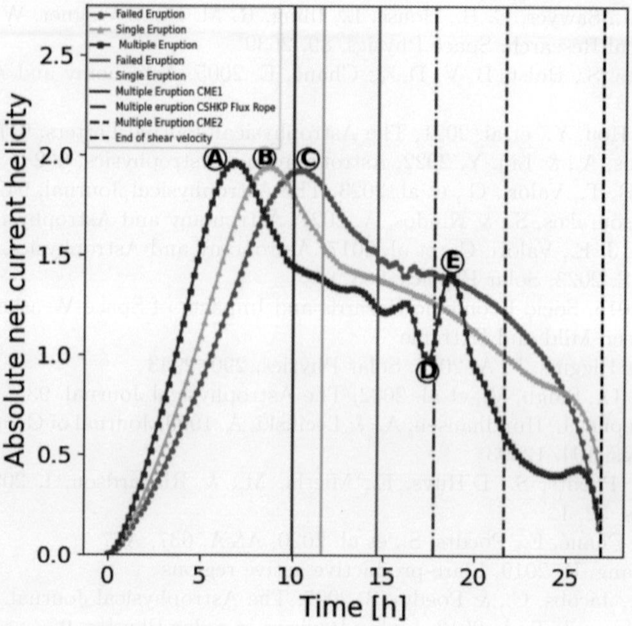

Figure 2. Temporal evolution of absolute net current helicity across three scenarios: failed eruption (teal), single eruption (orange), and multiple eruptions (purple). Vertical lines denote flux rope formation times specific to each scenario. The black dotted vertical line indicates the end of the shear period. Time is measured from the onset of shear.

distinct eruption scenarios–failed, single, and multiple eruptions. This variability underscores the challenge of predicting eruptions under subtly varying physical conditions. In our study, we discovered that among all magnetic parameters, the rate of change of absolute net current helicity, in addition to its magnitude, can be a more critical parameter.

References

Antiochos, S. K. 1998, The Astrophysical Journal Letters, 502, L181

Antiochos, S. K., DeVore, C. R., & Klimchuk, J. A. 1999, The Astrophysical Journal, 510, 485

Bobra, M. & Couvidat, S. P. 2014, in AGU Fall Meeting Abstracts, Vol. 2014, SH21A–4092

Chané, E., Holst, B. V. D., Jacobs, C., Poedts, S., & Kimpe, D. 2006, Astronomy and Astrophysics, 447, 727

Chané, E., Poedts, S., & Holst, B. V. D. 2008, Astronomy and Astrophysics, 492

Gopalswamy, N. 2004, A Global Picture of CMEs in the Inner Heliosphere (Dordrecht: Springer Netherlands), 201–251

Gopalswamy, N., Akiyama, S., Yashiro, S., et al. 2020, in Effect of the Weakened Heliosphere in Solar Cycle 24 on the Properties of Coronal Mass Ejections, Vol. 1620 (IOP Publishing Ltd)

Gopalswamy, N., Akiyama, S., Yashiro, S., et al. 2014, Geophysical Research Letters, 41, 2673

Gopalswamy, N., Mikić, Z., Maia, D., et al. 2006, Space Science Reviews, 123, 303

Gopalswamy, N., Xie, H., Yashiro, S., & Usoskin, I. 2005, in 29th International Cosmic Ray Conference (ICRC29), Volume 1, Vol. 1, 169

Hapgood, M. 2017, Space Weather, 2399–2891 (IOP Publishing)

Hosteaux, S., Chané, E., Decraemer, B., Talpeanu, D.-C., & Poedts, S. 2018, Astronomy & Astrophysics, 620, A57

Hosteaux, S., Chané, E., & Poedts, S. 2019, Astronomy and Astrophysics, 632

Hundhausen, A. J., Sawyer, C. B., House, L., Illing, R. M. E., & Wagner, W. J. 1984, Journal of Geophysical Research: Space Physics, 89, 2639
Jacobs, C., Poedts, S., Holst, B. V. D., & Chané, E. 2005, Astronomy and Astrophysics, 430, 1099
Li, T., Chen, A., Hou, Y., et al. 2021, The Astrophysical Journal Letters, 917, L29
Liokati, E., Nindos, A., & Liu, Y. 2022, Astronomy and Astrophysics, 662
Liu, Y., Welsch, B. T., Valori, G., et al. 2023, The Astrophysical Journal, 942, 27
Moraitis, K., Patsourakos, S., & Nindos, A. 2021, Astronomy and Astrophysics, 649
Pariat, E., Leake, J. E., Valori, G., et al. 2017, Astronomy and Astrophysics, 601
Raju, H. & Das, S. 2023, Solar Physics, 298, 96
Schrijver, C. J. 2015, Socio-Economic Hazards and Impacts of Space Weather: The Important Range between Mild and Extreme
Schrijver, C. J. & Higgins, P. A. 2015, Solar Physics, 290, 2943
Sinha, S., Gupta, O., Singh, V., et al. 2022, The Astrophysical Journal, 935, 45
St. Cyr, O., Burkepile, J., Hundhausen, A., & Lecinski, A. 1999, Journal of Geophysical Research: Space Physics, 104, 12493
Talpeanu, D.-C., Poedts, S., D'Huys, E., Mierla, M., & Richardson, I. 2022, Astronomy & Astrophysics, 32, 1
Talpeanu, D.-C., Chané, E., Poedts, S., et al. 2020, A&A, 637, A77
Toriumi, S. & Wang, H. 2019, Flare-productive active regions
van der Holst, B., Jacobs, C., & Poedts, S. 2007, The Astrophysical Journal, 671, L77
Webb, D. F. & Howard, T. A. 2012, Living Reviews in Solar Physics, 9, 3
Yashiro, S., Gopalswamy, N., Michalek, G., et al. 2004, Journal of Geophysical Research: Space Physics, 109 [https://agupubs.onlinelibrary.wiley.com/doi/pdf/10.1029/2003JA010282]
Zhang, H. 2001, Formation of current helicity and emerging magnetic flux in solar active regions
Zuccarello, F. P., Bemporad, A., Jacobs, C., et al. 2012a, Astrophysical Journal, 744
Zuccarello, F. P., Jacobs, C., Soenen, A., et al. 2009, Astronomy and Astrophysics, 507, 441
Zuccarello, F. P., Meliani, Z., & Poedts, S. 2012b, The Astrophysical Journal, 758, 117
Zuccarello, F. P., Pariat, E., Valori, G., & Linan, L. 2018, The Astrophysical Journal, 863, 41

Solar and Stellar Coronal Mass Ejections
Proceedings IAU Symposium No. 388, 2026
N. Gopalswamy, O. Malandraki, A. Vidotto & W. Manchester, eds.
doi:10.1017/S1743921325000602

The Temperature Difference between Eruptive and Confined Solar Flares

Seiji Yashiro[1,2], Nat Gopalswamy[1], Sachiko Akiyama[1,2] and Pertti A. Mäkelä[1,2]

[1]The Catholic University of America, 620 Michigan Avenue NE, Washington, DC 20064, USA

[2]NASA Goddard Space Flight Center, 8800 Greenbelt Road Greenbelt, MD 20771, USA

Abstract. Eruptive and confined flares are distinguished by their coronal mass ejection (CME) association. From 1996 January to 2023 December, we found 1,748 flares with definite CME association (eruptive flares) and 7,288 flares that lacked CMEs (confined flares) using the coronagraph observations by SOHO/LASCO and STEREO/SECCHI. Flare temperatures, which are measured from the GOES short/long wavelength ratio, increase with the peak X-ray intensity. Average maximum temperature is 10 MK at C3.0 and 24 MK at X10 level. Kay et al. (2003) examined a total of 69 eruptive and confined flares and found that eruptive flares tend to have lower temperatures than the confined ones. By analyzing a 100 times larger data set, we confirmed that result and found that, for a given X-ray flare intensity, the average maximum temperature of eruptive flares is approximately 0.9 MK cooler than that of the confined ones.

Keywords. Solar Flares, Coronal Mass Ejections, Flare Temperature

1. Introduction

A solar flare is an explosion in the solar atmosphere observed as a sudden flash of electromagnetic radiation emitted from the heated plasma. A coronal mass ejection (CME) is an eruption of magnetized plasma from the Sun into the interplanetary medium. Both phenomena are thought to be different manifestations of the same process (Harrison 1995; Jing et al. 2005). The flare-CME relationship has been investigated extensively using the data from Large Angle and Spectrometric Coronagraph (LASCO; Brueckner et al. 1995) on the Solar and Heliospheric Observatory (SOHO; Domingo et al. 1995), which has observed more than 38,000 CMEs since 1996 (Gopalswamy et al. 2024, in this issue). Zhang et al. (2001) and Yashiro et al. (2006) respectively examined the temporal and the spatial relationship between them and confirmed that the observations are consistent with the magnetic reconnection model developed by Carmichael (1964), Sturrock (1966), Hirayama (1974), and Kopp and Pneuman (1976). Using the acronym of the five scientists, this flare-CME model is called the CSHKP model.

Although the CSHKP flare-CME model is widely accepted as the standard model, there is no one-to-one correspondence between the flares and the CMEs. Statistical studies showed that about 10% X-class and half of M-class flares are not associated with CMEs (Yashiro et al. 2006; Gopalswamy et al. 2009). By their CME association, solar flares can be classified into eruptive and confined flares. Generally speaking, eruptive flares are large and long-lived; confined flares are compact and impulsive (e.g., Yashiro and Gopalswamy 2009). Indeed, eruptive flares tend to have lower temperatures than the confined ones based on a study of 69 eruptive and confined flares by Kay et al. (2003). We revisit the flare temperature issue by analyzing a much larger data set.

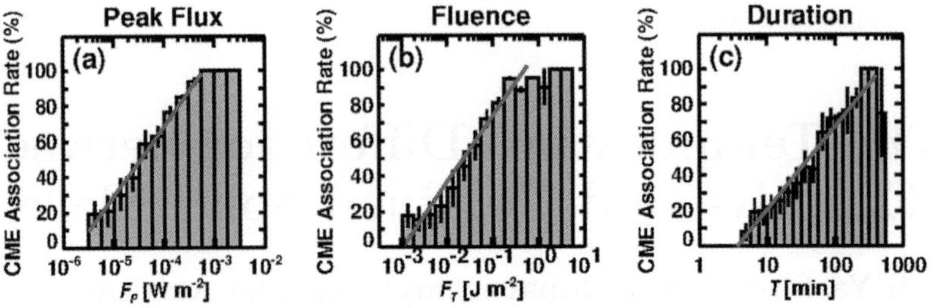

Figure 1. CME association rates as a function of the X-ray peak flux (a), fluence (b), and duration (c).

2. CME Associations of Solar Flares

In order to examine the CME associations of solar flares, we collect flare with GOES soft X-ray size C3.0 and above. From 1996 January to 2023 December, more than twelve thousand solar flares have been observed. We examine their CME associations using coronagraph images from SOHO/LASCO and Sun Earth Connection Coronal and Heliospheric Investigation (SECCHI; Howard et al. 2008) on the Solar Terrestrial Relation Observatory (STEREO; Kaiser et al. 2008) mission. The flare-CME associations are examined by viewing both flare and CME movies in the SOHO/LASCO CME Catalog (`https://cdaw.gsfc.nasa.gov/CME_list/`; Yashiro et al. 2004). Eruptive surface signatures, such as filament eruptions and coronal dimmings, help ascertain the associations (see Yashiro et al. 2006, for details). We find 1,748 flares with definite CME association (eruptive flares) and 7,288 flares that lacked CMEs (confined flares). We can not determine with confidence whether the remaining flares had the CME association or not because of coronagraph data gaps or the flares had obscure eruptive signatures.

Figure 1 shows the CME association rate as a function of X-ray peak flux (a), fluence (b), and duration (c). All flare parameters are taken from the Edited Event Reports compiled by NOAA Space Weather Prediction Center (`https://www.swpc.noaa.gov/products/solar-and-geophysical-event-reports/`). The error bars are determined from the uncertain flare-CME pairs, i.e., the lower (upper) limit are determined by assuming that all uncertain flare-CME pairs were false (true). As previous studies showed, the CME association rate of the flares increases with the peak X-ray intensity, X-ray fluence, and duration.

3. Temperatures of Solar Flares with and without CMEs

The top panels of Figures 2 show the GOES X-ray light curves in the two wavelength bands (long; 1-8 Å, and short; 0.5-4 Å) for an eruptive flare (left) and for a confined flare (right). By assuming an isothermal flare plasma, the ratio of the two channels can provide the flare temperature (Thomas et al. 1985; White et al. 2005). The background is subtracted to avoid the influence of non-flaring plasma in other active regions. The bottom panels of Figure 2 show the time profile of the flare temperature. In both flare types, the flare temperature increases quickly as the flare progresses and reaches the maximum during the impulsive (or rising) phase of the flare. For each flare we obtain two representative temperatures, namely, the maximum temperature during the impulsive phase and the temperature at the peak in the long wavelength band. We repeat the analysis for 1,748 eruptive flares and 7,288 confined flares.

Figure 3a shows a scatter plot of X-ray peak flux and the flare maximum temperature. The flare temperatures increase with the peak X-ray intensity. Average maximum temperature is 10 MK at C3.0 and 24 MK at X10 level. The blue and red dots represent the data points for confined and eruptive flares, respectively. The squares and error

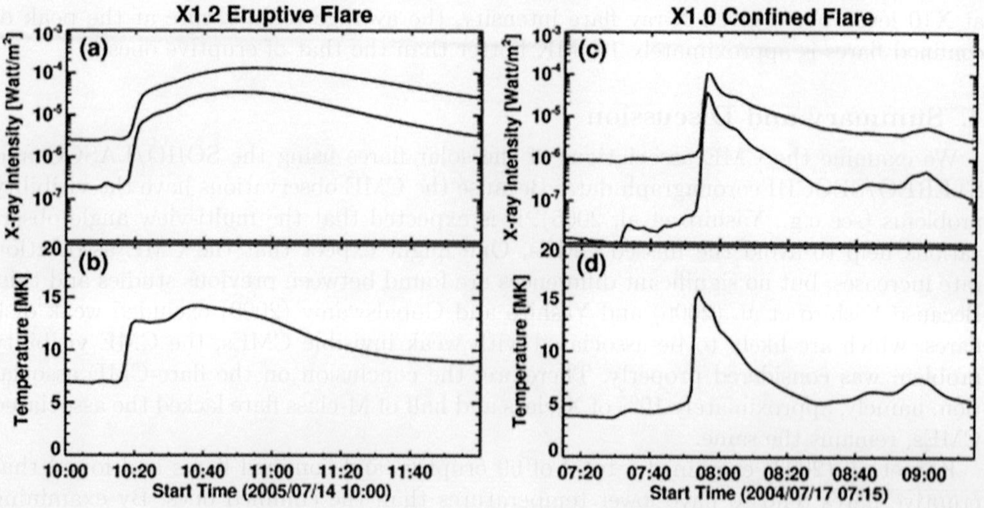

Figure 2. GOES X-ray light curve in long and short wavelength bands (top) and the flare temperature derived from the intensity ratio of the two bands (bottom). Left panels are for an X1.2 eruptive flare and right panels for an X1.0 confined flare, respectively.

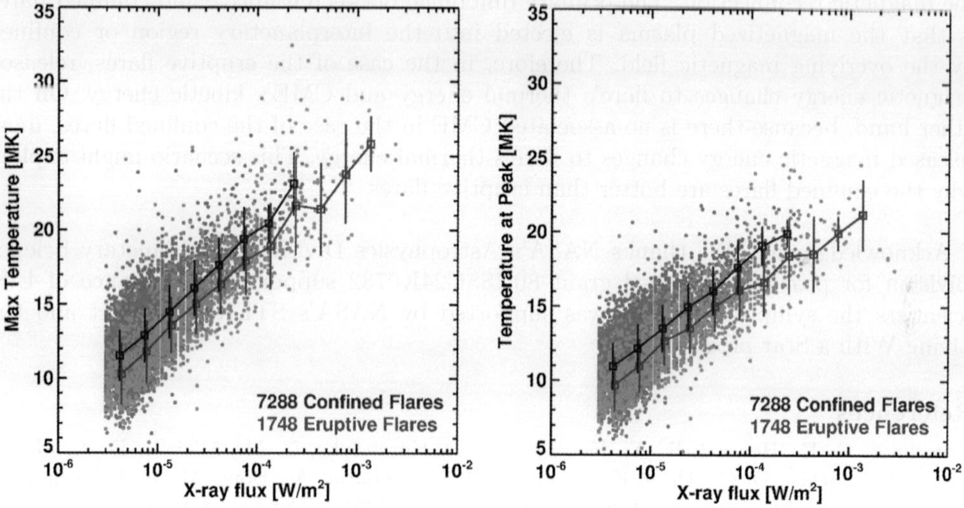

Figure 3. (a) A scatter plot of the peak X-ray flux and the maximum flare temperature of the confined (blue) and the eruptive (red) flares. The square symbols are the average maximum temperatures, and the vertical bars represent the standard deviation of the distribution. (b) Same as panel (a) but for the flare temperature at the peak.

bars show the average temperatures and the standard deviations. The blue and the red points highly overlap, but confined flares have higher average temperatures than those of the eruptive ones at all overlapping intervals. For a given X-ray flare intensity, the average maximum temperature of confined flares is approximately 0.9 MK hotter than the eruptive ones.

Figure 3b is the same as Fig. 3a but for the flare temperature at the peak. All data points move down compared to Fig. 3a, but the difference between confined and eruptive flares remains. The average flare temperature at the peak is 6.5 MK at C3.0 and 16 MK at X10 level. For a given X-ray flare intensity, the average temperature at the peak of confined flares is approximately 1.0 MK hotter than the that of eruptive ones.

4. Summary and Discussion

We examine the CME associations of the solar flares using the SOHO/LASCO and STEREO/SECCHI coronagraph data. Because the CME observations have the visibility problems (see e.g., Yashiro et al. 2005), it is expected that the multi-view angle observations help to avoid the missed CMEs. One might expect that the CME association rate increases, but no significant differences are found between previous studies and this. Because Yashiro et al. (2006) and Yashiro and Gopalswamy (2009) excluded weak disk flares, which are likely to be associated with weak invisible CMEs, the CME visibility problem was considered properly. Therefore, the conclusion on the flare-CME association, namely, approximately 10% of X-class and half of M-class flare lacked the associated CMEs, remains the same.

Kay et al. (2003) examined a total of 69 eruptive and confined flares and found that eruptive flares tend to have lower temperatures than the confined ones. By examining 100 times larger data set (1,748 eruptive and 7,288 confined flares), we confirm their result: for a given X-ray flare intensity, the average maximum temperature of confined flares is higher than the eruptive ones.

It is believed that mechanisms of the eruptive and confined flares are the same, i.e., the magnetic reconnection. The primary difference between eruptive and confined flares is that the magnetized plasma is ejected into the interplanetary region or confined by the overlying magnetic field. Therefore, in the case of the eruptive flares, released magnetic energy changes to flare's thermal energy and CME's kinetic energy. On the other hand, because there is no associated CME in the case of the confined flares, most released magnetic energy changes to flare's thermal energy. This scenario might explain why the confined flares are hotter than eruptive flares.

Acknowledgments. We thanks NASA's Astrophysics Division and Planetary Science Division for providing a travel grant 80NSSC24K0732 supporting attendance of U.S. scientists the symposium. N.G. was supported by NASA's STEREO project and the Living With a Star program.

References

Brueckner, G. E., Howard, R. A., Koomen, M. J., Korendyke, C. M., Michels, D. J., Moses, J. D., Socker, D. G., Dere, K. P., Lamy, P. L., Llebaria, A., Bout, M. V., Schwenn, R., Simnett, G. M., Bedford, D. K., and Eyles, C. J. (1995). The Large Angle Spectroscopic Coronagraph (LASCO). *Solar Phys.*, 162(1-2), 357–402.

Carmichael, H. (1964). A Process for Flares. In *NASA Special Publication*, volume 50, page 451.

Domingo, V., Fleck, B., and Poland, A. I. (1995). The SOHO Mission: an Overview. *Solar Phys.*, 162(1-2), 1–37.

Gopalswamy, N., Akiyama, S., and Yashiro, S. (2009). Major solar flares without coronal mass ejections. In Gopalswamy, N. and Webb, D. F., editors, *Universal Heliophysical Processes*, volume 257, pages 283–286.

Gopalswamy, N., Michałek, G., Yashiro, S., Mäkelä, P., Akiyama, S., Xie, H., and Vourlidas, A. (2024). The SOHO LASCO CME Catalog – Version 2. *arXiv e-prints*, page arXiv:2407.04165.

Harrison, R. A. (1995). The nature of solar flares associated with coronal mass ejection. *A&A*, 304:585.

Hirayama, T. (1974). Theoretical Model of Flares and Prominences. I: Evaporating Flare Model. *Solar Phys.*, 34(2), 323–338.

Howard, R. A., Moses, J. D., Vourlidas, A., Newmark, J. S., Socker, D. G., Plunkett, S. P., Korendyke, C. M., Cook, J. W., Hurley, A., Davila, J. M., Thompson, W. T., St Cyr, O. C., Mentzell, E., Mehalick, K., Lemen, J. R., Wuelser, J. P., Duncan, D. W., Tarbell, T. D., Wolfson, C. J., Moore, A., Harrison, R. A., Waltham, N. R., Lang, J., Davis, C. J., Eyles, C. J., Mapson-Menard, H., Simnett, G. M., Halain, J. P., Defise, J. M., Mazy, E., Rochus, P., Mercier, R., Ravet, M. F., Delmotte, F., Auchere, F., Delaboudiniere, J. P., Bothmer, V., Deutsch, W., Wang, D., Rich, N., Cooper, S., Stephens, V., Maahs, G., Baugh, R., McMullin, D., and Carter, T. (2008). Sun Earth Connection Coronal and Heliospheric Investigation (SECCHI). *Space Sci. Revs.*, 136(1-4), 67–115.

Jing, J., Qiu, J., Lin, J., Qu, M., Xu, Y., and Wang, H. (2005). Magnetic Reconnection Rate and Flux-Rope Acceleration of Two-Ribbon Flares. *ApJ*, 620(2), 1085–1091.

Kaiser, M. L., Kucera, T. A., Davila, J. M., St. Cyr, O. C., Guhathakurta, M., and Christian, E. (2008). The STEREO Mission: An Introduction. *Space Sci. Revs.*, 136(1-4), 5–16.

Kay, H. R. M., Harra, L. K., Matthews, S. A., Culhane, J. L., and Green, L. M. (2003). The soft X-ray characteristics of solar flares, both with and without associated CMEs. *A&A*, 400:779–784.

Kopp, R. A. and Pneuman, G. W. (1976). Magnetic reconnection in the corona and the loop prominence phenomenon. *Solar Phys.*, 50(1), 85–98.

Sturrock, P. A. (1966). Model of the High-Energy Phase of Solar Flares. *Nature*, 211(5050), 695–697.

Thomas, R. J., Starr, R., and Crannell, C. J. (1985). Expressions to Determine Temperatures and Emission Measures for Solar X-Ray Events from Goes Measurements. *Solar Phys.*, 95(2), 323–329.

White, S. M., Thomas, R. J., and Schwartz, R. A. (2005). Updated Expressions for Determining Temperatures and Emission Measures from Goes Soft X-Ray Measurements. *Solar Phys.*, 227(2), 231–248.

Yashiro, S., Akiyama, S., Gopalswamy, N., and Howard, R. A. (2006). Different Power-Law Indices in the Frequency Distributions of Flares with and without Coronal Mass Ejections. *ApJ* (Letters), 650(2), L143–L146.

Yashiro, S. and Gopalswamy, N. (2009). Statistical relationship between solar flares and coronal mass ejections. In Gopalswamy, N. and Webb, D. F., editors, *Universal Heliophysical Processes*, volume 257, pages 233–243.

Yashiro, S., Gopalswamy, N., Akiyama, S., Michalek, G., and Howard, R. A. (2005). Visibility of coronal mass ejections as a function of flare location and intensity. *Journal of Geophysical Research (Space Physics)*, 110(A12), A12S05.

Yashiro, S., Gopalswamy, N., Michalek, G., St. Cyr, O. C., Plunkett, S. P., Rich, N. B., and Howard, R. A. (2004). A catalog of white light coronal mass ejections observed by the SOHO spacecraft. *Journal of Geophysical Research (Space Physics)*, 109(A7), A07105.

Zhang, J., Dere, K. P., Howard, R. A., Kundu, M. R., and White, S. M. (2001). On the Temporal Relationship between Coronal Mass Ejections and Flares. *ApJ*, 559(1), 452–462.

Solar and Stellar Coronal Mass Ejections
Proceedings IAU Symposium No. 388, 2026
N. Gopalswamy, O. Malandraki, A. Vidotto & W. Manchester, eds.
doi:10.1017/S1743921324001686

Magnetic evolution of active regions: formation and eruption of magnetic flux ropes

P. Vemareddy

Indian Institute of Astrophysics, II Block, Koramangala, Bengaluru-560 034, India
email: vemareddy@iiap.res.in

Abstract. Magnetic flux ropes (FRs) are twisted structures appearing on the sun, predominantly in the magnetically concentrated regions. These structures appear as coronal features known as filaments or prominences in Hα observations, and as sigmoids in X-ray, EUV observations. Using the continuous vector magnetic field observations from *Helioseismic and Magnetic Imager* onboard *Solar Dynamics Observatory*, we study the evolution of the magnetic fields in the active regions (ARs) to understand the conditions of twisted flux formation. While ARs emerge and evolve further, flux motions such as shearing and rotation are efficient mechanisms to form twisted flux ropes. Magnetic helicity quantifies the twisted magnetic fields and helicity injection through photosphere leads to its accumulation in the corona. Therefore, coronal helicity accumulation leads to twisted FR formation and its eruption. The magnetic helicity injection is seen to evolve distinctly in the regions of flux rope formation and eruption. The ARs that are associated with eruptive activity are observed with helicity injection predominantly with one sign over a period of a few days. The ARs that inject helicity with a changing sign are unlikely to form twisted FRs because coronal helicity during the period of one sign of injected helicity gets cancelled by the opposite sign of injection in the later period. As a result, the coronal field reconfigures from shared to potential structure. For a given AR, the upper limit of helicity that could cause a CME eruption is not yet understood, which can be the subject of future studies of ARs. Magnetic reconnection plays a crucial role in both the initiation and driving of FR eruptions after their formation. Data-driven simulations of the AR evolution provide more insights on the flux rope formation and its onset of eruption.

Keywords. active regions, magnetic fields, magnetic flux ropes, magnetic helicity, coronal mass ejections (CMEs)

1. Introduction

Coronal mass ejections (CMEs) are large-scale magnetized plasma structures emanating from the solar atmosphere. CMEs occur more frequently and are most intense around the solar maximum. Fast CMEs drive interplanetary shocks, especially their propagation toward Earth, which was established to be the cause of the most severe geomagnetic disturbances on Earth (e.g., Gosling et al. 1991; Webb et al. 2000). Geo-magnetic storms are disturbances in Earth's magnetosphere and can have a significant impact on both ground- and space-based technological systems. Therefore, it is of scientific and technological interest to understand the complete picture of CMEs, such as their origin from the source regions, evolution, and sun-to-earth propagation (Webb and Howard 2012).

In white-light observations, the CMEs are seen with a three-part structure morphology; viz. the leading edge, cavity, and core. These features of the CMEs are attributed to the compressed plasma in front of a flux rope (FR), a cavity, and a bright filament or prominence surrounding the cavity (Illing and Hundhausen 1985). In this description, the magnetic configuration of the CME is a FR with a helical field that is wound around the central axis. A statistical study by Vourlidas et al. (2013) suggested that at least 40% of CMEs observed by space-borne instruments have a clear FR structure. The in situ counterparts of the CMEs are known as Interplanetary Coronal Mass Ejections (ICMEs) which comprise magnetic clouds (MCs) of large-scale, organized magnetic structures (Burlaga et al. 1981). The MCs are characterized by a smoothly rotating field of enhanced field strength from the background solar wind, low proton temperature, and low proton β. This characterization supports the FR topology to the in situ magnetic field and the MC is interpreted to be part of large scale bent FR extending from the Sun into interplanetary space (Burlaga 1991).

On the other hand, because the CMEs are linked to eruptions of solar features such as prominences, filaments, and sigmoids, the origin of the twisted FR must be in the source regions of these features. Therefore, to understand the origins of the CMEs on the Sun, it is very fundamental to study the physical mechanisms of the formation and eruption of FRs in the source regions.

Magnetic fields play a fundamental role in the structure and dynamics of the solar corona. Driven by continuous plasma motions at foot points, the coronal field is stressed to build up electric current, and magnetic free energy. Such magnetic fields explode as CMEs/flares to release the free energy into thermal and kinetic energies (Shibata and Magara 2011). Sheared and twisted magnetic structures represent a non-potential magnetic field with free magnetic energy. Using the unprecedented, uninterrupted imaging observations from the Solar Dynamics Observatory (SDO; Pesnell et al. 2012), in this proceeding paper, we study the magnetic evolution of the active regions (ARs) to comprehend the formation of sheared/twisted magnetic field and its subsequent eruption. **Having connection to the CME flux rope, we refer both sheared and twisted magnetic fields as twisted magnetic structures or FRs, although they are different from the modeling point of view.** This paper is organized as follows: In section 2, as an exemplary study, we showed the formation of twisted magnetic structure during the evolution of AR 11158. The characteristic evolution of ARs that is very likely to form the **sheared/twisted magnetic field** is studied by evaluating the helicity injection and its coronal accumulation in section 3. The triggering mechanisms of the eruptions are outlined in Section 4, followed by a brief summary of the paper in Section 5.

2. Magnetic evolution of the ARs and the twisted structures

Magnetic field evolution is better studied with continuous, high-cadence observations of the full disk magnetic fields. Routine vector magnetic field observations (`hmi.sharp.cea_720s` data product) are available from the Helioseismic and Magnetic Imager (HMI; Schou et al. 2012) onboard SDO. As an example, we study the AR 11158, which has been highly flare-productive since its emergence on February 11, 2011 and has been the subject of several studies (e.g., Schrijver et al. 2011; Vemareddy et al. 2012a; Sun et al. 2012; Vemareddy et al. 2015; Green et al. 2022). Its disk passage was well captured by multiple instruments, especially SDO. Fig 1a shows the map of vector magnetic fields of the AR 11158 on February 14 at 10:00 UT. The major sunspot polarities are labelled (positive as P1, P2, P3 and negative as N1, N2, N3) and the regions of interest are enclosed in rectangular boxes (R1 and R2). During the evolution, the polarities underwent shear motions predominantly in sub-region R1 and rotational motions in sub-region R2. As a result, the horizontal field vectors are stressed in these regions. Such horizontal fields are mostly aligned parallel to the local polarity inversion line (PIL). The extent of stress is quantified by the shear angle, which is measured with respect to the horizontal field vectors of the potential field (PF). The potential field is the minimum

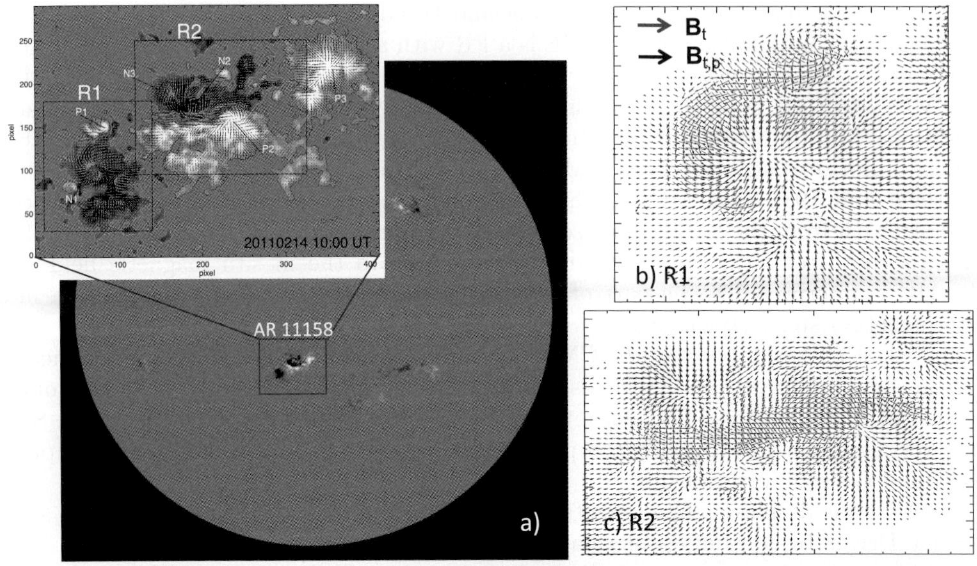

Figure 1. A typical distribution of non-potential magnetic fields in an AR. a) HMI full-disk observation of a line-of-sight magnetic field. Inset is the vector magnetic field map of AR 11158 at 10:00 UT on February 14, 2011. The horizontal field vectors in red (green) are overplotted on the vertical component of the magnetic field with ±150G contours. Major sunspot polarities are marked as P/N* within the rectangular regions of interest R1 and R2 (sub-regions) for further reference. The blue solid curves represent the strongly sheared (with shear angle greater than 45°) PILs. The axis units are in pixels of 0.5 arcsec. b-c) Horizontal magnetic fields (blue) in sub-regions R1 and R2, respectively. Horizontal magnetic field vectors (black) derived from potential field extrapolation are overplotted to show the non-potentiality of the observed fields with respect to the flux motions in the AR.

energy stage of the magnetic field and is calculated with the B_z component as a Neumann boundary condition. In Figure 1(b-c), we show the horizontal vectors of both observed and potential magnetic fields in subregions R1 and R2. It can be evidently noticed that the sheared field mostly spread around the PIL.

The buildup of magnetic non-potentiality during the AR evolution is monitored by parameters such as magnetic shear, force-free parameter (α_{av}), vertical current (J_z), helicity injection rate (dH/dt), magnetic free- energy (E_f), etc. These parameters are found to have good correspondence with the rotating sunspot N1 in R1, as reported by Vemareddy et al. (2012a). Moreover, they found that the free energy exhibits a step-down decrease at the onset of the flares from sub-region R2.

Since regular measurements of coronal magnetic fields are not available, one extrapolates the observed photospheric vector field into the corona using the force-free approximation to the coronal field (Wiegelmann 2004). In this model, the force-free parameter α varies spatially due to the presence of a sheared field, therefore known as non-linear force-free fields (NLFFFF). In Figure 2a, we display the vector magnetogram used to construct NLFFF. Figure 2b shows the rendered magnetic field of the NLFFF model, and is compared with the coronal plasma tracers observed in the AIA 171 Å image (Figure 2c). From the comparison, we see that the NLFFF model reproduced structures similar to the coronal plasma tracers, and would be a promising coronal magnetic field model. In addition, the magnetic field at the core of the AR (blue rectangular box) is highly twisted, as displayed in Figure 2d. Due to strong electric currents in the twisted field, the core field region is co-spatial with the bright plasma emission. The electric

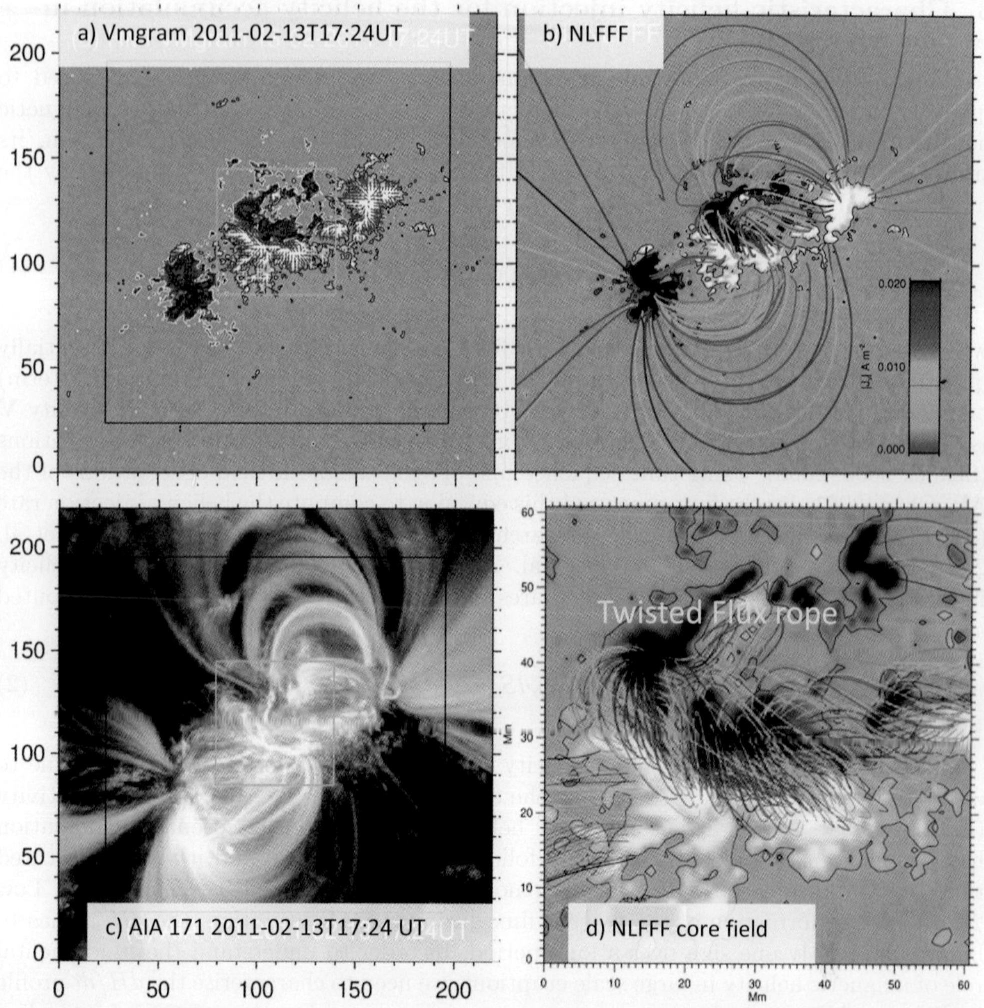

Figure 2. Magnetic field structure of AR 11158. a) vector magnetogram used as the bottom boundary for the coronal field extrapolation, b) NLFFF magnetic structure rendered on the vertical magnetic field map, c) image of the AR corona in AIA 171 Å as proxy of coronal magnetic fields for comparison, d) magnetic structure at the core of the AR (green rectangular box) indicating highly twisted flux system, appearing as bright plasma emission in 171 Å.

currents are non-neutralized in the regions of twisted flux (Török et al. 2014; Vemareddy 2019).

Thus, during the evolution of the AR, the flux motions drive the coronal field to non-potential form and form twisted FR. Shear and rotational motions are efficient energy storage mechanisms to account for CMEs and flares. The FRs thus formed are subject to erupt once they attain the critical limits of energy and helicity (Zhang and Low 2005). The eruption of the FRs is the only way to release the excess free magnetic energy, which is being converted to kinetic and thermal energy. An X2.2 flare associated with large CME and several C and M class flares occurred from subregion R2 of AR 11158 where shear motions are predominant. Some ARs were reported to launch successive CMEs due to the continuous process of FR formation and its eruption under a constant driver of shear and converging flux motions (Vemareddy 2017).

3. Characteristic helicity injection for the helicity accumulation in the AR corona

Magnetic helicity is a measure of magnetic twist and shear which can be used to quantify the complexity of the AR magnetic field. Due to conserved nature, magnetic helicity transfers from one volume to another. Especially in the case of the sun, its transport through photospheric surface in the open coronal volume is estimated by the mathematical expression derived by Berger and Field (1984),

$$\left.\frac{dH}{dt}\right|_S = 2\int_S (\mathbf{A}_P \bullet \mathbf{B}_t)\,\mathrm{V}_n dS - 2\int_S (\mathbf{A}_P \bullet \mathbf{V}_t)\,\mathrm{B}_n dS \qquad (1)$$

with the help of velocity ($\mathbf{V}$) and magnetic field $\mathbf{B}$ observations. This equation essentially quantifies the complexity generated by the flux motions, including emerging (first term) and shear motions (second term), as the ARs emerge and evolve further. The velocity $\mathbf{V}$ is derived by tracking the flux motions from time sequence magnetic field observations (Schuck 2005, 2008). Using time sequence line-of-sight magnetic field observations of the AR, Chae (2001) for the first time used this equation to compute the helicity injection rate (dH/dt), followed by several other researchers (e.g., Nindos et al. 2003; LaBonte et al. 2007; Park et al. 2010; Vemareddy et al. 2012b) to understand connection of helicity injection with the occurrence of large flares. Similarly, the energy injection is computed as

$$\left.\frac{dE}{dt}\right|_S = \frac{1}{4\pi}\int_S B_t^2 V_n dS - \frac{1}{4\pi}\int_S (\mathbf{B}_t \bullet \mathbf{V}_t)\, B_n dS \qquad (2)$$

Given the profiles of energy and helicity injections rates, one integrates over time to estimate their coronal budgets. Unlike the statistical studies for large flare productivity (LaBonte et al. 2007; Park et al. 2010), helicity injection and its coronal accumulation have implications to the FR formation followed by its eruption and it was conjectured that the CMEs are the only way to remove excess coronal helicity (Zhang and Low 2005). For the formation of the twisted flux during the AR evolution, the dH/dt has to be predominantly one sign over a long period. In order to understand the fundamental role of magnetic helicity in large scale eruptions, we need to characterise the dH/dt profile in different ARs producing eruptive or non-eruptive flares. For this purpose, we studied a few emerging ARs along with their generated activity.

Figure 3 shows magnetic evolution in emerging ARs 11560 and 11726 during their disk transit within ± 40 longitude. While magnetic polarities emerge and evolve, the dH/dt is predominant with a sign, so the coronal helicity is additive in time. Note the difference in injection rate in these two ARs because the flux motions (both emerging and shear) are different in different ARs. As a result, these ARs exhibit eruptive behaviour at some point in time. The time profiles of dH/dt are examined in several such ARs. In Figure 4, we plot the magnetic evolution in AR 12371 from where successive CME eruptions were launched over the course of 7 days of evolution. The dH/dt profile is negative all the time, then there could be instances when the accumulated helicity exceeds the critical value, and as a result, the FR eruption occurs intermittently as CMEs. Although the AR is in the decay phase, the core polarities are in shear, and converging motions, which form twisted flux along the PIL. **The pre-eruptive magnetic structure of NLFFF model reveals the existence of a twisted FR along the PIL of shearing polarities** (Vemareddy and Demóulin 2018).

There are also ARs that have different evolutions of magnetic fluxes. In Figure 5, we plot the magnetic evolution of AR 12257, which did not exhibit eruptive activity except for C-class flaring events. In this AR, the dH/dt changes sign from positive to negative

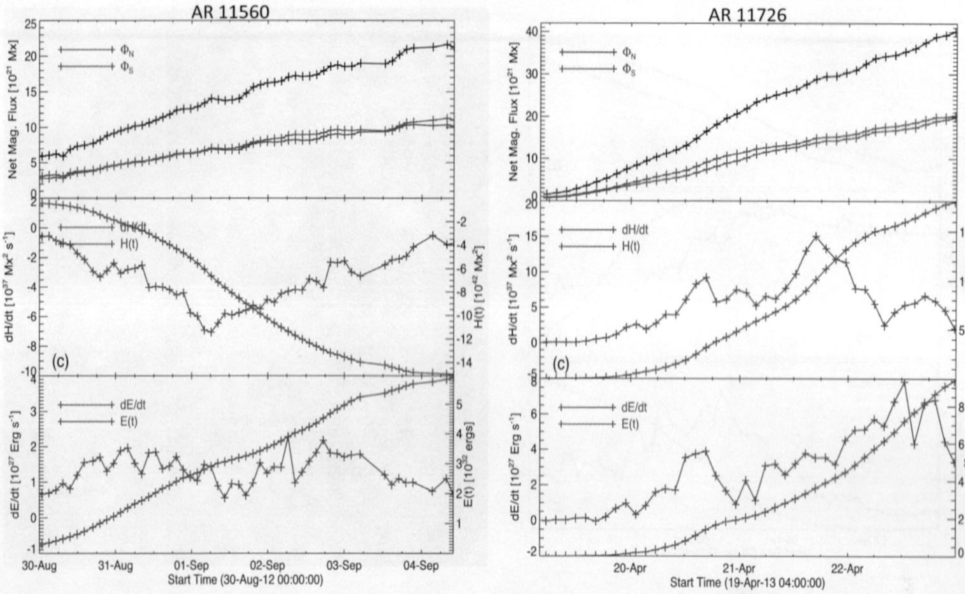

Figure 3. Time evolution of magnetic flux, helicity, and energy fluxes in ARs 11560 (left) and 11726 (right). Time-integrated coronal helicity and energy (accumulated) quantities are plotted with the y-axis scale on the right of each plot. AR 11560 evolves with a predominant negative sign of dH/dt, whereas AR 11726 has a positive sign. Due to continuous injection of one sign of helicity flux, both these ARs have exhibited eruptive behavior.

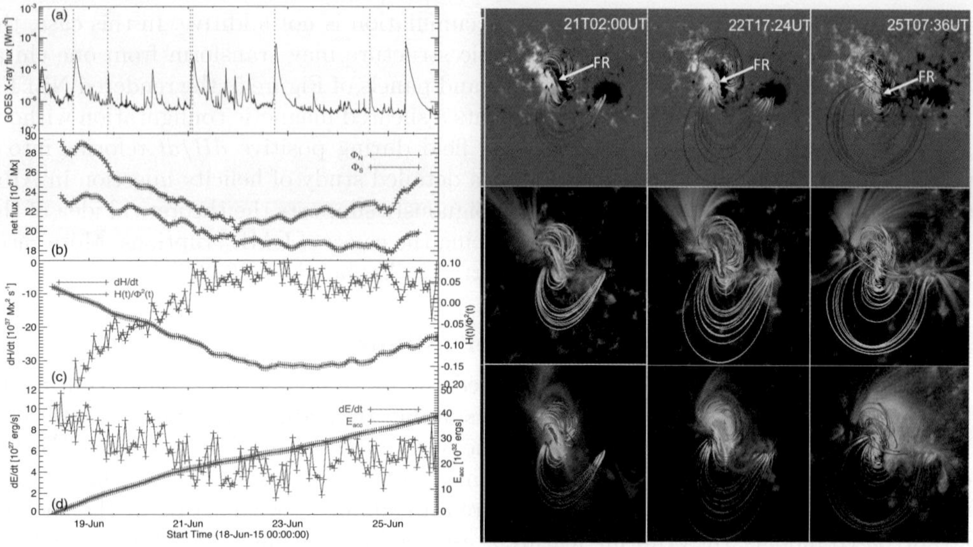

Figure 4. Magnetic evolution in successively erupting AR 12371. **left column panels:** time profiles of disk-integrated soft x-ray flux (a), net magnetic flux (b), dH/dt (c) and dE/dt (d). **right column panels:** Rendered NLFFF magnetic structure on B_z map and AIA 171Å and 94Å maps at different times. While helicity flux continuously accumulates with a predominant sign, the coronal magnetic fields store energy to form twisted flux along the sheared PIL.

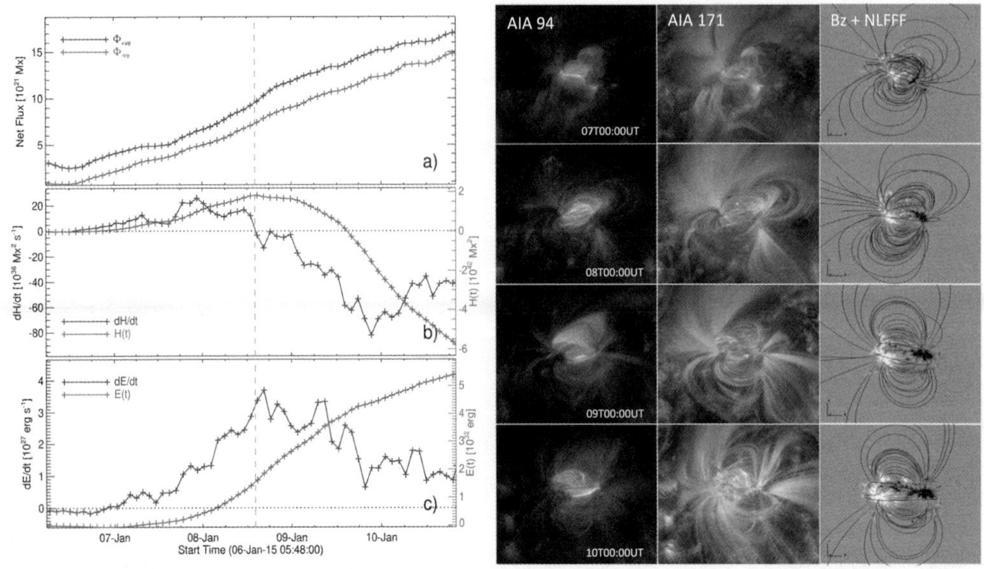

Figure 5. Magnetic evolution of the AR 12257 without exhibiting eruptive behavior. Left column panels: time evolution of net magnetic flux, the helicity injection rate (dH/dt), and the energy injection rate. Accumulated helicity, $H(t)$, and energy, $E(t)$, are plotted with their scales on the right-side axis. Vertical-dashed line marks the time (2015 January 08T14:00 UT) when dH/dt changes sign from positive to negative. Right column panels: Comparison of NLFFF magnetic structure with plasma tracers in EUV images in AIA 94, 171 Å images at different epochs of the AR evolution.

values, and as a consequence, the helicity accumulation is not additive. In this case, the twisted flux may not form, but the magnetic structure may transform from one chiral (handedness) form to another. In the right-hand panels of Figure 5, the rendered NLFFF magnetic structure at different epochs presents a sheared magnetic configuration without a twisted FR at the core. And the sheared field during positive dH/dt reforms into a potential field during negative dH/dt. Such a detailed study of helicity injection in ARs, supported by the coronal field model, unambiguously supports the theoretical idea of the conservation property of magnetic helicity being the cause of CME eruptions. More cases of opposite helicity injecting ARs were reported in Vemareddy (2022).

4. Triggering Mechanism of solar eruptions

When the AR has reached a state of stored energy configuration through continuous flux motions, its sudden release requires a suitable triggering mechanism. There have been several ideas explored for explaining the onset mechanism of the eruptions (Forbes et al. 2006; Moore and Sterling 2006; Parenti 2014). All these models are based on a pre-eruptive configuration, either as a sheared arcade or a twisted FR. In the sheared arcade based models, the erupting feature is assumed to be composed of the sheared and twisted field in the core of the AR, and then the internal/external runaway tether-cutting reconnection forms the FR and its subsequent eruption (Moore et al. 2001; Antiochos et al. 1999; Amari et al. 2003). On the other hand, FR based models invoke a twisted FR that could lose equilibrium due to ideal MHD instability such as kink or torus instability (Török et al. 2004). Because of the wide variety of dynamic processes occurring in complex ARs, however, it may be difficult to identify a particular triggering mechanism responsible for an observed eruption.

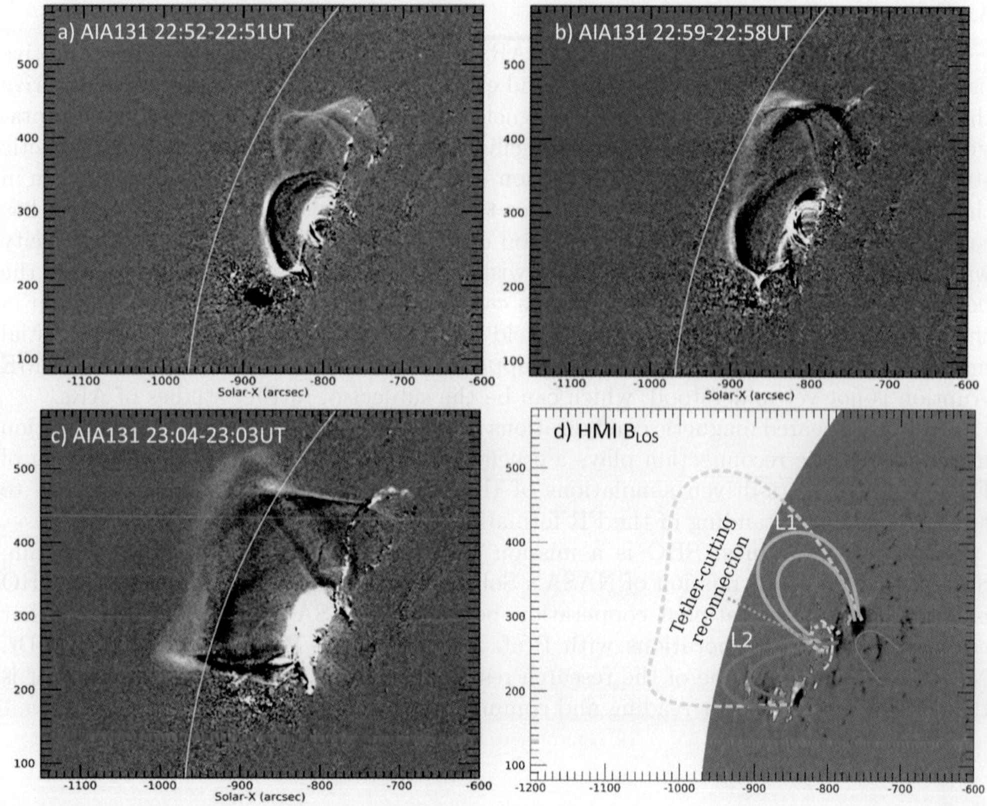

Figure 6. Evidence for tether-cutting reconnection as the initiation of the eruption. a-c) difference images of AIA 131 Å during the onset of the eruption. Two nearby loop structures, L1 and L2, first rise up, then reconnect at a coronal cross-point to form a single larger loop structure with flare loops below. b) A schematic illustration of the possible magnetic structure undergoing the tether-cutting reconnection.

Observational evidence for tether cutting reconnection forming the FR and triggering its eruption were reported in several studies (e.g., Vemareddy et al. 2012c; Liu et al. 2013; Xue et al. 2017; Chen et al. 2018; Vemareddy et al. 2022. Typically in this model, two adjacent opposite sheared loops cross each other over the PIL as shown in Figure 6(a-b). In top-view observations (e.g., Vemareddy 2017) they appear as two lobes of a sigmoid. When subject to a converging motion toward the PIL, these loops reconnect to form upward-rising twisted field lines, seen as continuous EUV structure in observations (Figure 6c). The schematic in Figure 6d better depicts this scenario, linking the foot points of the loops with the magnetic field observations. The reconnection also induces the flare emission underneath the FR and acts as a driving mechanism to further accelerate the FR. Similarly, there were observational studies that claimed direct evidence of ideal kink instability with the flux system underlying the prominence/filament body (e.g., Rust and Kumar 1996; Green et al. 2007; Gilbert et al. 2007; Vemareddy et al. 2017). After the FR reaches a certain height where the magnetic field decrease is critically steep, the torus instability plays a role in driving the eruption further (Vasantharaju et al. 2019; Aulanier et al. 2010). In both types of these mechanisms, reconnection plays a crucial role by transforming the stabilizing upper arcade into the twisted field surrounding the original core field.

5. Summary

We study the magnetic evolution of the ARs which is the central subject of the eruptive behavior of the sun. While ARs emerge and evolve further, flux motions constantly drive the coronal magnetic field to a stored magnetic energy configuration. Shearing and rotational motions of the fluxes are especially efficient mechanisms to form twisted magnetic structures in the form of FRs. FR formation is a consequence of helicity accumulation in the corona (Zhang and Low 2005), and for that, helicity injection from the surface has to be predominantly one sign over a period of a few days. The ARs that inject helicity with changing signs are unlikely to form twisted FRs because coronal helicity during the period of one sign of injected helicity gets cancelled by the opposite sign of injection in the later period. As a result, the coronal field reconfigures itself from sheared to potential magnetic structure. For a given AR, the upper limit of helicity that could cause a CME eruption is not yet understood, which can be the subject of further studies of ARs.

The FR or sheared magnetic configurations are basic ingredients in most of the eruption models. Magnetic reconnection plays a crucial role in both the initiation and driving of FR eruptions. Data-driven simulations of the AR evolution provide more insights to advance our understanding of the FR formation and its onset of eruption.

Acknowledgements: SDO is a mission of NASA's Living With a Star Program; STEREO is the third mission of NASA's Solar-Terrestrial Probes program; and SOHO is a mission of international cooperation between the ESA and NASA. The author acknowledges the collaborations with Prof. Jie Zhang, Dr. Pascal Demöulin, and Dr. Nat Gopalswamy on some of the research results discussed in this paper. The author is thankful to the referee for reading and commenting on this draft.

References

Amari, T., Luciani, J. F., Aly, J. J., & et al. 2003, Coronal Mass Ejection: Initiation, Magnetic Helicity, and Flux Ropes. I. Boundary Motion-driven Evolution. *ApJ*, 585, 1073–1086.

Antiochos, S. K., DeVore, C. R., & Klimchuk, J. A. 1999, A Model for Solar Coronal Mass Ejections. *ApJ*, 510, 485–493.

Aulanier, G., Török, T., Démoulin, P., & DeLuca, E. E. 2010, Formation of Torus-Unstable Flux Ropes and Electric Currents in Erupting Sigmoids. *ApJ*, 708, 314–333.

Berger, M. A. & Field, G. B. 1984, The topological properties of magnetic helicity. *Journal of Fluid Mechanics*, 147, 133–148.

Burlaga, L. F., Hundhausen, A. J., & Zhao, X.-P. 1981, The coronal and interplanetary current sheet in early 1976. *J. Geophys. Res.*, 86, 8893–8898.

Burlaga, L. F. E. 1991,. *Magnetic Clouds*, 152.

Chae, J. 2001, Observational Determination of the Rate of Magnetic Helicity Transport through the Solar Surface via the Horizontal Motion of Field Line Footpoints. *ApJ*, 560(1), L95–L98.

Chen, H., Duan, Y., Yang, J., Yang, B., & Dai, J. 2018, Witnessing Tether-cutting Reconnection at the Onset of a Partial Eruption. *ApJ*, 869(1), 78.

Forbes, T. G., Linker, J. A., Chen, J., Cid, C., & et al. 2006, CME Theory and Models. *Space Sci. Rev.*, 123, 251–302.

Gilbert, H. R., Alexander, D., & Liu, R. 2007, Filament Kinking and Its Implications for Eruption and Re-formation. *Solar Phys.*, 245, 287–309.

Gosling, J. T., McComas, D. J., Phillips, J. L., & Bame, S. J. 1991, Geomagnetic activity associated with earth passage of interplanetary shock disturbances and coronal mass ejections. *J. Geophys. Res.*, 96(A5), 7831–7839.

Green, L. M., Kliem, B., Török, T., van Driel-Gesztelyi, L., & Attrill, G. D. R. 2007, Transient Coronal Sigmoids and Rotating Erupting Flux Ropes. *Solar Phys.*, 246, 365–391.

Green, L. M., Thalmann, J. K., Valori, G., Pariat, E., Linan, L., & Moraitis, K. 2022, Magnetic Helicity Evolution and Eruptive Activity in NOAA Active Region 11158. *ApJ*, 937(2), 59.

Illing, R. M. E. & Hundhausen, A. J. 1985, Observation of a coronal transient from 1.2 to 6 solar radii. *J. Geophys. Res.*, 90, 275–282.

LaBonte, B. J., Georgoulis, M. K., & Rust, D. M. 2007, Survey of Magnetic Helicity Injection in Regions Producing X-Class Flares. *ApJ*, 671(1), 955–963.

Liu, C., Deng, N., Lee, J., Wiegelmann, T., Moore, R. L., & Wang, H. 2013, Evidence for Solar Tether-cutting Magnetic Reconnection from Coronal Field Extrapolations. *ApJ*, 778(2), L36.

Moore, R. L. & Sterling, A. C. 2006, Initiation of Coronal Mass Ejections. *Washington DC American Geophysical Union Geophysical Monograph Series*, 165, 43.

Moore, R. L., Sterling, A. C., Hudson, H. S., & Lemen, J. R. 2001, Onset of the Magnetic Explosion in Solar Flares and Coronal Mass Ejections. *ApJ*, 552, 833–848.

Nindos, A., Zhang, J., & Zhang, H. 2003, The Magnetic Helicity Budget of Solar Active Regions and Coronal Mass Ejections. *ApJ*, 594(2), 1033–1048.

Parenti, S. 2014, Solar Prominences: Observations. *Living Reviews in Solar Physics*, 11.

Park, S.-h., Chae, J., & Wang, H. 2010, Productivity of Solar Flares and Magnetic Helicity Injection in Active Regions. *ApJ*, 718(1), 43–51.

Pesnell, W. D., Thompson, B. J., & Chamberlin, P. C. 2012, The Solar Dynamics Observatory (SDO). *Solar Phys.*, 275, 3–15.

Rust, D. M. & Kumar, A. 1996, Evidence for Helically Kinked Magnetic Flux Ropes in Solar Eruptions. *ApJ*, 464, L199.

Schou, J., Scherrer, P. H., Bush, R. I., Wachter, R., & et al. 2012, Design and Ground Calibration of the Helioseismic and Magnetic Imager (HMI) Instrument on the Solar Dynamics Observatory (SDO). *Solar Phys.*, 275, 229–259.

Schrijver, C. J., Aulanier, G., Title, A. M., Pariat, E., & Delannée, C. 2011, The 2011 February 15 X2 Flare, Ribbons, Coronal Front, and Mass Ejection: Interpreting the Three-dimensional Views from the Solar Dynamics Observatory and STEREO Guided by Magnetohydrodynamic Flux-rope Modeling. *ApJ*, 738, 167.

Schuck, P. W. 2005, Local Correlation Tracking and the Magnetic Induction Equation. *ApJ*, 632, L53–L56.

Schuck, P. W. 2008, Tracking Vector Magnetograms with the Magnetic Induction Equation. *ApJ*, 683, 1134–1152.

Shibata, K. & Magara, T. 2011, Solar Flares: Magnetohydrodynamic Processes. *Living Reviews in Solar Physics*, 8, 6.

Sun, X., Hoeksema, J. T., Liu, Y., & et al. 2012, Evolution of Magnetic Field and Energy in a Major Eruptive Active Region Based on SDO/HMI Observation. *ApJ*, 748, 77.

Török, T., Kliem, B., & Titov, V. S. 2004, Ideal kink instability of a magnetic loop equilibrium. *A&A*, 413, L27–L30.

Török, T., Leake, J. E., Titov, V. S., & et al. 2014, Distribution of electric currents in solar active regions. *ApJ*, 782, L10.

Vasantharaju, N., Vemareddy, P., Ravindra, B., & Doddamani, V. H. 2019, Finding the Critical Decay Index in Solar Prominence Eruptions. *ApJ*, 885(1), 89.

Vemareddy, P. 2017, Successive homologous coronal mass ejections driven by shearing and converging motions in solar active region noaa 12371. *ApJ*, 845, 59.

Vemareddy, P. 2019, Degree of electric current neutralization and the activity in solar active regions. *MNRAS*, 486(4), 4936–4946.

Vemareddy, P. 2022, Nature of helicity injection in non-erupting solar active regions. *MNRAS*, 516(1), 158–166.

Vemareddy, P., Ambastha, A., & Maurya, R. A. 2012,a On the Role of Rotating Sunspots in the Activity of Solar Active Region NOAA 11158. *ApJ*, 761a, 60.

Vemareddy, P., Ambastha, A., Maurya, R. A., & Chae, J. 2012,b On the Injection of Helicity by the Shearing Motion of Fluxes in Relation to Flares and Coronal Mass Ejections. *ApJ*, 761b, 86.

Vemareddy, P. & Demóulin, P. 2018, Study of Three-dimensional Magnetic Structure and the Successive Eruptive Nature of Active Region 12371. *ApJ*, 857, 90.

Vemareddy, P., Démoulin, P., Sasikumar Raja, K., Zhang, J., Gopalswamy, N., & Vasantharaju, N. 2022, Eruption of the EUV Hot Channel from the Solar Limb and Associated Moving Type IV Radio Burst. *ApJ*, 927(1), 108.

Vemareddy, P., Gopalswamy, N., & Ravindra, B. 2017, Prominence Eruption Initiated by Helical Kink Instability of an Embedded Flux Rope. *ApJ*, 850(1), 38.

Vemareddy, P., Maurya, R. A., & Ambastha, A. 2012,c Filament Eruption in NOAA 11093 Leading to a Two-Ribbon M1.0 Class Flare and CME. *Solar Phys.*, 277c, 337–354.

Vemareddy, P., Venkatakrishnan, P., & Karthikreddy, S. 2015, Flux emergence in the solar active region NOAA 11158: the evolution of net current. *Research in Astronomy and Astrophysics*, 15, 1547.

Vourlidas, A., Lynch, B. J., Howard, R. A., & Li, Y. 2013, How Many CMEs Have Flux Ropes? Deciphering the Signatures of Shocks, Flux Ropes, and Prominences in Coronagraph Observations of CMEs. *Solar Phys.*, 284, 179–201.

Webb, D. F., Cliver, E. W., Crooker, N. U., Cry, O. C. S., & Thompson, B. J. 2000, Relationship of halo coronal mass ejections, magnetic clouds, and magnetic storms. *J. Geophys. Res.*, 105, 7491–7508.

Webb, D. F. & Howard, T. A. 2012, Coronal Mass Ejections: Observations. *Living Reviews in Solar Physics*, 9(1), 3.

Wiegelmann, T. 2004, Optimization code with weighting function for the reconstruction of coronal magnetic fields. *Solar Phys.*, 219, 87–108.

Xue, Z., Yan, X., Yang, L., Wang, J., & Zhao, L. 2017, Observing Formation of Flux Rope by Tether-cutting Reconnection in the Sun. *ApJ*, 840(2), L23.

Zhang, M. & Low, B. C. 2005, The Hydromagnetic Nature of Solar Coronal Mass Ejections. *ARA&A*, 43, 103–137.

Solar and Stellar Coronal Mass Ejections
Proceedings IAU Symposium No. 388, 2026
N. Gopalswamy, O. Malandraki, A. Vidotto & W. Manchester, eds.
doi:10.1017/S1743921325000304

On the Onset Mechanism for Solar Coronal Jets, and Implications for the Onset Mechanism for CME-Producing Eruptions

Alphonse C. Sterling[1], **Ronald L. Moore**[2,1] **and Navdeep K. Panesar**[3,4]

[1]NASA Marshall Space Flight Center, Huntsville, AL, USA

[2]The University of Alabama in Huntsville, Huntsville, AL, USA

[3]Bay Area Environmental Research Institute, Moffett Field, CA, USA

[4]Lockheed Martin Solar and Astrophysics Laboratory, Palo Alto, CA, USA

Abstract. Large-scale solar eruptions often include ejection of a filament, a solar flare, and expulsion of a coronal mass ejection (CME). Unravelling the magnetic processes that build up the free energy for these eruptions and trigger that energy's release in the eruption is a continuing challenge in solar physics. Such large-scale eruptions are comparatively infrequent, with the moderate level ones (say, GOES M-class events) occurring perhaps once every few days on average during active-activity times, and much less frequently during quieter times. In contrast, solar coronal jets, which are long (∼50,000 km), narrow (less than about 10,000 km), transient (∼10—20 min) plasma spires with bright bases and that are seen in soft X-rays and EUV, occur much more frequently, likely several hundred times per day independent of large-scale solar activity level. Recent studies indicate that coronal jets are small-scale versions of large-scale eruptions, often produced by eruption of a small-scale "minifilament," that results in a "miniflare" analogous to a larger typical solar flare, and that sometimes produces a CME analogue (a "narrow CME" or "white-light jet"). Under the assumption that jets are small-scale eruptions, their higher occurrence frequency and faster build-up evolution reveals perhaps fundamental aspects of all eruptions that are not as easy to discern in the more-complex magnetic environment and the slower build up to the larger eruptions. Therefore, the study of coronal jets can provide insights into the onset mechanism of CME-producing large-scale eruptions.

Keywords. Sun: filaments, prominences, flares, magnetic fields, CMEs

1. Overview of Solar Corona Jets

Solar coronal jets are long, narrow geyser-like ejections usually observed in soft X-ray (SXR) or EUV coronal images (Shibata et al. 1992; Raouafi, et al. 2016; Shen 2021; Schmieder, 2022; Sterling et al. 2023). They reach ∼50,000 km, have widths ∼8000 km, and lifetimes ∼10 min (Savcheva et al. 2007). (Fig. 1.)

Jets apparently result from eruptions of small-scale filaments (minifilaments) (Fig. 2). This minifilament forms and erupts from a neutral line formed by a magnetic minority polarity patch surrounded by majority polarity. A jet bright point (JBP) often forms at the minifilament-eruption site (Sterling et al. 2015). (Fig. 3.) Wyper et al. (2017) present numerical simulations of the basic process.

The small-scale eruptions making jets are analogous to the eruptions that make CMEs (Fig. 4); the erupting minifilament and JBP in the jet case correspond to erupting large-scale filaments and a solar flare in the CME-eruption case. We make a distinction between whether the erupting minifilament produces a "jet" or a "CME," based on whether the

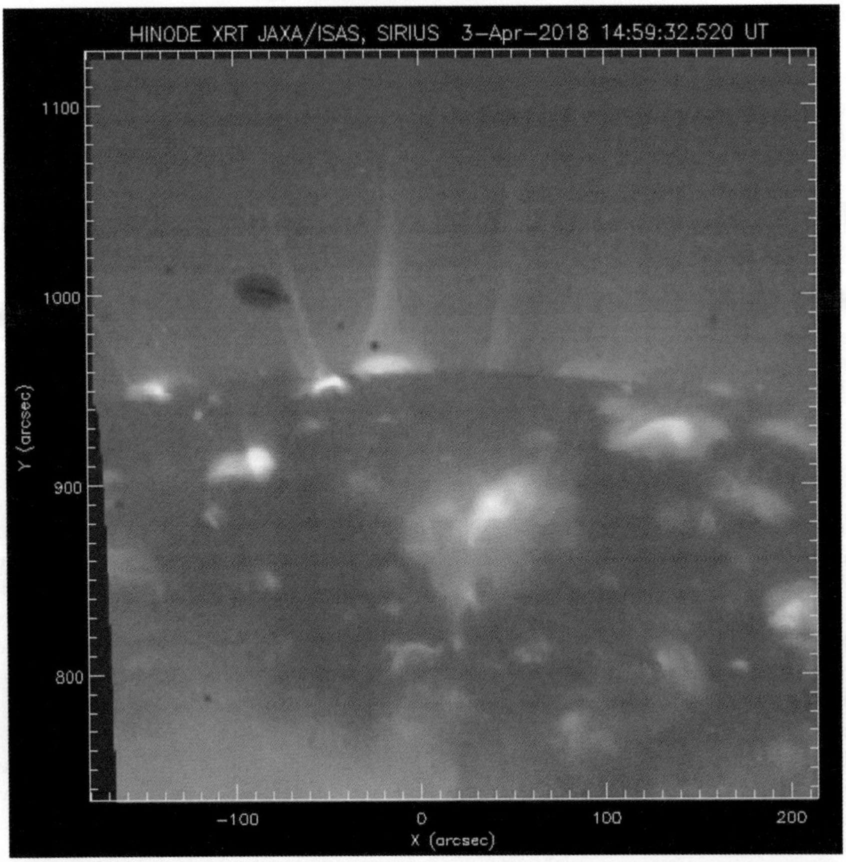

Figure 1. The north pole region of the Sun in SXRs, from the X-Ray Telescope (XRT) on the Hinode spacecraft, with date and time as in the top label. Vertical outward ejections are coronal jets. (From Sterling et al. 2022, which gives more details.)

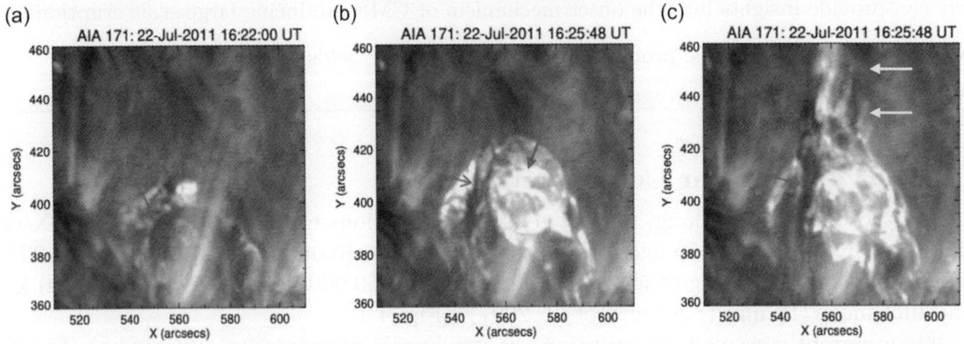

Figure 2. Jet closeup from SDO/AIA 171Å channel, showing an erupting minifilament (green arrows), likely JBP (red), and the jet spire (yellow). (Sterling 2021; Shen et al. 2012)

eruption results in the flux rope losing its flux-rope nature due to reconnection with the ambient coronal field. In the case of a jet, one part of the erupting flux rope completely reconnects with the external coronal field, so that eventually only one end of the erupting structure remains tied to the photosphere. In the case of a CME, the erupting flux rope has enough flux during its eruption so that part of the flux rope escapes out into

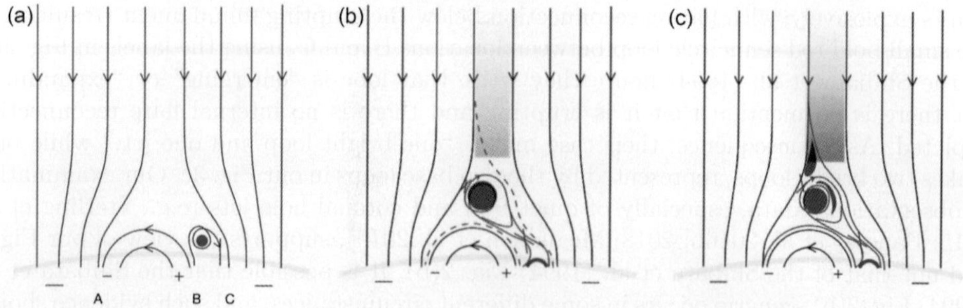

Figure 3. Schematic picture of how jets work (Sterling et al. 2015, 2018), showing magnetic field lines before (black) and after (red) reconnection, erupting minifilament (blue), JBP (red semicircle), the jet spire (orange), and magnetic polarities (+, -). See Sterling et al. (2015) for details.

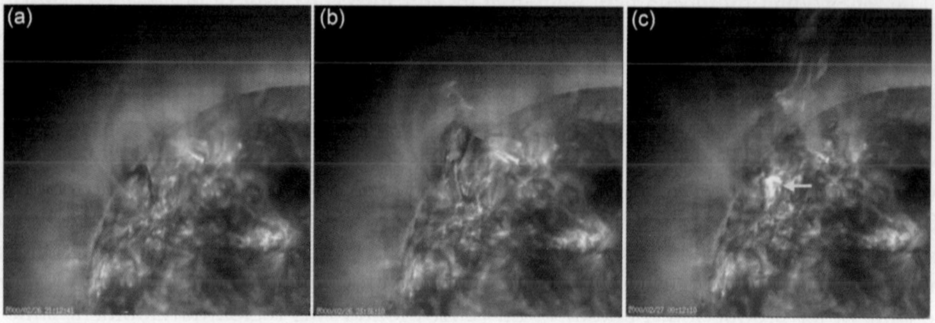

Figure 4. Typical large-scale solar eruption. A solar filament (orange arrow) erupts outwards, in many cases eventually producing a CME. A typical solar flare occurs in its wake (teal). Jets are apparetly small-scale versions of these large eruptions, with erupting minifilaments and JBPs analogous to erupting filaments and flares, respectively. This eruption was analyzed in Sterling et al. (2004).

the heliosphere, with both ends of that flux rope remaining tied to the photosphere. This point is also discussed in Sterling et al. (2023). It is, however, also possible for the eruptions that cause jets to trigger destabilization of a coronal loop to create an erupting-loop CME, perhaps via the method suggested in Panesar et al. (2016a).

Other studies indicate that the jet-producing minifilament eruption is – at least in many cases – caused by magnetic flux cancelation in the photosphere (e.g., Panesar et al. 2016b, 2018; Mcglasson et al. 2019; Muglach 2021). Sometimes these jet-base regions can be small-scale ephemeral regions, as discussed in some detail in Moore et al. 2022. But not all jets originate from such ephemeral regions that emerge as bipoles, and in fact in some cases the two opposite-polarity patches come from different, widely separated emergence locations, with one of the polarities migrating over and canceling with the other, resulting in the jet (e.g., Adams et al. 2014). In other cases, one polarity of the canceling neutral line can be the pole of an emerging ephemeral region, while the polarity with which it cancels is a pre-existing patch that developed independent of the emerging bipole (e.g., Panesar et al. 2018; Muglach 2021). Schrijver (2010) discussed eruptions from ephemeral regions that produce CMEs; although similar, our jets differ from those eruptions in two ways: (1) jets are different from CMEs, as discussed above; and (2), not all of the jet-origin locations are EFRs, as discussed in this paragraph.

A figure in Shibata et al. (1994) (their Fig. 2(b)) shows a proposed setup for jets with similarities to that in our Fig. 3, but there are key differences: In ours the base magnetic lobe that moves outward from the solar surface between our Fig. 3a and 3b erupts explosively, with flaring reconnection below the erupting miniflament (resulting in the small bold red semicircle loop between locations B and C, using the labels in Fig. 3a). In the Shibata et al. (1994) figure they write that lobe is "emerging" or "expanding," but there is no mention that it is erupting and there is no internal flare reconnection depicted. As a consequence, their case makes "one bright loop and one jet," while ours makes two bright loops, represented by the two base loops in our Fig. 3c. Our examination of observational data, especially of quiet Sun and coronal hole jets (e.g., Sterling et al. 2015; Panesar et al. 2016b, 2018; Mcglasson et al. 2019), supports the view of our Fig. 3 and not that of the Shibata et al. (1994) Fig. 2(b). It is possible that the Shibata et al. (1994) Fig. 2(b) scenario occurs in some different circumstances, and such evidence should be presented if found. (The Shibata et al. 1994 Fig. 2(c) case would represent a CME instead of a jet, in our view.)

2. Coronal Jets and CME-producing Eruptions

The jet-production mechanism may give insight into the onset of large-scale eruptions. One study (Sterling et al. 2018) indicates that large-scale eruptions originating from magnetically isolated locations also result from flux cancelation, analogous to jets. That study selected two active regions that could be followed on the solar disk from the time of emergence until the time when they produced an eruption. Those two regions were both comparatively small ones that had a build-up time of about five days prior to eruption; larger regions typically evolve longer than two weeks (and so rotate onto or off the Earth-facing solar disk) before suddenly releasing the free energy that drives the resulting eruption. Because the magnetic regions that produce jets are much smaller than even "small" active regions, they progress much faster from the time of flux emergence to the time of jet production; Panesar et al. (2018) found this region-development time for jets to be from about two hours to two days prior to jet occurrence. Similarly Moore et al. (2022) found that ephemeral regions, some of which produced jets, lasted under two days.

If, as we argue above, eruptions that produce jets are small-scale versions of larger eruptions, then we can possibly learn more about the lead up to and onset of eruptions from the eruptions that produce jets than from the larger eruptions. This is because, in the case of jets, we can often use uninterrupted magnetogram coverage – for about two days or less – of the evolution of the magnetic field from its initial emergence until jet-producing-eruption onset. In contrast, it is harder to find large-scale eruptions that build up quickly enough – i.e. over less than or about two weeks – for comparable studies of the pre-eruption magnetic flux evolution leading up to eruptions making typical flares/CMEs. This opens the possibility of investigating general eruption ideas by statistically studying large numbers of jets, provided such studies can be carried out with adequate spatial resolution and cadence and at suitable wavelengths (Sterling et al. 2023).

On the smaller size scale, the jet mechanism appears to produce features called "jetlets" (Raouafi & Stenborg 2014; Panesar et al. 2018, 2019; Kumar et al. 2022; Sterling et al. 2023). That same mechanism might also produce spicules, although this is more speculative (Sterling et al. 2016, 2020).

3. Summary & Discussion

Solar coronal jets apparently result from small-scale eruptions analogous to CME-producing large-scale solar eruptions, with the minifilament eruption, JBP, and jet spire corresponding to a filament eruption, flare, and CME.

Jets are much more common than large-scale eruptions (several 100/day vs. $\sim$1/day), and they develop more quickly ($\sim$1-50 hrs vs. days/weeks (e.g., Yashiro et al. 2004; Chen 2011; Panesar et al. 2017; Sterling et al. 2018).

If jets and large eruptions work the same way, studies of the former can complement studies of the latter because of jets' abundance and faster evolution (Sterling et al. 2023). Sterling et al. (2018) provide examples of two small CME-producing active regions, where the eruptions apparently result from processes analogous to those that cause jets. By further clarifying and confirming the onset mechanism for the comparatively frequent jets, we potentially can gain insight into the onset mechanism for the larger-scale eruptions also.

Moreover, CMEs on other stars likely occur via the same mechanism as CMEs from the Sun (e.g., Veronig 2021). Therefore, exploration of among the smallest eruptive features on the Sun (the small-scale minifilament/flux-rope eruptions that make coronal jets) potentially can provide clarification for far-away stellar eruptions also.

The authors received funding from the Heliophysics Division of NASA's Science Mission Directorate through the Heliophysics Supporting Research (HSR, grant No. 20-HSR20_2-0124) Program, and the Heliophysics Guest Investigators program. ACS and RLM were also supported through the Heliophysics System Observatory Connect (HSOC, grant No. 80NSSC20K1285) Program, and A.C.S. received support from the NASA/MSFC Hinode satellite Project Office. N.K.P. received additional support through a NASA SDO/AIA grant. We acknowledge the use of AIA data. AIA is an instrument onboard SDO, a mission of NASA's Living With a Star program.

References

Adams, M., et al. 2014, ApJ, 783, 11
Chen, P. F. 2011, LRSP, 8, 1
Kumar, P., et al. 2022, ApJ, 933, 21
McGlasson, R., et al. 2019, ApJ, 882, 16
Moore, R., et al. 2022, ApJ, 933, 12
Muglach, K. 2021, ApJ, 909, 133
Panesar, N., et al. 2016a, ApJ, 822L, 23
Panesar, N., et al. 2016b, ApJ, 862L, 7
Panesar, N., et al. 2017, ApJ, 844, 131
Panesar, N., et al. 2018, ApJ, 853, 189
Panesar, N. K., et al. 2019, ApJL, 887, L8
Raouafi, N. E., & Stenborg, G. 2014, ApJ, 787, 118
Raouafi, N., et al. 2016, SSRv, 201, 1
Savcheva, A., et al. 2007, PASJ 59, 771
Schmieder, B 2022, FrASS, 9, 820183
Schrijver, C. 2010, ApJ, 710, 1480
Shen et al. 2012, ApJ, 745, 164
Shen, Y. 2021, Proceedings of The Royal Society A, 477, 217
Shibata, K., et al. 1992, PASJ, 44, L173
Shibata, K., et al. 1994, ApJ, 431, 51
Sterling, A., & Moore, R. 2004, ApJ, 613, 1221
Sterling, A., et al. 2015, Nature, 523, 437
Sterling, A., & Moore, R. L.. 2016, ApJL, 828, L9
Sterling, A., et al. 2018, ApJ, 864, 68
Sterling, A., et al. 2020, ApJL, 893, L45
Sterling, A. C. 2021, in Solar Physics and Solar Wind, ed. N. E. Raouafi & A. Vourlidas, Vol. 258 (New York: Wiley), 221
Sterling, A., et al. 2022, ApJ, 927, 127
Sterling, A., et al. 2023, FrASS, 10, 49
Wyper, P., et al. 2017, Nature, 544, 452

Veronig, A., et al. 2021, Nature Astronomy, 5, 697
Yashiro, S., et al. 2004, JGRA, 109, A07105

Q & A

Philippe L. Lamy: In our 2019 review paper (Lamy, P., et al. 2019, Space Sci. Rev., 215, 39) we argue that CMEs at large arise from closed-field coronal regions at both large and small size scales. Your presentation appears consistent with our argument.

A. C. Sterling: I largely agree with your assessment, but with some caveats on the creation of the smallest CMEs. We have argued that on the size scale of jets and smaller there is a physical difference between jets and CMEs. In our jet picture (Fig. 3), the erupting flux rope carrying the erupting minifilament is totally consumed through reconnection with the open field (Figs. 3b and 3c), with the twist of the initially twisted minifilament/flux-rope unravelling onto the open field, and propagating out along the field as an Alfvenic pulse (Shibata, K. and Uchida, Y. 1986, Solar Phy. 178, 379; Sterling, A. & Moore, R. 2020, ApJ, 896, L18). Thus, in that jet case, a jet spire forms in the corona, and a "white-light jet" or "narrow CME" (e.g., Wang, Y.-M., et al. 1998, ApJ, 508, 899; Moore et al. 2015, ApJ, 806, 11) can form in the outer corona rather than a *bona fide* CME. Such *bona fide* CMEs, on the other hand, are generally thought of as having a flux-rope core. In order to make such a small-scale CME from a jet-like magnetic setup of Fig. 3, an erupting small-scale flux rope would have to be robust enough to survive as a flux rope the interchange reconnection of Figs. 3b and 3c. So in summary: below some (rather ill-defined) cutoff size the erupting minifilament/flux-rope will make a coronal jet and a narrow CME, and above that cutoff size a portion of the erupting minifilament/flux-rope will escape as a flux rope – perhaps severely reduced in flux content – and potentially form the core of a very small CME. There would be some sort of mixture of the features for erupting miniflament/flux-ropes of size close to that cutoff size. We discuss some of these points further in Sterling et al. (2023). Additionally, some CMEs might be triggered via the mechanism discussed in Panesar et al. (2026a), whereby the twist of a growing jet's field gets transferred to the base of a coronal loop, and subsequently destabilizes that loop causing it to erupt.

Solar and Stellar Coronal Mass Ejections
Proceedings IAU Symposium No. 388, 2026
N. Gopalswamy, O. Malandraki, A. Vidotto & W. Manchester, eds.
doi:10.1017/S1743921324001406

Can Small-scale Eruptions in the Quiet Sun Propagate into the Higher Solar Corona?

Rui Wang[1,2]

[1]State Key Laboratory of Solar Activity and Space Weather, National Space Science Center, Chinese Academy of Sciences, Beijing 100190, People's Republic of China
email: rwang@swl.ac.cn

[2]School of Astronomy and Space Sciences, University of Chinese Academy of Sciences, Beijing 100049, People's Republic of China

Abstract. Recent research has revealed that small-scale coronal mass ejections (CMEs) in the quiet Sun exhibit similarities in their formation mechanisms and eruption characteristics when compared to large-scale solar eruptions. The occurence of small-scale CMEs increases exponentially as their size decreases. Thus, the role of small-scale CMEs in solar wind formation should not be underestimated. Can small-scale eruptions in the quiet Sun propagate into the higher solar corona? I analyze a typical small-scale eruption in the quiet Sun and reconstruct a flux-rope (FR) structure. My results indicate that torus instability of the FR can be triggered at a relatively low critical height of ∼8 Mm. This suggests that the quiet Sun's FRs can easily overcome the confinement of the low coronal magnetic field and enter the higher corona. However, further investigation is still needed to determine if the large-scale coronal mangetic fields above constrain these small-scale eruptions.

Keywords. Coronal mass ejections, Quiet sun, Solar magnetic fields, Solar filaments

1. Introduction

Solar eruptive activities, such as flares, eruptive filaments, coronal mass ejections (CMEs), and coronal jets, exhibit a broad spectrum of energy scales. The observed solar eruptions encompass a vast range of energy levels, spanning at least a factor of 10^8 (Schrijver et al. 2012). Large, highly energetic "X-class" eruptions can release energy substantially exceeding $\sim 10^{33}$ erg. However, there are smaller-scale eruptions, such as the recent popular "campfires" events (Panesar et al. 2021, 2023), small-scale filament eruptions (Chen et al. 2020; Wang et al. 2023), and small-scale eruptions originating from coronal bright points (Mou et al. 2018). Although individual events release lower energy and mass than large eruptive events, Aschwanden et al. (2000) and Schrijver et al. (2012) suggest that the frequency of lower-energy events exhibits exponential growth. This implies that these low-energy events are more numerous and that the energy released by these small-scale eruptive events serves as a significant source for coronal heating. Additionally, an important question arises regarding the fate of their mass after the eruption. Does it return to the solar surface or become part of the higher corona and even interplanetary space, eventually becoming a component of the solar wind?

Jackson & Howard (1993) and Webb & Howard (1994) propose that the annualized mass flux of CMEs in relation to the solar wind is at most 16%. Lamy et al. (2017) further indicate that this ratio is limited to a maximum of 6%. However, it is important to note that the mass of these recorded CMEs typically falls within the range of $\sim 10^{14}$–10^{15} g. If the aforementioned small-scale eruptions can successfully get into the higher corona or interplanetary space, their considerably higher occurrence frequency indicates that they would make a significant contribution to the mass of the solar wind. Therefore, my study focuses on analyzing small-scale eruptive events occurring in quiet-Sun regions to investigate their potential for propagating into the higher corona.

2. Data Anlysis and Results

A small-scale eruption with coronal dimming (Wang et al. 2023) occurred near the central meridian observed by Solar Orbiter (Müller et al. 2020) and Solar Dynamics Observatory (Pesnell et al. 2012) on 2020 May 20 21:20 UT. STEREO-A was nearly in quadrature with Solar Orbiter, providing a vantage point to observe the ascending process of the eruption. Figure 1 presents the reconstructed flux-rope (FR) structure and the ambient magnetic field with the magnetohydrostatic method of Zhu & Wiegelmann (2018) which is considered more appropriate for extrapolation of low-coronal structures. To convert the disambiguated full-disk vector magnetic field series "hmi.B_720s" from native CCD coordinates to cylindrical equal area heliographic coordinates (Hoeksema et al. 2014), the "bvec2cea.pro" routine from the SSW packages is utilized. This routine employs a radial-acute method (Hoeksema et al. 2014) to resolve the 180° azimuthal uncertainty in the transverse field direction.

Figure 1a demonstrates the presence of a perfectly sigmoid FR, resembling FRs found in active regions. A magnetic dip is identified between the positive and negative polarities, indicating that a cancellation process is taking place. Additionally, an emergence-cancellation process of the positive polarity is observed along the polarity inversion line (PIL), which is closely associated with the FR's formation (Wang et al. 2018, 2022). Figure 1b depicts the surrounding magnetic fields encompassing the FR. Some selected magnetic field lines associated with the FR (yellow), which connect to a more distant region of the solar surface, mingle with the surrounding magnetic field lines (grey). Notably, a magnetic null point is presented above the FR, while open magnetic fluxes are rooted in its vicinity. This magnetic field configuration facilitates the escape of the FR from the confining fields and resembles the configurations illustrated in small-scale jet eruptions driven by mini-filaments (Sterling et al. 2015; Wang et al. 2023).

The torus instability of the flux rope in the lower corona (below ~70 Mm) is assessed using the decay index, defined as $n = -\partial ln\|\mathbf{B}_p\|/\partial ln R$ (Kliem & Török 2006), where $\|\mathbf{B}_p\|$ represents the magnitude of the external poloidal field and R denotes the height of the flux rope axis's apex above the photosphere. Figure 3a shows the critical height contours above the FR when $n = 1.5$, the maximum of which is below 22 Mm around the FR. Figure 3b shows a critical height of ~8 Mm above the PIL (within the yellow quadrilateral), indicating favorable conditions for the escape of a CME from its source region.

3. Discussion

This study reveals that eruptions in the quiet-Sun regions can occur with weaker confining magnetic fields, characterized by predominantly open magnetic fields around. This results in a relatively weak constraint on the near-surface magnetic fields. The decay index reveals that FRs in the quiet-Sun region indeed have a lower critical height for eruption. However, observations by Wang et al. (2023) indicate a slow ascent velocity of only ~5.4 km s^{-1} for such small-scale CME structures, suggesting a limited kinetic energy. As they rise to a height of about 300 Mm, they will encounter the streamer belt structures above, which can further confine these small CMEs. An attempt is made to connect observations from the solar surface to 1 AU to track the trajectory of the

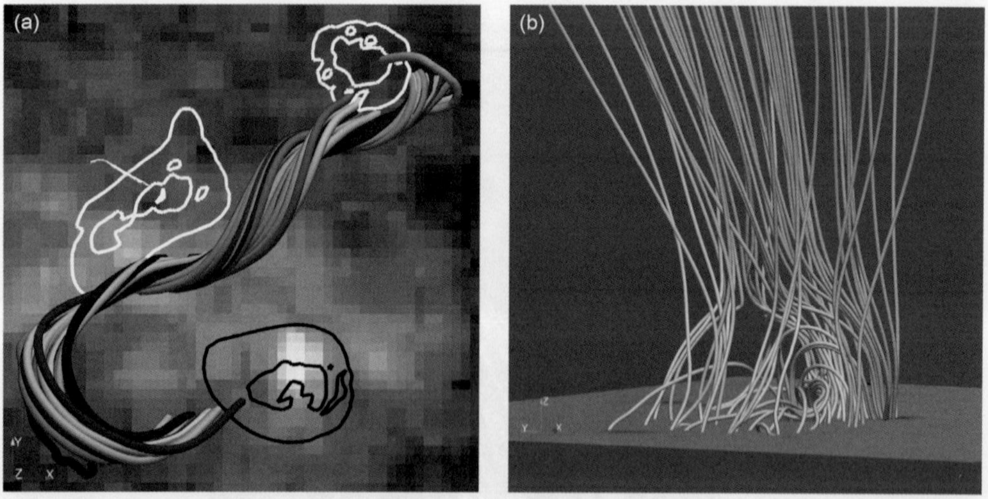

Figure 1. (a) Reconstructed flux-rope structure with the EUV background at 304 Å. The positive (white) in [20, 50] G and negative (black) fluxes in [-20, -50] G are overplotted. (b) Surrounding magnetic fields (grey) around the flux rope (yellow).

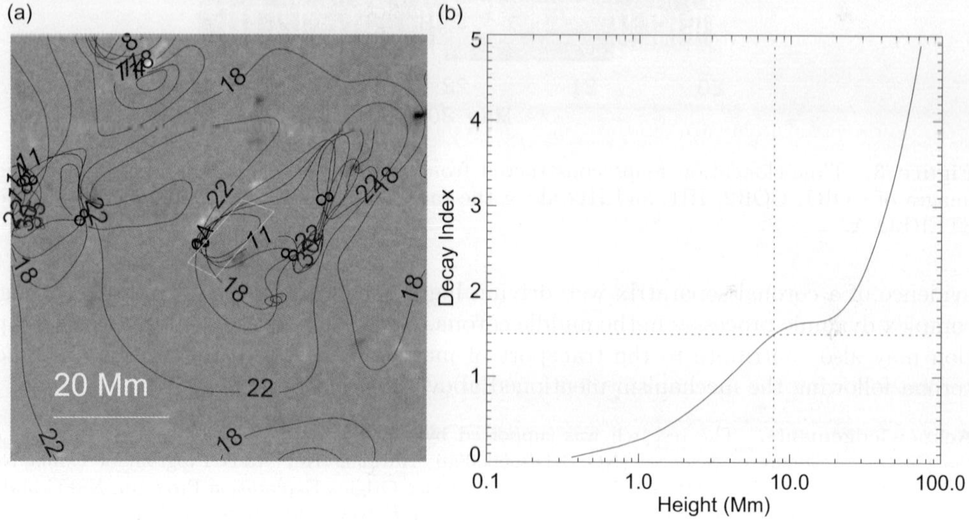

Figure 2. Torus instability of the flux rope. (a) Color contours correspond to a decay index n = 1.5 at each height (in Mm). (b) Average decay index above the PIL.

small-scale CME (as shown in Figure 4). Unfortunately, usable inner coronal data from STEREO-A/COR1 are lacking. The J-map plot in Figure 4 displays multiple front edges of the CME leading tracks in COR2, making it difficult to identify the specific front edge associated with this small CME. Nonetheless, determining whether quiet-Sun eruptions are confined or eruptive remains a challenge. In this study, numerous narrow blobs were observed moving outward in the STEREO-A/COR2 view. The table provided in Wang et al. (2023) presents potential small-scale CME eruption events, which motivate future work in tracking similar CMEs to obtain direct evidence of their entry into the higher corona. However, it is worth noting that mass and magnetic free energy could be transferred from the lower corona to the middle or higher corona, potentially serving as a source for the solar wind. Chitta et al. (2023) have provided direct observational

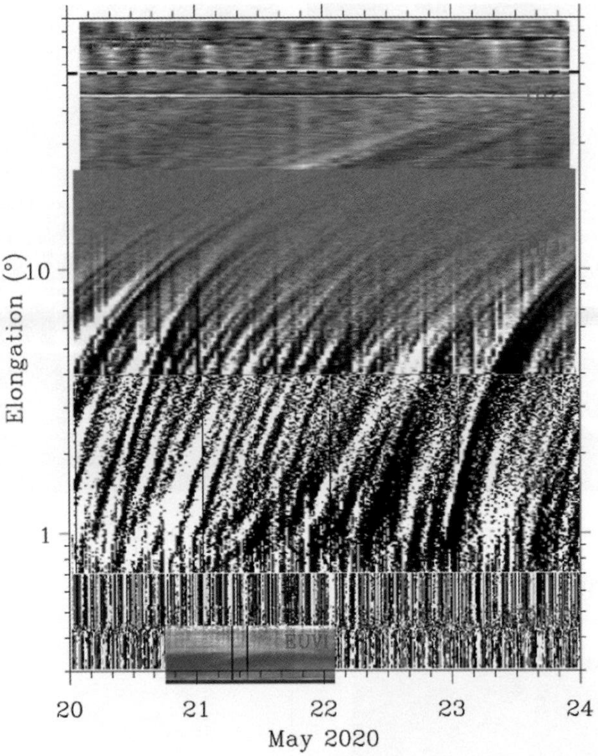

Figure 3. Time-elongation maps constructed from the EUVI image and running difference images of COR1, COR2, HI1, and HI2 along the radial direction of the CME propagation for STEREO-A.

evidence of a coronal separatrix web driving highly structured slow solar wind through complex dynamic processes in the middle corona. The observed and analyzed CME eruption may also contribute to the transport of mass and energy to the middle or higher corona following the mechanism mentioned above.

Acknowledgements. The research was supported by the Strategic Priority Research Program of the Chinese Academy of Sciences (NO. XDB0560000), National Key R&D Program of China No. 2022YFF0503800 and No. 2021YFA0718600, China's Space Origins Exploration Program, NSFC under grants 12073032, and the Specialized Research Fund for State Key Laboratories of China.

References

Aschwanden, M. J., Tarbell, T. D., Nightingale, R. W., et al. 2000, Time Variability of the "Quiet" Sun Observed with TRACE. II. Physical Parameters, Temperature Evolution, and Energetics of Extreme-Ultraviolet Nanoflares. *ApJ*, 535(2), 1047.

Chen, H., Hong, J., Yang, B., et al. 2020, High-resolution Chromospheric Observations of a Solar Minifilament: Formation and Destabilization. *ApJ*, 902(1), 8.

Chitta, L. P., Seaton, D. B., Downs, C., et al. 2023, Direct observations of a complex coronal web driving highly structured slow solar wind. *Nat. Astron.*, 7, 133.

Hoeksema, J. T., Liu, Y., Hayashi, K., et al. 2014, The Helioseismic and Magnetic Imager (HMI) Vector Magnetic Field Pipeline: Overview and Performance. *Sol. Phys.*, 289, 3483.

Jackson, B. V. & Howard, R. A. 1993, A CME Mass Distribution Derived from SOLWIND Coronagraph Observations. *Sol. Phys.*, 148(2), 359.

Kliem, B. & Török, T. 2006, Torus Instability. *Phys. Rev Lett.*, 96(25), 255002.

Lamy, P., Floyd, O., Quémerais, E., et al. 2017, Coronal mass ejections and solar wind mass fluxes over the heliosphere during solar cycles 23 and 24 (1996–2014). *J. Geophys. Res.: Space Phys.*, 122(1), 50.

Mou, C., Madjarska, M. S., Galsgaard, K., et al. 2018, Eruptions from quiet sun coronal bright points-i. observations. *A&A*, 619, A55.

Müller, D., St. Cyr, O. C., Zouganelis, I., et al. 2020, The Solar Orbiter mission. Science overview. *A&A*, 642, A1.

Panesar, N. K., Hansteen, V. H., Tiwari, S. K., et al. 2023, Solar orbiter and sdo observations, and a bifrost magnetohydrodynamic simulation of small-scale coronal jets. *ApJ*, 943(1), 24.

Panesar, N. K., Tiwari, S. K., Berghmans, D., et al. 2021, The magnetic origin of solar campfires. *ApJL*, 921(1), L20.

Pesnell, W. D., Thompson, B. J., & Chamberlin, P. C. 2012, The Solar Dynamics Observatory (SDO). *Sol. Phys.*, 275, 3.

Schrijver, C. J., Beer, J., Baltensperger, U., et al. 2012, Estimating the frequency of extremely energetic solar events, based on solar, stellar, lunar, and terrestrial records. *J. Geophys. Res.: Space Phys.*, 117(A8), A08103.

Sterling, A. C., Moore, R. L., Falconer, D. A., et al. 2015, Small-scale filament eruptions as the driver of X-ray jets in solar coronal holes. *Nature*, 523(7561), 437.

Wang, R., Liu, Y. D., Hoeksema, J. T., et al. 2018, Roles of Photospheric Motions and Flux Emergence in the Major Solar Eruption on 2017 September 6. *ApJ*, 869(2), 90.

Wang, R., Liu, Y. D., Yang, S., et al. 2022, Buildup of the Magnetic Flux Ropes in Homologous Solar Eruptions. *ApJ*, 925(2), 202.

Wang, R., Liu, Y. D., Zhao, X., et al. 2023, Observations of mini coronal dimmings caused by small-scale eruptions in the quiet sun. *ApJL*, 952(2), L29.

Webb, D. F. & Howard, R. A. 1994, The solar cycle variation of coronal mass ejections and the solar wind mass flux. *J. Geophys. Res.*, 99(A3), 4201.

Zhu, X. & Wiegelmann, T. 2018, On the Extrapolation of Magnetohydrostatic Equilibria on the Sun. *ApJ*, 866(2), 130.

Solar and Stellar Coronal Mass Ejections
Proceedings IAU Symposium No. 388, 2026
N. Gopalswamy, O. Malandraki, A. Vidotto & W. Manchester, eds.
doi:10.1017/S1743921324001704

EUV Wave and Coronal Seismology

Pooja Devi[1], Ramesh Chandra[1], Arun Kumar Awasthi[2], Brigitte Schmieder[3,4] and Reetika Joshi[5,6]

[1]Department of Physics, DSB Campus, Kumaun University, Nainital 263001, India

[2]Space Research Centre, Polish Academy of Sciences, Bartycka 18A, 00-716 Warsaw, Poland

[3]LIRA, Observatoire de Paris, CNRS, Université de Paris, 5 place Janssen, 92290 Meudon, France

[4]Centre for mathematical Plasma Astrophysics, Dept. of Mathematics, KU Leuven, 3001 Leuven, Belgium

[5]Rosseland Centre for Solar Physics, University of Oslo, P.O. Box 1029 Blindern, N-0315 Oslo, Norway

[6]Institute of Theoretical Astrophysics, University of Oslo, P.O. Box 1029 Blindern, N-0315 Oslo, Norway
Email: setiapooja.ps@gmail.com

Abstract. We present an investigation of the Extreme-Ultraviolet (EUV) wave linked to the flare that occurred on 28 October 2021, along with the associated coronal loop oscillation and type II radio burst. The EUV wave was observed by multi-viewpoint with Solar Dynamics Observatory and Solar Terrestrial Relations Observatory – A. The associated coronal mass ejection (CME) was observed by Large Angle and Spectrometric Coronagraph (LASCO) as well by COR1 coronagraph. From the multi-view observation, we found that the EUV wave is propagated ahead of the connected CME. The coronal magnetic field measurement was performed by the coronal loop oscillations as well by the associated m-type II radio burst observations. We found the magnetic field strength values computed by both methods are consistence and are in the range of $\approx$ 5 to 10 G.

Keywords. EUV wave, Coronal Mass Ejection, Loop Oscillations, m-Type II Radio Burst

1. Introduction

During the campaign of Hα observation in 1960, Moreton and Ramsey (1960) found the large scale moving disturbances on the solar chromosphere. Later these disturbances were named as “Moreton waves”. Their reported speed is in the range of 500 to 2000 km s^{-1} (see reviews by Warmuth 2015; Chen 2016). The generation of Moreton waves is due to the fast mode magnetohydrodynamic (MHD) shock sweeping into the chromosphere. Thompson et al. (1998) observed similar propagating features in 195 Å wavelength with Extreme-ultraviolet Imaging Telescope (EIT; Delaboudinière et al. 1995) onboard Solar and Heliospheric Observatory (SOHO; Domingo et al. 1995) and named them as “EIT waves”. Because of their clear visibility in Extreme-ultraviolet (EUV) wavelengths, they are popularly known as “EUV waves” and their speed ranges from a few hundreds to multiple of thousand km s^{-1}.

Initially, these waves were observed with low-cadence data of SOHO and Solar Terrestrial Relations Observatory (STEREO) satellites (Thompson et al. 1999; Zhukov and Auchère 2004; Zhukov et al. 2009) and were proposed to be the fast-mode MHD waves. However, after analyzing the high spatio-temporal data of Solar Dynamics

Observatory (SDO), two components (slow, speed is as low as 10 km s^{-1} and fast (speed can be up to multiple of 1000 km s^{-1}) as well as stationary fronts of this phenomena were reported (Chen and Wu 2011; Asai et al. 2012; Chandra et al. 2016, 2018, 2021). The observed features can be explained by the hybrid model proposed by Chen et al. (2002, 2005). According to this model, a piston-driven shock forms along the expanding flux rope during its eruption, sweeping the whole solar surface. The legs of the shock produce fast-mode EUV waves. On the other hand, the slower (or nonwave) component, forms due to successive opening or stretching of the magnetic field lines straddling over the erupting flux rope.

When the coronal mass ejection (CME) propagates out of the solar surface, it produces a shock wave and it is believed that this shock wave could be responsible for the EUV wave disturbances (Biesecker et al. 2002). There are studies, where people try to look into the association between the EUV waves and CMEs and it is found that these two physical phenomena are closely associated (Chen 2009; Dai et al. 2010; Schmieder et al. 2013; Nitta et al. 2014; Muhr et al. 2014). Another evidence of the shock wave is the type II radio bursts observed in coronal heights and can be associated with the EUV waves.

When the fast component of the EUV wave encounters the surrounding magnetic structures, these structures can become unstable. These magnetic structures can be filament or coronal loops. As a result of this instability, these structures can oscillate and be used for the coronal seismology (Warmuth and Mann 2005; Ballai 2007; Guo et al. 2015; Luna et al. 2017; Fulara et al. 2019; Shen et al. 2019; Devi et al. 2022).

In this article, we present an EUV wave from 28 October 2021, associated with a large GOES X1.0 eruptive solar flare, nearby loop oscillations, and a type II radio burst. Section 2 covers the observational data sources and the results. The summary of the article is provided in Section 3.

2. Observations

The EUV wave event of 28 October 2021 was observed by several space based telecopes and its counterpart i.e., Moreton wave, by ground based instruments. We analyse the data of Atmospheric Imaging Assembly (AIA; Lemen et al. 2012) onboard Solar Dynamics Observatory (SDO; Pesnell et al. 2012) and Global Oscillation Network Group (GONG; Harvey et al. 2011) from Earth view and by Extreme-ultraviolet Imager (EUVI) and coronagraphs (COR1) onboard ahead spacecraft of Solar Terrestrial Relations Observatory (STEREO–A; Kaiser et al. 2008) from a different view angle. The radio emission due to this event were observed by e-Callisto spectrographs at Birr Castle, Ireland (BIR) and also by ORFEES (Hamini et al. 2021) at Nançay Radio Observatory. The EUV wave event is accompanied by one of the major GOES X1.0 class two ribbon flare and filament eruption, which produced a ground level enhancement (GLE). The origin of the activities was from NOAA AR 12887, which was located at S26W07 on 28 October 2021. The studies have been done on different activities produced by this event which include: different features of the EUV and Moreton wave (Hou et al. 2022; Devi et al. 2022), solar energetic particles (Kouloumvakos et al. 2024) and related radio bursts (Klein et al. 2022), and GLE (Mavromichalaki et al. 2022; Papaioannou et al. 2022; Mishev et al. 2024). Along with the observations, data-driven simulations of the event are performed by Guo et al. (2023).

2.1. *Morphology of EUV Wave and associated CME*

The EUV wave appeared on the solar disk ∼ 15:25 UT in a circular shape. The propagating EUV wave was promptly visible in the northward and fainter in the southward direction. Figure 1 displays base difference snapshots of the wave in different wavelengths.

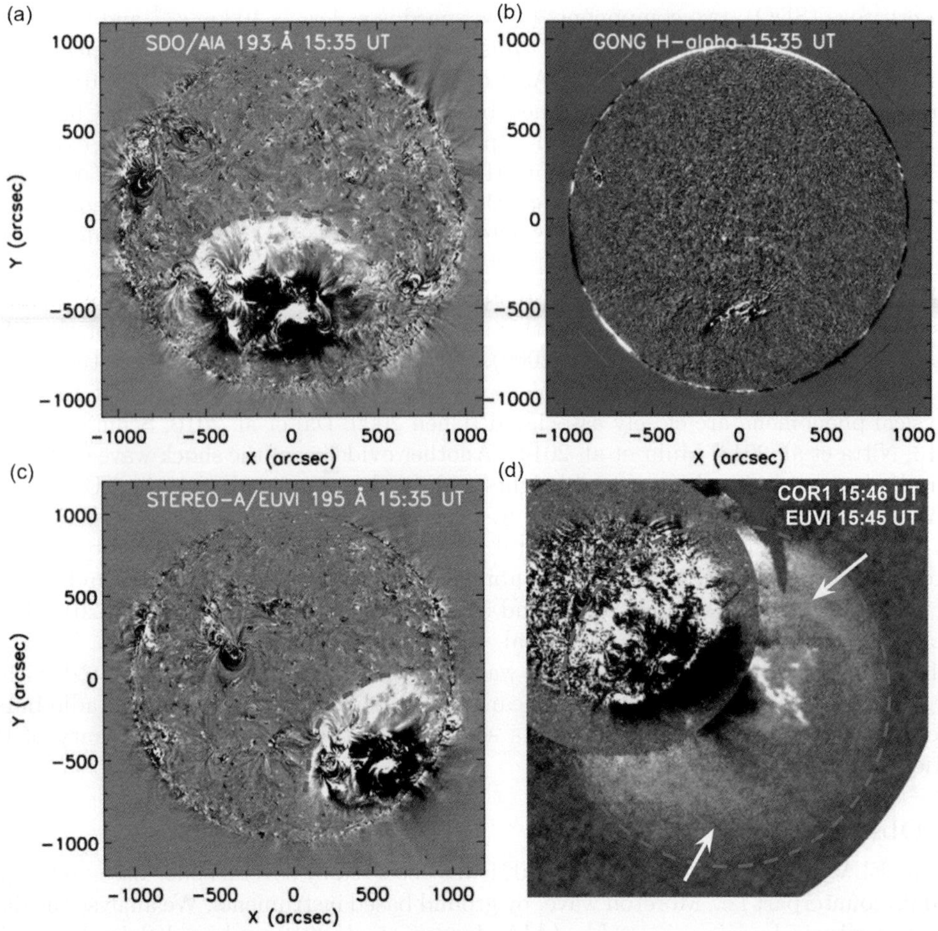

Figure 1. Panel (a, c): Snapshots of the EUV waves observed in AIA 193 Å, and STEREO-A 195 Å. Panel b: Hα Observation of GONG, where Moreton wave is visible. panel (d): Overlay of CME observed by COR1 with the STEREO-A data. The red dashed curves in panels (a – c) and the circle in panel (d) are the fronts of the wave. The white arrows in panel (d) show the CME leading edge. (Panels (a) and (d) are adapted and recreated from Devi et al. 2022).

The kinematics of the wave were done by Devi et al. (2022) and Hou et al. (2022) using the time-distance technique. Two components (i.e., fast-mode and nonwave) of the EUV wave were observed. The speed of the nonwave component was found to be $\approx$ 280 km s^{-1} and fast-mode ranges from $\approx$ 550 to 720 km s^{-1}. The speed of the Moreton wave observed in Hα was found in the range of $\approx$ 310 to 540 km s^{-1}.

The EUV wave was associated with a halo CME of linear speed 1519 km s^{-1} and an acceleration of -61 m s^{-2} as reported by the LASCO CME catalog. The CME was also observed by the STEREO-A COR1 coronagraph. In the STEREO-A data, the event is observed close to the western limb, which provides us to an excellent opportunity to compare spatially these two phenomena. For this comparison, the image of EUVI is overlaid on the COR1 image as shown in Figure 1(d). The internal yellow part is the EUVI 195 Å image at 15:45 UT and the outer green part is the COR1 image at 15:46 UT. Assuming the spherical structure of the EUV wave, we fitted a circle at the outer edge of the EUV wave (shown by a red dashed circle) as suggested by Gopalswamy et al. (2013). The alignment between the EUVI and COR1 images evidenced that the EUV wave was

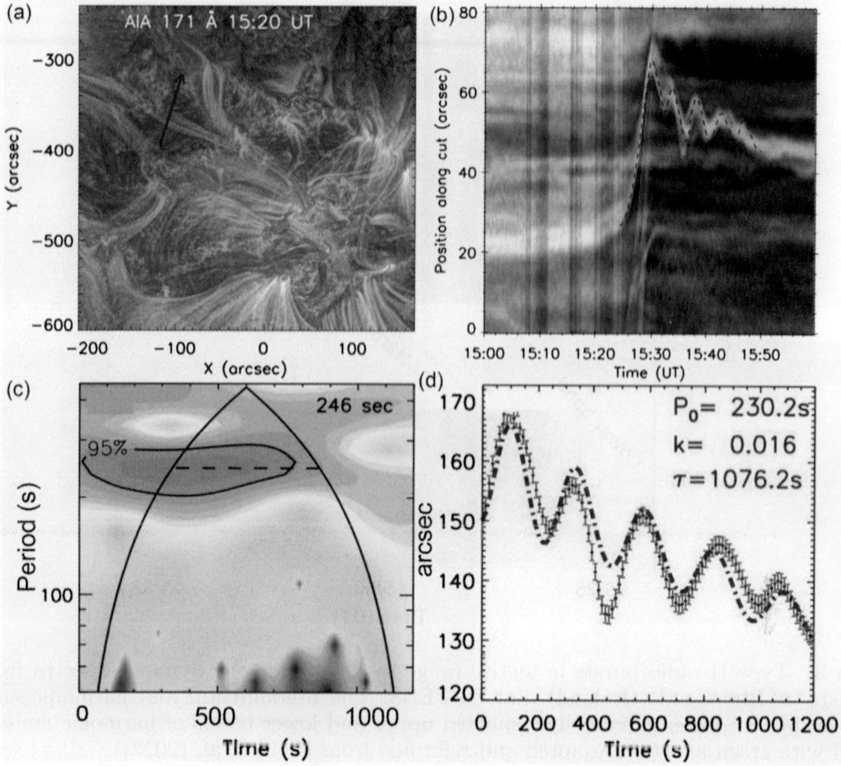

Figure 2. AIA 171 Å image with the clearly observed oscillating loop system is depicted in (a) panel. The direction of oscillation is in the blue arrow direction. The time-slice plot showing the oscillations is shown in (b) panel. The wavelet analysis along with the temporal evolution of oscillation (overlaid by dashed lines) is displayed in panel (c) and (d) panel of the figure. (Adapted from Devi et al. (2022))

traveling ahead of the CME. This result is confirmed by the data-driven simulation of Guo et al. (2023) (see their Figure 10).

2.2. *Loop Oscillations*

During the propagation of the EUV wave in the north direction, it encountered the neighboring coronal loops and triggered them to oscillate. The oscillations of two loop systems have been studied in detail by Devi et al. (2022). They computed the periods of loop oscillations using the wavelet method and by fitting a damped sinusoidal function given in their equation 1. The periods from both these methods are found to be consistent with each other and are in the range from 230 to 280 sec. They also calculated the densities inside (n_{in}) and outside (n_{out}) the oscillating loops using the Differential Emission Measure (DEM; Cheung et al. 2015) technique which are found in the range 1.6×10^8 – 2.1×10^8 cm^{-3} and 1×10^8 – 1.8×10^8 cm^{-3}, respectively. Oscillations in one of the loop systems along with its periodic analysis are presented in Figure 2. Further, Devi et al. (2022) calculated the coronal magnetic field strength (B) using the following equation (Roberts et al. 1984),

$$B = \frac{L}{P}\sqrt{8\pi\mu m_p n_{in}\left(1+\frac{n_{out}}{n_{in}}\right)} \qquad (1)$$

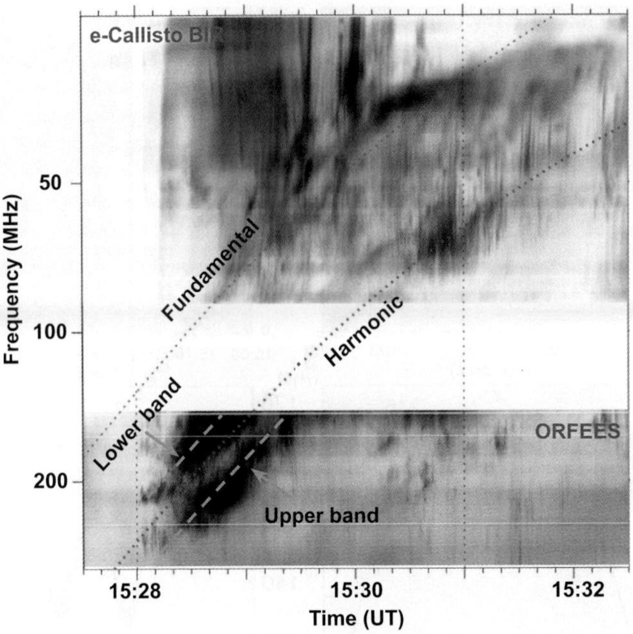

Figure 3. Type II radio bursts in metric range by combining the dynamic spectra from BIR (e-Callisto at Birr Castle, Ireland) and ORFEES. The fundamental and harmonic emissions are shown by red dotted curves. The splitted upper and lower bands of harmonic emission are denoted with green arrows. (Adapted and reformed from Klein et al. (2022))

where, μ is the average molecular weight for coronal abundances, and m_p is mass of proton. Putting all the above values along with the length (L) of oscillating loop (115 – 175 Mm), the B found to vary from $\approx$ 5.7 to 8.8 G.

2.3. *Type II Radio Burst*

The event was also associated with type II radio bursts in metric (m) and decameteric-hectometeric frequency range. The detailed study about the observed type II burst was investigated by Klein et al. (2022) (see their figures 2, 5, and 12). They computed the speed of m-type II burst in the range 1900 – 2600 km s^{-1}. The observed metric type II radio burst is presented in Figure 3. The type II radio burst shows the harmonic and fundamental structures. The splitting in the harmonic structure is observed in ORFEES dynamic spectra. Such splitting in the harmonic band is due to the consequence of the plasma emission of the upstream and downstream shock regions and it was initially interpreted by Smerd et al. (1974). The frequency of observed upstream (f_u) and downstream (f_l) are 173 and 130 MHz, respectively. Using the parameters of harmonic band splitting of m-type II bursts Klein et al. (2022) calculated the Alfvén Mach number (M_A) which is 1.6. Now, using the parameters computed by them, we have calculated B by the following formula (Vasanth et al. 2014),

$$B = 5.1 \times 10^{-5} \times v_a \times f_l, \tag{2}$$

where the Alfvén speed ($v_a = v_s/M_A$) found to vary from 1187 to 1625 km s^{-1}. Inserting the values of v_a and f_l in equation 2, the computed B is found in the range $\approx$ 7 – 10 G, which is consistent with B calculated using the loop oscillation parameters (Section 2).

3. Summary

The EUV wave of 28 October 2021 showed both the fast-mode and nonwave components which can be explained by the hybrid model proposed by Chen et al. (2002). Our analysis revealed that the EUV wave travels in front of the CME leading edge implying that the EUV wave is driven by the CME and not by the flare pressure pulse (Vourlidas and Ontiveros 2009; Kwon et al. 2014; Kwon and Vourlidas 2017; Guo et al. 2023). An important characteristic of this EUV wave is its association with the neighbouring loop oscillations and the type II radio bursts. We have computed the coronal magnetic field using the loops oscillation parameters which are 5.7 – 8.8 G and by the m-type II radio observations harmonic band splitting, the computed value is about 7 – 10 G. The computed B from both these methods are consistent with each other and in the range of previous studies (Nakariakov and Ofman 2001; Aschwanden and Schrijver 2011; Guo et al. 2015; Su et al. 2018; Zhang et al. 2022).

References

Asai, A., Ishii, T. T., Isobe, H., et al. 2012, *ApJL*, 745(2), L18.
Aschwanden, M. J. & Schrijver, C. J. 2011, *ApJ*, 736(2), 102.
Ballai, I. 2007, *Solar Phys.*, 246(1), 177.
Biesecker, D. A., Myers, D. C., Thompson, B. J., et al. 2002, *ApJ*, 569(2), 1009.
Chandra, R., Chen, P. F., Fulara, A., et al. 2016, *ApJ*, 822(2), 106.
Chandra, R., Chen, P. F., Joshi, R., et al. 2018, *ApJ*, 863(2), 101.
Chandra, R., Démoulin, P., Devi, P., et al. 2021, *ApJ*, 922(2), 227.
Chen, P. F. 2009, *ApJL*, 698(2), L112.
Chen, P. F. 2016, *Washington DC American Geophysical Union Geophysical Monograph Series*, 216, 381.
Chen, P. F., Fang, C., & Shibata, K. 2005, *ApJ*, 622(2), 1202.
Chen, P. F., Wu, S. T., Shibata, K., & Fang, C. 2002, *ApJL*, 572(1), L99.
Chen, P. F. & Wu, Y. 2011, *ApJL*, 732(2), L20.
Cheung, M. C. M., Boerner, P., Schrijver, C. J., et al. 2015, *ApJ*, 807(2), 143.
Dai, Y., Auchère, F., Vial, J. C., et al. 2010, *ApJ*, 708(2), 913.
Delaboudinière, J. P., Artzner, G. E., Brunaud, J., et al. 1995, *Solar Phys.*, 162(1-2), 291.
Devi, P., Chandra, R., Awasthi, A. K., et al. 2022, *Solar Phys.*, 297(12), 153.
Domingo, V., Fleck, B., & Poland, A. I. 1995, *Solar Phys.*, 162(1-2), 1.
Fulara, A., Chandra, R., Chen, P. F., et al. 2019, *Solar Phys.*, 294(5), 56.
Gopalswamy, N., Xie, H., Mäkelä, P., et al. 2013, *Adv. Space Res.*, 51(11), 1981.
Guo, J. H., Ni, Y. W., Zhong, Z., et al. 2023, *ApJS*, 266(1), 3.
Guo, Y., Erdélyi, R., Srivastava, A. K., et al. 2015, *ApJ*, 799(2), 151.
Hamini, A., Auxepaules, G., Birée, L., et al. 2021, *Journal of Space Weather and Space Climate*, 11, 57.
Harvey, J. W., Bolding, J., Clark, R., et al. 2011, *AAS/Solar Physics Division Meeting*, 42, 17.45.
Hou, Z., Tian, H., Wang, J.-S., et al. 2022, *ApJ*, 928(2), 98.
Kaiser, M. L., Kucera, T. A., Davila, J. M., et al. 2008, *Space Sci. Revs*, 136(1-4), 5.
Klein, K.-L., Musset, S., Vilmer, N., et al. 2022, *A&A*, 663, A173.
Kouloumvakos, A., Papaioannou, A., Waterfall, C. O. G., et al. 2024, *A&A*, 682, A106.
Kwon, R.-Y. & Vourlidas, A. 2017, *ApJ*, 836(2), 246.
Kwon, R.-Y., Zhang, J., & Olmedo, O. 2014, *ApJ*, 794(2), 148.
Lemen, J. R., Title, A. M., Akin, D. J., et al. 2012, *Solar Phys.*, 275(1-2), 17.
Luna, M., Su, Y., Schmieder, B., et al. 2017, *ApJ*, 850(2), 143.
Mavromichalaki, H., Paschalis, P., Gerontidou, M., et al. 2022, *Universe*, 8(7), 378.
Mishev, A. L., Koldobskiy, S. A., Larsen, N., & Usoskin, I. G. 2024, *Solar Phys.*, 299(2), 24.
Moreton, G. E. & Ramsey, H. E. 1960, *Pub. Astron. Soc. Pac.*, 72(428), 357.
Muhr, N., Veronig, A. M., Kienreich, I. W., et al. 2014, *Solar Phys.*, 289(12), 4563.

Nakariakov, V. M. & Ofman, L. 2001, *A&A*, 372, L53.
Nitta, N. V., Liu, W., Gopalswamy, N., & Yashiro, S. 2014, *Solar Phys.*, 289(12), 4589.
Papaioannou, A., Kouloumvakos, A., Mishev, A., et al. 2022, *A&A*, 660, L5.
Pesnell, W. D., Thompson, B. J., & Chamberlin, P. C. 2012, *Solar Phys.*, 275(1-2), 3.
Roberts, B., Edwin, P. M., & Benz, A. O. 1984, *ApJ*, 279, 857.
Schmieder, B., Démoulin, P., & Aulanier, G. 2013, *Adv. Space Res.*, 51(11), 1967.
Shen, Y., Chen, P. F., Liu, Y. D., et al. 2019, *ApJ*, 873(1), 22.
Smerd, S. F., Sheridan, K. V., & Stewart, R. T. 1974 Newkirk, G. A., editor, *Coronal Disturbances*, 57, 389.
Su, W., Guo, Y., Erdélyi, R., et al. 2018, *Sci. Rep.*, 8, 4471.
Thompson, B. J., Gurman, J. B., Neupert, W. M., et al. 1999, *ApJL*, 517(2), L151.
Thompson, B. J., Plunkett, S. P., Gurman, J. B., et al. 1998, *Geophys. Res. Lett.*, 25(14), 2465.
Vasanth, V., Umapathy, S., Vršnak, B., et al. 2014, *Solar Phys.*, 289(1), 251.
Vourlidas, A. & Ontiveros, V. 2009, Ao, X. & Burrows, G. Z. R., editors, *American Institute of Physics Conference Series*, 1183, 139.
Warmuth, A. 2015, *Living Reviews in Solar Physics*, 12(1), 3.
Warmuth, A. & Mann, G. 2005, *A&A*, 435(3), 1123.
Zhang, Q. M., Chen, J. L., Li, S. T., et al. 2022, *Solar Phys.*, 297(2), 18.
Zhukov, A. N. & Auchère, F. 2004, *A&A*, 427, 705.
Zhukov, A. N., Rodriguez, L., & de Patoul, J. 2009, *Solar Phys.*, 259(1-2), 73.

Solar and Stellar Coronal Mass Ejections
Proceedings IAU Symposium No. 388, 2026
N. Gopalswamy, O. Malandraki, A. Vidotto & W. Manchester, eds.
doi:10.1017/S1743921324001777

Comoving MHD simulations of the turbulent interplanetary evolution of a CME

Mattia Sangalli[1], **Andrea Verdini**[1], **Simone Landi**[1] **and Emanuele Papini**[2]

[1]Dipartimento di Fisica e Astronomia, Università degli Studi di Firenze
Via G. Sansone 1, 50019 Sesto Fiorentino, Italy
email: mattia.sangalli@unifi.it

[2]Istituto di Astrofisica e Planetologia Spaziali, Istituto Nazionale di Astrofisica
Via del Fosso del Cavaliere 100, 00133 Roma, Italy

Abstract. Coronal mass ejections (CMEs) are eruptive solar events with great influence on Earth. They often carry coherent magnetic fields in the form of magnetic clouds as they propagate through the heliosphere. Despite being large-scale structures, CMEs are subject to meso-scale dynamical processes such as turbulence and magnetic reconnection and exhibit fluctuations at many scales. A novel numerical approach is presented which separates the large-scale interplanetary motion of a CME from its local dynamics, investigating the effects of spherical expansion and turbulent fluctuations.

Keywords. Sun: coronal mass ejections – methods: numerical – magnetohydrodynamics (MHD) – turbulence

1. Introduction

Coronal mass ejections (CMEs) are large-scale releases of plasma and magnetic fields taking place in the outer layers of the Sun's atmosphere (Webb and Howard 2012). The "ejecta" are launched in the solar wind, becoming Interplanetary CMEs (ICMEs). ICMEs are often found to be complex multipart structures, possibly driving a shock wave (see e.g. Gosling et al. 1975; Sheeley et al. 1985), followed by a compressed and turbulent sheath (Kilpua et al. 2017) and embedding an enhanced magnetic field (see e.g. Klein and Burlaga 1982; Burlaga 1988). The "magnetic ejecta" can often be described as a magnetic flux rope, linked to the Sun at both extremes (see e.g. Burlaga 1988; Bothmer and Schwenn 1998).

CMEs are the most important drivers of geomagnetic storms (Gosling et al. 1991; Webb et al. 2000; Huttunen and Koskinen 2004; Zhang et al. 2007; Richardson and Cane 2012); in particular, their sheath regions with strong and rapidly variable plasma quantities are among the most geoeffective parts (Huttunen and Koskinen 2004; Kilpua et al. 2017).

Turbulence is found in both the background solar wind (Bruno and Carbone (2013)), the magnetic ejecta (e.g. Ruzmaikin et al. 1997; Leamon et al. 1998; Manoharan et al. 2000) and the sheath regions of ICMEs (e.g. Kennel et al. 1982; Kataoka et al. 2005; Kilpua et al. 2017). Moreover, estimates of coherence lengths for both magnetic ejecta and sheath regions are smaller than their average size (Lugaz et al. 2018; Ala-Lahti et al. 2020). Overall, this information points towards some important internal meso-scale dynamics occurring during the heliospheric motion of CMEs.

Global numerical simulations can reproduce complex heliospheric configurations including multiple CMEs and structured solar wind (see e.g. Manchester et al. (2017)

and references therein, or Allen et al. (2020) and Scolini et al. (2020) for more recent works) and provide unique tools for studying their evolution and space weather impacts.

In this work, we present a different numerical approach to the interplanetary evolution of a CME carrying a magnetic cloud: a semi-Lagrangian numerical model is used to decouple the internal dynamics from the large-scale heliospheric motion. The higher effective numerical resolution allows to investigate meso-scale processes such as turbulence and magnetic reconnection – taking place both inside and outside the ICME.

Here, we focus on defining an estimate for the ICME size and studying its heliospheric evolution under the influence of spherical expansion and turbulent fluctuations. The overall goal is to better understand the internal dynamics of ICMEs.

2. Methods and numerical setup

2.1. *Expanding Box Model equations*

The Expanding Box Model (EBM, Grappin et al. 1993; Grappin and Velli 1996; Rappazzo et al. 2005; Dong et al. 2014) follows the evolution of a parcel of plasma in the spherically expanding solar wind. For simplicity, we assume a constant mean wind velocity $\mathbf{U}_0 = U_0 \hat{\mathbf{e}}_r$ transporting the parcel radially outward from the Sun. The large-scale motion of the wind and the internal dynamics are thus effectively decoupled.

Our parcel will contain an ICME with an embedded magnetic cloud, described as a magnetic flux rope. The ambient medium will be considered as unmagnetized. The heliocentric position of the parcel of plasma evolves as $R(t) = R_0 + U_0 t$, where $R(t = 0) = R_0$ is the starting position, and U_0 is the constant parcel/ICME velocity. We can also define a normalized heliocentric position as

$$a(t) = \frac{R(t)}{R_0} = 1 + \varepsilon t \tag{1}$$

where $\varepsilon = \mathrm{d}a/\mathrm{d}t = U_0/R_0$ is the expansion rate (inverse of the expansion time $t_{\mathrm{exp}} = R_0/U_0$).

In a spherically expanding flow, the parcel expands as a unit element of spherical volume would: its radial dimension does not vary, whereas the transverse ones increase linearly with the heliocentric distance R. A coordinate transformation to comoving and expanding cartesian coordinates ($x' = x - U_0 t$, $y' = y/a(t)$, $z' = z/a(t)$) is performed. The spherical expansion thus reduces to the presence of anisotropic prefactors in the gradients: $\nabla' = (\partial/\partial x', \partial/\partial y', \partial/\partial z') = (\partial/\partial x, 1/a(t)\cdot\partial/\partial y, 1/a(t)\cdot\partial/\partial z)$. The velocity field is also explicitly decoupled: the mean flow is subtracted from the fixed-coordinate velocity field $\mathbf{U}$, focusing on the velocity fluctuations $\mathbf{u} = \mathbf{U} - \mathbf{U_0}$. Finally, several conservation laws (mass, magnetic flux, angular momentum) together with the anisotropic expansion imply anisotropic scalings of the physical fields with R:

$$\rho \propto R^{-2};\ B_x \propto R^{-2},\ B_{y,z} \propto R^{-1};\ u_x \propto \text{const.},\ u_{y,z} \propto R^{-1};\ T \propto R^{-2(\gamma-1)}, \tag{2}$$

where ρ, T and $\mathbf{u}$ denote density, temperature and fluid velocity, $\mathbf{B}$ is the magnetic field, P is the pressure, obtained from density and temperature via the equation of state for ideal gases, and $\gamma = 5/3$ is the adiabatic index. The following are the EBM (ideal) MHD evolution equations in adimensionalized form (from now on, the symbols (x, y, z) and ∇ will refer to local coordinates, dropping the primes for ease of reading):

$$\frac{\partial \rho}{\partial t} = -\nabla\cdot(\rho\mathbf{u}) - 2\rho\frac{\dot{a}}{a} \tag{3}$$

$$\frac{\partial \mathbf{u}}{\partial t} = -\mathbf{u}\cdot\nabla\mathbf{u} - \frac{1}{\rho}\nabla P + \frac{1}{\rho}(\nabla\times\mathbf{B})\times\mathbf{B} - \frac{\dot{a}}{a}\mathcal{P}_\perp\mathbf{u} \tag{4}$$

$$\frac{\partial \mathbf{B}}{\partial t} = \nabla \times (\mathbf{u} \times \mathbf{B}) + \frac{\dot{a}}{a} (\mathcal{P}_\perp - 2\mathcal{I})\, \mathbf{B} \tag{5}$$

$$\frac{\partial T}{\partial t} = -\mathbf{u} \cdot \nabla T - (\gamma - 1) T\, (\nabla \cdot \mathbf{u}) - 2(\gamma - 1)\frac{\dot{a}}{a} T \tag{6}$$

where $\mathcal{I}$ is the 3×3 identity matrix and $\mathcal{P}_\perp = \mathrm{diag}(0,1,1)$ is the projector on the transversal directions. Space, time, mass density, magnetic field, velocity and temperature are measured respectively in units of L^0, t^0, ρ^0, B^0, u^0 and T^0, where $u^0 = c_A^0 = B^0/\sqrt{4\pi\rho^0}$, $t^0 = L^0/u^0$ and $T^0 = m_p (u^0)^2/2k_B$. Since the expansion rate ε is an additional dimensional parameter with respect to the usual MHD equations, defined through U_0 and R_0, its adimensionalized counterpart is related to the velocity and length units (i.e., also to the magnetic field and density ones)†.

2.2. *Magnetic flux rope equilibrium configuration*

We consider a locally cylindric magnetic flux rope in the ICME. This assumption allows us to neglect the dynamics along the flux rope axis, i.e. $\partial/\partial z \equiv 0$. We then consider a 2D cut normal to the flux rope axis (r-θ plane, i.e. the x-y plane) while retaining all the 3 components of vector fields (2.5D geometry). We define an ideal 2.5D "magnetic pinch" equilibrium configuration which is static, stationary and axially symmetric around the flux rope axis. This means that $\mathbf{B} = B_z(r)\hat{\mathbf{e}}_z + B_\theta(r)\hat{\mathbf{e}}_\theta$ and $\partial \mathbf{u}/\partial t \equiv 0$. We choose a profile for both $B_\theta(r)$, $B_z(r)$ and $\rho(r)$, thus imposing $T(r)$ by inverting equation (4) to account for the ultimate stability of the configuration. In particular, we choose the profiles $B_z(r) = B_{ax}/\cosh(r/\Delta)$, $B_\theta(r) = B_{tw}(r/\Delta)^4 \exp(4(1 - r/\Delta))$ and $\rho(r) = \rho_{bg} + \rho_{in}(1 - \tanh((r - 2\Delta)/(\Delta/10))/2$. This means that the magnetic field has an axial (out-of-plane) component of width $\sim L_{CME} = 2\Delta$ which gradually rises to have a peak B_{ax} on the axis, and an azimuthal (in-plane) component which describes a twist condensed in a ring peaking at B_{tw} at distance $\sim \Delta$ from the axis; an overdensity of width $\sim L_{CME}$ and intensity ρ_{in} is also present around the axis. The chosen profiles imply an excess magnetic pressure, compensated by imposing a $T(r)$ which drops from T_{bg} for $r \gg \Delta$ to a $T_{in} < T_{bg}$ for $r \leq \Delta$, accordingly to the inversion of eq. (4). This consistently describes a denser and colder filament-like structure together with a twisted magnetic flux rope in static and stationary equilibrium.

2.3. *Superposition of fluctuations*

In order to study the effect of turbulence on the interplanetary evolution of this equilibrium magnetized ICME, we superpose pseudo-alfvenic random fluctuations to the initial stable equilibrium. We choose Fourier coefficients for $\delta B_{x,y}$ and $\delta u_{x,y}$ so to have a flat energy spectrum for $k_{CME} < k < 4k_{CME}$, where $k = ({k_x}^2 + {k_y}^2)^{1/2}$ and $k_{CME} = 2\pi/L_{CME}$, L_{CME} being the initial CME radial width. The fluctuations are normalized to a chosen root mean square value δB_{rms}.

2.4. *Numerical methods and setup of simulations*

We assume all the fields to be periodic in the local coordinates (x, y) and invariant in the axial (out-of-plane) coordinate z. Equations (3-6) are solved in a 2.5D domain $L_x \times L_y$ discretized on a fixed grid $(N_x \times N_y)$ using a pseudospectral code based on the one described in Papini et al. (2021). Fourier transforms are used to compute spatial derivatives via the FFTW library (Frigo and Johnson 2005), and a third order Runge-Kutta

† Note that the adimensionalized ε is actually the ratio between the unit time t^0 (the characteristic Alfvén time) and the expansion time t_{exp}.

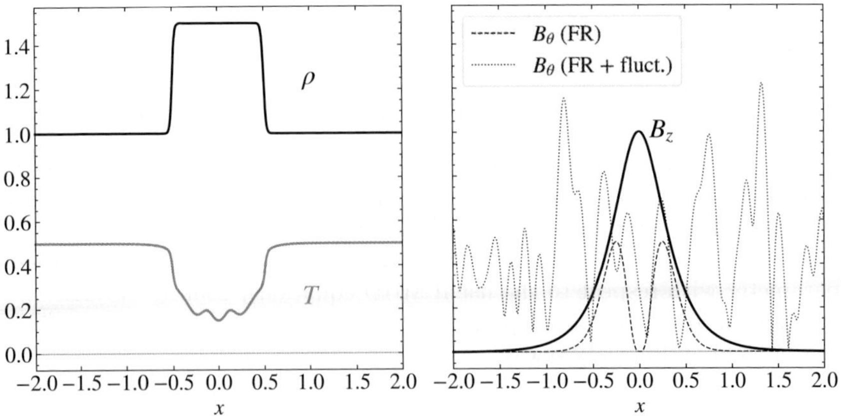

Figure 1. One-dimensional profiles along x at $t=0$. **Left panel**: profiles of nondimensional $\rho(x)$ (black) and $T(x)$ (grey). **Right panel**: profiles of nondimensional $B_z(x)$ (solid) and $B_\theta(x)$, the latter both without (dashed) and with (dotted) superposed fluctuations (runs I and T, respectively).

method is used for time advancement. A passive scalar is evolved together with the physical fields

$$\frac{\partial s}{\partial t} = -\mathbf{u} \cdot \nabla s, \tag{7}$$

initially defined as a scalar field equal to 1 inside the CME and 0 outside. The numerical equations for $\mathbf{u}$, $\mathbf{B}$, T all have physical dissipative terms; an artificial dissipative term is added to equation (7) in order to ensure numerical stability. The details of the numerical equations will be described in a dedicated paper (Sangalli *et al.*, in preparation).

The initial configuration is chosen with $\rho_{\rm bg}=1$, $\rho_{\rm in}=\rho_{\rm bg}/2$, $B_{\rm ax}=1$, $B_{\rm tw}=B_{\rm ax}/2$, $\Delta=0.5$, and $T_{\rm bg}=0.5$. Furthermore, we choose $R_0=5$ and $U_0=4$, meaning $\varepsilon=4/5$.

Here we present two simulations: run I ("ideal") with just the equilibrium ICME subject to geometrical expansion, and run T ("turbulent") with superposed fluctuations $\delta u_{\rm rms}=\delta B_{\rm rms}=0.5$. Their initial 1-D profiles along x for ρ, T and $\mathbf{B}$ are shown in fig. 1. The following differences are to be noted: run I is simulated in a 10×10 domain on a 512^2 grid, whereas for run T a 4×4 domain and a 1024^2 grid are used.

To study the evolution of the ICME size, we compute a time-dependent weighted position based on the passive scalar s (effective center of mass) and its dispersion σ (effective size); the values of $\sigma_{x,y}$ are used to estimate the evolution of radial and transversal sizes of the CME.

3. Results

3.1. *Ideal simulation*

The results of the ideal simulation can be seen in figure 2, where the distance-evolution of the passive scalar (fig. 2a) and the $\sigma_{x,y}(R)$ trends (fig. 2c) are depicted. An increase in the x size of the passive scalar (that is, a radial expansion of the ICME) is already apparent in the 2D maps. A quantitative confirmation comes from the $\sigma_x(R)$ monotonic increase; conversely, in $\sigma_y(R)$ we observe a departure from the "kinematic" domain stretching. Qualitatively, we can attribute both these behaviours to the effect of magnetic tension: we expect it to increase in the y direction and decrease in the x direction because of the geometrical stretching of the (initially circular) field lines. On the other hand, a departure from equilibrium is also expected due to anisotropy: the initial magnetic pinch

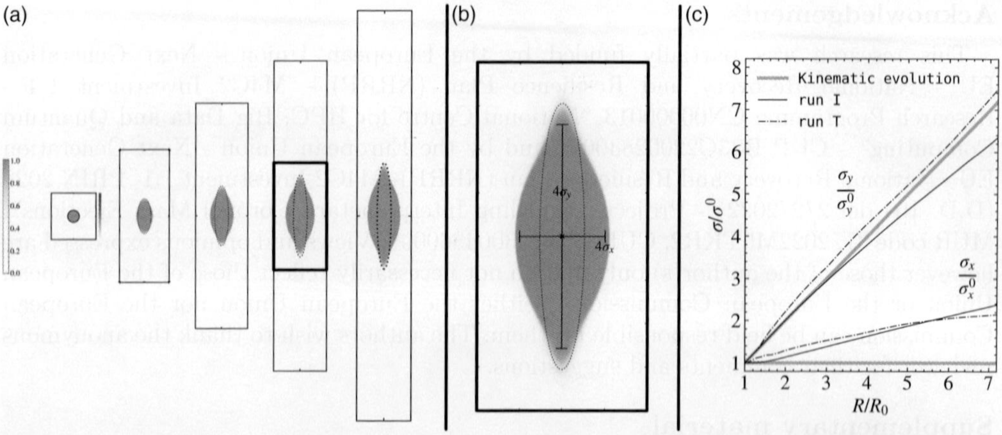

Figure 2. **(a)**: evolution of the passive scalar (color-coded) with $R(t)$ for run I; the snapshots show the evolution from $R = R_0$ to $R \simeq 7R_0$; the dotted closed line represents the purely geometrical expected ("kinematic") behavior. **(b)**: A close-up for $R \sim 6R_0$ showing the comparison between the actual passive scalar (color-coded), the expected kinematic behavior (dotted black line) and the size estimates $\sigma_{x,y}$. **(c)**: evolution of $\sigma_{x,y}(R)$ for the two runs I and T; the thick grey line represents the kinematic evolution, the solid black line represents the "ideal" run and the dash-dotted black line represents the "turbulent" run.

configuration was isotropic; however, the spherical expansion (both gradients and fields) is intrinsically anisotropic.

3.2. *Turbulent simulation*

The presence of random fluctuations developing into turbulence is by itself a perturbation to the equilibrium. However, this also has an effect on the dynamics described above. We can thus decouple the effects of turbulence into:

- a disordered increase in $\sigma_{x,y}$ because of the shearing and distortion of the plasma by turbulent eddies;
- a weakening of the coherent magnetic flux rope field because of magnetic reconnection, meaning less effective magnetic tension.

We can find these effects in the $\sigma(R)$ trends shown in fig. 2c: initially σ_y rises more than in the ideal case; a similar effect can be seen in σ_x, even though a second decreasing phase is also observed.

4. Conclusions

Our 2.5D simulations in a spherically expanding geometry show that a magnetized ICME in isotropic equilibrium does not follow the expected kinematic stretching. The system is driven out of equilibrium by the time-dependent expansion and a dynamical response arises: the radial size increases, as often observed in both data and models, and the transverse size increases slower than spherically. Superposing turbulent fluctuations on such a system has a twofold effect: on the one hand, turbulent eddies distort and shear the CME, resulting in a greater size; on the other hand, turbulence causes magnetic reconnection, weakening the magnetic equilibrium that otherwise would resist the geometrical expansion (along y) and produce radial expansion (along x). An upcoming paper will be dedicated to a deeper understanding of the internal dynamics driving both the radial expansion and the transversal resistance to expansion, also exploring the relevant parameters' space (Sangalli *et al.*, in preparation).

Acknowledgements

This research was partially funded by the European Union - Next Generation EU - National Recovery and Resilience Plan (NRRP) - M4C2 Investment 1.4 - Research Programme CN00000013 "National Centre for HPC, Big Data and Quantum Computing" - CUP B83C22002830001 and by the European Union - Next Generation EU - National Recovery and Resilience Plan (NRRP)- M4C2 Investment 1.1- PRIN 2022 (D.D. 104 del 2/2/2022) - Project "Modeling Interplanetary Coronal Mass Ejections", MUR code 31. 2022M5TKR2, CUP B53D23004860006. Views and opinions expressed are however those of the author(s) only and do not necessarily reflect those of the European Union or the European Commission. Neither the European Union nor the European Commission can be held responsible for them. The authors wish to thank the anonymous reviewer for their comments and suggestions.

Supplementary material

To view supplementary material for this article, please visit https://doi.org/10.1017/S1743921324001777

References

Ala-Lahti, M. *et al.* 2020, *Journal of Geophysical Research (Space Physics)*, 125(9), e2020JA028002.
Allen, R. C. *et al.* 2020, *ApJS*, 246(2), 36.
Bothmer, V. & Schwenn, R. 1998, *Annales Geophysicae*, 16(1), 1–24.
Bruno, R. & Carbone, V. 2013, *Living Reviews in Solar Physics*, 10(1), 2.
Burlaga, L. F. 1988, *J. Geophys. Res.*, 93(A7), 7217–7224.
Dong, Y., Verdini, A., & Grappin, R. 2014, *ApJ*, 793(2), 118.
Frigo, M. & Johnson, S. G. 2005, *IEEE Proceedings*, 93, 216.
Gosling, J. T. *et al.* 1975, *Sol. Phys.*, 40(2), 439–448.
Gosling, J. T., McComas, D. J., Phillips, J. L., & Bame, S. J. 1991, *J. Geophys. Res.*, 96(A5), 7831–7839.
Grappin, R. & Velli, M. 1996, *J. Geophys. Res.*, 101(A1), 425–444.
Grappin, R., Velli, M., & Mangeney, A. 1993, *Phys. Rev. Lett.*, 70(14), 2190–2193.
Huttunen, K. & Koskinen, H. 2004, *Annales Geophysicae*, 22(5), 1729–1738.
Kataoka, R. *et al.* 2005, *Geophys. Res. Lett.*, 32(12), L12103.
Kennel, C. F. *et al.* 1982, *J. Geophys. Res.*, 87(A1), 17–34.
Kilpua, E., Koskinen, H. E. J., & Pulkkinen, T. I. 2017, *Living Reviews in Solar Physics*, 14(1), 5.
Klein, L. W. & Burlaga, L. F. 1982, *J. Geophys. Res.*, 87(A2), 613–624.
Leamon, R. J., Smith, C. W., & Ness, N. F. 1998, *Geophys. Res. Lett.*, 25(14), 2505–2508.
Lugaz, N. *et al.* 2018, *ApJL*, 864(1), L7.
Manchester, W. *et al.* 2017, *Space Sci. Rev.*, 212(3-4), 1159–1219.
Manoharan, P. K., Kojima, M., Gopalswamy, N., Kondo, T., & Smith, Z. 2000, *ApJ*, 530(2), 1061–1070.
Papini, E. *et al.* 2021, *ApJL*, 917(1), L12.
Rappazzo, A. F., Velli, M., Einaudi, G., & Dahlburg, R. B. 2005, *ApJ*, 633(1), 474–488.
Richardson, I. G. & Cane, H. V. 2012, *Journal of Space Weather and Space Climate*, 2, A01.
Ruzmaikin, A., Feynman, J., & Smith, E. J. 1997, *J. Geophys. Res.*, 102(A9), 19753–19760.
Scolini, C. *et al.* 2020, *ApJS*, 247(1), 21.
Sheeley, N. R., Jr. *et al.* 1985, *J. Geophys. Res.*, 90(A1), 163–176.
Webb, D. F. *et al.* 2000, *J. Geophys. Res.*, 105(A4), 7491–7508.
Webb, D. F. & Howard, T. A. 2012, *Living Reviews in Solar Physics*, 9(1), 3.
Zhang, J. *et al.* 2007, *Journal of Geophysical Research (Space Physics)*, 112(A10), A10102.

Solar and Stellar Coronal Mass Ejections
Proceedings IAU Symposium No. 388, 2026
N. Gopalswamy, O. Malandraki, A. Vidotto & W. Manchester, eds.
doi:10.1017/S1743921324001571

Rotation of flux ropes in the low corona

Brigitte Schmieder[1,2], Anwesha Maharana[1], Jin Han Guo[1,3], Luis Linan[1] and Stefaan Poedts[1,4]

[1]Centre for mathematical Plasma Astrophysics, Dept. of Mathematics, KU Leuven, 3001 Leuven, Belgium

[2]LIRA, Observatoire de Paris, CNRS, Université de Paris, 5 place Janssen, 92290 Meudon, France

[3]University of Nanjing, Nanjing, China

[4] Institute of Physics, University of Curie-Skłodowska, Lublin, Poland

email: brigitte.schmieder@obspm.fr

Abstract. Eruptions of filaments are defined by different parameters, specially, sigmoid handedness and direction of the eruption, which are important parameters for forecasting the geoeffectiveness of consequent interplanetary coronal mass ejection (ICME) or magnetic cloud. Solar filaments often exhibit rotation and deflection during eruptions, which would significantly affect the geoeffectiveness of the coronal mass ejections (CMEs). Therefore, understanding the mechanisms that lead to such lateral displacement of filaments is a great concern to space weather forecasting. Two case studies are discussed. Firstly, the events of September 8 and September 10 2014, were analyzed from the Sun to the Earth. The numerical heliospheric simulation EUHFORIA shows that the handedness of the EUV sigmoid deduced from coronagraph observations was different from the tilt of the ICME at 1 au, suggesting a rotation of the CME in the low corona. A potential undetected low coronal rotation led to erroneous space weather prediction. The second event concerns a filament observed on August 20 2021, which underwent a rotation of 73 degrees during its eruption, implying a significant lateral drifting of the filament material. A data-constrained magnetohydrodynamic (MHD) simulation confirms such a rotation. These two studies reinforce the idea that CMEs are subjected to more significant rotation and deflection in the low corona than during their journey in the heliosphere.

Keywords. Coronal Mass Ejection, MHD simulations

---

1. Introduction

Geoeffective event prediction is now possible using MHD simulations (e.g. EUHFORIA). It relies on the orientation of the B_z component of the ejected magnetic flux rope from the Sun, frequently assimilated to a coronal mass ejection (CME). A forecasting limitation arises from the evolution of CME structures during their propagation and their interaction with the solar wind and other CMEs. The interaction of CMEs may lead to severe geoeffective events, as demonstrated by Shen et al. (2018), Scolini et al. (2020), and Koehn et al. (2022). The lack of multiple in situ crossings through CMEs makes predicting its global behaviour through reconstruction using data from a single point challenging.

A second limitation, such as CME rotation, deflection, and deformation between their solar source and their consequent ICME at 1au, is even more intriguing. Some previous works have indicated that the rotation of CME flux ropes mainly happens in the low corona due to the complicated magnetic field environment. For example, Shiota et al.

(2010) found that the rotation of CME flux ropes due to reconnection happens within 5 solar radius. Lynch et al. (2009) found that the rotation angle of the CME flux rope reaches 50 degrees when it propagates to a distance of 3.5 solar radii. A rotation up to 56 degrees of a newly built magnetic flux rope by helicity injection was found in the April 23 2023 geo-effective event (Vemareddy 2024). Kliem et al. (2012) highlighted the possibility of extensive low coronal rotation up to even more than 100 degrees by combining twist and shear-driven rotation, the latter being dominant in the lower corona. MHD simulations from the solar surface to 1 au revealed that the rotation mainly occurs within 15 solar radii (Regnault et al. 2023). The boundary conditions of the European Heliospheric FORecasting Information Asset (EUHFORIA Pomoell & Poedts 2018) are given at 0.1au and it does not account the CME evolution in the corona below 0.1 au. Although the white-light coronagraph images help reconstruct the CMEs in the middle and upper corona, they are insufficient to derive the magnetic field configuration. This leaves us to rely on the source region proxies for guessing the magnetic field configuration for prediction purposes, being agnostic to the dominant low coronal dynamics.
Data-driven MHD simulation seems to be a good tool to understand better what is happening to a flux rope during its eruption (Guo et al. 2023a,b, 2024). In this paper, we present two case studies – the first paper concerns a complex set of CMEs, showing that EUHFORIA can model the correct magnetic field profiles resulting from the interaction between two CMEs by assuming substantial CME rotation in the low corona, to justify the consistency of the CME orientation at 0.1 au and at 1 au (Maharana et al. 2023). The second case study shows that data-driven MHD simulation can explain the difference in orientation between the initial eruptive filament and the consequent CME. In that case, multiple reconnections of flux ropes have been found explaining the observed orientation discrepancy (Guo et al. 2023b) (Section 3).

2. MHD simulation of the September 2014 events

For the first event, we investigate the earthward evolution of two successive CMEs that erupted from the same active region, AR 12158, on September 8, 2014, and September 10, 2014, respectively. This active region was reported to be continuously shearing and deflecting during its evolution (Vemareddy et al. 2016). The first CME (hereafter CME1) was not recorded in any ICME catalogues. However, we noticed a shock signature at Earth and interpreted its atypical magnetic field signatures, which pointed to a potential glancing blow at Earth. The second CME (hereafter CME2) was predicted to be a direct hit and cause a major storm due to a strong negative B_z component. This prediction, made by the space weather community, considered only the source region magnetic field signatures of CME2 to interpret the orientation of the incoming flux rope. Surprisingly, a long-lasting positive B_z passed through Earth, with a brief period of negative B_z in the sheath region of the CME that drove the storm. So, instead of a major storm driven by the magnetic cloud, a moderate storm was caused due to the sheath region's magnetic field. Hence, the geoeffectiveness of different CME2 sub-structures was mispredicted.

The main objective of this study was to understand the change in the CME2 orientation that would have led to the erroneous geoeffectiveness prediction at 1 au. It could be because of the rotation or deformation of CME2 in the lower corona or during its heliospheric propagation. Or it could also be because CME2 impacted Earth with its flank, which gave rise to the in situ magnetic field signatures different from what was predicted, assuming a direct hit based on insights from the early remote-sensing observations. We characterised CME2 in three phases, first at the Sun, then in the corona and finally at Earth. We constrained the magnetic axis orientation of the erupting flux rope using the polarity inversion line (PIL) proxy from the HMI magnetogram observations of the active region. The chirality of the flux rope, i.e., the sign of helicity, is determined using the

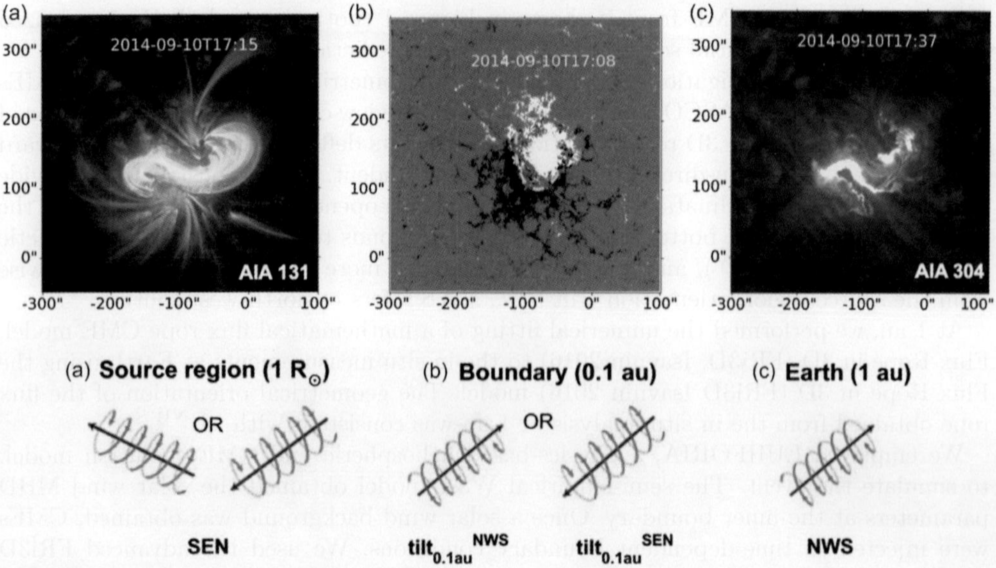

Figure 1. Sigmoid observed on September 9 2014, in AR 12158: Top panels: (a) by AIA in 131 Å (b) by HMI, (c) by AIA in 304 Å. The red line indicates the tilt of the active region, and the green dashed line is the polarity inversion line. Bottom panels: Schematic representation of the CME2 orientation inferred from different observational proxies at different locations. (a) Close to 1 $R_\odot$, based on the analysis of the source region (sigmoid); (b) close to 0.1 au, based on the 3D reconstruction of the white-light images, and (c) at 1 au, based on the in situ observations. Adapted from Maharana et al. (2023).

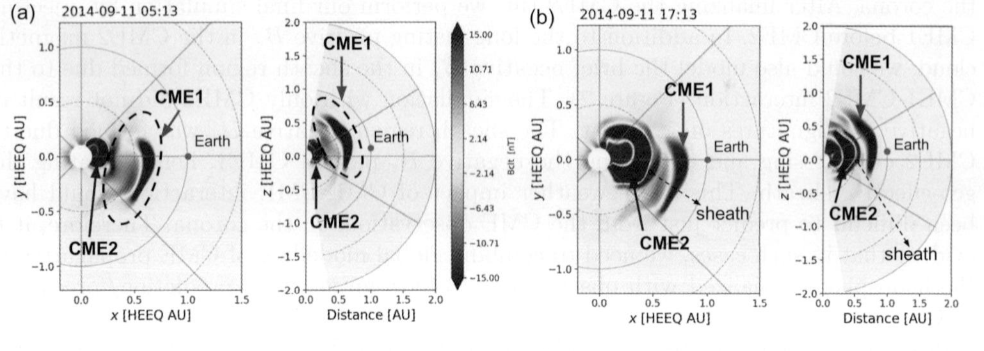

Figure 2. Evolution of B_z for the CME-CME interaction event of September 2014 using EUHFORIA. The co-latitudinal component in the spherical coordinate system is B_{clt}, equivalent to $-B_z$ on the ecliptic plane. The red and blue spectra of the colour bar correspond to positive and negative B_z, respectively. Each sub-figure shows the view of the equatorial (X-Y) and the meridional (X-Z) planes at a particular time mentioned at its top. (a) CME2 is shown evolving behind CME1 in the early stage of interaction; (b) CME2 compresses the trailing negative B_z part of CME1, creating the geo-effective sheath ahead of itself. Adapted from Maharana et al. (2023).

handedness of the EUV sigmoid (Figure 1 top panel a). Determining the PIL orientation of this eruption was not straightforward. It was more of a curved topology than a straight line, as shown in Figure 1 (top panel b). As our heliospheric simulations require us to adhere to a single orientation, we compromised and chose an average straight-line south-east-north (SEN) configuration as defined by Bothmer & Schwenn (1998). The schematic representation of the CME front is shown in Figure 1 (bottom panel a). We recognized this as one of the potential sources of errors in the prediction.

Continuing our investigation, we performed a 3D geometric reconstruction of the CMEs in the coronagraph of LASCO and STEREO-B field of view close to 0.1 au. The position of CME2 suggested by the 3D reconstruction indicated its deflection in the northwest-ward direction making the non-direct encounter at Earth evident. Such reconstructions provide only the geometrical inclination of the flux rope, which opened up two possibilities for the magnetic axis (Figure 1 bottom panels). One corresponds to the source region magnetic axis orientation (tilt^{SEN}), and the other was rotated more than 200° counter-clockwise from the source region orientation (tilt^{NWS}, NWS refers to north-west-south).

At 1 au, we performed the numerical fitting of a mathematical flux rope CME model, Flux Rope in 3D (FRi3D, Isavnin 2016) to the in situ measurements at Earth using the Flux Rope in 3D (FRi3D Isavnin 2016) model. The geometrical orientation of the flux rope obtained from the in situ analysis at 1 au was consistent with tilt^{NWS}.

We employed EUHFORIA, a physics-based heliospheric and CME evolution model, to simulate the event. The semi-empirical WSA model obtained the solar wind MHD parameters at the inner boundary. Once a solar wind background was obtained, CMEs were injected as time-dependent boundary conditions. We used the advanced FRi3D model with a global CME geometry to capture the flank encounter of CME2. Finally, the CMEs self-consistently evolved in the solar wind background in the heliospheric domain and forecasting was done. We first performed two simulations just for CME2, initializing it with tilt^{SEN} and tilt^{NWS}, respectively.

With tilt^{SEN}, a negative B_z signature was obtained at Earth opposite to what was observed. With tilt^{NWS}, prolonged positive B_z was reproduced at Earth. Hence, our numerical experiments suggested that low corona rotation was in fact one of the key factors that could not be considered due to the lack of magnetic field observations in the corona. After finalizing the CME2 tilt, we perform our final simulation by injecting CME1 before CME2. In addition to the long-lasting positive B_z in the CME2 magnetic cloud, we could also model the brief negative B_z in the sheath region formed due to the CME1-CME2 interaction (Figure 2). The simulation with only CME2 did not result in negative B_z signatures on its front. The sheath region negative B_z was formed due to CME2 compressing and enhancing the negative B_z part of CME1, hence creating the geo-effective sheath. This space weather impact of CME-CME interaction would have been difficult to predict just from the CME observations in the corona. Therefore, it is evident that in such cases, we need to combine global modelling of CME propagation in the heliosphere initialized with observations to have an accurate prediction/forecast at 1AU.

In summary, we successfully reproduced the magnetic field components at 1 au for the early September 2014 storm by modelling both CME1 and CME2. We identify the correct orientation of the erupting flux rope at 0.1 au, i.e., the EUHFORIA simulation boundary. It is different from the sigmoid at the source region, and we proposed a possible rotation below 0.1 au, essential for initializing the CME2 in the simulations. We find that the interaction between CME1 and CME2 is responsible for producing the geoeffectiveness in the sheath. Finally, we highlight the importance of 3D MHD modelling of CME evolution in understanding the geoeffectiveness due to the different sub-structures in CME-CME interaction events.

3. Flux rope rotation in the corona

In the previous section, we show that the orientation of the main flux rope at 1 au does not correspond to the source region tilt. A rotation of the FR has to be considered in the low corona. We were tempted to use our data-driven method to see how the sigmoid

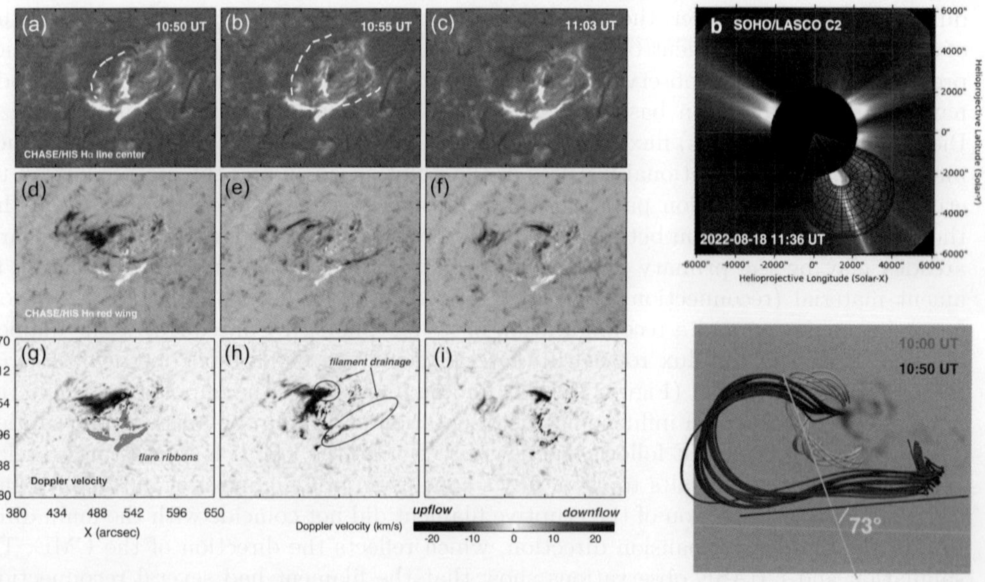

Figure 3. Observation of the ejection of a filament and change of orientation by CHASE; top panels: intensity in Hα, middle panels in the Hα wing, bottom panels: Doppler shifts. In panels a and b, the spine of the filament is indicated by a dashed line; in panel h, the oval indicates the drainage of the filament towards its drifting footpoint. The CME is on the top right panel, and in the bottom right panel, from the MHD simulation, the initial flux rope is green, and the final flux rope is violet. 73 degrees is the estimated rotation angle of the filament. Adapted from Guo et al. (2023b).

changed orientation in the low corona. However, this case is very complex and difficult to model, with the interaction of two CMEs and a background solar wind approximated with WSA. We chose a more simple case, the second case.

Solar filaments often exhibit rotation and deflection during eruptions, which would significantly affect the geoeffectiveness of CMEs. Therefore, understanding the mechanisms that lead to such a rotation and lateral displacement of filaments is a great concern to space weather forecasting. In fact there are three candidates responsible for the rotation of a CME flux rope in the low corona: (1) kink instability (Rust & Kumar 1996); (2) lateral Lorentz force induced by sheared/toroidal magnetic fields (Isenberg & Forbes 2007; Kliem et al. 2012); and (3) magnetic reconnection (Shiota et al. 2010; Guo et al. 2023b). The rotation due to kink instability mainly occurs in twisted flux ropes. In this scenario, the twist of the eruptive flux rope will release to the right due to helicity conservation. Hence, the rotation direction is consistent with the sign of helicity (positive helicity corresponds to clockwise rotation). As for the effects of sheared/toroidal magnetic fields, they can produce a lateral Lorentz force acting upon the two legs of the flux rope, which can easily lead to large-angle rotation, as validated in Kliem et al. (2012). Unlike the effects of kink instability, the sheared magnetic fields can also result in a large-angle rotation of a sheared arcade, even if the twist number is lower than 1. Regarding the last candidate, the magnetic reconnection between the eruptive flux rope and the ambient sheared arcades can directly change the connectivity of their magnetic field lines, potentially leading to rotation. The second event showing a strong rotation of flux rope in the low corona belongs to this category.

We examine an intriguing filament eruption event observed by the Chinese Hα Solar Explorer (CHASE) and the Solar Dynamics Observatory (SDO) on August 18 2022 (Guo et al. 2023b). The filament, which evolves into a CME, exhibits significant lateral drifting during its rising. Moreover, the orientation of the CME flux rope axis deviates from that of the pre-eruptive filament observed in the source region. To investigate the physical processes behind these observations, we perform a data-constrained magnetohydrodynamic (MHD) simulation based on MPI-AMRVAC code after inserting a Regularized Biot-Savart laws (RBSLs) flux rope (Titov et al. 2018). The numerical model reproduces many prominent observational features in the eruption, including the morphology of the eruptive filament, eruption path, and flare ribbons. The simulation results reveal that the magnetic reconnection between the flux rope leg and neighbouring low-lying sheared arcades may be the primary mechanism responsible for the lateral drifting of the filament material (reconnection aa-rf). Such a reconnection geometry leads to flux rope footpoint migration and a reconfiguration of its morphology. Consequently, the filament material hosted in the flux rope drifts laterally, and the CME flux rope deviates from the pre-eruptive filament (Figure 3). This finding underscores the importance of external magnetic reconnection in influencing the orientation of a flux rope axis during eruption. The direction of the CME follows the new direction of the FR. It is interesting to relate the MHD simulation results to the study's conclusion of Chikunova et al. (2023). They conclude that the direction of the eruptive filament did not coincide with the main direction of the dimming expansion direction, which reflects the direction of the CME. The simulation and CHASE observations show that the filament had several reconnections before the CME escaped.

4. Conclusion

During filament/flux rope eruption, many reconnections with surrounding magnetic fields (arcade top or flux rope leg) may occur as suggested by the 3D standard model (Aulanier et al. 2010). It was demonstrated that in such a model, three types of reconnection with the arcades, with the preexisting flux and in the legs of the flux rope (aa-fr, ar-fr, fr-fr) exist (Aulanier & Dudík 2019). Such reconnection implies the rotation of the flux rope during its ejection in the low corona. We show two cases: one case where we suspect that there is a rotation of the FR before its journey in the heliosphere using EUHFORIA simulation, and the second case, the effective rotation in the low corona suspected in the CHASE observations and confirmed by an MHD simulation (MPI-AMRVAC). Rotation of erupting features in the low corona is important for space weather prediction (Vemareddy 2024). Therefore, knowledge of the CME magnetic field is crucial for deriving the correct orientation of the erupting flux rope in the low corona and, therefore, for propagating it further in the heliospheric models for space weather forecasting purposes.

Supplementary material

To view supplementary material for this article, please visit https://doi.org/10.1017/S1743921324001571

References

Aulanier, G., & Dudík, J. 2019, *A&A*, 621, A72
Aulanier, G., Török, T., Démoulin, P., & DeLuca, E. E. 2010, *ApJ*, 708, 314
Bothmer, V., & Schwenn, R. 1998, Annales Geophysicae, 16, 1
Chikunova, G., Podladchikova, T., Dissauer, K., Veronig, A. M., Dumbović, M., Temmer, M., & Dickson, E. C. M. 2023, *A&A*, 678, A166
Guo, J. H., et al. 2024, *ApJ*, 961, 140
Guo, J. H., et al. 2023a, *ApJs*, 266, 3
Guo, J. H., et al. 2023b, *ApJ*, 956, 119
Isavnin, A. 2016, *ApJ*, 833, 267

Isenberg, P. A., & Forbes, T. G. 2007, *ApJ*, 670, 1453
Kliem, B., Török, T., & Thompson, W. T. 2012, *Solar Phys.*, 281, 137
Koehn, G. J., Desai, R. T., Davies, E. E., Forsyth, R. J., Eastwood, J. P., & Poedts, S. 2022, *ApJ*, 941, 139
Lynch, B. J., Antiochos, S. K., Li, Y., Luhmann, J. G., & DeVore, C. R. 2009, *ApJ*, 697, 1918
Maharana, A., Scolini, C., Schmieder, B., & Poedts, S. 2023, A&A, 675, A136
Pomoell, J., & Poedts, S. 2018, Journal of Space Weather and Space Climate, 8, A35
Regnault, F., Strugarek, A., Janvier, M., Auchère, F., Lugaz, N., & Al-Haddad, N. 2023, *A&A*, 670, A14
Rust, D. M., & Kumar, A. 1996, *ApJL*, 464, L199
Scolini, C., et al. 2020, *ApJs*, 247, 21
Shen, C., Xu, M., Wang, Y., Chi, Y., & Luo, B. 2018, *ApJ*, 861, 28
Shiota, D., Kusano, K., Miyoshi, T., & Shibata, K. 2010, *ApJ*, 718, 1305
Titov, V. S., Downs, C., Mikić, Z., Török, T., Linker, J. A., & Caplan, R. M. 2018, *ApJL*, 852, L21
Vemareddy, P. 2024, *ApJ*, 961, 199
Vemareddy, P., Cheng, X., & Ravindra, B. 2016, *ApJ*, 829, 24

Solar and Stellar Coronal Mass Ejections
Proceedings IAU Symposium No. 388, 2026
N. Gopalswamy, O. Malandraki, A. Vidotto & W. Manchester, eds.
doi:10.1017/S1743921324001789

Filament eruption deflection and associated CMEs

K. Koleva[1], R. Chandra[2], P. Duchlev[1] and P. Devi[2]

[1]Space Research and Technology Institute, Bulgarian Academy of Sciences, Sofia, Bulgaria
[2]Department of Physics, DSB Campus, Kumaun University, Nainital

Abstract. We present the observations of a quiescent filament eruption and its deflection from the radial direction. The event occurred in the southern solar hemisphere on 2021 May 9 and was observed by the Atmospheric Imaging Assembly (AIA) on board the Solar Dynamics Observatory (SDO), by the STEREO -A Observatory and GONG. Part of the filament erupted in the west direction, while major part of the filament deviated towards east direction. LASCO observed a very weak CME towards the west direction where it faded quickly. Moreover, the eruption was associated with CME observed by STEREO A COR1 and COR2. Our observations provide the evidence that the filament eruption was highly non-radial in nature.

Keywords. Filament Eruption, Coronal Mass Ejection, deflection

1. Introduction

Solar filament eruptions are one of the most studied phenomena in the field of solar physics. Filaments can erupt fully or partially and these erupted material can be observed in form of coronal mass ejections (CMEs) (Webb *et al.* (1976); Munro *et al.* (1979); Gopalswamy (2000, 2003); Schmieder *et al.* (2013); Seki *et al.* (2021), and references cited therein). There are cases of the deflection of CME during its propagation near to Sun or in the interplanetary space (for example: Zuccarello *et al.* (2012); Koleva *et al.* (2024). The study of the deflection of filament eruption and CMEs can be a crucial study to understand the geoeffectiveness of solar eruptions.

In this work we present the study of a sigmoid filament eruption that occurred in the southern solar hemisphere on 2021 May 9. The kinematics and morphology of the event were investigated by Koleva *et al.* (2021). Here we focused on the associated CME and its offset from the radial propagation.

The used data and observations from two points of view are introduced in Section 2. Section 3 provides our interpretation of the observations.

2. Data and Observations

The prominence eruption occurred at the southern solar hemisphere between 9:30 and 11:00 UT on 2021 May 9.

For the present study, we used the data from the Atmospheric Imaging Assembly (AIA: Lemen *et al.* 2012) on board the Solar Dynamics Observatory (SDO: Pesnell *et al.* 2012) in 304 Å. For a different viewpoint of the event, we used data from the Sun Earth Connection and Heliospheric Investigations (SECCHI: Howard *et al.* 2008) instruments on board the Solar Terrestrial Relations Observatory Ahead (STEREO-A: Kaiser *et al.* 2008) spacecraft. More specifically, data from the Extreme Ultraviolet Imager (EUVI) in

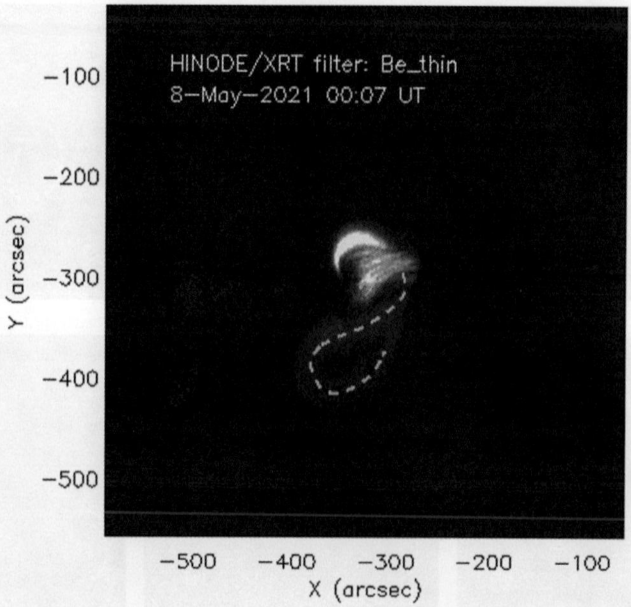

Figure 1. HINODE/XRT observation on 2021 May, 8 at 00:07 UT. The blue dotted line traces the sigmoid structure.

the 304 Å channel, the STEREO-A COR1 inner coronagraph (Thompson *et al.* 2003), and the COR2 outer coronagraph are used in the study.

We used also data obtained by the XRT (Golub *et al.* 2007) onboard HINODE (Kosugi *et al.* 2007) to follow the filament evolution one day before the eruption.

Before the eruption, the sigmoid filament was located in a plage region close to the disk center and laid along the S-shaped magnetic polarity inversion line. A weak X-ray sigmoid was observed in Hinode/XRT Al and Be filters along the filament channel one day before the eruption. In Figure 1 the sigmoid on 8 May 2024, observed in the Be filter image of HINODE/XRT instrument is presented.

2.1. *AIA/SDO FOV*

The eruption began at around 09:30 UT in the AIA field-of-view (FOV). Part of the filament erupted in the west direction, while major part of the filament deviated towards east direction. Before deviation from the radial direction, the erupted filament undergoes a twisting motion.

The eruption evolution is shown in Figure 2 in AIA 304 Å channel. The radial path through the filament loop center at 10:20 UT is indicated by the green dotted line. The filament's loop is traced by the yellow dotted line. The eastward deflection is well observed.

2.2. *EUVI/STEREO A FOV*

The eruption was observed also by STEREO Ahead spacecraft close to the bottom right limb. At the time of observation the separation angle of the STEREO-A spacecraft with Earth was 52°. The filament eruption as observed by EUVI on board the STEREO A spacecraft is shown in Figure 3.

In the COR1 FOV the prominence eruption starts at 10:21 UT at Position angle (PA) 242° and left COR1 FOV at 12:01 UT at PA 247° (see Figure 4, left panel). The deflection

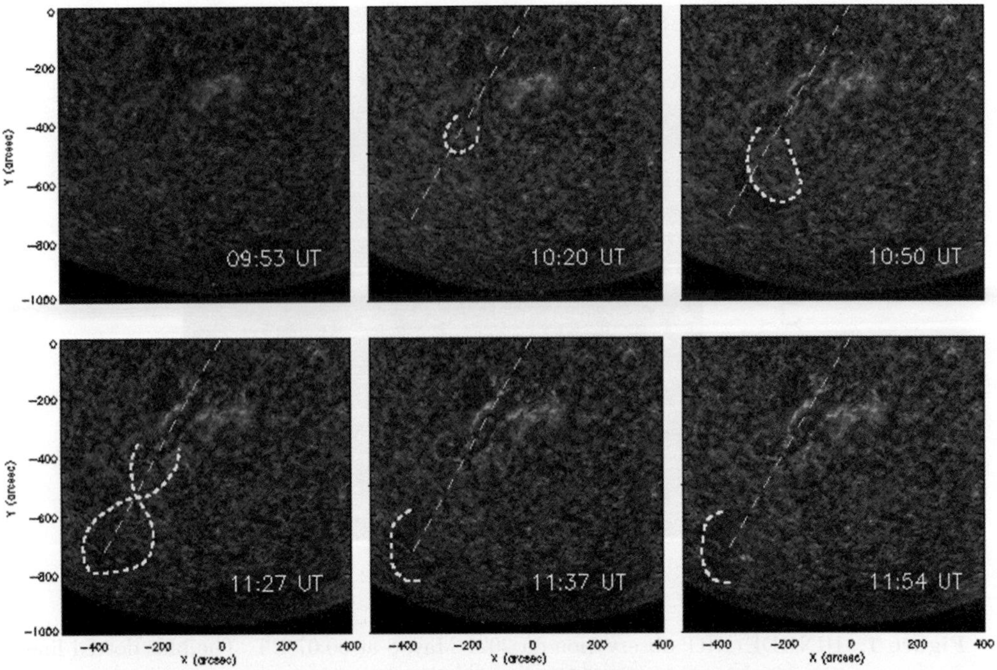

Figure 2. The eruption evolution as seen by AIA/SDO in 304 Å channel. The green dotted line shows the radial path through the filament loop center at 10:20 UT.

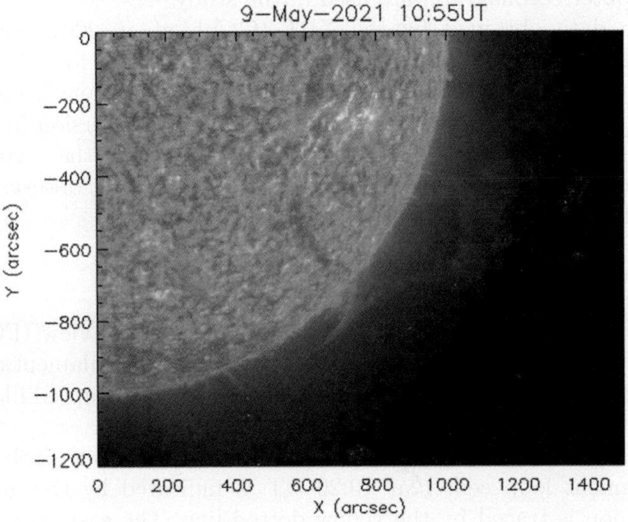

Figure 3. Filament eruption observed by EUVI/STEREO A in 304 Å channel at 10:55 UT.

stopped at about 1.94 solar radii (Figure 4, right panel). After that the core continues its propagation in C2 FOV without any visible deflection.

3. Results

We interpret our result as follows: part of the filament went into west direction in the AIA FOV and appeared as a weak CME in LASCO C2 coronagraph at around 13:25 UT. The major part of the erupted filament interact with the southern situated filament and

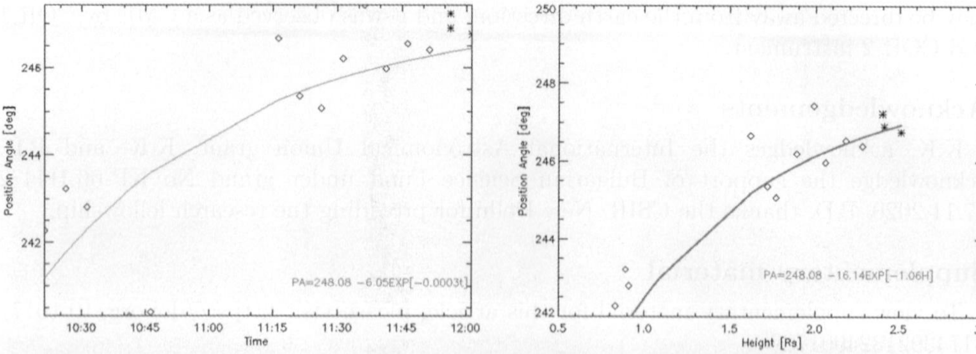

Figure 4. *left:* Variation of the position angle of CME core as a function of time. Diamonds and stars represent measurements from COR1 and COR2 coronagraphs of STEREO A, respectively. *right:* Prominence Position angle (PA) at various heliocentric distances. The solid line is the fit to the data points.

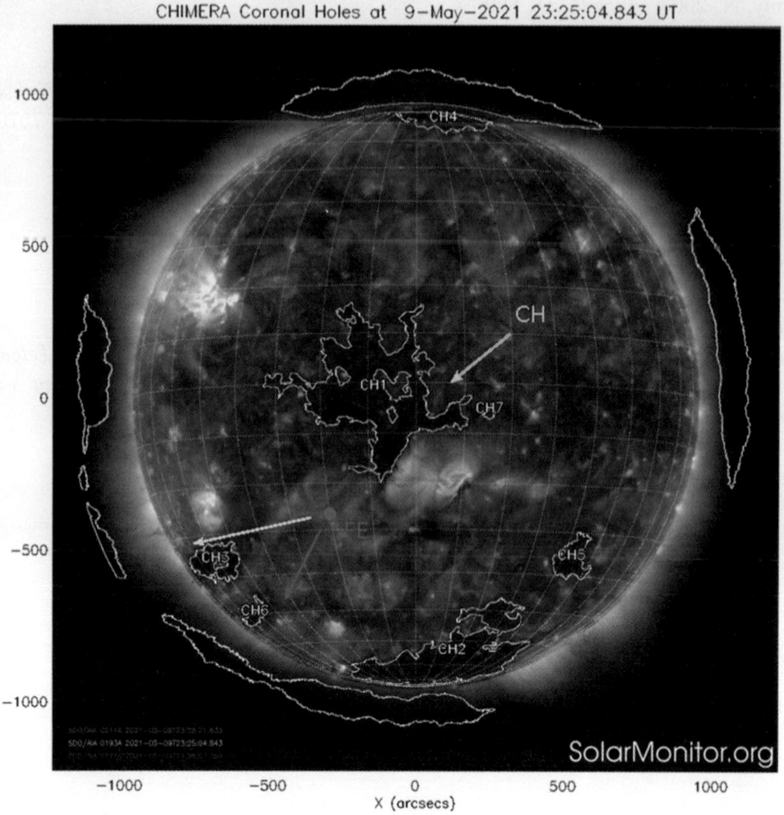

Figure 5. The composite image of three AIA channels: 211 Å, 193 Å and 171 Å at 9 May 2021. The eruption direction at 10:20 UT and 11:54 UT, is indicated by the red and yellow arrows, respectively.

deviated in the eastern direction. The eruption starts at a position angle at about 153°. After deviation the position angle of the prominence loop center was about 145°. In Figure 5 the eruption direction at 10:20 UT is indicated by the red arrow, while the yellow marks the eruption direction at 11:54 UT. The angle between the two arrows, which represents the deflection angle was approximately 8°. The deflected filament part may be directed away from the earth direction, and it was observed as a CME by COR 1 and COR 2 instrument.

Acknowledgements

K.K. acknowledges the International Astronomical Union grant. K.K. and P.D. acknowledge the support of Bulgarian Science Fund under grand No KP-06-H44/2 27.11.2020. P.D. thanks the CSIR, New Delhi for providing the research fellowship.

Supplementary material

To view supplementary material for this article, please visit https://doi.org/10.1017/S1743921324001789

References

Golub, L., Deluca, E., Austin, G., *et al.* 2007, *SoPhys*, 243, 63
Gopalswamy, N., Hanaoka, Y., Hudson, H. S., *et al.* 2000, *AdSpR*, 25, 1851
Gopalswamy, N. 2003, *AdSpR*, 31, 869
Howard, R. A., Moses, J. D., Vourlidas, A., *et al.* 2008, *SSRv*, 136, 67
Kaiser, M.L., Kucera, T.A., Davila, J.M., *et al.* 2008, *Space Sci. Rev.*, 136, 5
Koleva, K., Devi, P., Duchlev, P., *et al.* 2021, *Proceedings of the Thirteenth Workshop "Solar Influences on the Magnetosphere, Ionosphere and Atmosphere"*, ISSN 2367-7570, 34
Koleva, K., Gopalswamy, N., Devi, P., *et al.* 2024, *ApJ*, 966, 22
Kosugi, T., Matsuzaki, K., Sakao, T., *et al.* 2007, *SoPhys*, 243, 3
Lemen, J.R., Title, A.M., Akin, D.J., *et al.* 2012, *SoPhys*, 275, 17
Munro, R. H., Gosling, J. T., Hildner, E., *et al.* 1979, *SoPhys*, 61, 201
Pesnell, W.D., Thompson, B.J., Chamberlin, P.C., *et al.* 2012, *SoPhys*, 275, 3
Seki, Daikichi; Otsuji, Kenichi; Ishii, Takako T. *et al.* 2021, *EP&S*, 73, 58
Schmieder, B., Démoulin, P., Aulanier, G. 2013, *AdSpR*, 51, 1967
Thompson, W.T., Davila, J.M., Fisher, R.R., *et al.* 2003, *Innovative Telescopes and Instrumentation for Solar Astrophysics. Edited by Stephen L. Keil, Sergey V. Avakyan. Proceedings of the SPIE*, 4853, 1
Webb, D. F., Krieger, A. S., Rust, D. M. 1976, *SoPhys*, 48, 159
Zuccarello, F. P., Bemporad, A., Jacobs, C., *et al.* 2012, *ApJ*, 774, 66

Solar and Stellar Coronal Mass Ejections
Proceedings IAU Symposium No. 388, 2026
N. Gopalswamy, O. Malandraki, A. Vidotto & W. Manchester, eds.
doi:10.1017/S1743921324001236

Prominence material observed by SoloHI beyond 0.5 au

C. Mac Cormack[1,2], S. B. Shaik[3,4], P. Hess[4], R. Colaninno[4] and T. Nieves-Chinchilla[1]

[1]Heliospheric Physics Laboratory, Heliophysics Science Division, NASA Goddard Space Flight Center, 8800 Greenbelt Rd., Greenbelt, MD 20770, USA
email: ceciliamaccormack@gmail.com

[2]The Catholic University of America, Washington, DC 20064, USA

[3]George Mason University, Fairfax, VA 22006, USA

[4]U.S. Naval Research Laboratory, Washington, D.C., USA

Abstract. With the combination of observations from in-situ and high-resolution remote sensing instruments, the Solar Orbiter (SolO) mission has become a particularly valuable mission for studying the inner heliosphere. With a field of view (FOV) of 40° to the east of the Sun, the Solar Orbiter Heliospheric Imager (SoloHI) is one of the six remote sensing instruments on board the Solar Orbiter (SolO) spacecraft. SoloHI's high-resolution imaging observations of the heliosphere and its higher cadence than previous generations of heliospheric imagers make it a perfect candidate to perform and complement studies of CMEs evolution through the heliosphere. In this work, we present the first prominence material detected by SoloHI along with other remote sensing instruments. We report that as the CME propagates out in the heliosphere, the associated filament material reaches a heliocentric height of ~122.5 $R_\odot$ (~0.57 au).

Keywords. Prominences, coronal mass ejections (CMEs), heliosphere, Filament material, Sun: activity

1. Introduction

Coronal mass ejections (CMEs) are known for being one of the main drivers of heliospheric variability. CMEs exhibit highly structured magnetic fields that play a critical role in their dynamics and evolution. These magnetic structures can vary broadly in shape, ranging from confined loop eruptions to complex helical configurations that disrupt large regions of the interplanetary medium. In these structures, we can often identify three main parts: a bright leading edge, a dark cavity, and a bright core following the cavity (Howard *et al.* 1985; Illing and Hundhausen 1985). The brightness of the core is mostly related to the associated prominence (or filament) material erupted from the Sun. If it is dense enough, the bright material can be observed in its evolution through the white-light coronagraphs (Srivastava *et al.* 2000), and becomes weaker as long as it propagates in the heliosphere. Some prominences can withstand the forces and magnetic reconfiguration associated with the eruptions and retain their structure as they propagate out in the interplanetary medium. These prominences that survive appear bright in white-light images and were observed at different heights up to 1 au (Wood, Howard, and Linton 2016; Howard 2015a), and further (Howard 2015b), by the heliospheric imagers on board the Solar-Terrestrial Relations Observatory mission (STEREO; Kaiser *et al.* 2008).

Newer solar missions have integrated high-resolution instruments that are able to fill some pre-existing gaps by providing observations closer to the Sun and out of the ecliptic plane. The Solar Orbiter Heliospheric Imager (SoloHI; Howard *et al.* 2020), onboard the Solar Orbiter (SolO; Müller *et al.* 2020) mission, provides white-light images that allow us to track specific features detected before the CME erupts through the solar corona and the heliosphere. SoloHI is a single telescope made up of four distinct tiles with a typical field of view (FOV) of $5° - 45°$ off the east limb of the Sun relative to the Solar Orbiter. SoloHI has a spatial resolution comparable to that of STEREO HI-1 (HI-1; Eyles *et al.* 2009), but because the orbit of the spacecraft reaches heights to within a third of an AU, its effective resolution is significantly greater when the spacecraft is close to the Sun (Hess *et al.* 2023).

From the beginning of the mission until April 2023, SoloHI has detected 140 large-scale structure events, and they have been cataloged with complementary information about their whole evolution through the heliosphere. On 2022 September 23, SoloHI detected, for the first time, a prominence material that evolves in its entire FOV with the associated CME. The filament being observed through the FOV allows us to study the evolution of the CME as a whole and probe the underlying magnetic structures of the event with unprecedented image resolution beyond other available coronagraphs FOV. By combining SoloHI observations with other remote sensing data, along with the CME, we track this prominence material from its early evolution until it appears to completely diffused.

For examining the source region, we use extreme ultraviolet (EUV) images provided by the Extreme Ultraviolet Imager (EUI; Rochus *et al.* 2020) on board SolO, the Atmospheric Imaging Assembly (AIA; Lemen *et al.* 2012) on board the Solar Dynamic Observatory (SDO; Pesnell, Thompson, and Chamberlin 2012) from Earth's point of view (POV), and the Extreme Ultraviolet Imager (EUVI) on board STEREO-A (ST-A hereafter) of the Sun-Earth Connection Coronal and Heliospheric Investigation (SECCHI; Howard *et al.* 2020) suite of instruments. To understand the coronal evolution of the prominence material, we use the observations from the Large Angle Spectroscopic Coronagraph (LASCO; Brueckner *et al.* 1995) coronagraphs (C2 and C3) on board the Solar and Heliospheric Observatory (SOHO; Domingo, Fleck, and Poland 1995) mission and the COR1 and COR2 coronagraphs on board ST-A. To track the propagation of the CME through the heliosphere, we use SoloHI, HI-1, and HI-2 heliospheric imagers. In Section 2, we describe the main features detected by SoloHI for this event. In Section 3, we discuss the source region and the CME early evolution. In Section 4, we report observations in the upper corona and heliosphere and their comparison with the observations made by SoloHI in each case. Finally, conclusions are presented in Section 5.

2. 2022 September 23 event: SoloHI Observations

On 2022 September 23, prominence material was detected in the SoloHI FOV. Panel a of Figure 1 shows the spacecraft position at 16:00 UT during the detection of the CME in SoloHI. The black arrow denotes the longitudinal direction of the CME propagating at 68° of longitude. At the time of the detection of the CME in the SoloHI FOV, SolO was located at $\sim$0.49 au. The first appearance of the associated CME in the SoloHI FOV was around 15:52:41 UT. Mainly evolving in the ecliptic plane, the CME displays a well-defined leading edge and streamers on the sides. Although the CME evolves in the whole SoloHI FOV, the prominence material is only detected in the bottom tiles within the detector, as indicated in panel b on Figure 1. A well-defined cavity can be identified below the prominence material. In the Figure, we denote the position of the filament with a yellow arrow, and a curved line indicates the low-density cavity. A movie of the CME

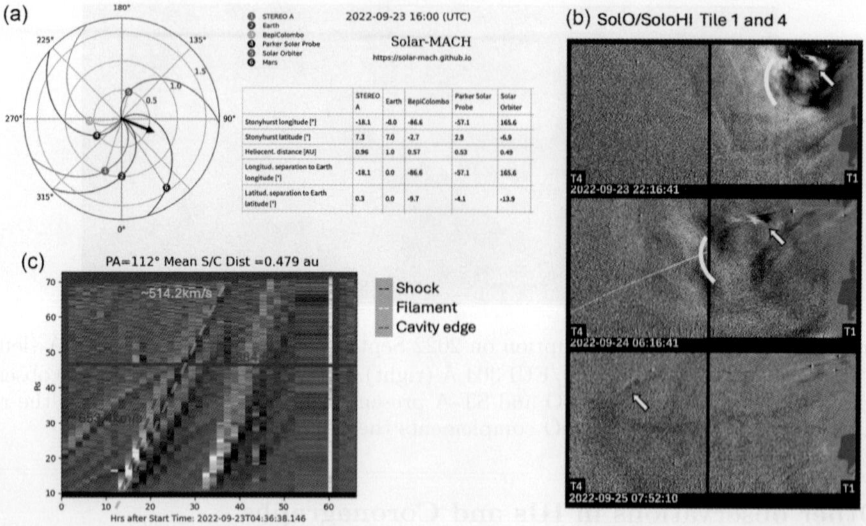

Figure 1. Panel a: Spacecraft position at the moments of the CME detection in SoloHI FOV. The black arrow indicates the longitudinal propagation of the CME. The details of the position of the spacecraft are also shown in this panel. Panel b: Bottom tiles of SoloHI observations on 2022 September 23 at three different times in the CME evolution. The filament material is denoted by a yellow arrow, and the cavity is indicated by the yellow line. Panel c: J-map of SoloHI observations from September 23 until September 25. Dashed violet, yellow, and red lines indicate the CME shock, filament material, and back part of the cavity, respectively. Projected velocities and estimated heights on the plane of the sky are marked.

evolution in the SoloHI FOV is attached to this article. The filament seems to vanish during its evolution in tile 4 on September 25 at 07:52:16 UT. Panel c of Figure 1 shows a J-map made with SoloHI observations from September 23 until September 25. The evolution of the mentioned structures can be clearly identified, and a projected velocity is estimated (as indicated by the values within the panel). Although it disappears during the initial evolution of the CME in the SoloHI FOV, the shock is slightly faster than the filament. In particular, we estimated the height of the filament on the plane of the sky, being detected up to 70 $R_\odot$ or around 37° of elongation in the SoloHI FOV. Knowing the angle of propagation of the CME, we can detect the filament up to the non-projected height of 62 $R_\odot$. Multi-viewpoint analysis can provide a better description of the propagation of the filament.

3. Source Region of the Eruption

The FOVs of SDO, STEREO-A, and SolO complement one another effectively for this event. In particular, the source can be observed from Earth's POV with the SDO mission to be NOAA active region AR13102 located at south 25° and west 67°. The filament starts its formation days before the eruption, but during September 22 and 23, it gets thicker and twisted, and finally erupts on 2022 September 23 at ~13:25:53 UT. Figure 2 shows the moment of the eruption detected from the POV of SDO/AIA, ST-A/EUVI, and SolO/EUI with the 304 Å band-pass wavelength. It is easy to correlate and identify the same features in the EUV images, such as the concave structure at the highest point and the brightening leg of the filament in the AIA and EUVI images, as pointed out in the Figure.

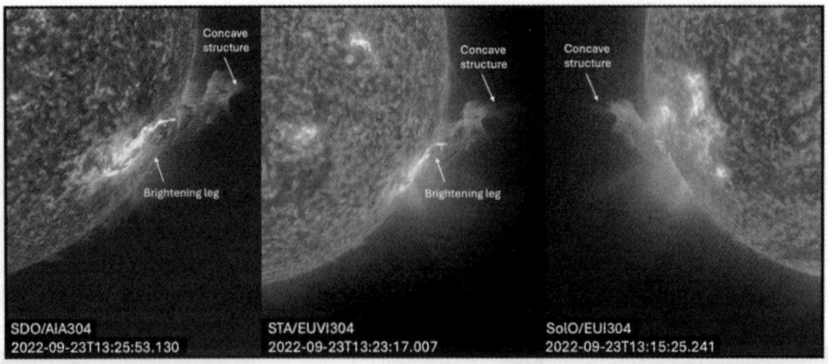

Figure 2. Observations of the eruption on 2022 September 23 by SDO/AIA 304 Å (left), ST-A/EUVI 304 Å (middle), and SolO/EUI 304 Å (right). The erupted material can be observed in each of the three POVs. While SDO and ST-A present similar observations due to the relative similarity of their perspectives, SolO complements them on the opposite side.

4. Other observations in HIs and Coronagraphs

Figure 3 shows the observations made by different remote sensing instruments on board ST-A and SOHO for the 2022 September 23 event. In all cases, the prominence material is denoted by a yellow arrow. The yellow line indicates the low-density cavity next to the prominence, and the green line indicates the leading edge of the associated CME. Panel a shows the first stage of the evolution from the FOV of ST-A/EUVI 304 Å at 13:23:17 UT. As we mentioned before, the eruption starts around 13:25:53 UT, and the filament material can be tracked through the entire EUVI FOV (up to 1.7 $R_\odot$). Panel b presents the eruption observed by SolO/EUI 304 Å at 14:30:25 UT. The image is flipped to provide a more direct comparison with the other viewpoints. With the high resolution and wide FOV of the EUI instrument, the filament can be tracked up to 3.8° at the west of the Sun (6.984 $R_\odot$ on the plane of the sky).

EUI observations can be compared with white-light ST-A/COR1 detection shown on panel c of Figure 3 at 14:31:18 UT. This is a complementary FOV between SolO and ST-A, and it can be seen that similar structures are identified with better resolution in EUI. The event appears in the COR1 FOV around 13:31:18 UT, and the leading edge exits the COR1 FOV at 15:36:18 UT. In the image, we detect a low-density cavity (pointed with the yellow line in the Figure) and the limits of the leading edge (green line). Panel d shows SOHO/LASCO-C2 observations at 15:36:07 UT. The prominence material seems to be more elongated and travels next to the low-density cavity detected in the COR1 observations. The leading edge of the CME is also denoted by green lines. The CME appears in the LASCO-C2 FOV around 14:00:07 UT, and it is no longer detected at 17:12:07 UT. Panel e of Figure 3 shows ST-A/COR2 observation at 17:38:43 UT, where we can identify the same structures denoted in LASCO-C2 observations. The CME appears at 13:53:43 UT in COR2 FOV, and the filament reaches the FOV limit at 21:23:43 UT. In panel f, we show the SOHO/LASCO-C3 observations at 17:42:07 UT. The elongated filament can be detected as long as the low-density cavity and leading edge.

The LASCO-C3 FOV overlaps with SoloHI and ST-A/HI-1, as shown in Figure 4. Panel a shows the prominence material observed by LASCO-C3 at 23:30:07 UT. It still appears as an elongated filament that travels next to the low-density cavity previously identified. The yellow square represents the SoloHI FOV, and the orange square represents the ST-A/HI-1 FOV. Panel b and c represent the CME detected in SoloHI (at 22:16:41 UT) and ST-A/HI-1 (at 23:08:00 UT), respectively. The yellow arrow points to the prominence material detected in all instruments. In both cases we can identify the

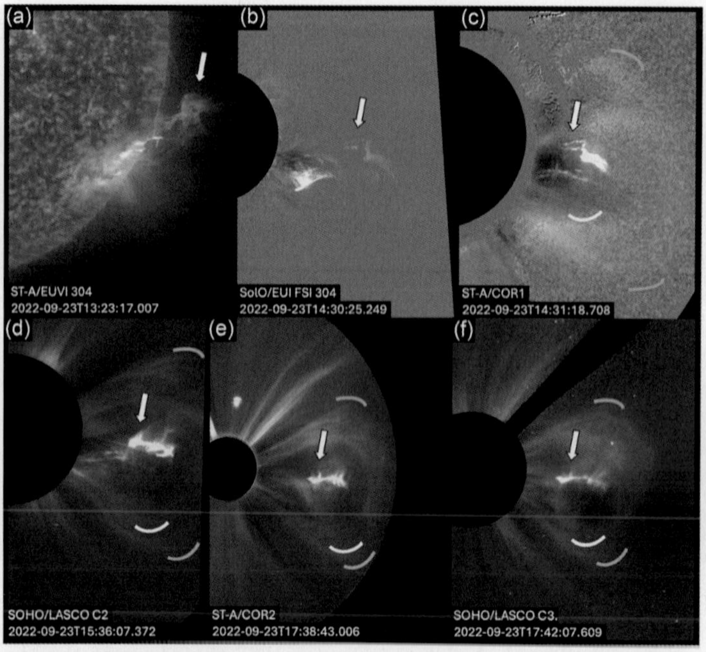

Figure 3. Evolution of the prominence material observed by different instruments on September 23, 2022. Panel a: ST-A/EUVI 304 Å observations at 13:23:17 UT. Panel b: Running difference image of SolO/EUI 304 Å observations at 14:30:25 UT. The image is flipped for an easier comparison with the other instruments' FOVs. Panel c: ST-A/COR1 observations at 14:31:18 UT (also running difference image). Panel d (f): SOHO/LASCO C2 (C3) observations at 15:36:07 UT (17:42:07 UT). Panel e: ST-A/COR2 observations of the prominence material at 17:38:43 UT. In all cases, a yellow arrow indicates the prominence material, and colored lines denote the limits of the cavity and the leading edge of the CME in the coronagraphs.

elongated prominence material in the lower limit of the FOVs. SoloHI allows us to better understand the internal structure of the CME as long as the shape of the prominence and the shock structure. While the ST-A/HI-1 image shows a clear front and internal structure, SoloHI reveals a clear V-shaped front and the circular flux rope still visible with the filament at the base in the first stage of the evolution. Moreover, in SoloHI movies, different structures in the elongated filament can still be identified, while the ST-A/HI-1 images are more blurred. In Hess *et al.* (2023), the authors perform a comparison between the observations made by SoloHI and ST-A/HI-1, highlighting the new structures that SoloHI can describe thanks to its improved resolution.

The event is also observed by ST-A/HI-2. The CME appears in HI-2 FOV on 2022 September 24 at 14:00 UT and becomes diffuse on September 25 at around 22:08 UT, as shown in Panel b on Figure 4. The structure seems to reach 35° of elongation on the ST-A/HI-2 plane of the sky. Combining this information with the location of the spacecraft at that moment, we conclude that the filament reached the height of $\sim$122.5 $R_\odot$ ($\sim$0.57 au).

5. Conclusions

We presented the first prominence material detected by SoloHI. The filament was observed at different stages of its evolution by different missions that conform to the heliophysics system of observatories. We compare the observations provided by coronagraphs and heliospheric imagers along its trajectory and track it up to a heliocentric distance of $\sim$122.5 $R_\odot$.

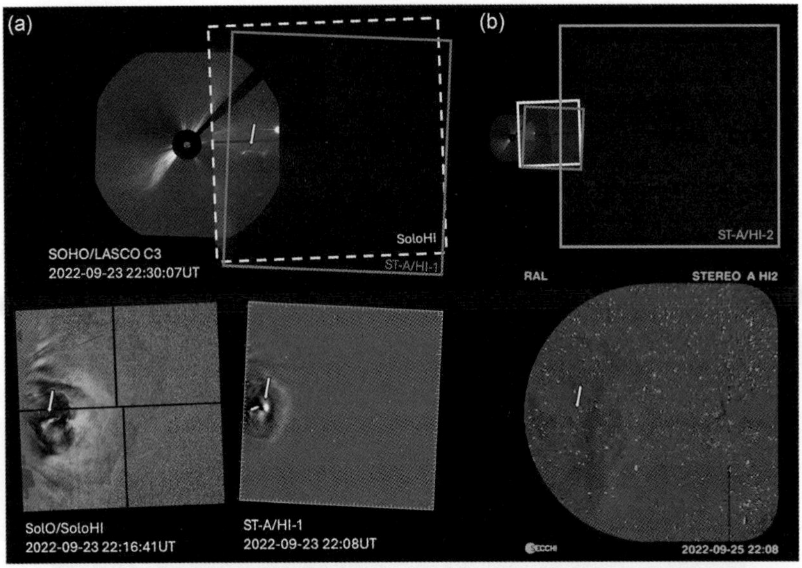

Figure 4. Panel a) Top: LASCO C3 observation of the CME on 2022 September 23 at 23:30:07 UT overlaid with the complementary FOVs of SolO/SoloHI (yellow square) and ST-A/HI-1 (orange square). The dashed line indicates that the plane of the sky for SoloHI is behind ST-A/HI-1 FOV. Bottom: CME detected in SoloHI (at 22:16:41 UT) and ST-A/HI-1 (at 22:08:00 UT), respectively. Panel b) Top: Same as top panel a but adding ST-A/HI-2 FOV. Bottom: Filament detected by ST-A/HI-2 at 22:08:00 UT on September 25, 2022. The yellow arrow points to the prominence material detected in all instruments.

With its high-resolution images and complementary FOV, SoloHI became crucial in the multi-view analysis not only for the filament tracking along the heliosphere but also in implementing models and further analyzing the associated CME. Since the loss of STEREO-B in 2014, a gap in the multi-viewpoint studies significantly limited the study of solar transients. The Solar Orbiter mission is playing an important role in filling this gap with new orbits closer to the Sun and, eventually, out of the ecliptic.

Although in the case presented in this study there were no in-situ signals of the CME detected on Earth or Mars, the variability of the Sun can have a strong impact on its surroundings and the heliosphere in general. The consequences of these phenomena have become increasingly relevant over the years with space exploration. The development of space-based technology and the upcoming NASA human explorations make understanding of space weather hazards a crucial task. The Solar Orbiter mission, and SoloHI in particular, is contributing key information for a better understanding of the evolution of large-scale structures in the heliosphere.

Supplementary material

To view supplementary material for this article, please visit https://doi.org/10.1017/S1743921324001236

References

Brueckner, G.E., Howard, R.A., Koomen, M.J., Korendyke, C.M., Michels, D.J., Moses, J.D., and, ...: 1995, *Solar Physics* **162**, 357. doi:10.1007/BF00733434.

Domingo, V., Fleck, B., and Poland, A.I.: 1995, *Solar Physics* **162**, 1. doi:10.1007/BF00733425.

Eyles, C.J., Harrison, R.A., Davis, C.J., Waltham, N.R., Shaughnessy, B.M., Mapson-Menard, H.C.A., and, ...: 2009, *Solar Physics* **254**, 387. doi:10.1007/s11207-008-9299-0.

Hess, P., Colaninno, R.C., Vourlidas, A., Howard, R.A., and Stenborg, G.: 2023, *Astronomy and Astrophysics* **679**, A149. doi:10.1051/0004-6361/202346907.

Howard, R.A., Vourlidas, A., Colaninno, R.C., Korendyke, C.M., Plunkett, S.P., Carter, M.T., and, ...: 2020, *Astronomy and Astrophysics* **642**, A13. doi:10.1051/0004-6361/201935202.

Howard, R.A., Sheeley, N.R., Michels, D.J., and Koomen, M.J.: 1985, *Journal of Geophysical Research* **90**, 8173. doi:10.1029/JA090iA09p08173.

Howard, T.A.: 2015a, *The Astrophysical Journal* **806**, 175. doi:10.1088/0004-637X/806/2/175.

Howard, T.A.: 2015b, *The Astrophysical Journal* **806**, 176. doi:10.1088/0004-637X/806/2/176.

Howard, R.A., Moses, J.D., Vourlidas, A., Newmark, J.S., Socker, D.G., Plunkett, S.P., and, ...: 2008, *Space Science Reviews* **136**, 67. doi:10.1007/s11214-008-9341-4.

Illing, R.M.E. and Hundhausen, A.J.: 1985, *Journal of Geophysical Research* **90**, 275. doi:10.1029/JA090iA01p00275.

Kaiser, M.L., Kucera, T.A., Davila, J.M., St. Cyr, O.C., Guhathakurta, M., and Christian, E.: 2008, *Space Science Reviews* **136**, 5. doi:10.1007/s11214-007-9277-0.

Lemen, J.R., Title, A.M., Akin, D.J., Boerner, P.F., Chou, C., Drake, J.F., and, ...: 2012, *Solar Physics* **275**, 17. doi:10.1007/s11207-011-9776-8.

Müller, D., St. Cyr, O.C., Zouganelis, I., Gilbert, H.R., Marsden, R., Nieves-Chinchilla, T., and, ...: 2020, *Astronomy and Astrophysics* **642**, A1. doi:10.1051/0004-6361/202038467.

Pesnell, W.D., Thompson, B.J., and Chamberlin, P.C.: 2012, *Solar Physics* **275**, 3. doi:10.1007/s11207-011-9841-3.

Rochus, P., Auchère, F., Berghmans, D., Harra, L., Schmutz, W., Schühle, U., and, ...: 2020, *Astronomy and Astrophysics* **642**, A8. doi:10.1051/0004-6361/201936663.

Srivastava, N., Schwenn, R., Inhester, B., Martin, S.F., and Hanaoka, Y.: 2000, *The Astrophysical Journal* **534**, 468. doi:10.1086/308749.

Wood, B.E., Howard, R.A., and Linton, M.G.: 2016, *The Astrophysical Journal* **816**, 67. doi:10.3847/0004-637X/816/2/67.

Solar and Stellar Coronal Mass Ejections
Proceedings IAU Symposium No. 388, 2026
N. Gopalswamy, O. Malandraki, A. Vidotto & W. Manchester, eds.
doi:10.1017/S1743921324001315

On Using Gradient Dynamic Spectra (GraDS) to Study Type-II Solar Radio Bursts

Puja Majee[1], Devojyoti Kansabanik[2]† and Divya Oberoi[1]

[1]National Centre for Radio Astrophysics, Tata Institute of Fundamental Research, S. P. Pune University Campus, Pune, India, 411007

[2]Cooperative Programs for the Advancement of Earth System Science, University Corporation for Atmospheric Research, Boulder, CO, USA

Abstract. Solar type-II radio bursts are coherent plasma emissions arising from magnetohydrodynamic shocks produced by either coronal mass ejections (CMEs) or flares. Type-II bursts sometimes show split-band emissions in the dynamic spectrum. When these split-band emissions come from regions just upstream and downstream of the shock, type-II band-splitting can be used as an important tool for estimating magnetic fields at the shock front. Earlier studies have shown that only ~20% of the type-IIs show morphologically similar split-bands. Imaging studies can unambiguously identify such instances, though they remain very rare. Here we suggest a useful approach to augment dynamic spectra based studies by also examining the Gradient Dynamic Spectra (GraDS) of type-II emission. We also verified the conclusions of this approach against those from an imaging study.

Keywords. Coronal Mass Ejections, Shocks, Solar Radio Burst, Radio Imaging

---

1. Introduction

Type-II solar radio bursts are plasma emissions generated by magnetohydrodynamic (MHD) shocks driven by energetic eruptions from the Sun, such as flares and coronal mass ejections (CMEs). The MHD shocks accelerate electrons to high speeds which excite Langmuir waves, eventually resulting in plasma emissions at the fundamental (F) and harmonic (H) of the local plasma frequency (McLean & Labrum 1985). Type-II emissions have been observed from metric (m-), decametric (Dm-) to kilometric (km-) wavelengths arising over a wide range of coronal heights and interplanetary medium. Association with shocks of large-scale solar eruptions makes studying type-IIs extremely attractive from the perspective of understanding space weather, particularly for determining solar energetic particle (SEP) events. Several studies have concluded that metric type-IIs mostly originate from CME-driven shocks (e.g., Gopalswamy 2006; Kumari et al. 2023) and hence can be used to understand the energetics, kinematics, and dynamics of the shocks produced by the CMEs.

Type-II bursts often appear as two (F and H) slowly drifting bands (with a typical drift rate $\lesssim 1$MH/s) in the dynamic spectrum. Each of these bands is often seen to be split into two or multiple parallel lanes, known as band-splitting and multi-lane emissions, respectively. Type-IIs can also show short-lived, fragmented fine structures resulting from the interaction between the shock and the inhomogeneities in the coronal density

† NASA Jack Eddy fellow hosted at the Johns Hopkins University Applied Physics Laboratory, USA

structures (Magdalenić et al. 2020). Despite being known for several decades, the origin of the band-splitting is still under debate. Several interpretations have been presented in this regard. One of the leading hypotheses, proposed by Smerd et al. (1974), is the so-called "upstream/downstream scenario". According to this interpretation, band-splitting occurs when shock-accelerated electrons travel both ahead and behind the shock front. They produce plasma emission at slightly different frequencies due to different electron densities in the upstream (ahead) and downstream (behind) regions of the shock front. Another interpretation is that the emissions originate from different parts of the shock that propagate through the coronal medium with different densities (McLean 1967). An attractive aspect of the upstream/downstream scenario is that it allows one to determine the shock properties (i.e. shock Mach number, density jump), ambient plasma properties (i.e. Alfvén velocity, local electron density), magnetic field strength even in the upper corona and inner heliosphere where few remote observational probes to estimate the physical properties of this magnetized plasma medium are available (Smerd et al. 1974, 1975).

In a study of coronal and upper coronal (UC)/interplanetary (IP) type-II bursts, Vršnak et al. (2001, 2002, 2004) have estimated the shock properties, Alfvén speed, magnetic field strength, and their evolution with coronal height using the band-splitting of type-II emissions under the assumption of upstream/downstream scenario. To avoid the possibility of the coincidence of two different emission lanes originating from different parts of the large extended shock front, the authors laid down some detailed criteria for a type-II emission to be regarded as band-splitting. These criteria are – 1. synchronized temporal and spectral evolution; 2. similar intensity variation 3. frequency ratio differs from 2. They imply a physical scenario where the two lanes originate from two closely spaced radio sources moving in tandem through coronal regions with similar conditions, except for slightly different electron densities. In a sample of 112 metric type-IIs and 139 of UC/IP type-IIs, the authors reported 20% of metric type-IIs and 17% of UC/IP type-IIs showing band-splitting emissions satisfying these criteria.

Many studies have used the band-splitting of type-II bursts to estimate the properties of shocks, Alfvén waves, and magnetic field strengths and try to understand the conditions for the occurrence of type-II bursts (e.g., Cho et al. 2011; Ramesh & Kathiravan 2022; Ndacyayisenga et al. 2023, etc.). However, establishing that a given type-II does indeed have band-splitting based on the detailed criteria given by Vršnak et al. (2001) has been done in a few studies, potentially leading to incorrect conclusions. While the dynamic spectra (DS) cannot provide any spatial information about the emission, a careful examination of the emission lanes can indeed help in exploring the possibility of the radio sources being located in the regions upstream and downstream of the shock.

Instead of visual identification of morphological similarities based on the dynamic spectrum, which can have significant human bias, we tried to develop a systematic approach of identifying the upstream/downstream scenario from the dynamic spectrum to reduce this human bias. Section 2 describes the observation and proposed method. Section 3 presents the results and discussion followed by the conclusion from this work in Section 5.

2. Observation and Methodology for Identification of Band-splitting Scenario

Here, we present a type-II solar radio burst from 28 September 2014 observed with the Murchison Widefield Array (MWA; Tingay et al. 2013) covering frequency range 80 – 133 MHz with a temporal and spectral resolution of 0.5 seconds and 40 kHz, respectively. A detailed spectroscopic snapshot imaging study of this type-II has been presented in Bhunia et al. (2023). To cover a larger spectral span, we use dynamic spectrum from

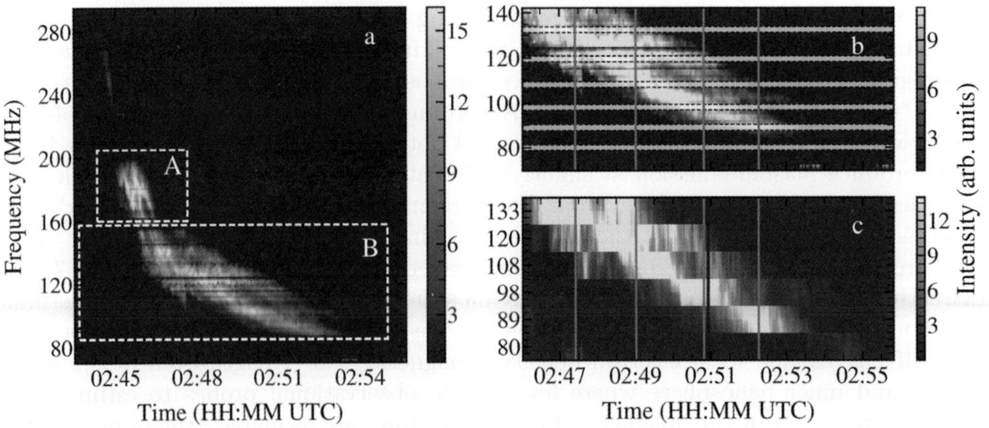

Figure 1. Left panel (a): The dynamic spectrum of the type-II harmonic emission captured by HiRAS-2 spectrograph. The two white dashed boxes (A and B) show the parts of the emission that have been examined in detail to distinguish emission lanes using gradient analysis. Top right panel (b): The lower part of the HiRAS-2 dynamic spectrum overlaps with the MWA observations in six spectral bands highlighted with white rectangles. Bottom right panel (c): The MWA DS in the six spectral bands. The red solid lines in the right panels indicate the four timestamps corresponding to four panels in Figure 3.

the Hiraiso Radio Spectrograph (HiRAS) (Kondo et al. 1995), which monitors the Sun over the wide frequency range from 25–2500 MHz. HiRAS comprises three independent antenna systems covering different spectral ranges (HiRAS-1: 25-70 MHz; HiRAS-2: 50-550 MHz; HiRAS-3: 500-2500 MHz). Here, we have used HiRAS-2 which provides a temporal and spectral resolution of 1 second and 0.5 MHz, respectively. Figure 1(a) shows the HiRAS dynamic spectrum of the harmonic emission. To make the emission lanes stand out visually, boxcar smoothing has been done over 5 seconds and 2 MHz. Figure 1(b) and (c) zoom in to show the six spectral bands, each 2.56 MHz wide, observed by the MWA, and the MWA DS respectively.

Identifying morphological similarities between type-II emission lanes can sometimes be difficult due to varying intensities across the lanes and the intrinsic complexity of the emission, in addition to limitations arising from SNR and loss of data due to flagging. Based on the idea that the locus of an emission lane should correspond to the local maximum in the dynamic spectrum, which would, in turn, correspond to a zero crossing in the gradient of the dynamic spectrum, irrespective of the strength of the emission. Hence, in order to systematically identify morphological similarities across lanes, we also examine the dynamic spectrum of the gradient of intensity alongside the usual DS. Gradient across time is taken for individual spectral channels to produce the gradient DS (GraDS). As shown in 2, the local maxima, i.e. the transition between the positive (red) to negative (blue) gradients indicate the peak of the emission. Green curves connecting the local maxima have been drawn to make the emission lanes easier to follow. The drift rate and morphology of the harmonic emission changes across the burst. Hence, we divide the emission into two parts, marked by the white dashed boxes in Figure 1(a), and carry out the gradient analysis separately.

3. Results and Comparison with Spectroscopic Imaging using the MWA

As shown in Figure 2, the GraDS unambiguously marks the transition from the rising (red) to the falling (blue) part of the dynamic spectrum. It systematically identifies the

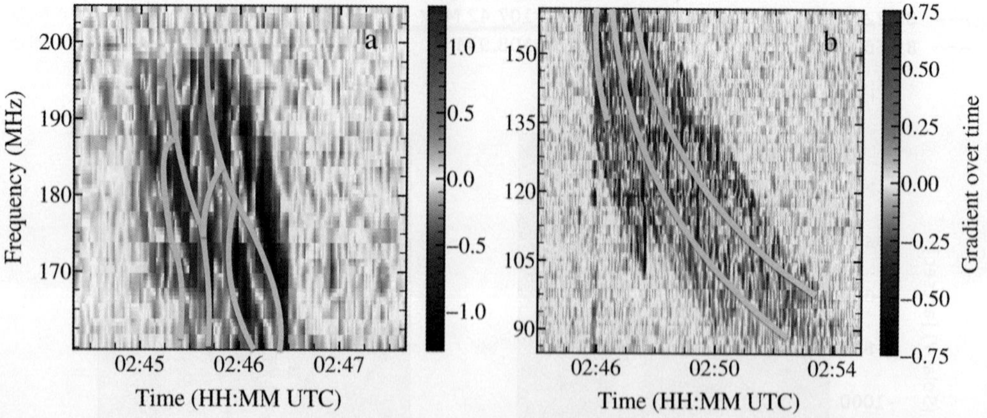

Figure 2. The GraDS of the harmonic emissions, marked by boxes A and B in Figure 1(a), are shown in the left (a) and right (b) panels, respectively. The green curves mark the positive-to-negative transitions indicating the peaks of emissions.

spectro-temporal location of the peaks of the emission lanes even when identifying them directly from the dynamic spectrum is not straightforward. Panels (a) and (b) of Figure 2 show the GraDS of the regions marked by the upper (A) and lower (B) white dashed boxes in Figure 1(a). The locations of these peaks of the emission lanes are marked by green curves in Figure 2 to guide the eye. Figure 2(a) starts with two well-formed parallel lanes, they become non-parallel and branch as the type-II proceeds.

The GraDS of the type-II marked box B in Figure 1(a) is shown in Figure 2(b). Three large-scale positive-to-negative transition regions are identified in the earlier part of the emission. However, in the later part, only two lanes can be identified with similar frequency drifts. It also shows numerous positive-to-negative transitions confined to small spectral and temporal extents, suggesting the presence of multiple narrow-band short-lived fine structures. A comparison of the two panels of Figure 2 brings out the remarkable difference in the nature of emission between them.

The above exercise shows that this type-II emission does not show band-splitting with morphologically similar emission lanes. Rather it has a complex morphology and for some part has emission lanes that can not be identified clearly. While the harmonic emission marked with box A in Figure 1(a) shows a clear signature for the presence of multiple lanes, the emission marked by box B in Figure 1(a) does show two lanes after the first few minutes. The presence of multiple lanes in the earlier part of the emission suggests that the corresponding sources likely lie at different parts of the shock front. The multiple lanes with different frequency drift rates also suggest that the multiple sites of type-II emission are likely moving with different shock speeds. The presence of two bands with similar drift rates after about 02:49 UTC, however, does imply that the emission is coming from regions with similar shock speeds.

In their study of this type-II, Bhunia et al. (2023) used the visually estimated similarity of the two lanes to regard this as a case of split-band emission. Their imaging analysis, however, revealed radio sources corresponding to the two lanes to be located far apart and moving in different directions with different speeds. This led them to conclude that this event is inconsistent with expectations from the upstream-downstream scenario. Having started from the hypothesis of band-splitting, Bhunia et al. (2023) limited themselves to study the location of radio sources corresponding to the peaks of the two split bands. As shown in Figure 2, the emission shows complex morphology, and we have just presented

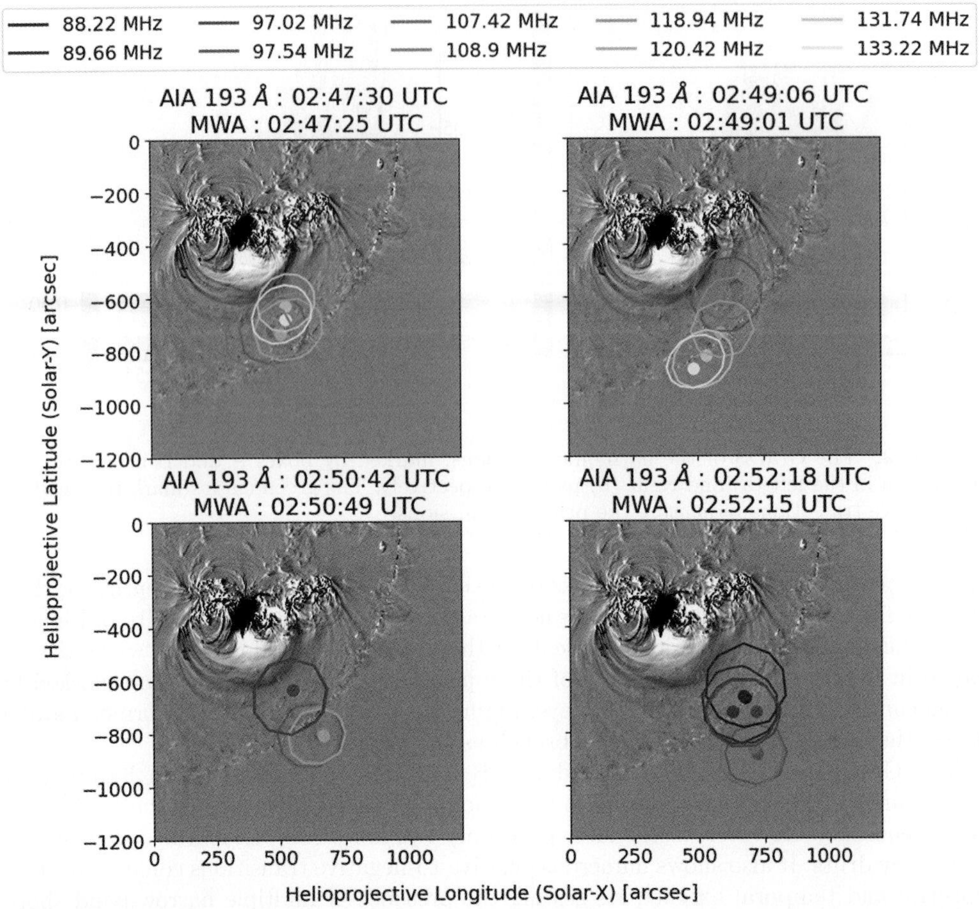

Figure 3. Each of the panels shows locations of the type-II source at different frequencies obtained from MWA spectroscopic snapshot images, overlaid on the AIA base difference images at 193 Å at nearby times. Sources at different radio frequencies are shown in different colors and times corresponding to different panels are marked in Figure 1(c). The contours and circles correspond to the 90% of peak intensity and the peak of the best-fit Gaussian used to model the source in MWA radio images. Uncertainties in the location of the peak of the Gaussian source are ∼1arcsec and are too small to be visible in this plot.

evidence for type-II emission to be coming from different parts of the shock. It would, hence, be interesting to investigate the locations of the sources corresponding to the full spectral extent of the emission across time. This exercise can potentially shed light on the type-II source regions as well as their evolution with time. Figure 3 shows the locations of the type-II radio sources obtained by spectroscopic snapshot imaging using the MWA. Due to the high signal-to-noise ratio (SNR), the astrometric accuracy of the location of these sources is much smaller than the point-spread-function (PSF) of the instrument. Astrometric accuracy is given as, $\frac{PSF}{2\times SNR}$, which is ~ 1 arcsecond, and the separation between the sources at different frequencies is much larger than the astrometric accuracy. This clearly shows that the type-II sources at different frequencies are spatially separated and confirms that they could not arise from the upstream-downstream scenario. The locations of these sources do not seem to be random but rather follow well-defined spectral trends, which are currently under investigation.

4. Conclusion

Type-II solar radio bursts are signatures of shocks. Type-II band-splitting satisfying the upstream-downstream scenario (Vršnak et al. 2001, 2002, 2004) is a powerful tool for estimating magnetic fields at shock locations. However, band-splitting is not commonly observed in type-IIs. Hence, it is important to carefully verify that it is indeed the case for the event under consideration before interpreting it as one. Naturally, incorrectly identifying type-IIs to have split-band emission is likely to lead to incorrect estimates of shock strengths and magnetic fields.

While spectroscopic imaging observations offer the most robust way to unambiguously distinguish between the upstream-downstream and multiple sources on the shock front scenarios, such observations are comparatively rare. In this article, we present a systematic approach to augment the usual study of DS to identify morphological similarities of type-II emission lanes by also examining the GraDS. Our analysis, based on this approach, found this type-II emission does not follow the upstream-downstream scenario, without relying on imaging and confirms the imaging-based results from (Bhunia et al. 2023). This confirmation establishes the usefulness of the GraDS-based morphological identification of type-II band-splitting and we hope for wider use of this approach in the community for identifying split-band type-IIs. Our imaging analysis also shows that, for the type-II studied here, the emission at different frequencies comes from a spatially extended region.

Acknowledgements

We thank the National Institute of Information and Communications Technology (NICT) for providing the HiRAS data publicly. This scientific work makes use of the Murchison Radio-astronomy Observatory (MRO), operated by the Commonwealth Scientific and Industrial Research Organisation (CSIRO). We acknowledge the Wajarri Yamatji people as the traditional owners of the Observatory site. Support for the operation of the MWA is provided by the Australian Government's National Collaborative Research Infrastructure Strategy (NCRIS), under a contract to Curtin University administered by Astronomy Australia Limited. We acknowledge the Pawsey Supercomputing Centre, which is supported by the Western Australian and Australian Governments. P.M. and D.O. acknowledge the support of the Department of Atomic Energy, Government of India, under the project no. 12-R&D-TFR-5.02-0700. D.K. acknowledges the support by the NASA Living with a Star Jack Eddy Postdoctoral Fellowship Program, administered by UCAR's Cooperative Programs for the Advancement of Earth System Science (CPAESS) under award #80NSSC22M0097. D.K. thanks the organizers of the IAU Symposium 388 for providing travel support to attend the symposium and present this work on behalf of the first author.

References

Bhunia, S., Carley, E. P., Oberoi, D., & Gallagher, P. T. 2023, Astronomy & Astrophysics, 670, A169, doi: 10.1051/0004-6361/202244456

Cho, K. S., Bong, S. C., Moon, Y. J., et al. 2011, A&A, 530, A16, doi: 10.1051/0004-6361/201015578

Gopalswamy, N. 2006, Geophysical Monograph Series, 165, 207, doi: 10.1029/165GM20

Kondo, T., Isobe, T., Igi, S., Watari, S., & Tokumaru, M. 1995, Journal of the Communications Research Laboratory, 42, 111

Kumari, A., Morosan, D. E., Kilpua, E. K. J., & Daei, F. 2023, Astronomy & Astrophysics, 675, A102, doi: 10.1051/0004-6361/202244015

Magdalenić, J., Marqué, C., Fallows, R. A., et al. 2020, ApJ, 897, L15, doi: 10.3847/2041-8213/ab9abc

McLean, D. 1967, Publications of the Astronomical Society of Australia, 1, 47

McLean, D. J., & Labrum, N. R. 1985, Solar radiophysics: studies of emission from the sun at metre wavelengths

Ndacyayisenga, T., Uwamahoro, J., Uwamahoro, J. C., et al. 2023, EGUsphere, 2023, 1, doi: 10.5194/egusphere-2023-201

Ramesh, R., & Kathiravan, C. 2022, ApJ, 940, 80, doi: 10.3847/1538-4357/ac9c65

Smerd, S. F., Sheridan, K. V., & Stewart, R. T. 1974, in Coronal Disturbances, ed. G. A. Newkirk, Vol. 57, 389

Smerd, S. F., Sheridan, K. V., & Stewart, R. T. 1975, Astrophys. Lett., 16, 23

Tingay, S. J., Goeke, R., Bowman, J. D., et al. 2013, PASA, 30, e007, doi: 10.1017/pasa.2012.007

Vršnak, B., Aurass, H., Magdalenić, J., & Gopalswamy, N. 2001, A&A, 377, 321, doi: 10.1051/0004-6361:20011067

Vršnak, B., Magdalenić, J., Aurass, H., & Mann, G. 2002, A&A, 396, 673, doi: 10.1051/0004-6361:20021413

Vršnak, B., Magdalenić, J., & Zlobec, P. 2004, A&A, 413, 753, doi: 10.1051/0004-6361:20034060

Solar and Stellar Coronal Mass Ejections
Proceedings IAU Symposium No. 388, 2026
N. Gopalswamy, O. Malandraki, A. Vidotto & W. Manchester, eds.
doi:10.1017/S1743921325000559

The SOHO LASCO CME Catalog - Version 2

Nat Gopalswamy[1], **Grzegorz Michałek**[2], **Seiji Yashiro**[3,1], **Pertti Mäkelä**[3,1], **Sachiko Akiyama**[3,1], **Hong Xie**[3,1] **and Angelos Vourlidas**[4]

[1]NASA Goddard Space Flight Center, Greenbelt, MD, USA
email: `nat.gopalswamy@nasa.gov`

[2]Jagiellonian University, Kraków, Poland

[3]The Catholic University of America, Washington, DC, USA

[4]Johns Hopkins University Applied Physics Laboratory, Laurel, MD, USA

Abstract. This paper provides an update on the coronal mass ejection (CME) catalog maintained at the CDAW Data Center, NASA Goddard Space Flight Center (`http://cdaw.gsfc.nasa.gov/CME_list`). This is version 2 (v2) of the Catalog that has been made as the default version as of May 1, 2024. The new features of the Catalog v2 are (i) online measurement tool, (ii) combination JavaScript movies from the STEREO and Solar Dynamics Observatory (SDO) missions, and (iii) insertion of newly identified CMEs for the period 1996 to 2004. The CME identification was revisited resulting in a set of ∼3000 new CMEs added to the Catalog. A vast majority of these CMEs are weak and narrow. The resulting statistical properties of CMEs are not significantly different from those reported using version 1.

Keywords. coronal mass ejections

1. Introduction

The CDAW CME Catalog grew out of a coordinated data analysis workshop (CDAW) in 1999 organized to study the properties of CMEs associated with interplanetary radio bursts. During the course of analyzing these events, it was realized that creating a set of plots, movies, and tables for each CME is a convenient way of investigating CME-related phenomena. The CMEs are manually identified in the images obtained by the Large Angle and Spectrometric Coronagraph (LASCO, Brueckner et al. 1995) on board the Solar and Heliospheric Observatory (SOHO). The Catalog resides in the CDAW Data Center at the Goddard Space Flight Center (`http://cdaw.gsfc.nasa.gov`) and is open to the scientific community. The version 1 catalog was initially reported in Yashiro et al. (2004), which was expanded with new features as reported in Gopalswamy et al. (2009). The number of CMEs listed in the CDAW catalog is an order of magnitude larger than the total number of CMEs known from pre-SOHO coronagraphs (Gopalswamy et al. 2009; Howard et al. 2023). This catalog serves as an important reference to test automatic detection catalogs such as cactus (`https://www.sidc.be/cactus/`) and SEEDS (`http://spaceweather.gmu.edu/seeds/`).

2. Catalog

The main feature of the catalog is a 13-column html table (`http://cdaw.gsfc.nasa.gov/CME_list/index.html`). The first column gives the year of CME observations; the

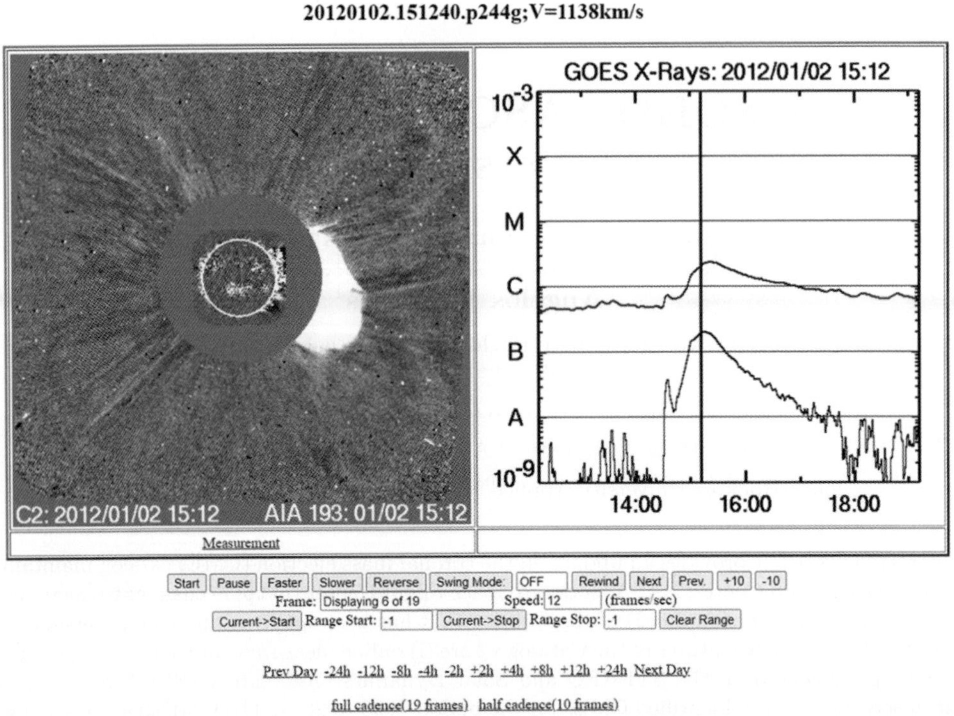

Figure 1. One frame in the JavaScript movie of the 2012 January 2 CME that first appeared in LASCO/C2 FOV at 15:12:40 UT. The movie is built using LASCO/C2 difference images with superposed EUV images from the Solar Dynamics Observatory (SDO, Pesnell et al. (2012)). The CME ID is given at the top as 20120102.151240.p244g. The numbers following p (for position angle) indicate the central position angle, while the suffix g is the first letter of the first name of the person who made the height-time measurement. Also given at the top is the CME speed (V=1138 km s^{-1}) obtained from a linear fit to the height-time measurements. The JavaScript movie can be controlled using the buttons given below the movie frame. The length of the movie can be extended to earlier or later times up to 24 hours. The movie can also be run with full or half cadence. The link "Measurement" takes to a new page, where h-t measurements can be made.

next 12 columns give links to monthly list of CMEs. The monthly list is also a html table with each row giving all available information for one CME. The information is in the form of movies, plots, digital data, and remarks. The first three columns provide a unique identification (ID) for each CME: date (YYYY/MM/DD), universal time (HH:MM:SS), and position angle (in degrees). The first column is linked to a JavaScript movie constructed from LASCO/C2 running difference images with superposed EUV difference images. Figure 1 shows one frame of the JavaScript movie. The unique ID for the CME is given above the movie frame, in the format YYYYMMDD.HHMMSS.pAAAf;V=xxxxkm/s (see Fig. 1) where p indicates that the succeeding three numbers give the central position angle of the CME in degrees; f indicates the first letter of the first name of the person who made the height-time measurement; V is the average speed of the CME within the coronagraph field of view (FOV). While the date is linked to the Javascript movie noted above, the time is linked to the height-time (h-t) data points as measured. The remaining columns in the monthly table are as described in the previous version (Gopalswamy et al. 2009).

2.1. *The measurement tool*

One of the new features of the catalog is the online tool using which the users can make their own measurements. The measurements can be made to by clicking on the "Measurement" link given directly below the images. The catalog gives routine h-t measurements made at the fastest-moving leading edge of a CME. These measurements do not distinguish whether the leading edge is a shock or flux rope. The user can measure sub features such as shock, flux rope, or prominence core of a CME. Measurements can be made at various position angles if needed. The measurement tool for radio dynamic spectra is mainly used for measuring drift rates of radio bursts. Based on the cursor position, in the measurement tool displays the pixel coordinates, distance from the Sun center in solar radii, the position angle, and the heliographic coordinates. Thus, one can find the solar source locations of eruptions by clicking on the disturbances observed in EUV images.

The measurement tool for the radio dynamic spectra gives the time and frequency corresponding to the cursor position. Thus, by clicking on features in the dynamic spectra, one can get a table of time vs. frequency data, which can be used to derive the drift rates. The radio dynamic spectra can be found under "Java Movie" in the column 12 of the catalog. For example, for the event in Fig. 1, clicking on "Java Movie" opens the link, `https://cdaw.gsfc.nasa.gov/CME_list/UNIVERSAL_ver2/2012_01/univ2012_01.html`. Under this link, one can find various combination movies including WAVES dynamic spectra. There are three movies that combine LASCO C2 and C3 images with Wind/WAVES dynamic spectra. Any of these can be used for making measurements on the dynamic spectra. A commonly used combination is "c2rdif_waves.html", which opens the movie, `https://cdaw.gsfc.nasa.gov/movie/make_javamovie.php?date=20120102&img1=lasc2rdf&img2=wwaves`. In the movie, the left frames are running difference images (c2rdif) from SOHO/LASCO while the right frame is the WAVES dynamic spectrum. The measurement feature in this movie can be used for both CME h-t and drift rate measurements. The java movie needs to be paused before clicking on the measurement tool at the WAVES frame containing the feature to be measured.

2.2. *STEREO movies*

Under the column "Movies, plots, & links" we have added a new set of daily movies displaying various combinations of images and spectra from the Solar Terrestrial Relations Observatory (STEREO, Kaiser et al. 2008) starting in November 2006, soon after the commissioning of the mission. EUV and coronagraph images are from the Sun Earth Connection Coronal and Heliospheric Investigation (SECCHI, Howard et al. 2008) and the radio dynamic spectra are from the STEREO/WAVES instrument (Bougeret et al. 2008). Figure 2 shows list of links to the JavaScript movies.

2.3. *Newly added CMEs*

CMEs were identified in the LASCO FOV by different people until 2004, beyond which a single person (co-author G. Michałek) identifies CMEs. In order to make the identification uniform throughout, G. Michałek revisited the 1996–2004 images to check if any CMEs were missing. This exercise resulted in 3072 CMEs over the 9-year period. Figure 3 shows the statistics of the newly identified CMEs. The top panel shows the annual number of CMEs in catalog v2. The second panel from the top shows all the newly identified CMEs amounting to hundreds of CMEs in each year. The number was high in the beginning of the mission when people were getting used to the new data set. Narrow CMEs (angular width $W < 30°$) constitute the largest population among the

[2008/03/24] -- 2008/03/25 -- [2008/03/26]

	All Day	Short Intervals					
STEREO A+B EUVI 171	0-24	0-4	4-8	8-12	12-16	16-20	20-24
STEREO A+B EUVI 195	0-24	0-4	4-8	8-12	12-16	16-20	20-24
STEREO A+B EUVI 284	0-24	0-4	4-8	8-12	12-16	16-20	20-24
STEREO A+B EUVI 304	0-24	0-4	4-8	8-12	12-16	16-20	20-24
STEREO A+B COR1	0-24	0-4	4-8	8-12	12-16	16-20	20-24
STEREO A+B COR1RD	0-24	0-4	4-8	8-12	12-16	16-20	20-24
STEREO A+B COR2	0-24	0-4	4-8	8-12	12-16	16-20	20-24
	All Day	Short Intervals					
STEREO A COR1 + EUVI 171	0-24	0-4	4-8	8-12	12-16	16-20	20-24
STEREO A COR1 + EUVI 195	0-24	0-4	4-8	8-12	12-16	16-20	20-24
STEREO A COR1 + EUVI 284	0-24	0-4	4-8	8-12	12-16	16-20	20-24
STEREO A COR1 + EUVI 304	0-24	0-4	4-8	8-12	12-16	16-20	20-24
STEREO A COR1 + WAVES	0-24	0-4	4-8	8-12	12-16	16-20	20-24
STEREO A COR1RD + EUVI 171	0-24	0-4	4-8	8-12	12-16	16-20	20-24
STEREO A COR1RD + EUVI 195	0-24	0-4	4-8	8-12	12-16	16-20	20-24
STEREO A COR1RD + EUVI 284	0-24	0-4	4-8	8-12	12-16	16-20	20-24
STEREO A COR1RD + EUVI 304	0-24	0-4	4-8	8-12	12-16	16-20	20-24
STEREO A COR1RD + WAVES	0-24	0-4	4-8	8-12	12-16	16-20	20-24
STEREO A COR2 + WAVES	0-24	0-4	4-8	8-12	12-16	16-20	20-24
	All Day	Short Intervals					
STEREO B COR1 + EUVI 171	0-24	0-4	4-8	8-12	12-16	16-20	20-24
STEREO B COR1 + EUVI 195	0-24	0-4	4-8	8-12	12-16	16-20	20-24
STEREO B COR1 + EUVI 284	0-24	0-4	4-8	8-12	12-16	16-20	20-24
STEREO B COR1 + EUVI 304	0-24	0-4	4-8	8-12	12-16	16-20	20-24
STEREO B COR1 + WAVES	0-24	0-4	4-8	8-12	12-16	16-20	20-24
STEREO B COR1RD + EUVI 171	0-24	0-4	4-8	8-12	12-16	16-20	20-24
STEREO B COR1RD + EUVI 195	0-24	0-4	4-8	8-12	12-16	16-20	20-24
STEREO B COR1RD + EUVI 284	0-24	0-4	4-8	8-12	12-16	16-20	20-24
STEREO B COR1RD + EUVI 304	0-24	0-4	4-8	8-12	12-16	16-20	20-24
STEREO B COR1RD + WAVES	0-24	0-4	4-8	8-12	12-16	16-20	20-24
STEREO B COR2 + WAVES	0-24	0-4	4-8	8-12	12-16	16-20	20-24

Figure 2. Table showing the set of new STEREO movies added to the catalog. The movies use various combinations of EUV images at 171, 195, 284, and 304 Å, inner (COR1) and outer (COR2) coronagraph images, and WAVES dynamic spectra from STEREO Ahead (A) and Behind (B). Movies can be displayed for the whole day (0-24) or in 4-hour segments. The date of the movies is given at the top with options to display from the previous day or next day. Text in blue indicate links. Greyed out texts indicate data gaps.

newly added CMEs, followed by regular CMEs ($30° \leq W < 120°$). Only 25 new wide CMEs ($W \geq 120°$) were found as shown in the bottom panel. Among the wide CMEs, only one was a halo CME.

Interestingly, most of the wide CMEs are generally faint. Figure 4 shows the only newly identified wide CME in the year 2003. Clearly the CME is very faint and can easily be missed. The CME is actually visible mainly in the difference images. The CME has an ID: 20030120.005405.p360g;V=233km/s indicating that the CME is also very slow.

To see the effect of adding the new CMEs, we have compared the distributions of speed, width, and acceleration in Figure 5. After the new CMEs are added, the average speed decreases by 27 km s^{-1} and the average angular width decreases by 2 degrees.

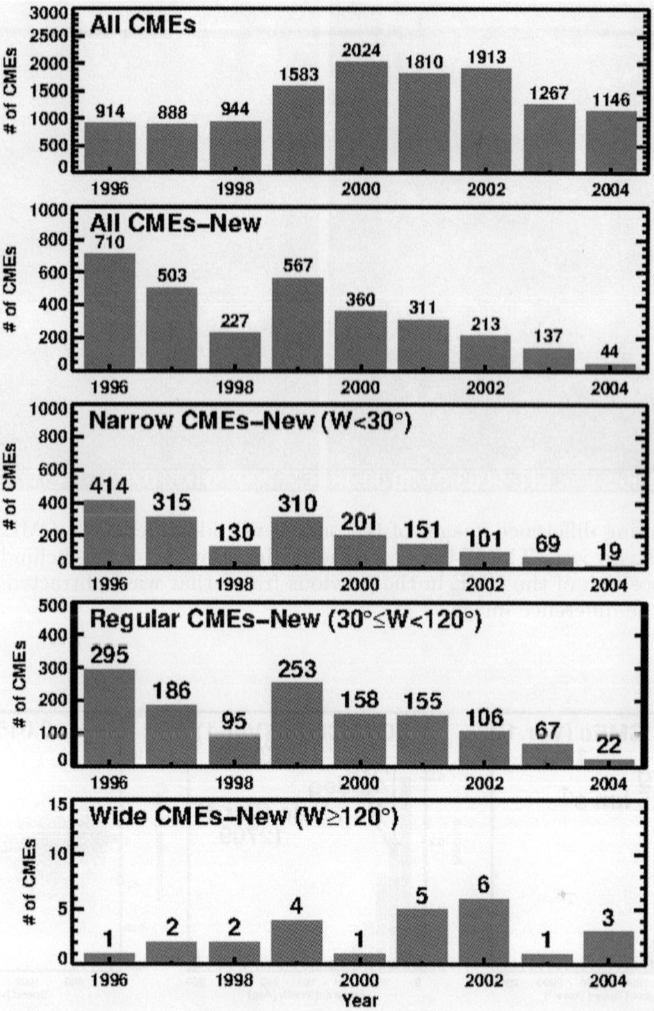

Figure 3. All CMEs in the catalog v2 over the period 1996–2004 (top) and the newly added CMEs (lower 4 panels). The number of CMEs in each year are indicated on the bars of the plots. The only missing CME in the year 2000 was a faint halo CME that occurred on 2000 January 2. This is the only newly added halo CME for the entire interval (1996 to 2004).

The acceleration within the LASCO C2 FOV remains the same. Thus, the newly added CMEs did not affect the statistics.

Figure 6 shows distributions similar to those in Figure 5 except that they include all CMEs measured until the end of 2023 (includes CMEs from cycles 23, 24, and the rise phase of cycle 25). Once again, the effect of inclusion of the new CMEs is insignificant. However, we do see the effect of weak solar cycles 24 and 25: the average values are smaller because of the higher abundance of weaker CMEs in cycles 24 and 25 than in cycle 23.

3. Summary

As of May 1, 2024 the CDAW Data Center has released the SOHO/LASCO CME catalog v2. The new catalog has three major changes: (i) we have incorporated online tools that can be used on various images in the Javascript moes and radio dynamic

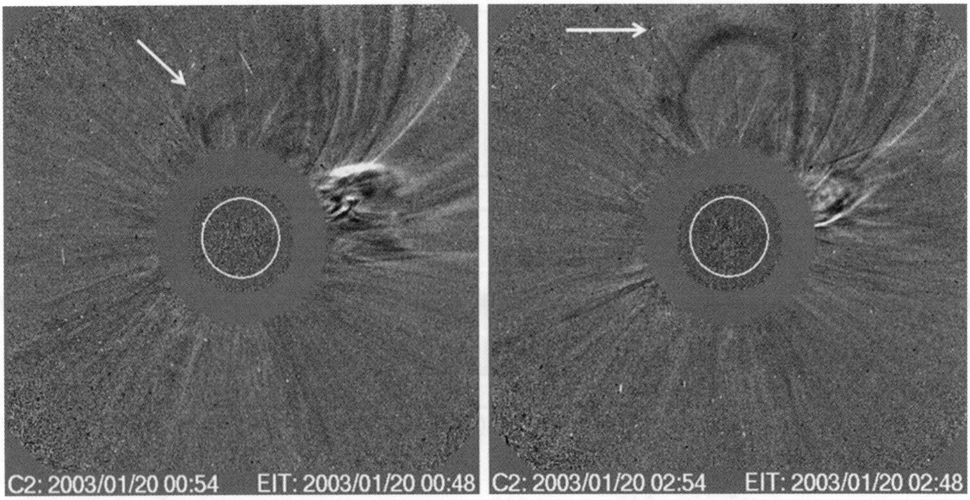

Figure 4. Running difference images of the only newly identified wide CME in the year 2003 (pointed to by the arrows). The CME is extremely faint. Dark structure behind the leading edge represents the position of the CME in the previous frame that was subtracted from the current frame to make the difference image.

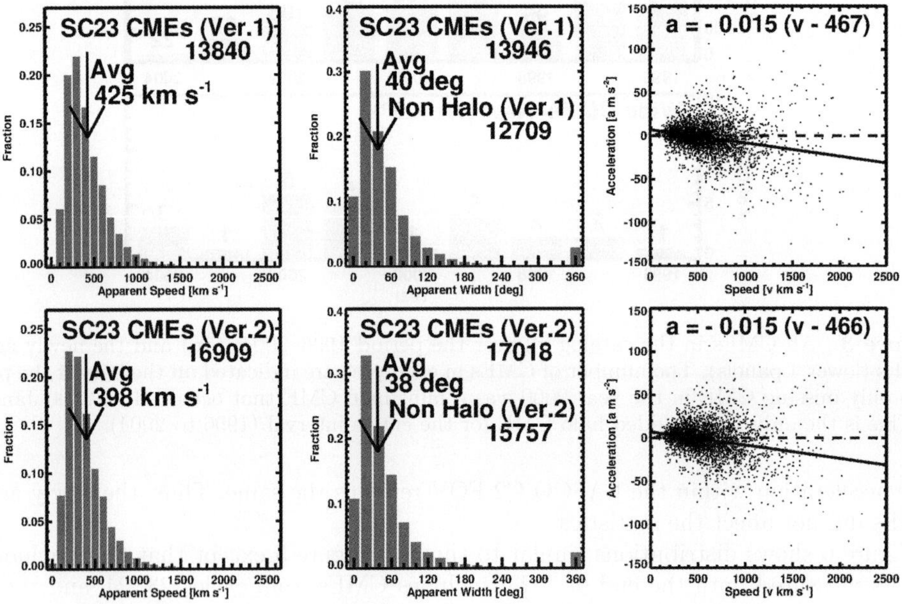

Figure 5. Speed, width, and acceleration distributions of CMEs in cycle 23 from catalog v1 (top) and the same parameters including newly added CMEs (bottom). The number of CMEs in the interval and the averages of the distributions are shown on the plots. The acceleration information is presented as a scatter plot between CME speed and the acceleration within the LASCO FOV. Note that this acceleration is typically the residual acceleration. The main acceleration peaks generally closer to the Sun (below the LASCO C2 occulting disk. Equations of the regression line is shown on the plots. E.g., a = 0.015 (v-466), where a is the acceleration in m s^{-2} and v is the speed in km s^{-1}. Only CMEs that have five or more height-time measurements are included in the acceleration plot. CMEs with feature quality $\geq$1 are used in the speed and width distributions (i.e., ill-defined events are excluded).

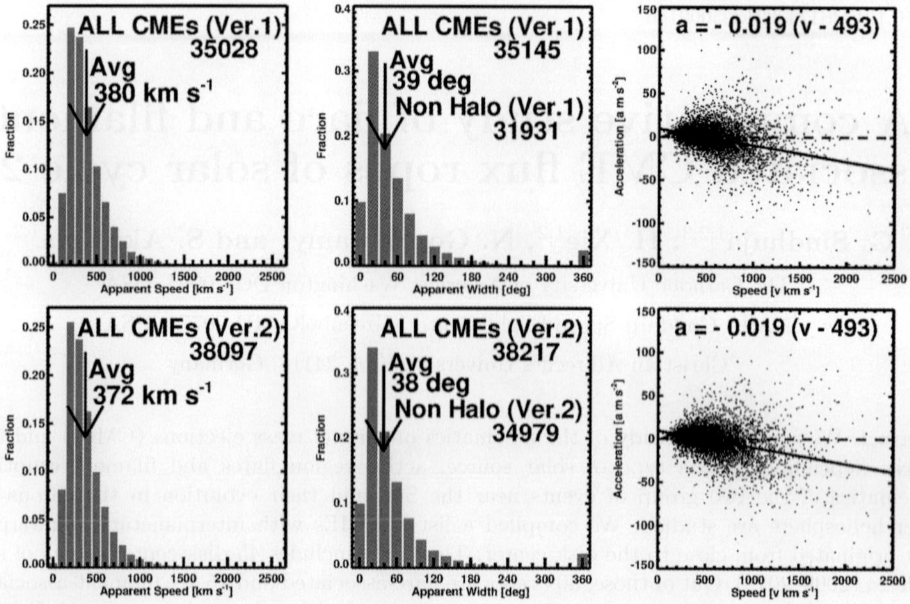

Figure 6. Same as Figure 5, but the full data sets are used (1996 to 2023, inclusive)

spectra, (ii) we have included 29 Javascript movies that combine various images and dynamic spectra from the STEREO mission, and (iii) we revisited SOHO/LASCO data to identify missing CMEs resulting in 3072 new CMEs. Most of the newly included CMEs are weak and narrow. Inclusion of the new CMEs did not have much impact on the statistical properties of CMEs. The v1 catalog resides at `https://cdaw.gsfc.nasa.gov/CME_list/index_ver1.html`

We thank the SOHO, STEREO, SDO, Wind, GOES, Yohkoh/SXT, and Nobeyama radioheliograph teams for their contribution in making this Catalog possible.

Supplementary material

To view supplementary material for this article, please visit https://doi.org/10.1017/S1743921325000559

References

Bougeret, J. L., Goetz, K., Kaiser, M. L., et al. 2008, *SSRv*, 136, 487, doi: 10.1007/s11214-007-9298-8

Brueckner, G. E., Howard, R. A., Koomen, M. J., et al. 1995, *SoPh*, 162, 357, doi: 10.1007/BF00733434

Gopalswamy, N., Yashiro, S., Michalek, G., et al. 2009, *EM&P*, 104, 295, doi: 10.1007/s11038-008-9282-7

Howard, R. A., Vourlidas, A., & Stenborg, G. 2023, *FrASS*, 10, 1264226, doi: 10.3389/fspas.2023.1264226

Howard, R. A., Moses, J. D., Vourlidas, A., et al. 2008, *SSRv*, 136, 67, doi: 10.1007/s11214-008-9341-4

Kaiser, M. L., Kucera, T. A., Davila, J. M., et al. 2008, *SSRv*, 136, 5, doi: 10.1007/s11214-007-9277-0

Pesnell, W. D., Thompson, B. J., & Chamberlin, P. C. 2012, *SoPh*, 275, 3, doi: 10.1007/s11207-011-9841-3

Yashiro, S., Gopalswamy, N., Michalek, G., et al. 2004, *JGR*, 109, A07105, doi: 10.1029/2003JA010282

Solar and Stellar Coronal Mass Ejections
Proceedings IAU Symposium No. 388, 2026
N. Gopalswamy, O. Malandraki, A. Vidotto & W. Manchester, eds.
doi:10.1017/S1743921325000146

A comparative study of flare and filament associated CME flux ropes of solar cycle 24

G. Sindhuja[1,2,3]**, H. Xie**[1,2]**, N. Gopalswamy**[2] **and S. Akiyama**[1,2]

[1]The Catholic University of America, Washington DC 20064, USA

[2]NASA Goddard Space Flight Center, Greenbelt, MD 20771, USA

[3]Christian Albrechts University, Kiel, 24118, Germany

Abstract. We report on a study of the kinematics of coronal mass ejections (CMEs) and their shocks depending on the type of solar source: active region flares and filament eruptions. Kinematics of the two group of events near the Sun and their evolution in the corona and inner heliosphere are studied. We compiled a list of CMEs with interplanetary counterparts that originated from close to the disk center. The study includes 45 disk center events of solar cycle 24 (2009–2019). Out of those, 30 events are flare-associated and 15 are filament-associated. Among the 30 flare-associated events, 8 are also associated with a shock. Among the 15 filament-associated events only one event is associated with a shock. We measure the CME flux rope (FR) speed along with the shock speed and the associated reconnection flux (RC). We found that the CME speed and the RC flux are about 30% more for shock-associated events compared without a shock. Similarly, the CME speed and the RC flux are about 50% more for flare-associated events compared to the filament-associated events.

Keywords. Flare, Filaments, Shock, Reconnection flux, CME flux rope kinematics

1. Introduction

CMEs and flares are thought to be different consequences of instabilities and reconnection of coronal magnetic fields (e.g, reviews by Fletcher et al. (2011), Shibata et al. (2011), Wiegelmann et al. (2014)). In general, CMEs may occur without flares, and flares may occur without CMEs. However, the association rate is a steeply increasing function of the GOES flare class, and in the strongest events typically both occur together. In active regions the association between filament eruptions and flares is more clearly visible (e.g., Jing et al. (2004); Chandra et al. (2010)) as compared with quiescent filament eruptions Gosain et al. (2009). In the Quiescent filament eruption, the emissions are mostly too weak because of the weaker magnetic field Forbes (2000). Quiescent filaments are from long-time decayed active regions with weak magnetic fields and usually give rise to slow CMEs. For the CMEs from the quiescent filaments, the velocity is inclined to be smaller, with a range of 500-1400 km/s. Most of them lack detectable GOES soft X-ray (SXR) 1-8 Å flux.

In this work we compare the flare and filament associated CME flux ropes (FR) of solar cycle 24. In addition, we study the events that are associated with a shock. Fast CMEs drive MHD shocks. For space weather purposes, large gradual solar eruptive events (SEPs) from CME-driven shocks are more important Gopalswamy et al. (2001). Therefore, we study their kinematics and the magnetic flux associated with the corresponding active region.

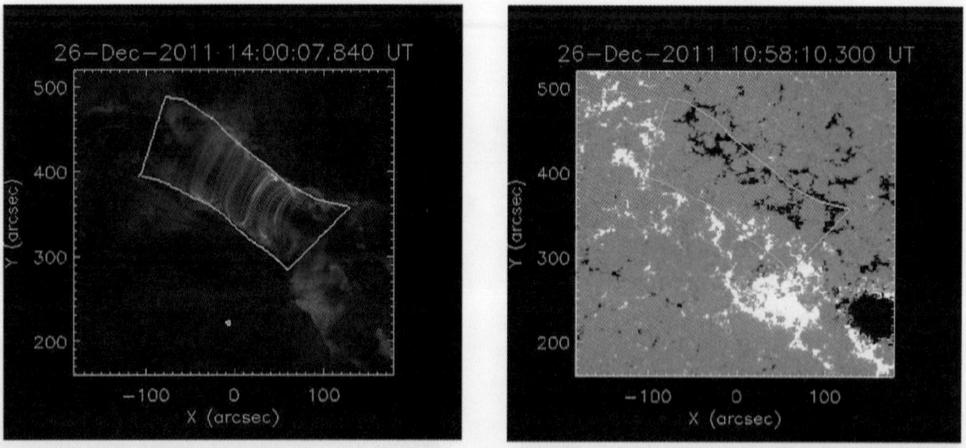

Figure 1. (left) PEA during the 26 December 2011 eruption observed in an SDO/AIA EUV image at 193 Å with the boundaries of the arcade marked by blue lines. (right) SDO/HMI LOS magnetogram with the PEA overplotted.

In Section 2, we discuss about the data selection and the methods used for this study. In Section 3, we elaborate the results and discussions.

2. Data selection and methods

There are 45 disk center events of solar cycle 24 (2009–2019). Only eruptions occurring from within ± 45° in longitude of the disk center were considered. Out of those, 30 are flare-associated and 15 are filament-associated. Among the 30 flare-associated events, are also associated with a shock. Among the 15 filament-associated events only one is associated with a shock.

2.1. *Post eruption arcade method*

Qiu et al (2007) determined the RC flux as the photospheric magnetic flux underlying the area swept up by the flare ribbons on one side of the polarity inversion line. Gopalswamy et al. (2017) introduced the post eruption arcade (PEA) method to measure the reconnected flux in solar eruptions. According to this method, a single EUV image (e.g, 193 Å) from the Atmospheric Imaging Assembly (AIA, Lemen et al. 2012) on board the Solar Dynamic Observatory (SDO, Pesnell et al. 2012) is chosen during the decay phase of a flare, when the PEA appears to be constant with time. We then use the line-of-sight (LOS) magnetogram of the SDO's Helioseismic Magnetic Imager (HMI, Schou et al. 2012) to estimate the reconnection (RC) flux, where the RC flux is defined as half of the magnetic flux underlying the PEA. We employ this method to extract the RC flux for the events we investigate (see Figure 1 for an example showing the PEA of the 2011 December 26 event superposed on a magnetogram obtained near the time of the eruption). To quantify the uncertainties in the RC flux, we repeated the selection of PEA for 10 instances in a single EUV image and calculated the mean and standard deviation of the RC flux.

2.2. *GCS modelling and shock analysis*

We obtain the three-dimensional (3D) speed of CMEs and shocks in the corona and interplanetary medium by fitting a Graduated Cylindrical Shell ((GCS, Thernisien 2011)) FR to a CME and a spheroid to a shock Olmedo et al. (2013). Close to the Sun, we use

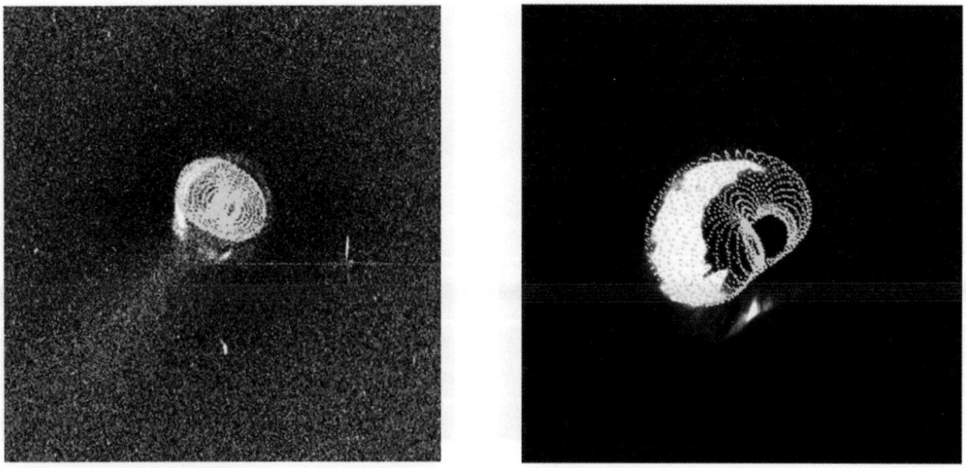

Figure 2. ((Left) SOHO/LASCO C3 coronagraph difference image with the superposed flux rope (green). (Right) the same but a STEREO-A COR2 difference image is used.

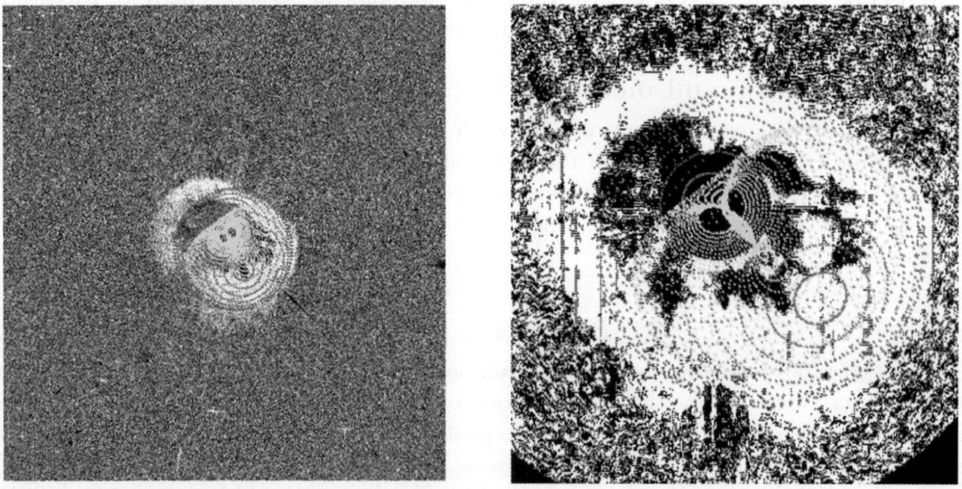

Figure 3. ((Left) SOHO/LASCO C3 coronagraph difference image with the flux rope (green) and the shock (red) are superposed. (Right) the same but a STEREO-A COR2 difference image is used.

EUV images from the SDO and the Solar Terrestrial Relations Observatory (STEREO) in the forward modeling. In the coronagraph field of view, we use images from STEREO and the Solar and Heliospheric Observatory (SOHO) missions. We adjust the initial test parameters like latitude, longitude and tilt angle of the source region, half angle, aspect ratio and height of the GCS FR until the fit is best match the white-light images. By doing this we derive the geometrical and kinematic properties of the CME FR, including FR radius R_0, aspect ratio $\Lambda = 1/2\ (R_{tip} - R_0)/R_0$ where R_{tip} is the heliocentric distance to the tip of the FR, tilt angle, half angle, and the CME speed. Figure 2 illustrates the GCS fitting: the FR (green wire) is overlaid on SOHO/LASCO (left) and STEREO-A COR2 difference images (right). In figure 3 we show the GCS fitting of the FR with a shock. The green and red wire meshes are the fitted flux rope and shock, respectively.

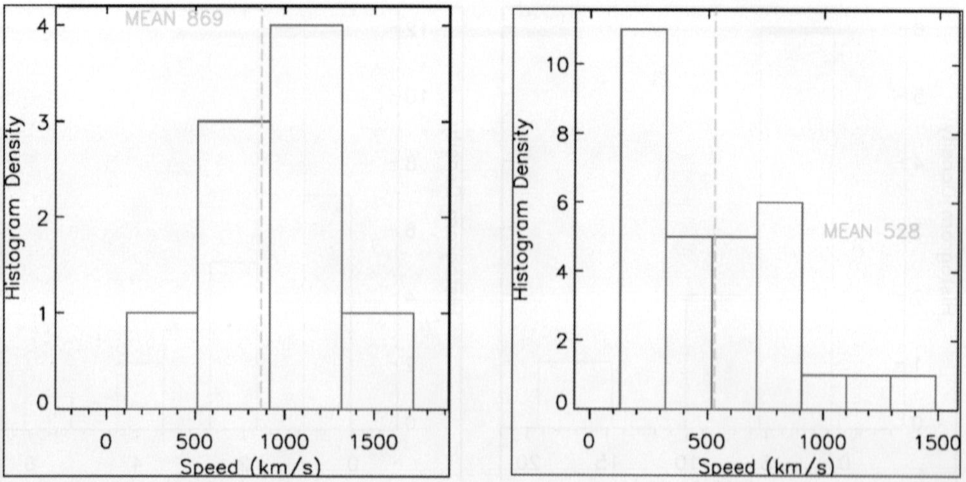

Figure 4. Distribution of GCS speed with shock (left). Distribution of GCS speed of CME FR associated without a shock (red)

3. Results and discussion

We compiled a list of 45 CMEs from cycle 24 with interplanetary counterparts that originated from close to the disk center, which include 30 flare-associated events and 15 filament-associated events. We measure the CME FR speed along with the shock speed and the associated RC flux.

3.1. *Comparison of CME FR associated with shock and without shock*

Among the 9 shock associated events, 8 are flare-associated that occurred on 2011 February 15, 2012 July 12, 2013 March 15, 2013 April 11, 2014 February 12, 2014 September 10, 2015 November 04 and 2015 December 28. Only one event that occurred on 2018 July 04 is filament-associated. Figure 4 (left) shows the GCS speed of the events that are associated with shock. The mean value is 869 km/s and mean of the corresponding shock speeds is 925 km/s. The right panel shows the distribution of GCS speed of the events that are associated without a shock. The mean of the FR speeds is 528 km/s. The Figure 5 (left) shows the distribution of the RC fluxes of shock associated events. The mean of the distribution is 5.74 x 10^{21} Mx compared with the 36 events not associated with shock (right). The mean value is 1.71 x 10^{21} Mx. Qiu et al (2007) studied 9 events and found a strong relationship between the magnetic cloud (MC) flux and the total reconnection flux for the nine events studied. The relationships between these physical quantities suggest that the magnetic flux and twist of interplanetary magnetic flux ropes (as observed in MCs) are significantly influenced by low-corona magnetic reconnection during CME eruptions.

We see that the speed and RC flux of shock-associated events are about 30% greater than those of events without a shock.

3.2. *Comparison of CME FR associated with flare and filament eruptions*

There are 30 flare-associated events and 15 filament-associated events. Among the flare events the 2012 June 14 associated with an M1.9 class flare and had the highest GCS speed (1295 km/s). The highest RC flux is 1.47 x 10^{22} Mx that occurred on 2012 July 12 associated with an X1.4 class flare. The lowest GCS speed is 285 km/s for 2012 October 09 flare events and the lowest value of RC flux is 3.83 x 10^{20} Mx for the 2011 July 01

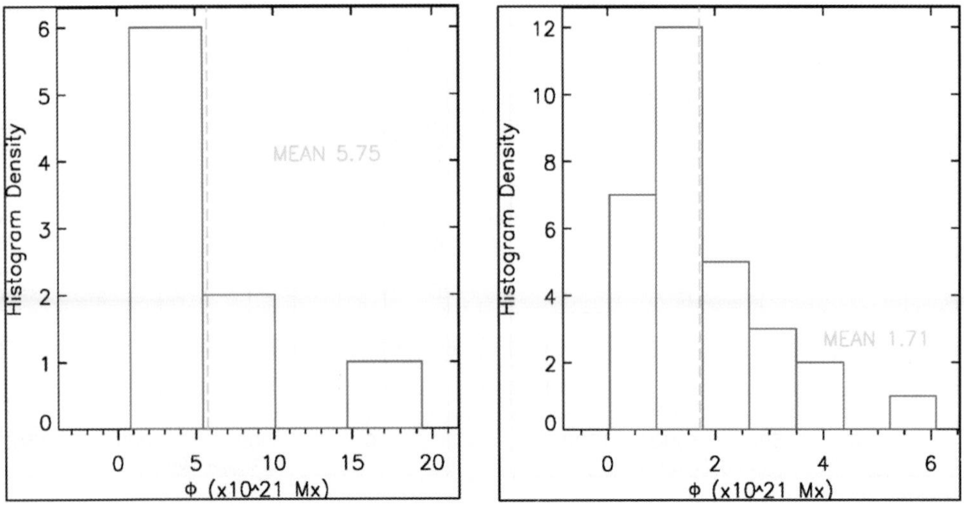

Figure 5. Distributions of RC fluxes of events with shock and events without shock along with their mean values mentioned in green

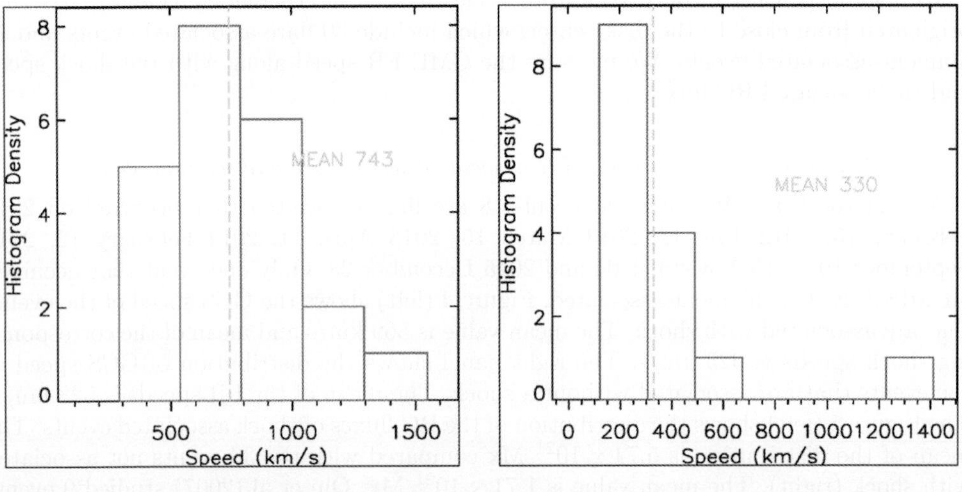

Figure 6. We have also shown the distribution of CME speed for Flare-associated (left) and filament-associated events (right). The green dotted line depict the mean of the distribution in each figure.

event. For the filament-associated events the highest GCS speed is 1228 km/s for the 2012 May 12 event and the highest RC flux is 5.23 x 10^{21} Mx for the 2009 January 21 event. The lowest GCS speed is 121 km/s for the 2010 October 26 event and the lowest RC flux is 3.3 10^{19} Mx for the 2009 September 25 event.

Figure 6 shows the distribution of the GCS speed of the flare and filament. Figure 7 shows the acceleration of the flare and filament associated events. The mean values are 1.59 km/s^2 and 0.52 km/s^2 for the two categories, respectively. Correspondingly, RC fluxes of flare and filament events are shown in Figure 8 with their mean values being 2.75 x 10^{21} Mx for flare events and 1.49 x 10^{21} Mx for filament events. The main findings are,

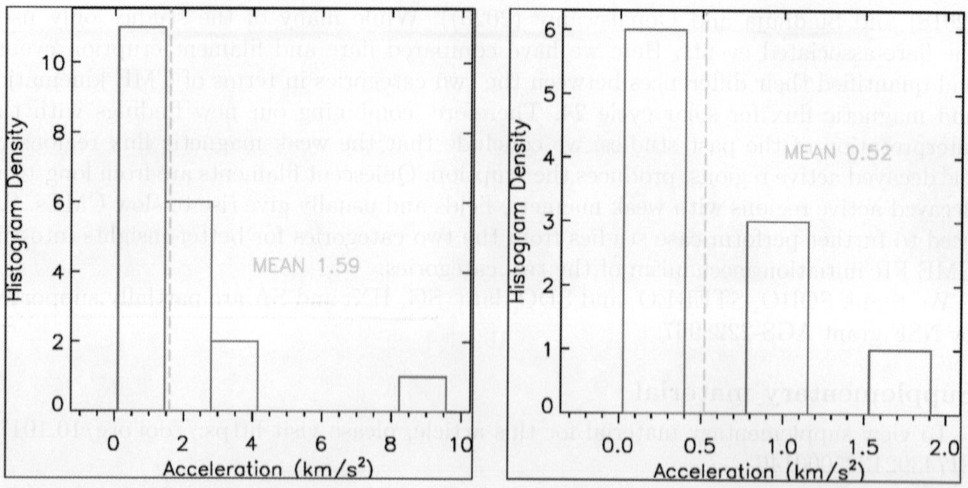

Figure 7. We have also measured the acceleration of the CME for flare-associated CMEs (left) and Filament-associated CMEs (right). The green dotted line depict the mean of the distribution in each figure.

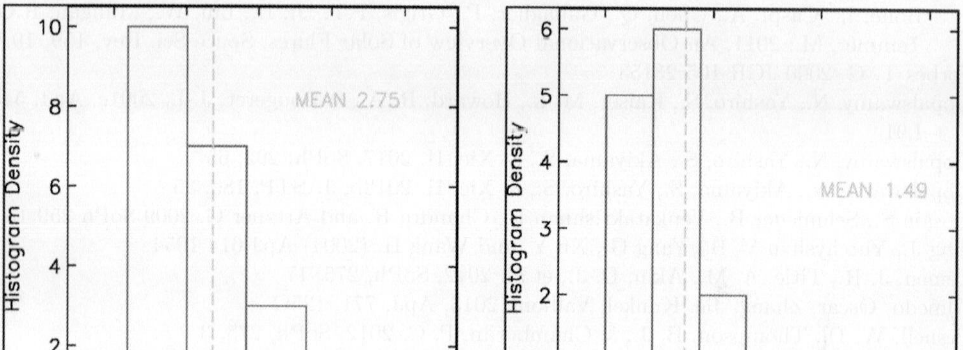

Figure 8. Distributions of RC fluxes of flare-associated events (left), filament-associated events (right), with their mean values mentioned in green

- The flare and filament associated events have a mean GCS speed of 743 km/s and 342 km/s, respectively. The average filament associated CME speed is about 55% smaller.
- The mean RC flux is 2.75 x 10^{21} Mx and 1.49 x 10^{21} Mx for flare and filament associated events, respectively. The RC flux of filament-associated events is about 45% smaller.
- We also measure the CME FR acceleration and found that the acceleration values of the flare-associated events are 75% more than the filament-associated events.

From the past studies we understood that the larger the flare the more the RC flux. This has been shown by many authors Gopalswamy et al. (2018), Veronig and Polanec (2015), Sindhuja et al. (2019),Sindhuja and Gopalswamy (2020). The RC flux is also known to be correlated with the CME speed and kinetic energy (Gopalswamy et al.

(2018) and Sindhuja and Gopalswamy (2020)). While many of the studies only used the flare-associated events. Here we have compared flare and filament eruption events and quantified their differences between the two categories in terms of CME kinematics and magnetic flux for solar cycle 24. Therefore, combining our new findings with the interpretation of the past studies, we conclude that the weak magnetic flux regions or the decayed active regions, produces the eruption. Quiescent filaments are from long-time decayed active regions with weak magnetic fields and usually give rise to slow CMEs. We need to further perform case studies from the two categories for better insights into the CME FR initiation mechanism of the two categories.

We thank SOHO, STEREO, and SDO data. SG, HX, and SA are partially supported by NSF grant AGS-2228967.

Supplementary material

To view supplementary material for this article, please visit https://doi.org/10.1017/S1743921325000146

References

Chandra R., Pariat E., Schmieder B., Mandrini C. H. and Uddin W. 2010 SoPh 261 127
Fletcher, L., Dennis, B.R., Hudson, H.S., Krucker, S., Phillips, K., Veronig, A., Battaglia, M., Bone, L., Caspi, A., Chen, Q., Gallagher, P., Grigis, P.T., Ji, H., Liu, W., Milligan, R.O., Temmer, M.: 2011, An Observational Overview of Solar Flares. Space Sci. Rev. 159, 19.
Forbes T. G. 2000 JGR 105 23153
Gopalswamy, N., Yashiro, S., Kaiser, M. L., Howard, R. A., & Bougeret, J.-L. 2001c, ApJ, 548, L91
Gopalswamy, N., Yashiro, S., Akiyama, S., & Xie, H. 2017, SoPh, 292, 65
Gopalswamy, N., Akiyama, S., Yashiro, S., & Xie, H. 2018b, JASTP, 180, 35
Gosain S., Schmieder B., Venkatakrishnan P., Chandra R. and Artzner G. 2009 SoPh 259 13
Jing J., Yurchyshyn V. B., Yang G., Xu Y. and Wang H. (2004) ApJ 614 1054
Lemen, J. R., Title, A. M., Akin, D. J., et al. 2012, SoPh, 275, 17
Olmedo, Oscar, Zhang, Jie, Kunkel, Valbona 2013, ApJ, 771, 125O
Pesnell, W. D., Thompson, B. J., & Chamberlin, P. C. 2012, SoPh, 275, 3
Qiu, J., Hu, Q., Howard, T. A., & Yurchyshyn, V. B. 2007, ApJ, 659, 758
Schmieder B., Démoulin P. and Aulanier G. 2013 AdSpR 51 1967
Schou, J., Borrero, J. M., Norton, A. A., et al. 2012, SoPh, 275, 327
Shibata, K., Magara, T.: 2011, Solar Flares: Magnetohydrodynamic Processes. Living Rev. Solar Phys. 8, 6.
Sindhuja, G., Srivastava, N., Veronig, A. M., & Pötzi, W. 2019, MNRAS, 482, 3744
Sindhuja, G., & Gopalswamy, N. 2020, The Astrophysical Journal, 889, 104
Thernisien, A. 2011, ApJS, 194, 33
Veronig, A. M., & Polanec, W. 2015, SoPh, 290, 2923
Wiegelmann, T., Thalmann, J.K., Solanki, S.K.: 2014, The magnetic field in the solar atmosphere. Astron. Astrophys. Rev. 22, 78.

Solar and Stellar Coronal Mass Ejections
Proceedings IAU Symposium No. 388, 2026
N. Gopalswamy, O. Malandraki, A. Vidotto & W. Manchester, eds.
doi:10.1017/S1743921324002229

Insights from the March 13, 2012, event with 110° longitudinal separation: Propagation and flux rope orientation

F. Carcaboso[1,2], C. Kay[3], M. Dumbović[4], M. A. Hidalgo[5], D. Lario[2], L. K. Jian[2], R. Gómez-Herrero[5], L. B. Wilson III[2], M. Temmer[6], S. G. Heinemann[7], T. Nieves-Chinchilla[2] and A. M. Veronig[6,8]

[1]NASA Postdoctoral Program Fellow, NASA Goddard Space Flight Center, Greenbelt, MD, USA

[2]Heliophysics Science Division, NASA Goddard Space Flight Center, Greenbelt, MD, USA

[3]The Johns Hopkins University Applied Physics Laboratory, Laurel, MD, 20723, USA

[4]Hvar Observatory, Faculty of Geodesy, University of Zagreb, Kaciceva 26, HR-10000, Zagreb, Croatia

[5]Universidad de Alcalá, (UAH), Plaza de San Diego s/n, 28801 Alcalá de Henares, Madrid, Spain

[6]Institute of Physics, University of Graz, Universitätsplatz 5, 8010 Graz, Austria

[7]Department of Physics, University of Helsinki, P.O. Box 64, 00014, Helsinki, Finland

[8]Kanzelhöhe Observatory for Solar and Environmental Research, University of Graz, Austria

Abstract. A very fast, broad, and inclined Coronal Mass Ejection (CME) occurred on March 13, 2012, and it was surrounded by three coronal holes. The particular configuration of the event induced that two days later, the interplanetary counterpart of the CME was captured by Earth-surrounding spacecraft and STEREO-A, which at that moment were separated 110° in longitude. In this proceeding, the propagation and magnetic topology of the CME is analysed by utilising the OSPREI propagation model and Hidalgo's analytical flux rope (FR) model respectively. This research complements previous findings provided in Carcaboso et al. (2024). The study reveals that the complexity of this event is such that, with the current version of the OSPREI model, we cannot fully explain the CME's dynamics. Magnetic topology analysis with Hidalgo's model indicates the CME had a FR orientation of approximately 30 degrees with respect to the ecliptic plane at 1 au, with Earth encountering a quasi-frontal FR and STEREO-A a flank crossing. The findings underscore the complexity of CME propagation and the importance of advanced modelling techniques in understanding CME propagation and topology.

Keywords. Coronal mass ejections (CMEs), stream interaction region (SIR), numerical model, analytical model, solar-terrestrial relations

1. Introduction

Coronal mass ejections (CMEs) are a huge release of plasma with its own magnetic field that can cover a wide angular extension in the heliosphere as analysed, for example, in the articles (Palmerio et al. 2024), with 4° separation in longitude; Lugaz et al. (2022), 55°; Liu et al. (2008), 74°; Rodríguez-García et al. (2021), 80°; Chen et al. (2022), 90°, and Richardson et al. (2002), 160°. It is often possible to study their evolution and interaction with other structures in the solar wind (e.g. Gopalswamy et al. 2009) if the available space missions equipped with the proper in situ plasma and magnetic field instruments intercept these structures at different locations. The interaction with other structures can lead the CMEs to deform, deflect, kink or rotate (Riley and Crooker 2004; He et al. 2018; Chen et al. 2019). The interaction with stream interaction regions (SIRs) originating from the high-speed streams (HSSs) from coronal holes (CHs) may also slower (Gopalswamy et al. 2009) or accelerate them (Geyer, Paul et al. 2023; Cargill et al. 1996; Vršnak 2001).

In order to analyse the propagation of CMEs in the interplanetary medium, the use of models is crucial. Some of these models are complex magnetohydrodynamic models such as ENLIL (Odstrcil 2003) or EUHFORIA (Pomoell and Poedts 2018), in which a CME with certain orientation, size and speed is injected. They are computationally expensive and require supercomputers to run. However, there are more simplistic models that can improve our understanding about the propagation of CMEs in the heliosphere. One of these models is the Open Solar Physics Rapid Ensemble Information (OSPREI, Kay et al. 2022), and its component Mostly Empirical Operational Wind with a High Speed Stream (MEOW-HiSS, Kay et al. 2023), which were used for this analysis.

There are also multiple analytical models aiming to infer the local and even global topology of the flux ropes (FRs) associated to CMEs in the heliosphere (e.g. Burlaga et al. 1981; Burlaga 1988; Farrugia et al. 1992; Romashets and Vandas 2003; Nieves-Chinchilla et al. 2019; Hidalgo 2016). Reproducing the global topology of these structures is a complex task, as normally there is a single point observation of structures whose cross-section at 1 au can reach up to 0.25 au (Burlaga et al. 1981). Among numerical models, there are two main groups: force-free and non-force-free models. Force-free models are generally simpler, as they assume the magnetic field unperturbed and typically as frozen structures. This is typically translated into less accurate topology of the FRs with the advantage of having less free parameters.

This proceeding complements the information of a previously published work, Carcaboso et al. (2024), using some modelling to understand the propagation and magnetic morphology of the CME. This event consists of a very fast and wide CME originating from the Sun on March 13, 2012. Three CHs were surrounding the active region where the CME originated, resulting in a particular propagation behaviour of this structure towards the different observatories. The CME was measured in situ at 1 au on March 15, 2012, by missions surrounding Earth (such as Cluster (Escoubet et al. 2001), Wind (Ogilvie and Desch 1997; Wilson III et al. 2021) and the Advanced Composition Explorer (ACE, Stone et al. 1998)) and Solar Terrestrial Relations Observatory Ahead (STEREO-A, Kaiser et al. 2008). The in situ signatures of these two locations –separated by 110° in longitude at that moment– were remarkably similar. In our previous work, a scenario was presented, in which Earth crossed the east flank of the CME, and STEREO-A intercepted the west flank.

This proceeding is divided into the following sections: Section 2, where the properties of the origin of the CME are presented, as well as the solar and interplanetary conditions; Section 3, where the results of some simulations for the propagation of the CME are presented; Section 4, where fits are computed for the in situ magnetic FR at the different locations; and Section 5, where a brief summary and conclusions of the study are provided.

2. CME Origin and Solar Wind Context

The CME was originated in a complex active region AR11429 was located at N18W64 as seen from Earth (catalogued by the National Oceanic and Atmospheric Administration, NOAA, Space Weather Prediction Center, SWPC) on March 13, 2012.

The filament was left handed with a polarity inversion line inclined 40° wrt the ecliptic plane (Palmerio et al. 2018). There was a solar M7.9 class flare in association with the release of the CME. Multiple catalogues and our own reconstruction based on the Geometrical Cyllindrical Shell model (Thernisien 2011) draw an approximate speed of 2,000 km/s with an average orientation of ~18° in latitude and ~55° in longitude, similar to the AR location (N18W64, Carcaboso et al. 2024). Also, no rotation of the structure was observed in the field of view of the available coronagraphs, and sustained an orientation of approximately -45° counterclockwise wrt to the ecliptic, similar to the inferred filament PIL. More information about the CME origin and the context of the conditions of the Sun and its heliosphere can be found in Carcaboso et al. (2024).

3. CME Propagation

The theoretical framework posit, due to the location of the CHs with respect to the active region where the CME originated, one SIR upstream and one downstream of the CME. The upstream SIR hit STEREO-A and was identified by the STEREO/MAG team, spanning from March 14, 2012, at 06:50 UT to March 15, 2012, at 03:12 UT. Given the velocities recorded during this period and the estimated speed of the CME, it is theorised that the interaction between these large-scale structures likely occurred around 0.3 astronomical units (au).

In our previous work, the Drag-Based Model (DBM, Vršnak et al. 2013) was used and it indicated that the drag experienced by the CME along the Sun–STEREO-A trajectory was approximately ~46% greater than that along the Sun–Earth trajectory. This influenced the CME's arrival time at 1 au and its structural deformation, observable in its time transit and orientation at the different locations. In order to understand the nature of this interaction and what forces were involved, the OSPREI model was utilised.

Using the GCS reconstruction parameters and no background HSSs, the OSPREI CME arrives at both locations in roughly a day, arriving slightly earlier at STEREO-A than at L1, which is the opposite of the observed case. Due to its pronounced inclination and latitude, the observed transit times can only be reproduced by lowering the CME latitude, changing its tilt and/or yaw, and reducing its speed. The hypothetical reconstruction of Carcaboso et al. (2024) also shows the CME with lower latitude and flatter inclination at 1 au than the GCS reconstructed.

At the current OSPREI version, only one HSS can be introduced. However, the GCS parameters still do not yield matching transit times when including a background HSS. From the different runs in the quiescent background, we find that the difference in transit time can be explained with a positive yaw angle of 11°, where the side of the CME closer to Earth is ahead of the side near STEREO-A. When initialising the CME with 0° yaw and including the HSS behind the CME (side closer to Earth than STEREO-A) then the reduced drag on this side introduces a yaw rotation.

The observed transit times can be reproduced if a low inclination CME at lower latitude and simulate the yaw rotation from the trailing HSS are used. Opposite effect is observed when including the upstream HSS, causing an earlier arrival at STEREO-A. For these reasons, it may be that this complex of an event is beyond the scope of the OSPREI and MEOW-HiSS system, though it generally remains somewhat inconclusive how the observed CME evolved from the orientation from the GCS coronal reconstruction to the orientation observed in situ.

4. Flux Rope Reconstruction

The magnetic orientation of the in situ ejecta at ACE and STEREO-A was determined using the non-force-free Hidalgo's analytical model (Hidalgo 2014). This model is able to

Table 1. Result of the parameters after the fitting process. ϕ corresponds to the local FR angle in the ecliptic plane, θ with respect to the ecliptic plane in Radial, Tangential Normal (RTN) coordinates. η corresponds to the circularity of the cross section (the smaller, more elliptical).

	ϕ [deg]	θ [deg]	η	Handedness
ACE	-80	151	11	Left
STEREO-A	11	31	8.1	Left

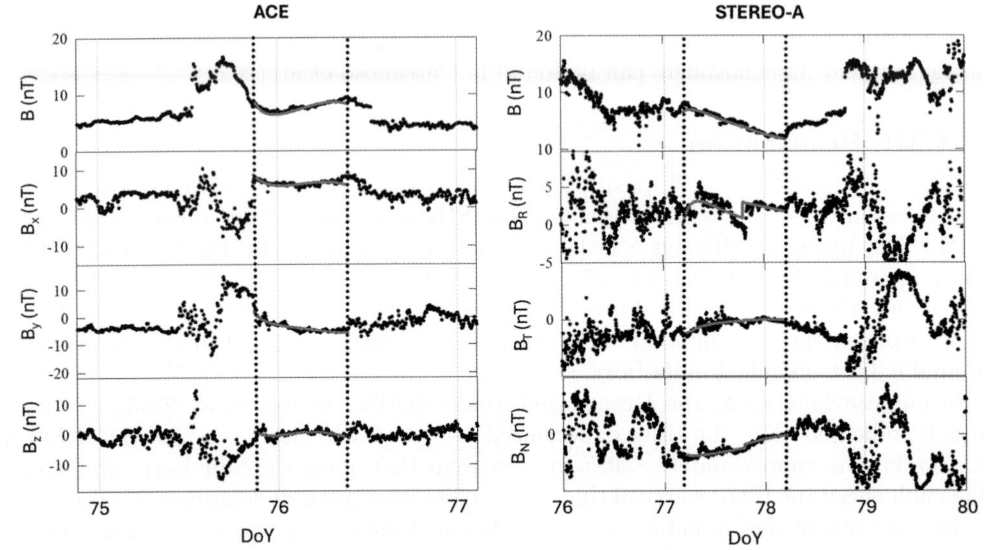

Figure 1. Magnetic field strength and its components in GSE (ACE, left) and RTN (STEREO-A, right). Red line corresponds to the results of the fit process using Hidalgo's model (Hidalgo 2014). Vertical dotted lines correspond to the limits of the FR fit.

provide a global topology and orientation of the magnetic structure out of the crossing of a single spacecraft. The model assumes a torus morphology with a variable cross section. Although we think that the structure underwent some deformation due to the interaction with the ambient solar wind, the morphology would still be close enough to a torus when reaching 1 au.

The interval for the fitting was from March 15, 2012, (day of year, DoY, 75) at 21:00 UT to March 15, 2012, at 09:46 UT for ACE and March 16, 2012, (DoY 76) at 04:48 UT to March 17, 2012, (DoY 77) at 05:08 UT for STEREO-A. The obtained parameters can be found in Table 1. Figure 1 shows the magnetic field in GSE for ACE (left) and in RTN for STEREO-A and the result of the fit (red line).

The angles obtained by the model indicate that the structure is inclined approximately 30° (θ) wrt the ecliptic plane at both ACE and STEREO-A. Also, it shows that Earth has a cross closer to the front part, while STEREO-A has a cross closer to the flank. STEREO-A presents a more circular cross section (η) in comparison to ACE. However, the shape of the cross section is hard to infer due to the particular flank transit. In both cases, a left handedness is obtained.

5. Summary and Conclusions

In this work we complement the information provided in Carcaboso et al. (2024) with the propagation and magnetic topology of the CME released on March 13, 2012, that impacted both Earth and STEREO-A on March 15, 2012, when they were separated approximately 110° in longitude. In order to do that, we used the propagation model

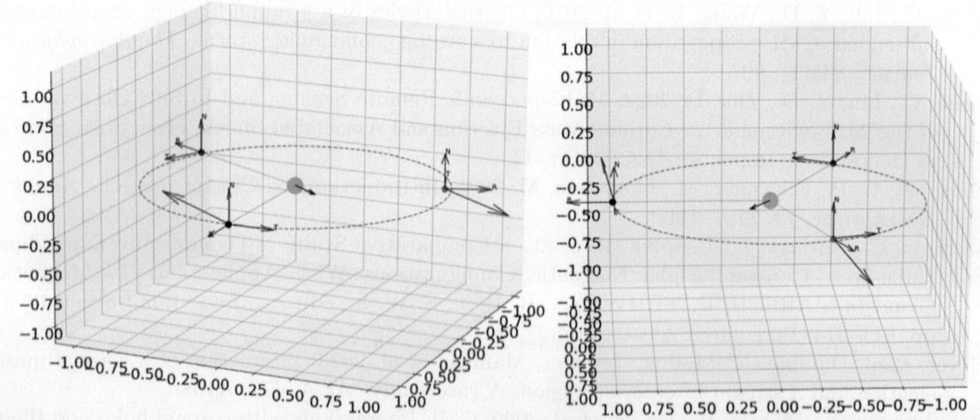

Figure 2. Orbital position of the spacecraft (red, STEREO-A; blue, STEREO-B; black, ACE) with their respective RTN coordinate reference system. Black arrow at the Sun represents the longitude where the active region was located at the moment of the eruption. Purple arrows show the axis orientation as derived from Hidalgo's model.

OSPREI (Kay et al. 2022) and the analytical FR model Hidalgo (Hidalgo 2014) respectively. By these means, we support the conclusions obtained in Carcaboso et al. (2024) and obtain the new following insights:

(1) The current version of OSPREI cannot reproduce the observed transit times using the GCS reconstruction as inputs. Whether this is due to an oversimplification in the model of the deflection, rotation, and deformation at play for this complicated event or an issue with the inputs requires further analysis.

(2) The magnetic obstacle of the CME had a FR with an orientation wrt the ecliptic plane of 30° at 1 au, with a quasi-frontal cross at ACE and a flank crossing at STEREO-A.

Acknowledgements. The authors acknowledge the open policy of the data for the different missions. We acknowledge use of NASA/GSFC's Space Physics Data Facility's CDAWeb service. FC acknowledges the financial support by an appointment to the NASA Postdoctoral Program at NASA Goddard Space Flight Center, administered by Oak Ridge Associated Universities through a contract with NASA, the support from the IAU, and the support of the Solar Orbiter mission. FC and LKJ thank the support of the STEREO mission. MD acknowledges the support by the Croatian Science Foundation under the project IP-2020-02-9893 (ICOHOSS). RGH acknowledges the financial support of the Spanish Ministerio de Ciencia Innovación y Universidades Project PID2019-104863RBI00/AEI/10.13039/501100011033. SGH acknowledges funding by the Austrian Science Fund (FWF): Erwin-Schrödinger fellowship J-4560. LKJ and TNC are supported by HGI-Open grant 80NSSC23K0447. This work has been possible thanks to the scientific discussions within the LASSOS-Goddard group.

References

Burlaga, L., Sittler, E., Mariani, F. et al. 1981, Magnetic loop behind an interplanetary shock: Voyager, Helios, and IMP 8 observations. *Journal of Geophysical Research*, 86(A8), 6673.

Burlaga, L. F. 1988, Magnetic clouds and force-free fields with constant alpha. *Journal of Geophysical Research: Space Physics*, 93(A7), 7217–7224.

Carcaboso, F., Dumbović, M., Kay, C. et al. 2024, Unveiling the journey of a highly inclined CME. Insights from the March 13, 2012, event with 110° longitudinal separation. *A&A*, 684, A90.

Cargill, P. J., Chen, J., Spicer, D. S. et al. 1996, Magnetohydrodynamic simulations of the motion of magnetic flux tubes through a magnetized plasma. *J. Geophys. Res.*, 101(A3), 4855–4870.

Chen, C., Liu, Y. D., Wang, R. et al. 2019, Characteristics of a gradual filament eruption and subsequent CME propagation in relation to a strong geomagnetic storm. *The Astrophysical Journal*, 884(1), 90.

Chen, C., Liu, Y. D., Zhu, B. 2022, Multispacecraft Remote Sensing and In Situ Observations of the 2020 November 29 Coronal Mass Ejection and Associated Shock: From Solar Source to Heliospheric Impacts. *ApJ*, 937(1), 44.

Escoubet, C. P., Fehringer, M., Goldstein, M. 2001, IntroductionThe Cluster mission. *Annales Geophysicae*, 19, 1197–1200.

Farrugia, C., Burlaga, L., Lepping, R. et al. a Comparative Study of Dynamically Expanding Force-Free, Constant-Alpha Magnetic Configurations With Applications To Magnetic CloudsIn MARSCH, E., SCHWENN, R. B. T. S. W. S., editors, *Solar Wind Seven* 1992, pp. 611–614. Pergamon, Amsterdam.

Geyer, Paul, Dumbovi´c, Mateja, Temmer, Manuela et al. 2023, Interaction of a coronal mass ejection and a stream interaction region: A case study. *A&A*, 672, A168.

Gopalswamy, N., Mäkelä, P., Xie, H. et al. 2009, CME interactions with coronal holes and their interplanetary consequences. *Journal of Geophysical Research (Space Physics)*, 114(A3), A00A22.

He, W., Liu, Y. D., Hu, H. et al. 2018, A stealth CME bracketed between slow and fast wind producing unexpected geoeffectiveness. *The Astrophysical Journal*, 860(1), 78.

Hidalgo, M. A. 2014, A Global Magnetic Topology Model for Magnetic Clouds. III. *The Astrophysical Journal*, 784(1), 67.

Hidalgo, M. A. 2016, A Global Magnetic Topology Model for Magnetic Clouds. IV. *The Astrophysical Journal*, 823(1), 3.

Kaiser, M. L., Kucera, T. A., Davila, J. M. et al. 2008, The STEREO Mission: An Introduction. *Space Science Reviews*, 136(1), 5–16.

Kay, C., Mays, M. L., Collado-Vega, Y. M. 2022, OSPREI: A Coupled Approach to Modeling CME-Driven Space Weather With Automatically Generated, User-Friendly Outputs. *Space Weather*, 20(4), e2021SW002914.

Kay, C., Nieves-Chinchilla, T., Hofmeister, S. J. et al. 2023, A Series of Advances in Analytic Interplanetary CME Modeling. *Space Weather*, 21(11), e2023SW003647. e2023SW003647 2023SW003647.

Liu, Y., Luhmann, J. G., Müller-Mellin, R. et al. 2008, A Comprehensive View of the 2006 December 13 CME: From the Sun to Interplanetary Space. *ApJ*, 689(1), 563–571.

Lugaz, N., Salman, T. M., Zhuang, B. et al. 2022, A Coronal Mass Ejection and Magnetic Ejecta Observed In Situ by STEREO-A and Wind at 55° Angular Separation. *The Astrophysical Journal*, 929(2), 149.

Nieves-Chinchilla, T., Jian, L. K., Balmaceda, L. et al. 2019, Unraveling the Internal Magnetic Field Structure of the Earth-directed Interplanetary Coronal Mass Ejections During 1995–2015. *Solar Physics*, 294(7), 89.

Odstrcil, D. 2003, Modeling 3-D solar wind structure. *Advances in Space Research*, 32(4), 497–506. Heliosphere at Solar Maximum.

Ogilvie, K. W., Desch, M. D. 1997, The wind spacecraft and its early scientific results. *Advances in Space Research*, 20(4-5), 559–568.

Palmerio, E., Carcaboso, F., Khoo, L. Y. et al. 2024, On the Mesoscale Structure of Coronal Mass Ejections at Mercury's Orbit: BepiColombo and Parker Solar Probe Observations. *ApJ*, 963(2), 108.

Palmerio, E., Kilpua, E. K. J., Möstl, C. et al. 2018, Coronal Magnetic Structure of Earthbound CMEs and In Situ Comparison. *Space Weather*, 16(5), 442–460.

Pomoell, J., Poedts, S. 2018, EUHFORIA: European heliospheric forecasting information asset. *Journal of Space Weather and Space Climate*, 8, A35.

Richardson, J. D., Paularena, K. I., Wang, C. et al. 2002, The life of a CME and the development of a MIR: From the Sun to 58 AU. *Journal of Geophysical Research: Space Physics*, 107(A4), SSH 1-1–SSH 1-9.

Riley, P., Crooker, N. U. 2004, Kinematic Treatment of Coronal Mass Ejection Evolution in the Solar Wind. *ApJ*, 600(2), 1035–1042.

Rodríguez-García, L., Gómez-Herrero, R., Zouganelis, I. et al. 2021, The unusual widespread solar energetic particle event on 2013 August 19-Solar origin and particle longitudinal distribution. *Astronomy & Astrophysics*, 653, A137.

Romashets, E. P., Vandas, M. 2003, Force-free field inside a toroidal magnetic cloud. *Geophysical Research Letters*, 30(20).

Stone, E. C., Frandsen, A., Mewaldt, R. et al. 1998, The advanced composition explorer. *Space Science Reviews*, 86(1-4), 1–22.

Thernisien, A. 2011, Implementation of the graduated cylindrical shell model for the three-dimensional reconstruction of coronal mass ejections. *Astrophysical Journal, Supplement Series*, 194(2), 33.

Vršnak, B. 2001, Deceleration of Coronal Mass Ejections. *Solar Phys.*, 202(1), 173–189.

Vršnak, B., Žic, T., Vrbanec, D. et al. 2013, Propagation of Interplanetary Coronal Mass Ejections: The Drag-Based Model. *Solar Phys.*, 285(1-2), 295–315.

Wilson III, L. B., Brosius, A. L., Gopalswamy, N. et al. 2021, A Quarter Century of Wind Spacecraft Discoveries. *Reviews of Geophysics*, 59(2), e2020RG000714. e2020RG000714 2020RG000714.

Solar and Stellar Coronal Mass Ejections
Proceedings IAU Symposium No. 388, 2026
N. Gopalswamy, O. Malandraki, A. Vidotto & W. Manchester, eds.
doi:10.1017/S174392132400111X

Observational Signatures of Stellar CMEs Associated with Superflares

Kosuke Namekata[1,2,3]

[1]The Hakubi Center for Advanced Research, Kyoto University, Kyoto 606-8302, Japan

[2]Department of Physics, Kyoto University, Kitashirakawa-Oiwake-cho, Sakyo-ku, Kyoto, 606-8502, Japan

[3]Division of Science, National Astronomical Observatory of Japan, NINS, Osawa, Mitaka, Tokyo, 181-8588, Japan

Abstract. The Kepler and TESS missions have unveiled that cool stars frequently host exoplanets. These exoplanets are subject to environmental changes due to flares and associated coronal mass ejections (CMEs) that may change their atmospheric dynamics and chemistry. However, substantial observational evidence of CMEs from G-M dwarfs remains scarce. Recent efforts to detect stellar CMEs have predominantly utilized methodologies derived from solar observations. Notably successful approaches have been the detection of Doppler shifts and coronal dimming associated with stellar flares, which are interpreted as strong indicators of stellar CMEs. Although the detection of type-II radio bursts remains elusive, candidates for type-IV radio bursts have been recently reported. Numerous indicators have been tested; however, each methodology presents its own set of issues and limitations, and a definitive answer remains elusive. This paper reviews the recent progress and proposes several future directions in CME observations of active stars.

Keywords. Flare stars, Stellar flares, Stellar coronal mass ejections, Optical flares, Stellar x-ray flares

1. Introduction

Solar and stellar flares, explosions on the surfaces observed from X-ray to radio wavelengths, are known to be driven by the conversion of magnetic to kinetic and thermal energies through the process of magnetic reconnection (Shibata & Magara 2011; Namekata et al. 2017). These solar flares often occur alongside eruptions of filaments or prominences and are frequently accompanied by coronal mass ejections (CMEs; Gopalswamy et al. 2003). These solar CMEs are capable of producing shockwaves and high-energy particles. Furthermore, solar active regions emit strong X-ray and extreme ultraviolet (EUV) radiation, influencing the ionospheric layers of planets. These magnetic phenomena on the Sun thus play a significant role in shaping planetary environments and affecting human activities (Cliver et al. 2022).

Recent observations by the Kepler and TESS missions have shown that F, G, K, and M dwarf stars are common hosts for exoplanets. These planets orbiting active stars are often exposed to intense flares, elevated XUV radiation, and strong stellar winds, potentially leading to significant atmospheric erosion and chemical composition changes (Linsky 2019; Airapetian et al. 2020). The magnetized plasma from stellar CMEs might even strip away the atmospheres of nearby exoplanets. Furthermore, it is suggested that high-energy particles from these stellar CMEs could trigger chemical processes in planetary atmospheres, forming molecules critical for habitability and the emergence of life, including greenhouse gases and precursors to prebiotic compounds (Airapetian et al. 2016). Consequently, the study and identification of stellar CMEs have become increasingly important for assessing the habitability of exoplanets.

The Kepler and TESS missions have also significantly enhanced our understanding of the statistical relationships between stellar characteristics–such as surface temperature, rotation period, and age–and flare properties, including their frequency and energy, over the past decade (Maehara et al. 2012; Davenport 2016). Kepler/TESS observations show young stars, cool M-dwarfs, and close-binary stars produce more frequent and energetic flares, called "superflares" (energy of more than 10^{33} erg), indicating a more pronounced effect on the environments of planets, particularly during the formation of habitable conditions (Maehara et al. 2012; Davenport 2016; Notsu et al. 2019; Okamoto et al. 2021). However, until recently, little was known about whether these superflares were accompanied by super-CMEs.

Based on the observed correlation between solar flares and CMEs, it is expected that stellar superflares could be associated with larger and more frequent CMEs. However, many active stars possess properties distinct from those of the Sun, such as different magnetic field strengths, magnetic topology, rotational velocities, and characteristics of stellar winds (e.g., Vidotto et al. 2014). These factors could potentially inhibit the occurrence of stellar CMEs (e.g., Alvarado-Gómez et al. 2018; Sun et al. 2022) or alter their observational signatures (Mullan & Paudel 2019; Alvarado-Gómez et al. 2020). Consequently, the mere presence of flares does not guarantee detectable stellar CMEs using the same methods applied to solar observations. This necessitates observational investigation to ascertain whether stellar CMEs are occurring and if they can be detected in a manner analogous to solar techniques.

In this paper, I review the current status of observational signatures of stellar CMEs in Section 2 and give recommendations for future studies in Section 3. Please also refer to the review papers by Osten & Wolk (2017); Leitzinger & Odert (2022); Namekata et al. (2022b).

2. Observational Signatures of Stellar CMEs

Observing stellar flares, characterized by brightening on the stellar surfaces, is relatively straightforward. However, detecting stellar CMEs is significantly more challenging and rare. Unlike the Sun, other stars cannot be spatially resolved with current technology, making it almost impossible to visually observe CMEs as they erupt. Additionally, the signal from a CME is often much weaker compared to the brightness of the stellar flare and the star itself, causing it to be easily lost in the noise of the observation.

Despite these difficulties, efforts have been made to leverage our understanding of solar flares to indirectly or directly identify evidence of stellar CMEs. For instance, Figure 1 presents a schematic overview of observational evidence from the Sun. There are also attempts to identify unique observational signatures that, while not well characterized in solar flares, could indicate the presence of large stellar CMEs. These methods and approaches, some of which have been successful while others remain unproven, are summarized in Table 1. The following discussion will cover the results obtained from these methods, the challenges faced, and potential improvements for future observations.

2.1. *Doppler shifts*

The Doppler shift of spectral lines offers a direct method to detect plasma motion. Differentiating these shifts from chromospheric evaporation/condensation and post-flare

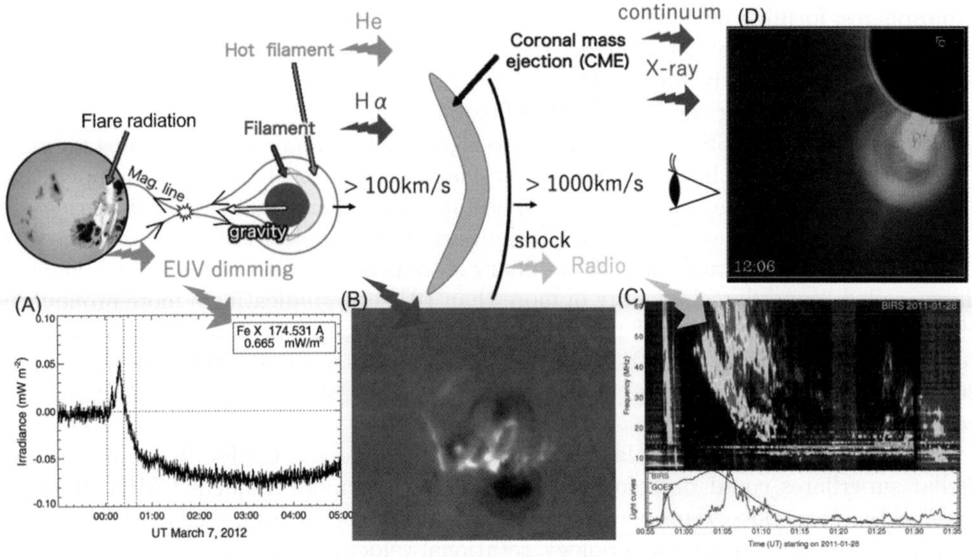

Figure 1. Schematic examples of detection methods of solar CMEs as a guide to stellar CMEs (adapted from Namekata et al. 2022b). (A) A coronal dimming is observed as a decrease in EUV, serving as an indirect indicator of CME (Harra et al. 2016, Courtesy of Harra et al.). (B) Doppler shifts from prominence/filament Eruptions (Image courtesy of Hida Observatory). (C) Type-II Radio Burst demonstrates a CME shockwave moving through space (Crosley et al. 2017, Courtesy of Crosley et al.). (D) Coronagraph: A CME captured in white light using a coronagraph on the Sun (Image courtesy of HAO/SMM C/P project team & NASA. HAO is part of the National Center for Atmospheric Research, supported by the National Science Foundation; DOI: 10.5065/D64J0CXB).

loops is difficult for stars that cannot be spatially resolved, but it remains a promising approach for identifying plasma motions linked to CMEs.

Numerous studies have reported blueshifted chromospheric lines during stellar flares, primarily in the optical (e.g., Houdebine et al. 1990; Gunn et al. 1994; Vida et al. 2016; Honda et al. 2018; Argiroffi et al. 2019; Vida et al. 2019; Maehara et al. 2021; Namekata et al. 2022c; Inoue et al. 2023; Namekata et al. 2024; Notsu et al. 2024; Leitzinger et al. 2024) but also in the UV spectrum (Hawley et al. 2007; Leitzinger et al. 2014). These observations have mostly focused on M-K dwarfs, although recent findings have extended to G dwarfs (Namekata et al. 2022c, 2024; Leitzinger et al. 2024) and RS CVn-type binary systems (Inoue et al. 2023). Blueshifted chromospheric lines are often interpreted as indicators of stellar prominence or filament eruptions, with some researchers suggesting that redshifted lines indicate backward eruptions. Figure 2 shows a recent report of blueshift absorption spectrum as a clear evidence of stellar filament eruptions from a young G dwarf EK Dra (Namekata et al. 2022c). Comparison with the-Sun-as-a-star analyses of solar filament eruptions suggest that the events occurring on the G-dwarf might be analogous to those observed on the Sun. On the other hand, in the case of M-dwarfs, all observed events show emission signatures, making it difficult to distinguish between the flare components and filament/prominence eruptions (see, the model calculations by Leitzinger et al. 2022).

In many cases, the observed blueshift velocities in chromospheric lines do not surpass the surface escape velocity of stars. This makes it difficult to definitively identify these events as escaping CMEs; therefore, these observations are not always considered conclusive evidence of CMEs. However, even on the Sun, prominence and filament eruptions associated with CMEs frequently occur at sub-escape velocities (e.g., Gopalswamy et al.

Table 1. List of Successful and Unsuccessful Indications of Stellar CMEs. (A) Reported but not well-justified. (B) Attempted but not detected. (C) Challenging with current techniques. ‡The number of events reported before June 15th, 2024. †Time-resolved spectroscopic observations are still limited while snapshot blueshifts have been many reported by Vida et al. (2019). References: [1] e.g., Houdebine et al. (1990); Gunn et al. (1994); Vida et al. (2016); Honda et al. (2018); Argiroffi et al. (2019); Maehara et al. (2021); Namekata et al. (2022c); Inoue et al. (2023); Namekata et al. (2024); Notsu et al. (2024); and many snapshot data from Vida et al. (2019) [2] Veronig et al. (2021) ; Loyd et al. (2022); Namekata et al. (2024), [3] Moschou et al. (2017); Moschou et al. (2019), [4] Zic et al. (2020); Mohan et al. (2024), [5] Golay et al. (2024), [6] Giampapa et al. (1982); Leitzinger et al. (2014) [7] See efforts by Crosley & Osten (2018a,b) [8] Report on possible SPI by Vedantham et al. (2020), but probably not CME-related [9] Reported on the Sun (Grechnev et al. 2008). §The signatures have been already attempted but not detected with stellar observations, or data exists in archive but has not been sufficiently explored.

Methods/Signatures	**Wavelength**	**Sun**	**Stars**	**No.‡**	**Ref.**	**Sect.**
Doppler Shifts	Optical/UV/X-ray	Yes	Yes	≥100†	[1]	2.1
Coronal Dimming	X-ray/EUV	Yes	Yes	23	[2]	2.2
Absorption Dimming	X-ray	?	Yes	7	[3]	2.3
Type IV Radio Burst	Radio	Yes	Yes	2	[4]	2.4
Moving Radio Source(A)	Radio	?	?	?	[5]	2.5
Pre-flare Dips(A)	Optical/UV	Yes	?	?	[6]	2.5
Type II Radio Burst(B)	Radio	Yes	?	?	[7]§	2.6
CME-related Star-Planet Interaction (e.g., Aurora)(B)	Radio/UV/Optical	Yes	?	?	[8]§	2.6
Negative Radio Burst(B)	Radio	Yes	?	?	[9]§	2.6
Interplanetary Scintillation(C)	Radio	Yes	?	?	–	2.7
Coronagraph(C)	–	Yes	?	?	–	

2003), implying that low velocity does not necessarily rule out the presence of CMEs. Additionally, some observations have reported velocities near (e.g., Namekata et al. 2022c, 2024) or exceeding the escape velocity (e.g., Houdebine et al. 1990; Vida et al. 2016, 2019; Inoue et al. 2023). The tendency to overlook the link between fast prominence/filament eruption candidates and potential stellar CMEs may be due to the short-lived nature of these events, complicating the tracking of their propagation into interplanetary space.

Doppler shifts have also been detected in X-rays flares (Argiroffi et al. 2019; Chen et al. 2022; Inoue et al. 2024), although distinguishing these shifts from chromospheric evaporation remains challenging. For example, Argiroffi et al. (2019) detected a Doppler shift of 90 km s^{-1} (below the escape velocity) only in lines corresponding to the steady corona temperature and interpreted this as indicative of a CME, supported by 1D loop calculations. Nevertheless, given the inherently three-dimensional nature of flares, relying on a one-dimensional model presents limitations. Moreover, no similar phenomena have been observed in solar CMEs.

The advantage of the Doppler method lies in its ability to leverage numerous observational examples to understand statistical properties and to derive plasma velocities and masses with fewer assumptions. It has been suggested that the mass of prominence eruptions/CMEs can be explained by solar prominence eruption/CME–flare energy scaling laws (e.g., Aarnio et al. 2012; Drake et al. 2013; Moschou et al. 2019). This implies that the underlying physics might be consistent with solar physics (Takahashi et al. 2016; Kotani et al. 2023). On the other hand, the kinetic energy of stellar CME candidates appears to fall below the solar empirical relationships. This discrepancy could be due to the suppression of kinetic energy by magnetic fields (e.g., Alvarado-Gómez et al. 2018) and/or the differences in velocities between CMEs and prominences (e.g., Namekata et al. 2022c). To resolve this issue, it is important to determine the kinetic energy using other methods (e.g., absorption dimming, Moschou et al. 2019).

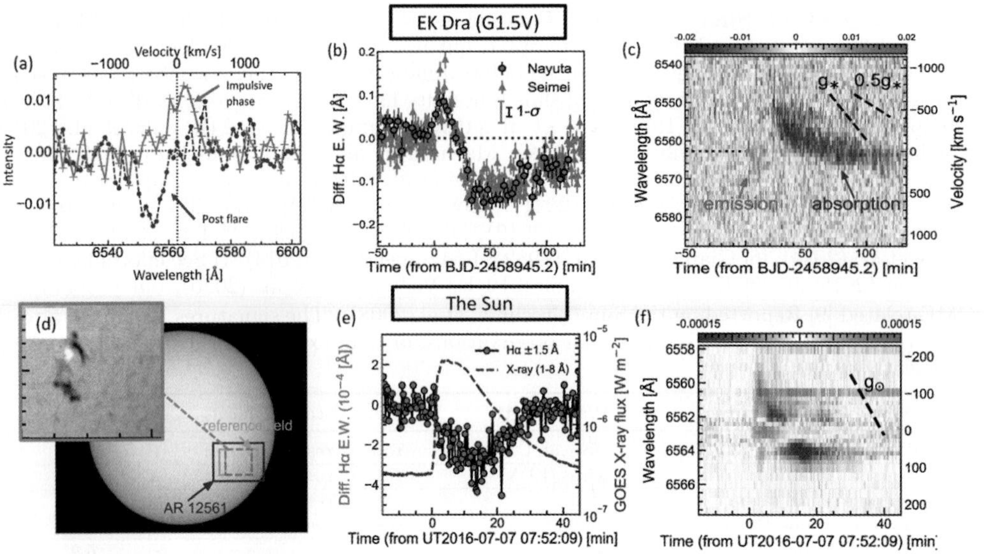

Figure 2. Example of Doppler Shifts Methods for Young Solar-type Star, EK Dra (G1.5V), and the Sun (reproduced from Namekata et al. 2022c). (a) Typical pre-flare-subtracted spectra during the superflare and filament eruption on EK Dra. (b) Light curve of Hα equivalent width of EK Dra (pre-flare level is subtracted). (c) Time evolution of pre-flare-subtracted spectra of Hα line of EK Dra. (d) A solar filament eruption in Hα. (e) Light curve of Hα equivalent width of the solar filament eruption in the Sun-as-a-star view. (f) Time evolution of pre-flare-subtracted spectra of Hα line of the Sun in the Sun-as-a-star view.

2.2. *Coronal Dimming (a.k.a. Mass-loss dimming, XUV dimming)*

Solar coronal dimming is a phenomenon where the coronal emission decreases following flares (see Figure 1). This dimming is believed to result from coronal plasma, initially confined by closed magnetic fields, escaping from the solar surface due to CMEs. Signals of coronal dimming have been observed in Sun-as-a-star EUV spectral lines (Harra et al. 2016), and its application to stellar observations has been proposed. Subsequently, Veronig et al. (2021) expanded on this idea and found that solar coronal dimming is frequently associated with CMEs, with a probability of more than 80 percent. Therefore, coronal dimming is one of the most promising indicators for detecting the occurrence of stellar CMEs.

Veronig et al. (2021) reported post-flare coronal dimming in X-ray and EUV wavelengths by utilizing the archival dataset of stellar flare observations on cool stars (see Figure 3). Later, Loyd et al. (2022) identified potential post-flare FUV dimming of Fe XII 1349 Å and Fe XXI 1354 Å emissions on ϵ Eri (K-type star) using Hubble data. More recently, Namekata et al. (2024) reported a potential X-ray dimming associated with gigantic prominence eruptions in Hα from young solar-type star EK Dra. These observations are considered promising evidence of CMEs. However, the number of such detections remains limited, and they depend on the definition of the quiescent flux level, which can vary over time due to several reasons.

A significant point is that the depth and slope of the dimming light curve can be measured from stellar light curves (Veronig et al. 2021). Solar studies have shown that the mass of a CME can be estimated from the depth of the dimming, and the velocity can be determined from its slope (Mason et al. 2016). Although these methods have yet to be applied to stars and require modeling of stellar atmospheres for their application, they

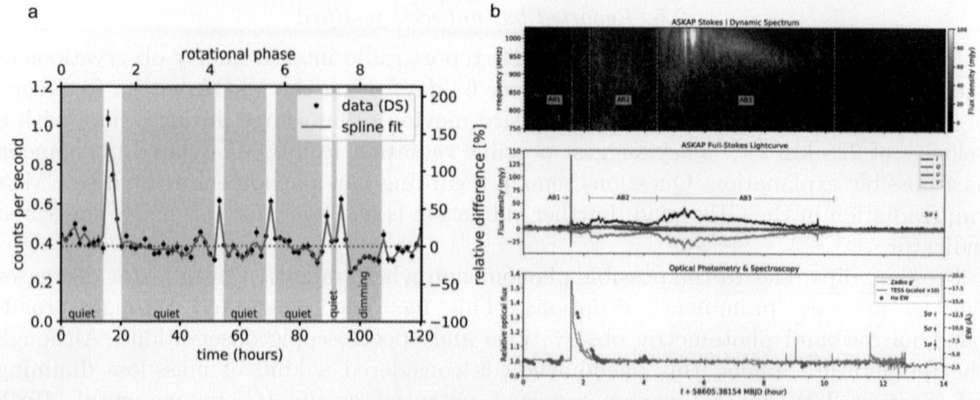

Figure 3. Example of Coronal dimming signature on the K-dwarf AB Dor (Vida et al. 2024; reproduced from Veronig et al. 2021, Courtesy of Vida et al.) and type-IV radio burst on the M-dwarf Proxima Centauri (Zic et al. 2020, Courtesy of Zic et al.).

are considered very promising for investigating the mass and velocity of stellar CMEs to complement Doppler methods.

2.3. *Absorption Dimming*

X-ray absorptions during stellar flares are also considered a potential signature of stellar CMEs. This is indirectly obtained as a result of X-ray spectral fitting with a parameter of hydrogen column density causing ISM absorption, N_H. The additional absorption during flares can indicate the presence of plasma obscuring the flaring radiation, likely due to stellar filament eruptions. Favata & Schmitt (1999) (and later Moschou et al. 2017) reported that a superflare on Algol exhibited continuous X-ray absorption, which gradually decayed following an inverse square law over time. They suggested that this phenomenon could be explained by expanding plasma above the star. Additional promising candidates for such signatures are summarized by Moschou et al. (2019). However, CME properties are model- and fitting-dependent, and without a direct solar counterpart, they may only be considered possible CME candidates. In the future, the validity of this method can be constrained by comparing it with other techniques.

2.4. *Type IV Radio Burst*

Type-II radio bursts are considered the most promising signature of stellar CME (shockwave) propagation, yet no such signature has been reported in publications (see Section 2.6). Alternatively, recent observations report type-IV radio bursts from nearby M dwarfs. Type-IV radio bursts come from high-energy electrons in the flux rope in CMEs and/or flaring loops. Zic et al. (2020) reported a (possible moving) type-IV radio burst from the M dwarf Proxima Centauri (see Figure 3) using ASKAP in Australia, which may indicate the presence of a stellar CME flux rope. Although distinguishing the radiation from CMEs and that from flaring loops is difficult, the possible drift in frequency can represent a possible CME origin. More recently, Mohan et al. (2024) reported a possible type-IV radio burst from the M dwarf AD Leo using GMRT in India. Currently, although there are few detections, the existing observations suggest the potential for an increase in detection rates. Future observations in the MHz band with instruments such as SKA, GMRT, and LOFAR are highly anticipated.

2.5. *Reported but not well-justified*

Moving Radio Source: Golay et al. (2024) report radio interferometry observations of the close-binary HR 1099 at six epochs over 63 days using the VLBA with a frequency of 22.2 GHz. They report a possible outward-moving radio source during a flare with a velocity of 228 km s^{-1}. They suggest possible radiation from CME-related phenomena as a possible explanation. Questions remain regarding how phenomena related to CMEs emit radiation in the GHz band. Further validation is necessary for this newly suggested indicator.

Pre-flare dips refer to the possible phenomenon where radiation from a star decreases due to pre-flare prominence eruptions. This has been suggested through broadband/narrowband photometric observations and spectroscopic observations. Although the temperature differs, this phenomenon is considered a kind of mass-loss dimming (cf. Section 2.2). Several papers reported potential events (Giampapa et al. 1982; Leitzinger et al. 2014), but the cause remains unknown.

2.6. *Attempted but not detected*

Type-II radio bursts are considered the most promising indicators of stellar CME shockwave propagation. There have been no publications for stars despite some efforts (Crosley & Osten 2018a,b). The absence of type-II radio bursts might suggest fewer or no CMEs on these stars. However, since other stars may have different coronal conditions compared to the Sun, the expected frequency of radio bursts might differ, or current radio telescopes may lack the necessary sensitivity (Alvarado-Gómez et al. 2020). Additionally, CMEs may occur without generating shockwaves (Mullan & Paudel 2019). Future observations with high sensitivity at low frequencies, using instruments such as LOFAR, SKA, or a potential lunar observatory, are required.

Negative radio bursts, akin to absorption dimming (Section 2.3) observed in radio GHz frequencies, have been reported on the Sun (Grechnev et al. 2008). This is thought to be caused by the eruptive prominence obscuring the stellar surface or bright active regions. Since these can be observed in the GHz band on the Sun, it may be possible to validate this phenomenon using existing data (e.g., VLA data archive).

Star-planet interactions (e.g., planetary auroras) have been gaining more interest and have been attempted (e.g., Vedantham et al. 2020). Although available signatures are limited, future observations may enable us to speculate on the occurrence of stellar CMEs in the future.

2.7. *Challenging with current techniques*

Interplanetary scintillation is also a phenomenon that represents solar CME propagation, which is detected by observing extrasolar objects with radio waves. If a CME crosses the line of sight between the object and Earth during observation, it scatters the radio waves from the object. This method is also discussed for stars, but it requires a sufficient angular diameter of CME propagation and a convenient background radio source.

3. Recommendations for Future Study

What can we do for detecting and characterizing stellar CMEs in the next decade? To date, we have maximized our efforts using existing telescopes and observational data across various wavelengths. As a result, there have been significant advancements over the past ten years. Based on our current understanding, I recommend focusing on the following four future directions over the next decade: increasing radio & UV data (Section

3.1), multi-wavelength observations (Section 3.2), reproducing observations with numerical models (Section 3.3), and solar studies extracting information from stellar data (Section 3.4).

3.1. *Increasing UV and Radio Observations*

UV and radio wavelengths are clearly lacking but crucial for observations. In the EUV, we can detect dimming (Veronig et al. 2021) and Doppler shifts (Xu et al. 2022) associated with CMEs, as observed in the Sun, making it an important wavelength for direct comparison with solar CME signature. To cover this wavelength, missions such as the ESCAPE mission have been proposed recently (France et al. 2019), and it is essential to promote these initiatives. Additionally, recent proposals suggest that certain coronal lines in the FUV might be suitable for detecting dimming (Loyd et al. 2022) and Doppler shifts associated with prominence eruptions. A unique aspect of the FUV is its ability to simultaneously capture planetary responses, such as atmospheric escape and auroras. Currently, we rely on the expensive Hubble Space Telescope for these observations, but there is already a wealth of observational data available, making an archive survey crucial. Furthermore, Japan's LAPYUTA mission and other new missions should be considered to enable more flexible and long-duration FUV observations.

Radio observations are critical for studying phenomena such as type-II, type-III, type-IV bursts, negative radio bursts, and star-planet interactions (auroras). In the solar system, many of these require observations in the MHz band, which are currently lacking for stellar observations. To address this gap, large-scale surveys using, e.g. LOFAR and GMRT, should be conducted, along with high-sensitivity surveys by SKA starting in the late 2020s. Additionally, there is great anticipation for observations by a potential lunar observatory.

3.2. *Multi-Wavelength Observations*

Currently, most observations of CMEs are limited to single-wavelength and single-signature detections. In the future, detecting CME signatures across multiple wavelengths and from various perspectives will not only provide a more comprehensive picture of CMEs but also enhance our understanding of related phenomena. This is a direction likely agreed upon by the entire community. For example, Namekata et al. (2024) provided the first multi-wavelength study linking various CME signatures from a single event. They observed a superflare on the young solar-type star EK Dra and detected an enormous prominence eruption with a mass of 10^{18-20} g. Simultaneous X-ray observations indicated possible coronal dimming (or its upper limit), revealing that the mass of the coronal dimming region was significantly smaller than that of the prominence eruption. While it is unclear how this comparison stands since such mass comparisons between prominence eruptions and dimming regions have not been extensively done for the Sun, the study succeeded in quantitatively assessing the spatial extent of these regions. Additionally, they conducted simultaneous spot and magnetic field mapping using DI/ZDI techniques. Such comprehensive datasets are expected to facilitate the modeling of stellar winds and CMEs, allowing for direct comparisons between numerical simulations and observations. Increasing such multi-perspective observational data will likely enhance our understanding of CME propagation in various stellar environments.

3.3. *Direct Comparison between Theoretical Model and Observations*

There are numerous observations and theoretical studies; however, direct comparisons with observations and numerical models remain limited. For a limited example,

Ikuta & Shibata (2024) conducted 1D simple modeling to explain the filament eruption shown in Figure 2, and suggest that the loop top expands with a velocity above the escape velocity although the observed velocity is below the escape velocity. Leitzinger et al. (2022) performed radiative transfer modeling of stellar prominences on M-dwarfs, exploring the prominence intensity. Close collaboration between theoretical and observational groups will be essential in the future.

3.4. *Multi-Aspect Sun-as-a-star Approach*

Solar observations serve as the best guide for characterizing stellar observations, as we learned its usefulness from recent Sun-as-a-star studies for eruptive events (e.g., Harra et al. 2016; Veronig et al. 2021; Namekata et al. 2022c,a; Xu et al. 2022; Otsu et al. 2022; Otsu & Asai 2024). However, more observations are needed. For example,

- Sun-as-a-star multi-wavelength analyses to complement stellar studies (Namekata et al. 2024). Recently, Otsu & Asai (2024) have initiated this effort.
- To verify whether phenomena observed in stars but not in the Sun, such as X-ray absorption dimming and X-ray blueshift from CMEs (Table 1), can indeed be seen in the Sun.
- Understanding and predicting the missing wavelength in stellar observations (i.e. EUV/FUV and radio) by available solar datasets (e.g., Xu et al. 2022 for the effort in EUV).

4. Summary

In the past decade, significant observations have been made for detecting and characterizing stellar CMEs, yet direct detection has not yet been achieved. Recent observations across various wavelengths have provided valuable insights, but many detections are limited to single-wavelength and single-signature methods. To achieve a more comprehensive understanding of CMEs and their impact on exoplanet habitability, a multifaceted approach combining multi-wavelength observations and numerical modeling is essential. This paper proposes future directions including the enhancement of UV and radio observations, conducting multi-wavelength analyses, fostering collaborations between theoretical and observational groups, and leveraging solar observations as a guide for stellar studies.

Acknowledgement The author was supported by JSPS KAKENHI Grant Nos. JP21J00316. The author thanks K. Shibata, H. Maehara, and Y. Notsu for fruitful advice. The author thanks P. Odert for providing the data of X-ray dimming from Proxima Cen. The author acknowledges ideas from the participants in the workshop organized by the W.M. Keck Institute for Space Studies.

References

Aarnio, A. N., Matt, S. P., & Stassun, K. G. 2012, ApJ, 760, 9, doi: 10.1088/0004-637X/760/1/9

Airapetian, V. S., Glocer, A., Gronoff, G., Hébrard, E., & Danchi, W. 2016, Nature Geoscience, 9, 452, doi: 10.1038/ngeo2719

Airapetian, V. S., Barnes, R., Cohen, O., et al. 2020, International Journal of Astrobiology, 19, 136, doi: 10.1017/S1473550419000132

Alvarado-Gómez, J. D., Drake, J. J., Cohen, O., Moschou, S. P., & Garraffo, C. 2018, ApJ, 862, 93, doi: 10.3847/1538-4357/aacb7f

Alvarado-Gómez, J. D., Drake, J. J., Fraschetti, F., et al. 2020, ApJ, 895, 47, doi: 10.3847/1538-4357/ab88a3

Argiroffi, C., Reale, F., Drake, J. J., et al. 2019, Nature Astronomy, 3, 742, doi: 10.1038/s41550-019-0781-4

Chen, H., Tian, H., Li, H., et al. 2022, ApJ, 933, 92, doi: 10.3847/1538-4357/ac739b

Cliver, E. W., Schrijver, C. J., Shibata, K., & Usoskin, I. G. 2022, Living Reviews in Solar Physics, 19, 2, doi: 10.1007/s41116-022-00033-8

Crosley, M. K., & Osten, R. A. 2018a, ApJ, 862, 113, doi: 10.3847/1538-4357/aacf02

—. 2018b, ApJ, 856, 39, doi: 10.3847/1538-4357/aaaec2

Crosley, M. K., Osten, R. A., & Norman, C. 2017, ApJ, 845, 67, doi: 10.3847/1538-4357/aa7edc

Davenport, J. R. A. 2016, ApJ, 829, 23, doi: 10.3847/0004-637X/829/1/23

Drake, J. J., Cohen, O., Yashiro, S., & Gopalswamy, N. 2013, ApJ, 764, 170, doi: 10.1088/0004-637X/764/2/170

Favata, F., & Schmitt, J. H. M. M. 1999, A&A, 350, 900. https://arxiv.org/abs/astro-ph/9909041

France, K., Fleming, B. T., Drake, J. J., et al. 2019, in Society of Photo-Optical Instrumentation Engineers (SPIE) Conference Series, Vol. 11118, UV, X-Ray, and Gamma-Ray Space Instrumentation for Astronomy XXI, ed. O. H. Siegmund, 1111808, doi: 10.1117/12.2526859

Giampapa, M. S., Africano, J. L., Klimke, A., et al. 1982, ApJl, 252, L39, doi: 10.1086/183715

Golay, W. W., Mutel, R. L., & Abbuhl, E. E. 2024, ApJ, 965, 86, doi: 10.3847/1538-4357/ad29fb

Gopalswamy, N., Shimojo, M., Lu, W., et al. 2003, ApJ, 586, 562, doi: 10.1086/367614

Grechnev, V. V., Uralov, A. M., Slemzin, V. A., et al. 2008, SoPh, 253, 263, doi: 10.1007/s11207-008-9178-8

Gunn, A. G., Doyle, J. G., Mathioudakis, M., Houdebine, E. R., & Avgoloupis, S. 1994, A&A, 285, 489

Harra, L. K., Schrijver, C. J., Janvier, M., et al. 2016, SoPh, 291, 1761, doi: 10.1007/s11207-016-0923-0

Hawley, S. L., Walkowicz, L. M., Allred, J. C., & Valenti, J. A. 2007, PASP, 119, 67, doi: 10.1086/510561

Honda, S., Notsu, Y., Namekata, K., et al. 2018, PASJ, 70, 62, doi: 10.1093/pasj/psy055

Houdebine, E. R., Foing, B. H., & Rodono, M. 1990, A&A, 238, 249

Ikuta, K., & Shibata, K. 2024, ApJ, 963, 50, doi: 10.3847/1538-4357/ad1ce6

Inoue, S., Maehara, H., Notsu, Y., et al. 2023, ApJ, 948, 9, doi: 10.3847/1538-4357/acb7e8

Inoue, S., Buz Iwakiri, W., Enoto, T., et al. 2024, arXiv e-prints, arXiv:2406.06940. https://arxiv.org/abs/2406.06940

Kotani, Y., Shibata, K., Ishii, T. T., et al. 2023, ApJ, 943, 143, doi: 10.3847/1538-4357/acac76

Leitzinger, M., & Odert, P. 2022, Serbian Astronomical Journal, 205, 1, doi: 10.2298/SAJ2205001L

Leitzinger, M., Odert, P., & Greimel, R. 2024, MNRAS, doi: 10.1093/mnras/stae1404

Leitzinger, M., Odert, P., Greimel, R., et al. 2014, MNRAS, 443, 898, doi: 10.1093/mnras/stu1161

Leitzinger, M., Odert, P., & Heinzel, P. 2022, MNRAS, 513, 6058, doi: 10.1093/mnras/stac1284

Linsky, J. 2019, Host Stars and their Effects on Exoplanet Atmospheres, Vol. 955, doi: 10.1007/978-3-030-11452-7

Loyd, R. O. P., Mason, J. P., Jin, M., et al. 2022, ApJ, 936, 170, doi: 10.3847/1538-4357/ac80c1

Maehara, H., Shibayama, T., Notsu, S., et al. 2012, Nature, 485, 478, doi: 10.1038/nature11063

Maehara, H., Notsu, Y., Namekata, K., et al. 2021, PASJ, doi: 10.1093/pasj/psaa098

Mason, J. P., Woods, T. N., Webb, D. F., et al. 2016, ApJ, 830, 20, doi: 10.3847/0004-637X/830/1/20

Mohan, A., Mondal, S., Wedemeyer, S., & Gopalswamy, N. 2024, A&A, 686, A51, doi: 10.1051/0004-6361/202347924

Moschou, S.-P., Drake, J. J., Cohen, O., Alvarado-Gomez, J. D., & Garraffo, C. 2017, ApJ, 850, 191, doi: 10.3847/1538-4357/aa9520

Moschou, S.-P., Drake, J. J., Cohen, O., et al. 2019, ApJ, 877, 105, doi: 10.3847/1538-4357/ab1b37

Mullan, D. J., & Paudel, R. R. 2019, ApJ, 873, 1, doi: 10.3847/1538-4357/ab041b

Namekata, K., Ichimoto, K., Ishii, T. T., & Shibata, K. 2022a, ApJ, 933, 209, doi: 10.3847/1538-4357/ac75cd

Namekata, K., Maehara, H., Honda, S., et al. 2022b, arXiv e-prints, arXiv:2211.05506, doi: 10.48550/arXiv.2211.05506

Namekata, K., Sakaue, T., Watanabe, K., et al. 2017, ApJ, 851, 91, doi: 10.3847/1538-4357/aa9b34

Namekata, K., Maehara, H., Honda, S., et al. 2022c, Nature Astronomy, 6, 241, doi: 10.1038/s41550-021-01532-8

Namekata, K., Airapetian, V. S., Petit, P., et al. 2024, ApJ, 961, 23, doi: 10.3847/1538-4357/ad0b7c

Notsu, Y., Maehara, H., Honda, S., et al. 2019, ApJ, 876, 58, doi: 10.3847/1538-4357/ab14e6

Notsu, Y., Kowalski, A. F., Maehara, H., et al. 2024, ApJ, 961, 189, doi: 10.3847/1538-4357/ad062f

Okamoto, S., Notsu, Y., Maehara, H., et al. 2021, ApJ, 906, 72, doi: 10.3847/1538-4357/abc8f5

Osten, R. A., & Wolk, S. J. 2017, in Living Around Active Stars, ed. D. Nandy, A. Valio, & P. Petit, Vol. 328, 243–251, doi: 10.1017/S1743921317004252

Otsu, T., & Asai, A. 2024, ApJ, 964, 75, doi: 10.3847/1538-4357/ad24ec

Otsu, T., Asai, A., Ichimoto, K., Ishii, T. T., & Namekata, K. 2022, ApJ, 939, 98, doi: 10.3847/1538-4357/ac9730

Shibata, K., & Magara, T. 2011, Living Reviews in Solar Physics, 8, 6, doi: 10.12942/lrsp-2011-6

Sun, X., Török, T., & DeRosa, M. L. 2022, MNRAS, 509, 5075, doi: 10.1093/mnras/stab3249

Takahashi, T., Mizuno, Y., & Shibata, K. 2016, ApJl, 833, L8, doi: 10.3847/2041-8205/833/1/L8

Vedantham, H. K., Callingham, J. R., Shimwell, T. W., et al. 2020, Nature Astronomy, 4, 577, doi: 10.1038/s41550-020-1011-9

Veronig, A. M., Odert, P., Leitzinger, M., et al. 2021, Nature Astronomy, 5, 697, doi: 10.1038/s41550-021-01345-9

Vida, K., Kriskovics, L., Oláh, K., et al. 2016, A&A, 590, A11, doi: 10.1051/0004-6361/201527925

Vida, K., Leitzinger, M., Kriskovics, L., et al. 2019, A&A, 623, A49, doi: 10.1051/0004-6361/201834264

Vida, K., Kővári, Z., Leitzinger, M., et al. 2024, Universe, 10, 313, doi: 10.3390/universe10080313

Vidotto, A. A., Gregory, S. G., Jardine, M., et al. 2014, MNRAS, 441, 2361, doi: 10.1093/mnras/stu728

Xu, Y., Tian, H., Hou, Z., et al. 2022, ApJ, 931, 76, doi: 10.3847/1538-4357/ac69d5

Zic, A., Murphy, T., Lynch, C., et al. 2020, ApJ, 905, 23, doi: 10.3847/1538-4357/abca90

Solar and Stellar Coronal Mass Ejections
Proceedings IAU Symposium No. 388, 2026
N. Gopalswamy, O. Malandraki, A. Vidotto & W. Manchester, eds.
doi:10.1017/S1743921324001133

High-velocity blueshifts of Hα and Fe XXV Heα lines during superflares of the RS CVn-type stars

Shun Inoue[1], Teruaki Enoto[1,2], Yuta Notsu[3,4], Hiroyuki Uchida[1], Kosuke Namekata[1], Hiroyuki Maehara[5], Satoshi Honda[6], Wataru Buz Iwakiri[7], Miki Kurihara[8,9], Masahiro Tsujimoto[9], Keiichi Namizaki[10], Takeshi Go Tsuru[1], Daisaku Nogami[10,11] and Kazunari Shibata[12,13]

[1]Department of Physics, Kyoto University, Kitashirakawa-Oiwake-cho, Sakyo-ku, Kyoto, 606-8502, Japan
email: inoue@cr.scphys.kyoto-u.ac.jp

[2]RIKEN Cluster for Pioneering Research, 2-1 Hirosawa, Wako, Saitama, 351-0198, Japan

[3]Laboratory for Atmospheric and Space Physics, University of Colorado Boulder, 3665 Discovery Drive, Boulder, CO 80303, USA

[4]National Solar Observatory, 3665 Discovery Drive, Boulder, CO 80303, USA

[5]Okayama Branch Office, Subaru Telescope, NAOJ, NINS, Kamogata, Asakuchi, Okayama, 719–0232, Japan

[6]Nishi-Harima Astronomical Observatory, Center for Astronomy, University of Hyogo, Sayo, Hyogo, 679–5313, Japan

[7]International Center for Hadron Astrophysics, Chiba University, Inage-ku, Chiba, 263-8522 Japan

[8]Department of Astronomy, Graduate School of Science, The University of Tokyo, 7-3-1 Hongo, Bunkyo-ku, Tokyo 113-0033, Japan

[9]Japan Aerospace Exploration Agency, Institute of Space and Astronautical Science, Chuo-ku, Sagamihara, Kanagawa 252-5210, Japan

[10]Department of Astronomy, Kyoto University, Kitashirakawa-Oiwake-cho, Sakyo-ku, Kyoto, 606–8502, Japan

[11]Astronomical Observatory, Kyoto University, Sakyo-ku, Kyoto, 606–8502, Japan

[12]Kwasan Observatory, Kyoto University, Yamashina, Kyoto, 607–8471, Japan

[13]School of Science and Engineering, Doshisha University, Kyotanabe, Kyoto, 610–0321, Japan

Abstract. In December 2020, we conducted simultaneous optical spectroscopic and photometric observations of the RS CVn-type star V1355 Ori with Seimei and TESS. In July 2023, we performed X-ray observations of the RS CVn-type star IM Peg with NICER. Then, we found a blueshifted excess component of Hα extending its velocity up to 760−1690 km s^{-1} during a superflare of V1355 Ori and a blueshift of the line center of the Fe XXV Heα line with its maximum Doppler velocity reaching 2200 ± 600 km s^{-1} during a superflare of IM Peg. These results suggest an upward-moving plasma during the flares, such as a prominence eruption, coronal mass ejection (CME) and chromospheric evaporation. The high-velocity blueshifts, which overwhelmingly exceed the escape velocity of the stars, will help elucidate whether such events follow existing theories and scaling laws on solar flares and CMEs even when the energy scale far exceeds solar cases.

Keywords. Stellar flares, Stellar coronal mass ejections, stars: RS CVn-type

---

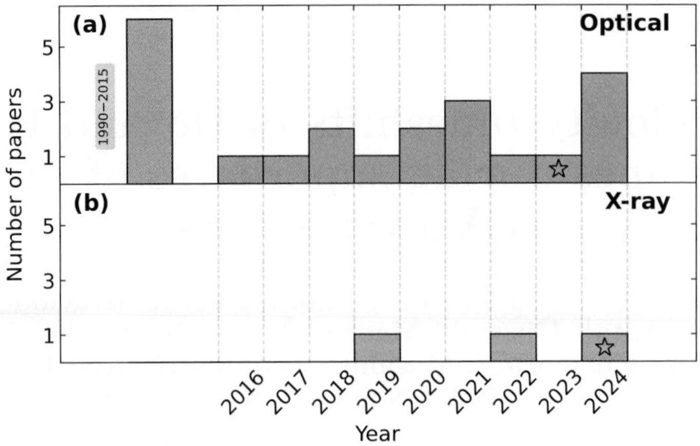

Figure 1. The histogram of the number of papers reporting blueshifts of (a) Balmer and (b) X-ray lines during stellar flares per year. The papers counted here are follows: Houdebine et al. (1990); Eason et al. (1992); Gunn et al. (1994); Crespo-Chacón et al. (2006); Fuhrmeister et al. (2008, 2011); Vida et al. (2016); Flores Soriano & Strassmeier (2017); Fuhrmeister et al. (2018); Honda et al. (2018); Vida et al. (2019); Muheki et al. (2020b,a); Koller et al. (2021); Maehara et al. (2021); Namekata et al. (2022b); Inoue et al. (2023); Namekata et al. (2024); Inoue et al. (2024b); Notsu et al. (2024); Leitzinger et al. (2024) for optical papers and Argiroffi et al. (2019); Chen et al. (2022); Inoue et al. (2024a) for X-ray papers. Open stars correspond to the results summarized in this proceeding (Inoue et al. 2023, 2024a).

1. Introduction

The Sun and stars suddenly release magnetic energy not only through radiation from flares but also through the motion of the plasma, called Coronal Mass Ejections (CMEs). While we cannot conduct spatially-resolved observations of stellar CMEs, we can observe the motion of them as Doppler shifts of the emission lines (e.g., Namekata et al. 2022a).

In the last ten years, the number of observations of the blue-shifted Balmer lines has greatly increased. Figure 1a shows the number of papers reporting blueshifts of Balmer lines during stellar flares per year. While there were only six reports in the 25 years from 1990 to 2015, some papers have been published every year since 2016. On the other hand, there were only two X-ray studies of detecting the blueshifts, excepting Inoue et al. (2024a), both of which were observed by Chandra HETG (Figure 1b). Furthermore, either in optical or X-ray studies, there are few blueshifts with a velocity exceeding the escape velocity of the star. Therefore, we cannot surely determine whether the erupted prominence eventually develops into a CME, escaping from the star. Some numerical studies consider the possibility that CMEs are suppressed by the strong magnetic fields on the stars (Alvarado-Gómez et al. 2018).

We detected the two superflares and high-velocity blueshifts were observed: Hα line on V1355 Ori and the Fe XXV Heα line on IM Peg. Both stars are classified as RS CVn-type stars, which are magnetically active and have been observed to produce large superflares (e.g., Tsuboi et al. 2016). We summarized the results of their observations in this proceeding.

2. Hα line on V1355 Orionis (Inoue et al. 2023)

2.1. *Observation*

We observed V1355 Orionis (K0-2IV+G1V) using KOOLS-IFU (Kyoto Okayama Optical Low-dispersion Spectrograph with optical-fiber Integral Field Unit;

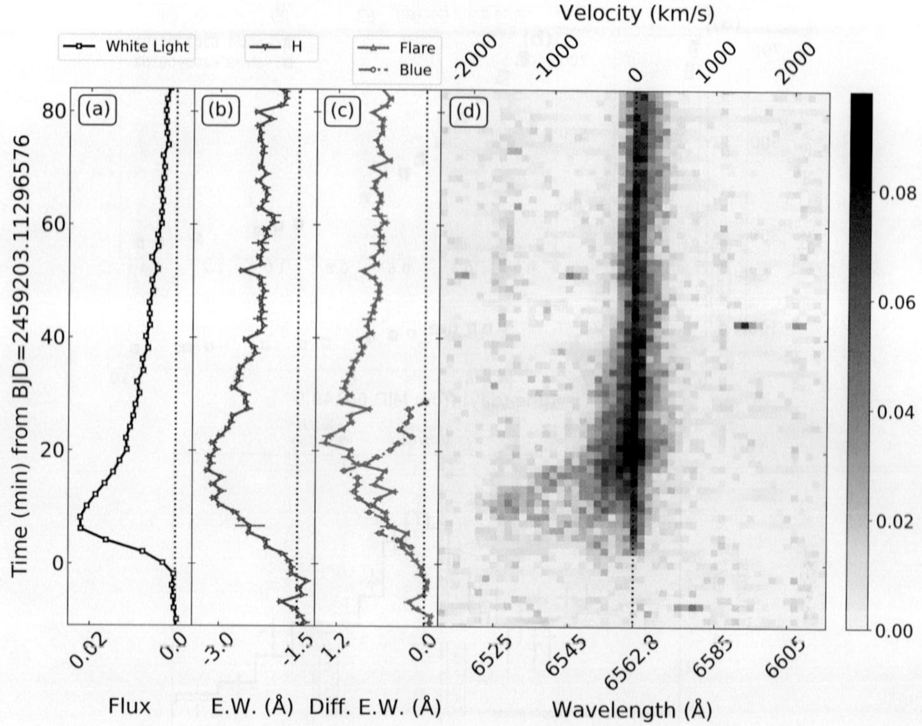

Figure 2. Light curves and Hα spectra of V1355 Ori during the 2020 December 19 taken from Inoue et al. (2023). (a) Normalized and detrended white-light light curve obtained by *TESS*. (b) Seimei Hα-line light curve. The negative equivalent width (Å) means the emission line flux in this light curve. (c) Hα-line light curve of the flare (pink) and blueshifted excess components (blue). (d) Time variation of the pre-flare-subtracted spectrum obtained by Seimei.

Matsubayashi et al. 2019), the spectrograph onboard the 3.8m Seimei telescope (Kurita et al. 2020). The observations were conducted in conjuction with the 2-min cadence photometric observations by Transiting Exoplanet Survey Satellite (TESS; Ricker et al. 2015) in Sector 34. KOOLS-IFU covers 5800-8000 Å, including Hα line (= 6562.8 Å), with wavelength resolution $R\,(=\lambda/\Delta\lambda)$ of ~ 2000.

2.2. *Result & Discussion*

We found a large superflare of V1355 Ori on December 19, 2020. The observed data were processed as described in Inoue et al. (2023). The superflare has the following characteristics:

(1) The white light and Hα flare lasted about 110 min and 210 min, respectively (Figure 2a, b, and c). Assuming that the flare spectrum is blackbody radiation with a temperature of $T_{\rm flare} = 10000$ K (Shibayama et al. 2013), the bolometric flare energy was estimated to be $E_{\rm bol} = 7.0 \times 10^{35}$ erg. Converting the equivalent width of the Hα to luminosity with the R-band magnitude (m_R; Cutispoto et al. 1995) and R-band Vega flux zero point per unit wavelength ($f_\lambda = 217.7 \times 10^{-11}$ ergs cm^{-2} sec^{-1} Å^{-1}; Bessell et al. 1998), the Hα flare energy was estimated to be $E_{\rm H\alpha} = 1.1 \times 10^{34}$ erg.

(2) For the first 30 min after the flare started, a high-velocity ($760 - 1690$ km s^{-1}) blueshift in the Hα emission line was observed confirming that the prominence

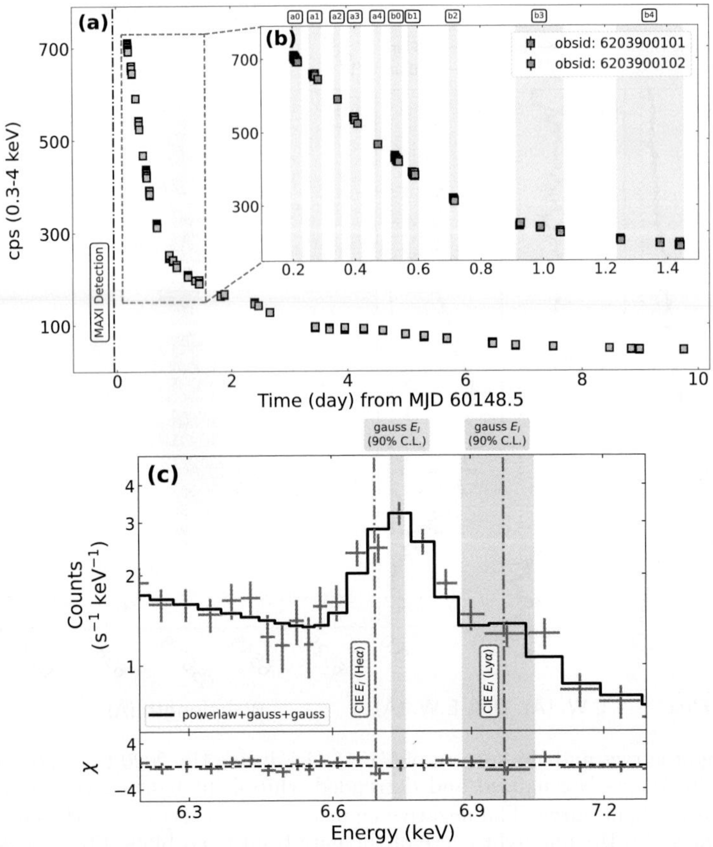

Figure 3. Light curves and X-ray spectrum of IM Peg during the superflare in July 2023 taken from Inoue et al. (2024a). (a) 64 s binned 0.3–4 keV light curve. The vertical is count rate (cps; counts s^{-1}). (b) Enlarged light curve during the first two ObsIDs 6203900101 (blue) and 6203900102 (green). a0–b4 shown above are interval numbers in our spectral analysis. (c) The interval b1 spectrum fitted with a phenomenological model (`powerlaw+gauss+gauss`). The vertical green and blue dashdot lines indicate the line center of the Fe Heα and Lyα emission lines at the rest frame in the CIE model, respectively. The green and blue shaded areas show the 90% error range of the line center of the two gaussians.

eruption occurred in association with the flare (Figure 2d). The velocity of the prominence was well above the escape velocity of 347 km s^{-1}.

(3) Using a method similar to that used by Maehara et al. (2021), the mass and kinetic energy of the prominence were estimated to be 9.5×10^{18} g $< M < 1.4 \times 10^{21}$ g and 4.5×10^{33} erg $< K < 1.0 \times 10^{37}$ erg, respectively. These results correspond to the value expected from the flare energy-mass scaling law that holds for solar CMEs (see Figure 8 in Inoue et al. 2023).

The fast prominence eruption detected in this study suggests that the massive plasma would certainly have flown outward from the star and developed into a CME. Furthermore, though the mass estimates make many uncertain assumptions and are highly indeterminate, the scaling law between the flare energy and prominence parameters indicate that the prominence eruption on V1355 Ori and solar prominence eruptions may have a common physical mechanism.

3. Fe XXV Heα line on IM Peg (Inoue et al. 2024a)

3.1. *Observation*

Monitor of All-sky X-ray Image (MAXI; Matsuoka et al. 2009) detected a large superflare, releasing 5×10^{37} erg in 2–10 keV, from IM Peg at 10:41 UT on 2023 July 23 (Iwakiri et al. 2023). We conducted follow up observations of the decay phase of the flare with Neutron Star Interior Composition ExploreR (NICER; Gendreau et al. 2016) from 16:52 UT on July 23 to 06:00 UT on August 2 (Figure 3a and b). The effective area and energy resolution of NICER are 600 cm^2 and 137 eV at 6 keV, respectively (Gendreau et al. 2012; Arzoumanian et al. 2014). The NICER data (ObsIDs: 6203900101–6203900111) were processed as described in (Inoue et al. 2024a). We extracted light curves, ObsID-averaged spectra, and time-resolved spectra of ObsIDs 6203900101 and 6203900102. We used `Xspec ver. 12.12.1` (Arnaud 1996) and `PyXspec ver. 2.1.0` (Gordon & Arnaud 2021) for our spectral analyses.

3.2. *Result & Discussion*

The superflare we observed in X-ray in July 2023 has the following characteristics:

(1) The Fe XXV Heα and Fe XXVI Lyα lines were clearly detected during the early phase of the flare decay, interval a0–a4 and b0–b4 (Figure 3b), which suggested the high temperature (> 50 MK) plasma (Smith et al. 2001).
(2) As shown in Figure 3c, the observed line center of the Fe XXV Heα during interval b1 was $E_l^{\text{gauss1}} = 6.74 \pm 0.01$ keV, which is $50 + 10$ eV larger than its lab-frame center (~ 6.69 keV). The corresponding Doppler velocity of the blueshift was -2200 ± 600 km s^{-1} and exceeded the escape velocity of the star (-230 km s^{-1}).
(3) We confirmed that the uncertainty of the NICER response function is limited to 8.7 eV (26 eV at 3σ, equivalent to 1200 km s^{-1}) using the NICER data of the supernova remnant Cas A. Then, we concluded that the -2200 ± 600 km s^{-1} ($(5 \pm 1) \times 10^{-2}$ keV) blueshift detected during the interval b1 was not due to the calibration uncertainty.
(4) There was a sign of reheating of the electron temperature from 60 MK to 100 MK one hour before the most pronounced blueshift (see Figure 4 in Inoue et al. 2024a, for details).

We consider that there are two possible cases for interpreting the blueshifted Fe XXV Heα line: a CME or chromospheric evaporation. While the velocity of solar CMEs ($-20 \sim -3000$ km s^{-1}: Webb & Howard 2012) can explain the observed Doppler velocity -2200 ± 600 km s^{-1}, the typical velocity of solar chromospheric evaporation is ~ -100 km s^{-1} (e.g., Doschek et al. 1980). On the other hand, the reheating one hour before the most pronounced blueshift can be well understood in the chromospheric evaporation interpretation. To know which interpretation is correct, we need the high resolution spectroscopy of the Fe line.

4. Summary & Future work

We detected the high-velocity blueshifts of Hα and Fe XXV Heα lines during superflares of V1355 Ori and IM Peg, respectively. These results suggest large CMEs can occur accompanied by superflares even on magnetically active stars.

The high resolution spectroscopy of X-Ray Imaging and Spectroscopy Mission (XRISM; Tashiro et al. 2020) with its microcalorimeter Resolve (Ishisaki et al. 2018) will greatly increase the number of the observations of X-ray blueshifts during stellar flares and deepen our understanding of the relationship between them and CMEs. For example, Resolve will enable us to distinguish whether the line itself is blue-shifted or non-shifted and blue-shifted components are added together in a case like the event of IM Peg. While observational studies of stellar CMEs in visible light have made significant progress in the last decade, the age of X-ray study of blueshifts is coming in XRISM era.

References

Alvarado-Gómez J. D., Drake J. J., Cohen O., Moschou S. P., Garraffo C., 2018, ApJ, 862, 93
Argiroffi C., et al., 2019, Nature Astronomy, 3, 742
Arnaud K. A., 1996, in Jacoby G. H., Barnes J., eds, Astronomical Society of the Pacific Conference Series Vol. 101, Astronomical Data Analysis Software and Systems V. p. 17
Arzoumanian Z., et al., 2014, in Takahashi T., den Herder J.-W. A., Bautz M., eds, Society of Photo-Optical Instrumentation Engineers (SPIE) Conference Series Vol. 9144, Space Telescopes and Instrumentation 2014: Ultraviolet to Gamma Ray. p. 914420, doi:10.1117/12.2056811
Bessell M. S., Castelli F., Plez B., 1998, A&A, 333, 231
Chen H., Tian H., Li H., Wang J., Lu H., Xu Y., Hou Z., Wu Y., 2022, ApJ, 933, 92
Crespo-Chacón I., Montes D., García-Alvarez D., Fernández-Figueroa M. J., López-Santiago J., Foing B. H., 2006, A&A, 452, 987
Cutispoto G., Pallavicini R., Kuerster M., Rodono M., 1995, A&A, 297, 764
Doschek G. A., Feldman U., Kreplin R. W., Cohen L., 1980, ApJ, 239, 725
Eason E. L. E., Giampapa M. S., Radick R. R., Worden S. P., Hege E. K., 1992, AJ, 104, 1161
Flores Soriano M., Strassmeier K. G., 2017, A&A, 597, A101
Fuhrmeister B., Liefke C., Schmitt J. H. M. M., Reiners A., 2008, A&A, 487, 293
Fuhrmeister B., Lalitha S., Poppenhaeger K., Rudolf N., Liefke C., Reiners A., Schmitt J. H. M. M., Ness J. U., 2011, A&A, 534, A133
Fuhrmeister B., et al., 2018, A&A, 615, A14
Gendreau K. C., Arzoumanian Z., Okajima T., 2012, in Takahashi T., Murray S. S., den Herder J.-W. A., eds, Society of Photo-Optical Instrumentation Engineers (SPIE) Conference Series Vol. 8443, Space Telescopes and Instrumentation 2012: Ultraviolet to Gamma Ray. p. 844313, doi:10.1117/12.926396
Gendreau K. C., et al., 2016, in den Herder J.-W. A., Takahashi T., Bautz M., eds, Society of Photo-Optical Instrumentation Engineers (SPIE) Conference Series Vol. 9905, Space Telescopes and Instrumentation 2016: Ultraviolet to Gamma Ray. p. 99051H, doi:10.1117/12.2231304
Gordon C., Arnaud K., 2021, PyXspec: Python interface to XSPEC spectral-fitting program, Astrophysics Source Code Library, record ascl:2101.014 (ascl:2101.014)
Gunn A. G., Doyle J. G., Mathioudakis M., Houdebine E. R., Avgoloupis S., 1994, A&A, 285, 489
Honda S., Notsu Y., Namekata K., Notsu S., Maehara H., Ikuta K., Nogami D., Shibata K., 2018, PASJ, 70, 62
Houdebine E. R., Foing B. H., Rodono M., 1990, A&A, 238, 249
Inoue S., Maehara H., Notsu Y., Namekata K., Honda S., Namizaki K., Nogami D., Shibata K., 2023, ApJ, 948, 9
Inoue S., et al., 2024a, arXiv e-prints, p. arXiv:2406.06940
Inoue S., et al., 2024b, PASJ, 76, 175
Ishisaki Y., et al., 2018, Journal of Low Temperature Physics, 193, 991
Iwakiri W., et al., 2023, The Astronomer's Telegram, 16152, 1
Koller F., Leitzinger M., Temmer M., Odert P., Beck P. G., Veronig A., 2021, A&A, 646, A34
Kurita M., et al., 2020, PASJ, 72, 48
Leitzinger M., Odert P., Greimel R., 2024, arXiv e-prints, p. arXiv:2406.01556
Maehara H., et al., 2021, PASJ, 73, 44
Matsubayashi K., et al., 2019, PASJ, 71, 102
Matsuoka M., et al., 2009, PASJ, 61, 999
Muheki P., Guenther E. W., Mutabazi T., Jurua E., 2020a, MNRAS, 499, 5047

Muheki P., Guenther E. W., Mutabazi T., Jurua E., 2020b, A&A, 637, A13
Namekata K., Maehara H., Honda S., Notsu Y., Nogami D., Shibata K., 2022a, arXiv e-prints, p. arXiv:2211.05506
Namekata K., et al., 2022b, Nature Astronomy, 6, 241
Namekata K., et al., 2024, ApJ, 961, 23
Notsu Y., et al., 2024, ApJ, 961, 189
Ricker G. R., et al., 2015, Journal of Astronomical Telescopes, Instruments, and Systems, 1, 014003
Shibayama T., et al., 2013, ApJS, 209, 5
Smith R. K., Brickhouse N. S., Liedahl D. A., Raymond J. C., 2001, ApJL, 556, L91
Tashiro M., et al., 2020, in den Herder J.-W. A., Nikzad S., Nakazawa K., eds, Society of Photo-Optical Instrumentation Engineers (SPIE) Conference Series Vol. 11444, Space Telescopes and Instrumentation 2020: Ultraviolet to Gamma Ray. p. 1144422, doi:10.1117/12.2565812
Tsuboi Y., et al., 2016, PASJ, 68, 90
Vida K., et al., 2016, A&A, 590, A11
Vida K., Leitzinger M., Kriskovics L., Seli B., Odert P., Kovács O. E., Korhonen H., van Driel-Gesztelyi L., 2019, A&A, 623, A49
Webb D. F., Howard T. A., 2012, Living Reviews in Solar Physics, 9, 3

Solar and Stellar Coronal Mass Ejections
Proceedings IAU Symposium No. 388, 2026
N. Gopalswamy, O. Malandraki, A. Vidotto & W. Manchester, eds.
doi:10.1017/S1743921324001145

Time-resolved spectroscopy of an active K-dwarf LQ Hydrae

Hiroyuki Maehara[1], Kosuke Namekata[2], Yuta Notsu[3,4], Shun Inoue[2], Kai Ikuta[5], Satoshi Honda[6], Daisaku Nogami[7] and Kazunari Shibata[8,9]

[1]Okayama Branch Office, Subaru Telescope, National Astronomical Observatory of Japan, NINS, Kamogata, Asakuchi, Okayama 719-0232, Japan
email: hiroyuki.maehara@nao.ac.jp

[2]Department of Physics, Kyoto University, Sakyo, Kyoto 606-8502, Japan

[3]Laboratory for Atmospheric and Space Physics, University of Colorado Boulder, 3665 Discovery Drive, Boulder, CO 80303, USA

[4]National Solar Observatory, 3665 Discovery Drive, Boulder, CO 80303, USA

[5]Department of Multidisciplinary Sciences, The University of Tokyo, 3-8-1 Komaba, Meguro, Tokyo 153-8902, Japan

[6]Nishi-Harima Astronomical Observatory, Center for Astronomy, University of Hyogo, Sayo, Hyogo 679-5313, Japan

[7]Department of Astronomy, Kyoto University, Sakyo, Kyoto 606-8502, Japan

[8]Kwasan Observatory, Kyoto University, Yamashina, Kyoto 607-8471, Japan

[9]School of Science and Engineering, Doshisha University, Kyotanabe, Kyoto 610-0321, Japan

Abstract. We carried out spectroscopic observations of an active K-dwarf LQ Hya using the 3.8-m Seimei telescope and 1.88-m telescope at Okayama Observatory simultaneously with the photometric observation by the Transiting Exoplanet Survey Satellite (TESS). During our observing run, we successfully detected a superflare with the bolometric energy of 10^{35} erg. The Hα line width significantly increased from ~ 3 Å to ~ 18 Å during the rising phase of the superflare and peaked at almost the same time as the flare peak time of optical continuum flux. In addition, the Hα line profile showed no line asymmetry during the flare. The absence of the line asymmetries in the Hα line suggests that the mass of the erupted plasma is smaller than 10^{17}–10^{18} g if we assume that the observed superflare was accompanied by the solar-like filament/prominence eruption with the velocity of ~ 500 km s^{-1}.

Keywords. Stellar flares, Stellar coronal mass ejections, K dwarf stars, Flare stars

1. Introduction

Flares are energetic explosions in the solar/stellar atmosphere caused by the rapid and sudden release of magnetic energy stored near the spots (e.g., Shibata & Magara 2011; Kowalski 2024). In the case of our Sun, flares are sometimes accompanied by mass ejections, such as filament/prominence eruptions (FEs/PEs) and coronal mass ejections (CMEs). By analogy with FEs/PEs on the Sun, stellar mass ejections are thought to be observed as the blue-shifted absorption/emission components in chromospheric lines (e.g., Odert et al. 2020). Recent optical spectroscopic observations of stellar flares found that some superflares can produce fast and massive FEs/PEs (e.g., G-dwarf EK Dra:

Namekata et al. 2022, 2024; RS CVn binary V1355 Ori: Inoue et al. 2023; M-dwarfs: Notsu et al. 2024). These massive and fast FEs/PEs are thought to be able to lead to "super-CMEs", and such massive CMEs associated with superflares can affect the habitability of exoplanets around active stars (e.g., Airapetian et al. 2016). According to Cuntz & Guinan (2016), early K-dwarfs are thought to be the most suitable for life-hosting planets based on evaluations of XUV radiations and stellar wind from the host star. However, it is unclear whether super-CMEs from K-dwarfs can affect exoplanetary habitability because the probable detections of stellar mass ejections associated with superflares are still limited to those on M-, G-dwarfs and RS CVn binaries, and no positive detection has been reported for K-dwarfs. In addition, the connection between stellar superflares and CMEs is still under debate. Therefore, we need more spectroscopic observations of stellar flares on various types of stars, including K-dwarfs, simultaneously with photometry. We carried out time-resolved spectroscopy of an active K-dwarf LQ Hya (spectral type K2V) simultaneously with the photometric observation by the Transiting Exoplanet Survey Satellite (TESS; Ricker et al. 2015).

2. Observation

We performed time-resolved spectroscopic observations of LQ Hya during the TESS Sector 35 (from 2021-02-09 to 2021-03-07) with the fiber-fed integral field spectrograph (KOOLS-IFU; Matsubayashi et al. 2019) mounted on the 3.8-m Seimei telescope Kurita et al. (2020) at Okayama Observatory, Kyoto University. We used the VPH683 grism, which covers 5800–8000 Å with a wavelength resolution of $R=\lambda/\Delta\lambda\sim 2000$ around the Hα line (6562.8 Å). The exposure time for our observations was 60 seconds and the typiacal signal-to-noise ratio is $\sim$200–300. We performed overscan and bias-pattern subtraction, spectral extraction, flat fielding, wavelength calibration and sky subtraction using a python script for the KOOLS-IFU data reduction† and PyRAF‡ to reduce the KOOLS-IFU data. We also performed time-resolved spectroscopy with the fiber-fed high-resolution spectrograph (HIDES-F; Kambe et al. 2013) mounted on the 188-cm telescope at Okayama on 2021-02-15 and 2021-02-17. The exposure time, wavelength resolution, and typical signal-to-noise ratio around the Hα line are 600 seconds, $\lambda/\Delta\lambda\sim 65000$, and ~ 30, respectively. Overscan/bias subtraction, gain adjustment, spectral extraction, flat fielding, wavelength calibration for HIDES data were performed using the "echelle" package in IRAF§.

3. Results

We simultaneously observed a superflare on 2021-02-15 with TESS, Seimei/KOOLS-IFU, and 188cm/HIDES-F. The light curve of the superflare is presented in Figure 1 (a). The equivalent width of the Hα line gradually increased and peaked at ~ 30 minutes after the peak timing of the white-light flare observed with TESS. We estimated the bolometric flare energy to be 1×10^{35} erg from the TESS light curve, assuming that the spectral energy distribution (SED) of flare continuum emission is the blackbody spectrum with the temperature of 10^4 K (e.g., Hawley & Fisher 1992, Shibayama et al. 2013). We also estimated the Hα flare energy to be 8×10^{32} erg using the Hα data obtained with the Seimei/KOOLS-IFU.

Figure 1 (b) shows the pre-flare subtracted dynamic spectra of the Hα line during the superflare on 2021-02-15. The Hα line width increased from ~ 3 Å to ~ 18 Å during the rising phase of flare, and peaked at almost the same time as the peak timing of the flare

† `http://www.o.kwasan.kyoto-u.ac.jp/inst/p-kools/reduction-201806/install_software.html`

‡ `https://iraf-community.github.io/pyraf.html`

§ `https://iraf-community.github.io/`

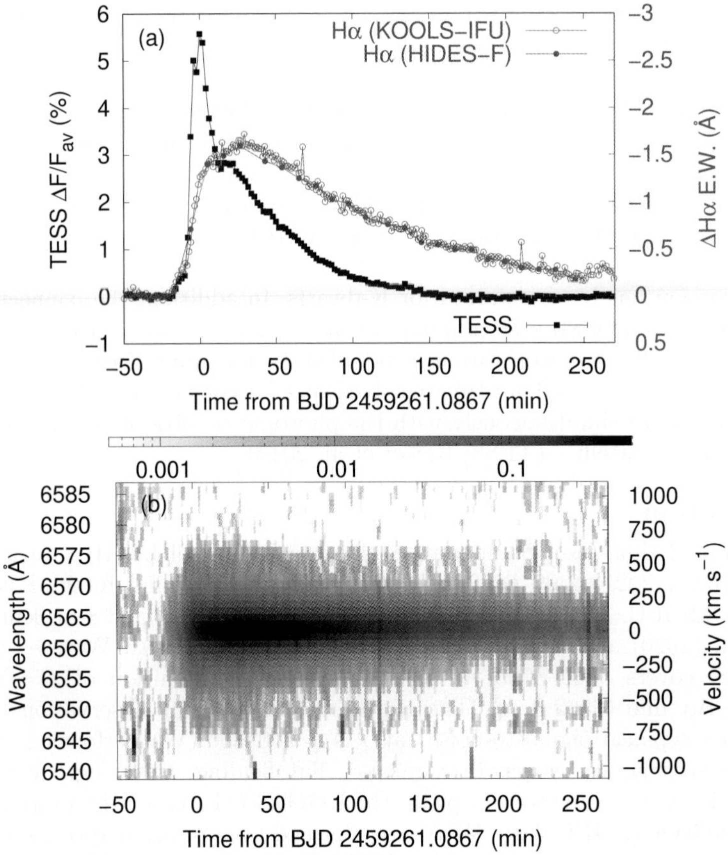

Figure 1. (a) Light curve of the superflare on 2021-02-15. Black filled squares indicate the TESS light curve of the flare. Red open and filled squares indicate the Hα light curves observed with Seimei/KOOLS and 1.88-m/HIDES-F, respectively. Horizontal axis represents the time from the flare peak BJD 2459261.0867 (t). Left- and right-vertical axis indicate the TESS flux difference and difference in the equivalent width (E.W.) of the Hα line relative to the pre-flare level, respectively. Pre-flare levels of TESS flux and Hα E.W. were determined from the data between $t = -50$ min and $t = -10$ min. (b) Pre-flare-subtracted dynamic spectra of the Hα line observed with Seimei/KOOLS. The pre-flare Hα line profile was derived by using the spectra from $t = -50$ min to $t = -10$ min.

light curve observed in the optical continuum. Similar Hα line broadening phenomena were observed around the peak timing of flares on M-dwarfs (e.g., Hawley & Pettersen 1991; Namekata et al. 2020). Namekata et al. (2020) suggested that the broadening phenomena in the Hα line during the flare can be interpreted by the Stark broadening or opacity broadening (see also section 10 in Kowalski 2024).

4. Discussion

As discussed in Odert et al. (2020), filament eruptions (FEs) on the stellar disk can be observed as the blue-shifted absorption components in the Hα line wing. Moreover, prominence eruptions (PEs) on the stellar limb can be observed as the blue- or red-shifted emission components (for Sun and solar-type stars: Otsu et al. 2022; Namekata et al. 2024). As shown in Figures 1 (b) and 2, the Hα line profile did not exhibit line asymmetry during this superflare. The upper limits of the equivalent width of blue-shifted absorption or emission components are estimated to be < 0.02 Å, if we assume that the

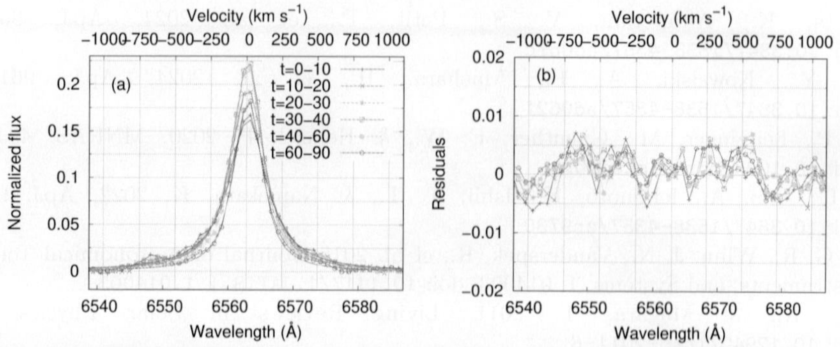

Figure 2. (a) Preflare-subtracted average profiles of the Hα line at $t = 0$–10 min, 10–20 min, 20–30 min, 30–40 min, 40–60 min, and 60–90 min. (b) Same as (a), but for the residual Hα line profiles after removing the symmetric component. The symmetric components for each spectrum were derived by fitting the preflare-subtracted profile with the Voigt function. Weak absorption and emission features in residual spectra can be explained by time variations of atmospheric line strength.

Doppler velocity and velocity dispersion of the blue-shifted component are ~ 500 km s^1 and ~ 300 km s^1, respectively. From this value, we estimated the upper limit of mass of the erupted materials. In the case of blue-shifted emission caused by the prominence eruption, the upper limit of the equivalent width of < 0.02 Å corresponds to the upper limit of the Hα luminosity of 6×10^{26} erg s^{-1} by adopting the stellar parameters from the Gaia DR3 Catalog (Gaia Collaboration et al. 2016, 2023). We estimated the upper limit of prominence mass to be $< 10^{18}$ g by using the same procedure in Maehara et al. (2021). For the blue-shifted absorption case caused by the filament eruption, we estimated the upper limit of filament mass by adopting the same manner as that of Namekata et al. (2022). If we assume that the observed superflare was accompanied by the solar-like filament eruption, the filament mass is smaller than $\sim 10^{17}$ g. These upper limits are within the scatter of the reported CME mass - flare energy relation for dMe flares (e.g., Notsu et al. 2024). Further spectroscopic observations for searching for stellar mass ejections associated with superflares are needed to clarify the CME association rate for superflares on K-dwarfs.

References

Airapetian, V. S., Glocer, A., Gronoff, G., Hébrard, E., & Danchi, W. 2016, Nature Geoscience, 9, 452, doi: 10.1038/ngeo2719

Cuntz, M., & Guinan, E. F. 2016, ApJ, 827, 79, doi: 10.3847/0004-637X/827/1/79

Gaia Collaboration, Prusti, T., de Bruijne, J. H. J., et al. 2016, A&A, 595, A1, doi: 10.1051/0004-6361/201629272

Gaia Collaboration, Vallenari, A., Brown, A. G. A., et al. 2023, A&A, 674, A1, doi: 10.1051/0004-6361/202243940

Hawley, S. L., & Fisher, G. H. 1992, ApJS, 78, 565, doi: 10.1086/191640

Hawley, S. L., & Pettersen, B. R. 1991, ApJ, 378, 725, doi: 10.1086/170474

Inoue, S., Maehara, H., Notsu, Y., et al. 2023, ApJ, 948, 9, doi: 10.3847/1538-4357/acb7e8

Kambe, E., Yoshida, M., Izumiura, H., et al. 2013, PASJ, 65, 15, doi: 10.1093/pasj/65.1.15

Kowalski, A. F. 2024, Living Reviews in Solar Physics, 21, 1, doi: 10.1007/s41116-024-00039-4

Kurita, M., Kino, M., Iwamuro, F., et al. 2020, PASJ, 72, 48, doi: 10.1093/pasj/psaa036

Maehara, H., Notsu, Y., Namekata, K., et al. 2021, PASJ, 73, 44, doi: 10.1093/pasj/psaa098

Matsubayashi, K., Ohta, K., Iwamuro, F., et al. 2019, PASJ, 71, 102, doi: 10.1093/pasj/psz087

Namekata, K., Maehara, H., Sasaki, R., et al. 2020, PASJ, 72, 68, doi: 10.1093/pasj/psaa051

Namekata, K., Maehara, H., Honda, S., et al. 2022, Nature Astronomy, 6, 241, doi: 10.1038/s41550-021-01532-8

Namekata, K., Airapetian, V. S., Petit, P., et al. 2024, ApJ, 961, 23, doi: 10.3847/1538-4357/ad0b7c

Notsu, Y., Kowalski, A. F., Maehara, H., et al. 2024, ApJ, 961, 189, doi: 10.3847/1538-4357/ad062f

Odert, P., Leitzinger, M., Guenther, E. W., & Heinzel, P. 2020, MNRAS, 494, 3766, doi: 10.1093/mnras/staa1021

Otsu, T., Asai, A., Ichimoto, K., Ishii, T. T., & Namekata, K. 2022, ApJ, 939, 98, doi: 10.3847/1538-4357/ac9730

Ricker, G. R., Winn, J. N., Vanderspek, R., et al. 2015, Journal of Astronomical Telescopes, Instruments, and Systems, 1, 014003, doi: 10.1117/1.JATIS.1.1.014003

Shibata, K., & Magara, T. 2011, Living Reviews in Solar Physics, 8, 6, doi: 10.12942/lrsp-2011-6

Shibayama, T., Maehara, H., Notsu, S., et al. 2013, ApJS, 209, 5, doi: 10.1088/0067-0049/209/1/5

Solar and Stellar Coronal Mass Ejections
Proceedings IAU Symposium No. 388, 2026
N. Gopalswamy, O. Malandraki, A. Vidotto & W. Manchester, eds.
doi:10.1017/S1743921324001121

Blue wing asymmetries in Balmer lines during mid M dwarf flares and possible stellar mass ejections

Yuta Notsu[1,2], Adam. F. Kowalski[1,2,3], Hiroyuki Maehara[4], Kosuke Namekata[5], Kenji Hamaguchi[6,7], Teruaki Enoto[5], Isaiah I. Tristan[1,2,3], Suzanne L. Hawley[8], James R. A. Davenport[8], Satoshi Honda[9], Kai Ikuta[10], Shun Inoue[5], Keiichi Namizaki[11], Daisaku Nogami[11] and Kazunari Shibata[12,13]

[1]Laboratory for Atmospheric and Space Physics, University of Colorado Boulder, Boulder, CO 80303, USA

[2]National Solar Observatory, Boulder, CO 80303, USA

[3]Department of Astrophysical and Planetary Sciences, University of Colorado Boulder, Boulder, CO 80305, USA

[4]Okayama Branch Office, Subaru Telescope, National Astronomical Observatory of Japan, NINS, Kamogata, Asakuchi, Okayama 719-0232, Japan

[5]Department of Physics, Kyoto University, Sakyo, Kyoto 606-8502, Japan

[6]CRESST II and X-ray Astrophysics Laboratory, NASA/GSFC, Greenbelt, MD 20771, USA

[7]Department of Physics, University of Maryland, Baltimore County, Baltimore, MD 21250, USA

[8]Department of Astronomy, University of Washington, Seattle, WA 98195, USA

[9]Nishi-Harima Astronomical Observatory, Center for Astronomy, University of Hyogo, Sayo, Hyogo 679-5313, Japan

[10]Department of Multidisciplinary Sciences, The University of Tokyo, Meguro, Tokyo 153-8902, Japan

[11]Department of Astronomy, Kyoto University, Sakyo, Kyoto 606-8502, Japan

[12]Kwasan Observatory, Kyoto University, Yamashina, Kyoto 607-8471, Japan

[13]School of Science and Engineering, Doshisha University, Kyotanabe, Kyoto 610-0321, Japan

Abstract. We conducted the 31-night time-resolved simultaneous optical spectroscopic and photometric observations of the mid M dwarf flare stars YZ CMi, EV Lac, and AD Leo, using APO 3.5m and SMARTS 1.5m telescopes. Among the 41 detected flares, 7 flares showed clear blue wing asymmetries in Balmer lines, with various correspondences in flare properties (e.g., flare durations are 20 min – 2.5 hours). The line-of-sight velocities of the blue-shifted components range from -73 to -122 km s^{-1}. Assuming that the blue-shifts were caused by prominence eruptions, the mass of upward moving plasma was estimated to be $10^{15} - 10^{19}$ g, which are roughly on the relation between flare energy and erupting mass expected from solar coronal mass ejections (CMEs). Although further investigations are necessary for understanding the observed various properties, these possible prominence eruptions on M-dwarfs could evolve into CMEs, assuming the similar acceleration mechanism with solar eruptions.

Keywords. Stellar flares, Stellar coronal mass ejections, M dwarf stars, Flare stars

1. Introduction

Flares are releases of magnetic energy in the solar/stellar atmosphere, and they have strong emissions from radio to X-rays (e.g., Shibata & Magara 2011; Kowalski 2024). Optical spectroscopic studies of solar and stellar flares have been carried out in order to understand the dynamics of plasma and the radiation mechanisms during flares. During some M dwarf flares, chromospheric line profiles show blue wing asymmetries (e.g., Houdebine et al. 1990; Vida et al. 2016, 2019; Fuhrmeister et al. 2018; Honda et al. 2018; Muheki et al. 2020; Maehara et al. 2021). Several recent studies have discussed the blue wing asymmetries from M-dwarfs can be related with stellar coronal mass ejections (CMEs) (Vida et al. 2019; Moschou et al. 2019). It is important to understand the properties of stellar CMEs in order to evaluate their effects not only on the mass and angular momentum loss of the star (e.g., Osten & Wolk 2015; Cranmer 2017; Odert et al. 2017), but also on the habitability of exoplanets (e.g., loss of atmosphere, atmospheric chemistry, radiation dose as described in Lammer et al. 2007; Segura et al. 2010; Airapetian et al. 2020). Attempts to detect stellar CMEs in various optical/UV/radio/X-ray methods (e.g., Crosley & Osten 2018; Zic et al. 2020; Veronig et al. 2021; Namekata et al. 2022b, 2024; Inoue et al. 2024a) have been extensively done in recent few years (see also Namekata et al. 2022a and Leitzinger & Odert 2022 for reviews). However, our understanding of blue and red asymmetries in chromospheric lines and their connections with stellar flares and CMEs is still limited by the low number of samples observed in time-resolved spectroscopy simultaneously with high-precision photometry. In order to investigate the connection between the line asymmetries in Balmer lines and the properties of flares more in detail, we conducted the time-resolved simultaneous optical spectroscopic and photometric observations of mid-M-dwarf flare stars, during the 31 nights (Notsu et al. 2024). The results of this observation campaign are summarized in this proceeding.

2. Observation

During the 31 nights over 2 yr (2019 January – 2021 February), we conducted time-resolved simultaneous optical spectroscopic and photometric observations of the three nearby mid-M-dwarf flare stars YZ CMi, EV Lac, and AD Leo. Spectroscopic observations were conducted using the ARC Echelle Spectrograph (ARCES; Wang et al. 2003) of the Apache Point Observatory (APO) 3.5 m telescope and the cross-dispersed, fiber-fed echelle CTIO HIgh ResolutiON (CHIRON) spectrogragh (Tokovinin et al. 2013) attached to the Cerro Tololo Inter-American Observatory (CTIO) Small and Moderate Aperture Research Telescope System (SMARTS) 1.5 m telescope. The wavelength resolution ($R = \lambda/\Delta\lambda$) and the spectral coverage of APO3.5m/ARCES are ~32000 and 3800–10000Å, respectively, while those of SMARTS1.5m/CHIRON are $R \sim 25000$ and 4500-8900Å, respectively. Photometric observations were obtained from the APO 0.5m telescope (Astrophysical Research Consortium Small Aperture Telescope, ARCSAT) and 1 m and 0.4 m telescopes of the Las Cumbres Observatory Global Telescope (LCOGT) network, while 5 nights were also covered with the observation window of TESS. For the 3 nights among the total 31 nights, we also conducted the soft X-ray spectroscopic observations with Neutron Star Interior Composition ExploreR (NICER). In this proceeding, we only summarize the results from the optical spectroscopic and photometric observations, and the results from the NICER X-ray data are described in Notsu et al. (2024).

3. Results and discussions

As a result of the 31 night simultaneous spectroscopic and photometric observations over two years, 41 flares were detected (29 flares on YZ CMi, 9 flares on EV Lac, and 3

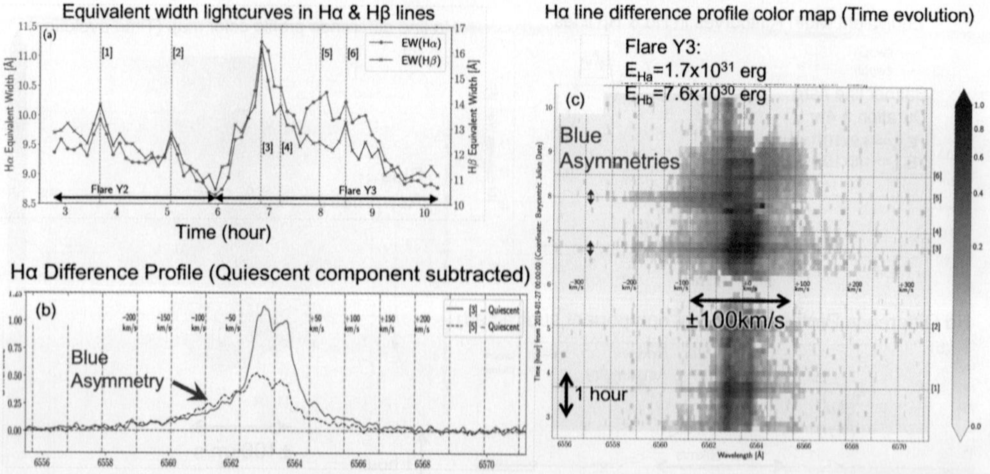

Figure 1. Lightcurves and spectra of a flare on YZ CMi (Flare Y3), which showed short-lived Hα blue wing asymmetries twice (20min×2) during the Hα flare over 4 hours. (a) Hα & Hβ equivalent width light curves of YZ CMi from APO 3.5m spectroscopic data. Red and Blue symbols correspond to Hα & Hβ EWs, respectively. The black double-headed arrows indicate the start and end time of two flares (Flares Y2 & Y3). The grey dashed lines with numbers ([1]–[6]) correspond to the time shown with the same numbers in (b)&(c). (b) Hα line profile differences from the quiescent phase at Time [3]&[5] during Flare Y3. (c) Time evolution of the Hα line profile during Flares Y2&Y3. The color map represents the line profile changes from the quiescent profile (cf. (b)). The blue double-headed arrows roughly represent the times when blue wing asymmetries were clearly seen.

flares on AD Leo). The energy ranges of the observed 41 flares are $10^{30} - 10^{32}$ erg in the Hα line, and $10^{30} - 10^{33}$ erg in u- & g-band continuum bands. Among the 41 flares, seven flares (4 flares on YZ CMi, 2 flares on EV Lac, and one flare on AD Leo) showed clear blue wing asymmetries in Hα line. There are various correspondences in flare properties as follows.

(i) The duration of the Hα blue wing asymmetries range from 20 min to 2.5 hours. As a notable example, a flare showed short-lived Hα blue wing asymmetries twice (20min×2) during the Hα flare over 4 hours (Figure 1). In contrast, a flare shown in Figure 2 showed the gradual shift from blue wing asymmetry to red wing asymmetry, during the Hα flare over 4.9 hours. See also Inoue et al. (2024b) for the appearance timing of blue asymmetries during flares.

(ii) Among the seven flares with blue wing asymmetries, two flares are categorized as candidate non white-light (NWL) flares and three flares are clearly white-light (WL) flares, while the remaining two flares do not have enough data coverage of simultaneous spectroscopic and photometric data to judge whether they are white-light or non white-light flares. For reference, among all the 41 flares, 4 flares are categorized as candidate NWL flares and 31 flares are clear WL flares, while the remaining 6 flares do not have enough data coverage of simultaneous spectroscopic and photometric data. These results can suggest that blue wing asymmetries of chromospheric lines can be commonly seen both during white-light and candidate non white-light flares.

(iii) All of the seven flares showed blue wing asymmetries also in the Hβ line, but there is a large variety in which other chromospheric lines showed blue wing asymmetries. For example, two flares showed blue wing asymmetries only in lower-order Balmer

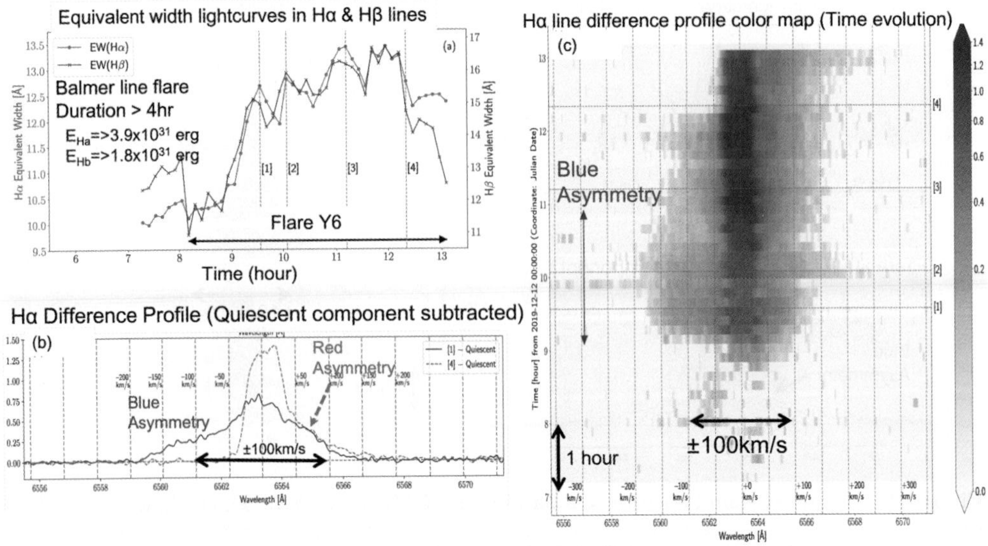

Figure 2. Lightcurves and spectra of a flare on YZ CMi (Flare Y6), which showed the gradual shift from blue wing asymmetry to red wing asymmetry, during the Hα flare over 4.9 hours. The lines and symbols are plotted in the same way as Figure 1.

lines (up to Hβ and Hγ lines, respectively). In contrast, the other two flares showed blue wing asymmetries in almost all the chromospheric lines (except for Ca II 8542 and Na I D1&D2, respectively). It is speculated that these differences can be caused by the differences of optical depth and line wing broadening physics, but observation-based modeling studies incorporating radiative transfer physics (e.g., Leitzinger et al. 2022) and comparison with solar flare data are necessary for further quantitative discussions.

(iv) The line-of-sight velocities of the blue wing excess components (blue wing asymmetries) are estimated to range from -73 to -122 km s^{-1}, and these are in the same range of solar prominence/filament eruptions. These velocity values (73–122 km s^{-1}) represent possible prominence eruptions of M-dwarfs and they are smaller than the escape velocities at the stellar surface ($\sim$600 km s^{-1} for YZ CMi, EV Lac, and AD Leo). However, this cannot simply lead to the conclusion that the plasma is not ejected from the star, as the blue-shift events only provide the lower limit of the velocities and as summarized in the following, based on the relevant discussions and similar interpretations in previous papers (e.g., Moschou et al. 2019; Vida et al. 2019; Maehara et al. 2021; Namekata et al. 2022b). Gopalswamy et al. (2003) showed that the average CME core velocity ($\sim$350 km s^{-1}) and average CME velocity ($\sim$610 km s^{-1}) are $\sim$4 and $\sim$8 times larger than that of the associated prominence eruptions ($\sim$80 km s^{-1}). This indicates that prominences with initial slow speeds are accelerated as they are lifted up and they evolve into CMEs. However, this indicates that if we assume similar acceleration mechanism would work, these prominence eruptions would be accelerated into $\sim$300 – 1000 km s^{-1}. This value is generally larger than the escape velocities at $\sim$2 – 3$R_{\rm star}$ ($\sim$300 – 450 km s^{-1}), and the prominence eruptions with the velocity of $\sim$100 km s^{-1} could evolve into CMEs (see also (vii) for the necessity of further investigations). Moreover, the observed blue-shift velocities are line-of-sight velocities, and the radial velocities can be larger considering the projection angle effect, which suggests

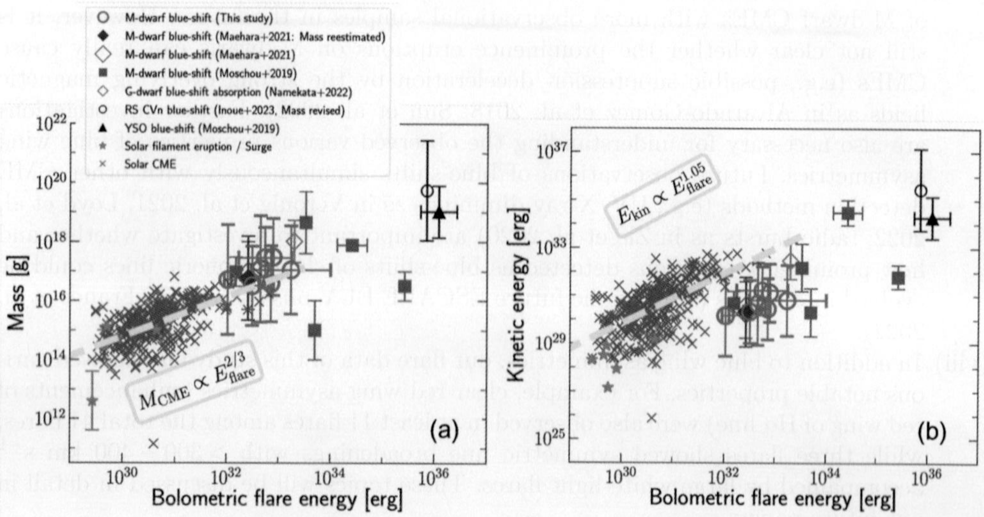

Figure 3. Mass and kinetic energy as a function of flare bolometric energy, for solar and stellar flares and prominence eruptions/CMEs (from Notsu et al. 2024). (a) Mass of prominence eruptions/CMEs. The cyan dashed line represents the relation: $M_{\rm CME} \propto E_{\rm flare}^{2/3}$ shown by Takahashi et al. (2016), which is fitted to the solar CME data points in this figure. (b) Kinetic energy of prominence eruptions/CMEs as a function of flare bolometric energy. The scaling relation denoted by the cyan dashed line ($E_{\rm kin} \propto E_{\rm flare}^{1.05}$) is obtained from Namekata et al. (2022b), which is also fitted to the solar CME data points in this figure.

that these prominence eruptions could evolve into CMEs with faster velocities. In addition, it is noted that red wing enhancements were observed during some flares with blue wing asymmetries (e.g., Figure 2), which indicates that some of the materials fell back to the stellar surface. This phenomenon is often observed in the case of solar filament/prominence eruptions even in the case that they evolve into CMEs (e.g., Otsu et al. 2022).

(v) Assuming the relation from the NLTE slab model calculation of solar prominences (Heinzel et al. 1994), the surface flux densities of the upward moving plasma causing blue-shifts are estimated from the luminosity ratio of blue wing asymmetry components in Hα and Hβ lines. Using these values, the erupted mass of the seven blue-shift (blue wing asymmetry) events are estimated to be $M_{\rm blue}^{\rm H\alpha} \sim 10^{15} - 10^{19}$ g. These estimated mass of the seven blue-shift events are roughly on the relation expected from solar CMEs, and are roughly on the same relation with other stellar events in the previous studies. This might suggest that these possible prominence eruptions on M-dwarfs could share a common underlying mechanism with solar filament/prominence eruptions/CMEs (i.e. magnetic energy release), although the large uncertainty of the mass estimation method should be considered. For more accurate evaluation, the model calculations of eruptive prominences in the M-dwarf stellar atmosphere (cf. Leitzinger et al. 2022; Ikuta & Shibata 2024) is important.

(vi) In contrast, the kinetic energies of the seven blue-shift events ($E_{\rm kin} \sim 10^{29} - 10^{32}$ erg) are roughly two orders of magnitude smaller than the the relation expected from solar CMEs, as also shown in previous studies. These small kinetic energies can be also understood if we assume the velocity difference/evolution of prominence eruptions and CMEs (see also the above discussions in (iv)).

(vii) The mass, velocity, and kinetic energy of the possible prominence eruptions of M-dwarfs in this study could eventually lead to understanding the statistical properties of M-dwarf CMEs with more observational samples in the future. However, it is still not clear whether the prominence eruptions on M-dwarfs can really cause CMEs (e.g., possible suppression/deceleration by the strong overlying magnetic fields as in Alvarado-Gómez et al. 2018; Sun et al. 2022). Further investigations are also necessary for understanding the observed various properties of blue wing asymmetries. Future observations of blue-shifts simultaneously with other CME detection methods (e.g., UV/X-ray dimmings as in Veronig et al. 2021, Loyd et al. 2022; radio bursts as in Zic et al. 2020) are important to investigate whether and how prominence eruptions detected as blue-shifts of chromospheric lines could be evolved into CMEs (see also the future ESCAPE EUV mission plan in France et al. 2022).

(viii) In addition to blue wing asymmetries, our flare data of this study also showed various notable properties. For example, clear red wing asymmetries (enhancements of red wing of Hα line) were also observed in at least 11 flares among the total 41 flares, while three flares showed symmetric line broadenings with $> 300 - 400$ km s^{-1} accompanied by large white-light flares. These topics will be discussed in detail in our future papers.

References

Airapetian, V. S., Barnes, R., Cohen, O., et al. 2020, International Journal of Astrobiology, 19, 136, doi: 10.1017/S1473550419000132

Alvarado-Gómez, J. D., Drake, J. J., Cohen, O., Moschou, S. P., & Garraffo, C. 2018, ApJ, 862, 93, doi: 10.3847/1538-4357/aacb7f

Cranmer, S. R. 2017, ApJ, 840, 114, doi: 10.3847/1538-4357/aa6f0e

Crosley, M. K., & Osten, R. A. 2018, ApJ, 856, 39, doi: 10.3847/1538-4357/aaaec2

France, K., Fleming, B., Youngblood, A., et al. 2022, JATIS, 8, 014006, doi: 10.1117/1.JATIS.8.1.014006

Fuhrmeister, B., Czesla, S., Schmitt, J. H. M. M., et al. 2018, A&A, 615, A14, doi: 10.1051/0004-6361/201732204

Gopalswamy, N., Shimojo, M., Lu, W., et al. 2003, ApJ, 586, 562, doi: 10.1086/367614

Heinzel, P., Gouttebroze, P., & Vial, J. C. 1994, A&A, 292, 656

Honda, S., Notsu, Y., Namekata, K., et al. 2018, PASJ, 70, 62, doi: 10.1093/pasj/psy055

Houdebine, E. R., Foing, B. H., & Rodono, M. 1990, A&A, 238, 249

Ikuta, K., & Shibata, K. 2024, ApJ, 963, 50, doi: 10.3847/1538-4357/ad1ce6

Inoue, S., Buz Iwakiri, W., Enoto, T., et al. 2024a, arXiv e-prints, arXiv:2406.06940, doi: 10.48550/arXiv.2406.06940

Inoue, S., Enoto, T., Namekata, K., et al. 2024b, PASJ, 76, 175, doi: 10.1093/pasj/psae001

Kowalski, A. F. 2024, Living Reviews in Solar Physics, 21, 1, doi: 10.1007/s41116-024-00039-4

Lammer, H., Lichtenegger, H. I. M., Kulikov, Y. N., et al. 2007, Astrobiology, 7, 185, doi: 10.1089/ast.2006.0128

Leitzinger, M., & Odert, P. 2022, Serbian Astronomical Journal, 205, 1, doi: 10.2298/SAJ2205001L

Leitzinger, M., Odert, P., & Heinzel, P. 2022, MNRAS, 513, 6058, doi: 10.1093/mnras/stac1284

Loyd, R. O. P., Mason, J. P., Jin, M., et al. 2022, ApJ, 936, 170, doi: 10.3847/1538-4357/ac80c1

Maehara, H., Notsu, Y., Namekata, K., et al. 2021, PASJ, 73, 44, doi: 10.1093/pasj/psaa098

Moschou, S.-P., Drake, J. J., Cohen, O., et al. 2019, ApJ, 877, 105, doi: 10.3847/1538-4357/ab1b37

Muheki, P., Guenther, E. W., Mutabazi, T., & Jurua, E. 2020, MNRAS, 499, 5047, doi: 10.1093/mnras/staa3152

Namekata, K., Maehara, H., Honda, S., et al. 2022a, arXiv e-prints, arXiv:2211.05506. https://arxiv.org/abs/2211.05506

—. 2022b, Nature Astronomy, 6, 241, doi: 10.1038/s41550-021-01532-8

Namekata, K., Airapetian, V. S., Petit, P., et al. 2024, ApJ, 961, 23, doi: 10.3847/1538-4357/ad0b7c

Notsu, Y., Kowalski, A. F., Maehara, H., et al. 2024, ApJ, 961, 189, doi: 10.3847/1538-4357/ad062f

Odert, P., Leitzinger, M., Hanslmeier, A., & Lammer, H. 2017, MNRAS, 472, 876, doi: 10.1093/mnras/stx1969

Osten, R. A., & Wolk, S. J. 2015, ApJ, 809, 79, doi: 10.1088/0004-637X/809/1/79

Otsu, T., Asai, A., Ichimoto, K., Ishii, T. T., & Namekata, K. 2022, ApJ, 939, 98, doi: 10.3847/1538-4357/ac9730

Segura, A., Walkowicz, L. M., Meadows, V., Kasting, J., & Hawley, S. 2010, Astrobiology, 10, 751, doi: 10.1089/ast.2009.0376

Shibata, K., & Magara, T. 2011, Living Reviews in Solar Physics, 8, 6, doi: 10.12942/lrsp-2011-6

Sun, X., Török, T., & DeRosa, M. L. 2022, MNRAS, 509, 5075, doi: 10.1093/mnras/stab3249

Takahashi, T., Mizuno, Y., & Shibata, K. 2016, ApJL, 833, L8, doi: 10.3847/2041-8205/833/1/L8

Tokovinin, A., Fischer, D. A., Bonati, M., et al. 2013, PASP, 125, 1336, doi: 10.1086/674012

Veronig, A. M., Odert, P., Leitzinger, M., et al. 2021, Nature Astronomy, 5, 697, doi: 10.1038/s41550-021-01345-9

Vida, K., Leitzinger, M., Kriskovics, L., et al. 2019, A&A, 623, A49, doi: 10.1051/0004-6361/201834264

Vida, K., Kriskovics, L., Oláh, K., et al. 2016, A&A, 590, A11, doi: 10.1051/0004-6361/201527925

Wang, S.-i., Hildebrand, R. H., Hobbs, L. M., et al. 2003, in SPIE, Vol. 4841, Instrument Design and Performance for Optical/Infrared Ground-based Telescopes, ed. M. Iye & A. F. M. Moorwood, 1145–1156, doi: 10.1117/12.461447

Zic, A., Murphy, T., Lynch, C., et al. 2020, ApJ, 905, 23, doi: 10.3847/1538-4357/abca90

Solar and Stellar Coronal Mass Ejections
Proceedings IAU Symposium No. 388, 2026
N. Gopalswamy, O. Malandraki, A. Vidotto & W. Manchester, eds.
doi:10.1017/S1743921325001577

Resolving the CME Confinement Paradox in the Highly Magnetic Environment of AB Doradus

D. Evensberget, K. M. Strickert and A. A. Vidotto

Leiden Observatory, The Netherlands
E-mail: evensberget@strw.leidenuniv.nl

Abstract. AB Doradus is a young, active solar-type star whose large-scale coronal magnetic field is 100–1000 times more intense than that of the Sun. Magnetic fields of such magnitude are believed to prevent stellar coronal mass ejections (CMEs) from erupting. It is, therefore, surprising that coronal dimming was recently observed in AB Doradus, as coronal dimming is typically associated with CMEs. We propose that CMEs in AB Doradus occur in high-latitude regions where parts of the coronal magnetic field may be open. We modelled the global coronal magnetohydrodynamics for twenty-one CMEs, injecting CMEs into high-latitude regions with both closed and open magnetic field configurations. Our models demonstrate that CMEs in areas with open magnetic fields are not obstructed by the overlying magnetic field, suggesting that magnetic confinement of CMEs may be less common in young solar-type stars than previously assumed. For full details of this study, we direct the reader to Strickert et al. (2024).

Keywords. Stellar magnetic fields, stellar coronal mass ejections

1. Introduction

Solar-type stars, which are often more magnetically active than the Sun, are expected to host frequent and energetic burst events. This is supported by the detection of stellar flares and super-flares (e.g., Maehara et al. 2012). Detections of stellar coronal mass ejections (CMEs) have, however, proved challenging (Moschou et al. 2019), although there is indirect evidence available in the form of Balmer line shifts (Vida et al. 2019) and blue-shifted X-ray emissions (Argiroffi et al. 2019). Another piece of indirect evidence of CMEs is *coronal dimming*, observed in X-ray, EUV, and far UV frequencies. Such dimming events follow most CME events in the Sun (Leitzinger et al. 2020).

In analogy with the Sun, it seems natural to consider whether dimming events could succeed stellar CMEs. In EUV and X-ray spectra of about 200 stars, Veronig et al. (2021) found thirteen stars that exhibited dimming after flare events. One of these stars is AB Doradus (HD36705) which is a young, rapidly rotating solar-type star aged about 120 Myr, a rotation period of 0.514 d, and a large-scale magnetic field in the kilogauss range (Donati et al. 1999; Hussain et al. 2007; Strassmeier 2009; Barenfeld et al. 2013).

Here we consider the 1994 dimming event of AB Doradus which lasted 13.3 h from the peak of the flare to the end of the dimming. The duration of the dimming event is comparable to the AB Doradus' rotation period. As AB Doradus rotation axis has an inclination of 60 deg (Donati et al. 1999), only a flare-dimming event occurring in a high-latitude region on the stellar surface would remain in observable for an entire stellar rotation. Such a high-latitude flare origin would also be consistent with the observed polar spots on AB Doradus (Collier Cameron and Unruh 1994).

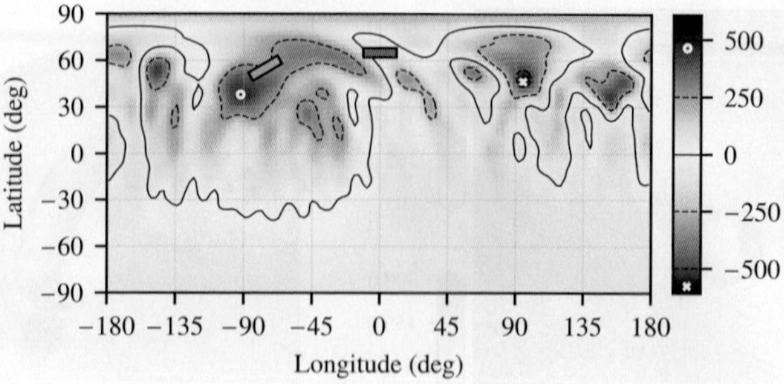

Figure 1. Surface magnetic field structure of AB Doradus, derived from spectropolarimetric observations in December 2002 and processed with Zeeman-Doppler imaging (Hussain et al. 2007). The green and magenta patches show different flux rope locations for the simulated CME models.

At first glance, the powerful kilogauss-range magnetic field of AB Doradus may seem to present an obstacle to CMEs escaping from the star, as powerful stellar magnetic field can prevent CMEs from erupting (Svestka and Cliver 1992; Drake et al. 2013, 2016). More specifically, powerful stellar magnetic field can prevent CMEs below an energy threshold from erupting, as Alvarado-Gómez et al. (2018) showed with a suite of numerical simulations. The suppression effect is, however, mainly present in the regions of closed magnetic field lines, where the magnetic tension force opposes the CME expansion. We therefore hypothesise that CMEs in AB Doradus may be more likely to erupt in high-latitude regions where the magnetic field is open. To test this hypothesis, we conducted a parametric study of CMEs in AB Doradus, varying the flux rope properties and the location of the flux rope on the stellar surface.

This text summarise the findings of our numerical study. For full details of this work, we refer the reader to Strickert et al. (2024).

2. Numerical modelling of CMEs

We model the corona, wind, and possible CMEs of AB Doradus using the Alfvén Wave Solar Model (AWSoM) (Sokolov et al. 2013; van der Holst et al. 2014) within the 3D MHD code BATS-R-US (Powell et al. 1999; Tóth et al. 2012). This model, initially developed for the solar wind, applies Alfvén waves to heat and accelerate the stellar wind and has been used for cool dwarf stars (e.g. Alvarado-Gómez et al. 2016; Ó Fionnagáin et al. 2021; Kavanagh et al. 2021; Evensberget et al. 2021, 2022, 2023; Vidotto et al. 2023). The model extends from the chromosphere, through the transition region, the stellar corona and into the wind region. We first find a steady-state solution for the stellar wind, and then inject a flux rope to simulate a CME event.

2.1. *Steady-state model*

We initialise our model with with AB Doradus' mass ($0.86\ M_{\odot}$), radius ($0.96\ R_{\odot}$), rotation period (0.514 d), and the December 2002 surface magnetic field map from Hussain et al. (2007), reconstructed using Zeeman-Doppler imaging (Semel 1989; Donati and Landstreet 2009). This map, shown in Figure 1, shows the large-scale features of the surface radial magnetic field. The radial field values range from -573 to 466 G, with an average field strength of 105 G.

The inner boundary of the model lies in the chromosphere, where we apply parameters similar to solar wind simulations (Evensberget et al. 2023). We apply a chromospheric electron density of $2 \times 10^{-11}\ \mathrm{cm}^{-3}$ and a temperature of 5×10^{4} K. Further model parameters are detailed in van der Holst et al. (2014). Our spherical grid has a smallest size

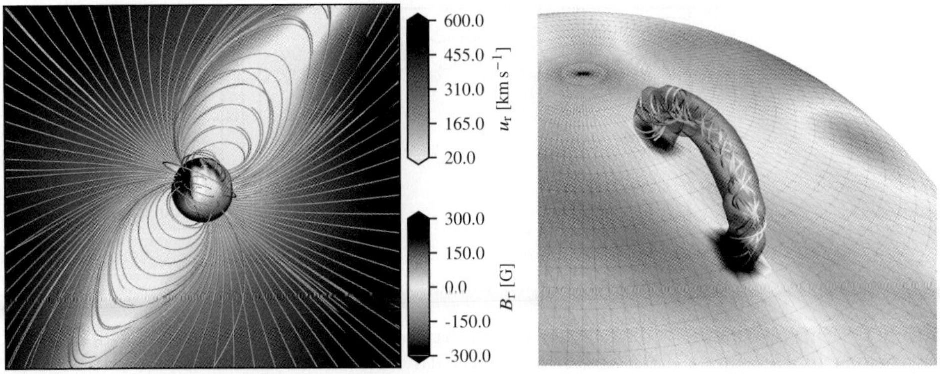

Figure 2. Left: Steady-state wind solution of AB Doradus using the Zeeman-Doppler imaging based map of Figure 1. The magnetic field lines are shown in grey. The background is colour-coded by radial wind velocity and the surface by radial magnetic field. Right: Initial flux rope configuration using the model of Titov and Démoulin (1999). Figures from Strickert et al. (2024).

of $1/256 R_\star$ near the surface and extends to $24\,R_\star$. The resulting steady-state wind solution, shown in Figure 2, provides the background for each of the simulated CME events. We compute a mass-loss rate of $10^{-12} M_\odot/\mathrm{yr}$, aligning with estimates from Cohen et al. (2010) and Jardine and Collier Cameron (2019). The wind shows an inclined dipole structure (see Figure 2), with speeds up to $1000\,\mathrm{km\,s^{-1}}$ in the polar regions.

2.2. *CME flux rope injection*

To initiate the CMEs, we add a magnetic flux rope anchored at the inner boundary of our simulations.We use the flux rope model from Titov and Démoulin (1999) (see Figure 2). The flux rope injection perturbs the background wind, often leading to a CME event. This approach follows similar models for solar and stellar CMEs (e.g., Jin et al. 2013; Alvarado-Gómez et al. 2018).

We conduct twenty-one CME simulations with varying flux rope parameters. The geometry of the flux rope torus (major radius $R_{\mathrm{maj}} = 128\,\mathrm{Mm}$, minor radius $R_{\mathrm{min}} = 20\,\mathrm{Mm}$, and depth $D = 1\,\mathrm{Mm}$) is constant. We place the flux rope at two different high-latitudes locations due to AB Doradus' polar active regions. In one of the locations the large-scale magnetic field is open ($-76°$ longitude, $55°$ latitude, $-30°$ orientation), while in the other location it is closed ($0°$ longitude, $65°$ latitude, east-west orientation). The locations and orientations are shown in Figure 1.

We vary the flux rope mass M_{FR} from 3.5×10^{14} to 3.5×10^{18} g, and central magnetic field B_{FR} from 100 to 1000 G. The current $I_{\mathrm{FR}} = 2cR_{\mathrm{maj}}B_{\mathrm{FR}}/4\pi$ is set to achieve the desired B_{FR}. We compute the flux rope's poloidal flux $\phi_{\mathrm{p}}^{\mathrm{FR}}$ and free magnetic energy $E_{\mathrm{free}}^{B} \propto B_{\mathrm{FR}}^2$ following Titov and Démoulin (1999); Roussev et al. (2003).

3. Overview of our simulated CMEs

As described in Section 2, we injected twenty-one different flux rope configurations into our background stellar wind model. Figure 3 shows fourteen simulated CMEs at a snapshot of one hour evolution. The surfaces shown, which track the CME evolution, comprise points where $n/n_{\mathrm{ss}} = 4$ between the simulated wind particle density and the steady-state stellar wind. The surface colour shows the relative speed of the CME compared to the background wind, while the model inner boundary (stellar surface) is coloured by the surface magnetic field strength.

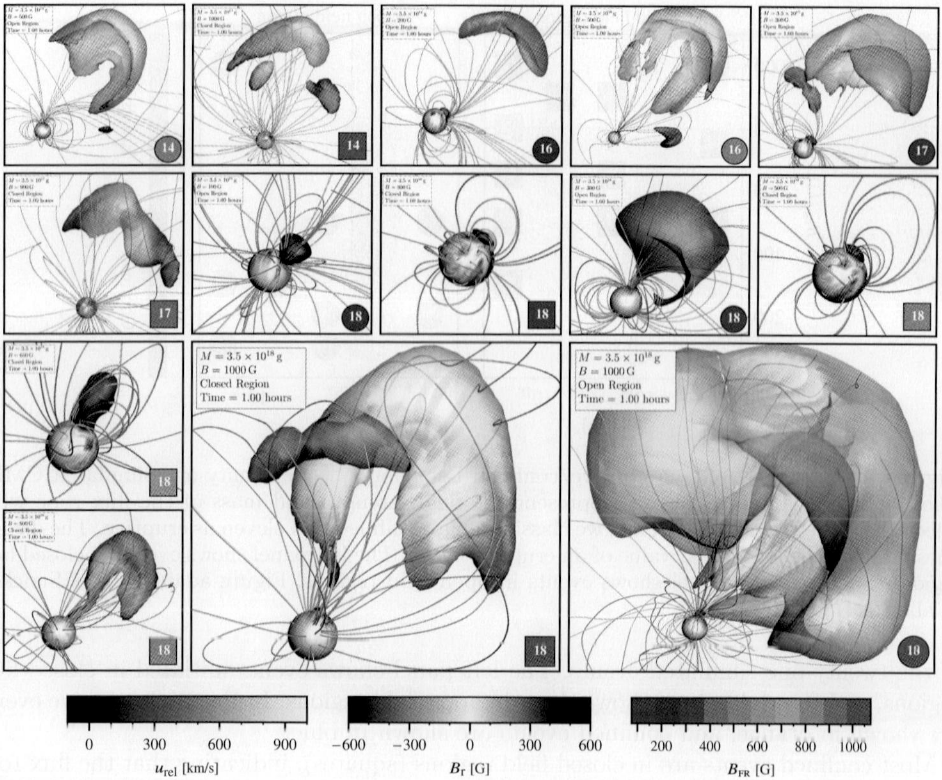

Figure 3. Snapshot of 14/21 CME models after one hour of evolution. The magnetic field of the initial flux rope B_{FR} (symbol colour) increases from top left to bottom right for each initial flux rope mass M_{FR} (cf. symbol text). Flux ropes located in open field lines (circle) evolve faster than those in closed field line regions (square) for the same B_{FR} and M_{FR}. Iso-surfaces are defined using a plasma density ratio of four and are colour-coded by their relative radial velocity compared to the local stellar wind value. Figure based on Strickert et al. (2024).

3.1. *Resulting CME shapes and speeds*

Several factors influence CME morphology. Fundamentally, the energy released from the flux rope must overcome the overlying magnetic tension to escape. Thus the placement (open/closed region) of the flux rope affects the energy needed for a CME to escape: Flux ropes in closed field regions require higher energy to break through compared to those in open field regions. For instance, model 18 has the most extended CME due to its energetic flux rope in an open field region, while model 18, with identical parameters but in a closed-field region, appears less extended and propagates more slowly.

We see cases of little CME evolution, such as model 18, where the flux rope energy is insufficient to overcome the magnetic tension. This model, with its weaker, 300G flux rope field is classified as confined. In contrast, model 18, with the same flux rope configuration but in an open field region, escapes.

Some eruptive CMEs leave behind a dense region of plasma. Models 16 and 17 show plasma remaining at the inner boundary, with some CME mass falling back to the star. These two events are still classified as eruptive, as the bulk of the CME mass has escaped.

3.2. *Confined versus eruptive CMEs*

We classify a CME behaviour as eruptive if it has a positive average velocity after one hour of evolution, otherwise we classify it as confined. Figure 4 summarises the behaviour

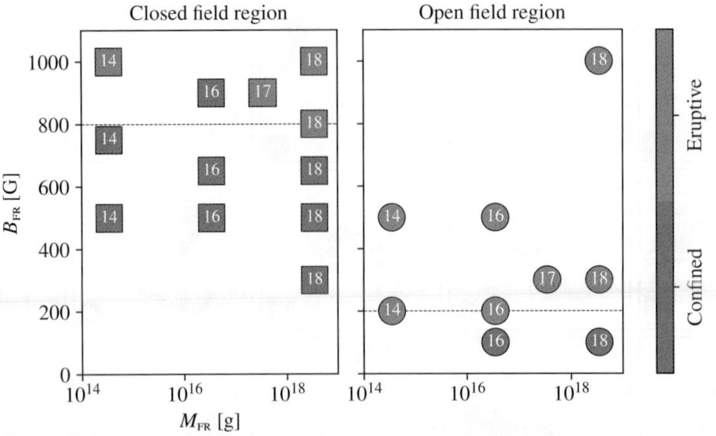

Figure 4. Summary of the eruptive/confined behaviour in our twenty-one simulated CMEs. The horizontal and vertical axes represent the field strength and mass of the flux rope when injected. Of the twenty-one events, we classify ten as confined and eleven as eruptive. The dashed lines indicate the lowest $B_{\rm FR}$ value of an eruptive event. The left panel shows events in closed field regions, and the right panel shows events in open field regions. Figure adapted from Strickert et al. (2024).

of the twenty-one simulated events. The left panel shows events initiated in closed field regions, and the right panel shows those in open field regions. In this plot eruptive events are shown in orange, and confined events are shown in blue.

Most confined events are in closed field regions (squares), indicating that the flux rope energy is too weak to break the overlying field tension. The two confined events in open field lines, 16 and 18, have weak flux rope magnetic fields of 100 G. Figure 4 shows that for a CME to erupt, higher magnetic field strength is required in closed field regions compared to open field regions. We see that no models with initial flux rope magnetic fields $B_{\rm FR} \lesssim 800$ G were eruptive in closed-field CMEs, while the threshold for open-field CMEs was $B_{\rm FR} \lesssim 200$ G.

4. Discussion

Active, flaring stars, would be expected to have more CMEs (Aarnio et al. 2012; Drake et al. 2013; Osten and Wolk 2015). but these have proven challenging to detect (Moschou et al. 2019; Namekata et al. 2022). Theoretical and numerical work (Svestka and Cliver 1992; Drake et al. 2013, 2016; Alvarado-Gómez et al. 2018; Sun et al. 2022) has, however, suggested that these may frequently be confined due to the presence of a strong overlying magnetic field and the accompanying magnetic tension. It was therefore of great interest to the authors to understand whether and how CMEs could escape from the highly magnetised environment of AB Doradus, where, recently, Veronig et al. (2021) reported the detection of coronal dimmings indicative of the presence of eruptive CMEs.

In this work, we investigate this apparent paradox by conducting a parametric study with twenty-one simulations of CME events on AB Doradus. The presence of polar spots in AB Doradus led us to inject our modelled CMEs in high-latitude regions ($\sim 60°$) of the star. Using three-dimensional magnetohydrodynamic modelling driven by an observation-based surface magnetic map obtain from Zeeman-Doppler imaging, we fixed the flux rope geometry, while varying the central magnetic field strength of the flux rope, the mass of the flux rope and also the region (open/closed field) where the flux rope was placed. The coronal magnetic field was based on a reconstructed magnetic map based on spectropolarimetric observations taken in December 2002 (Hussain et al. 2007).

We find that CMEs initiated in open regions are more likely to be eruptive compared to those initiated in closed regions. In our simulations, no models with initial flux rope magnetic fields under ~ 800 G were eruptive for closed-field CMEs. For open-field CME models, the limit was ~ 200 G.

We note that our model does not include flux rope emergence, as shown by the initial state of the injected flux rope in Figure 2. Our approach permits us to focus on the dynamics of already emerged flux rope in regions of open and closed large-scale magnetic fields. Further modelling work could include the emergence of flux ropes in the context of stellar CMEs.

5. Conclusion

Our work presents a way to reconcile the apparent paradox† that active stars like AB Doradus would have more confined CMEs (Alvarado-Gómez et al. 2018), and yet, observations show the presence of eruptive CMEs (Veronig et al. 2021). We demonstrate that CMEs can escape the stellar surface in stars with polar active regions, since magnetic confinement is comparatively weak in regions of open magnetic field lines.

It is also interesting to note that in our models, the most energetic CMEs expand considerably, including in the direction towards the equatorial plane, where potentially planets would be orbiting. Therefore, even though the CME originates in stellar polar regions, they would still affect the space weather environment around planets orbiting in the equatorial plane. The geometry of the large-scale magnetic field of the star, as well as the location of the CME initiation, are crucial factors in determining whether a CME will be confined or eruptive. For the full details of this work, please see Strickert et al. (2024).

Acknowledgements

This project has received funding from the European Research Council (ERC) under the European Union's Horizon 2020 research and innovation programme (grant agreement No 817540, ASTROFLOW). This work used the Dutch national e-infrastructure with the support of the SURF Cooperative using grant nos. EINF-2218 and EINF-5173. This research has made use of NASA's Astrophysics Data System. This work was carried out using the SWMF tools developed at The University of Michigan Center for Space Environment Modelling (CSEM) and made available through the NASA Community Coordinated Modelling Center (CCMC). This work has made use of the following additional numerical software and visualisation software: NumPy (van der Walt et al. 2011), SciPy (Virtanen et al. 2020), and Matplotlib (Hunter 2007).

References

Aarnio, A. N., Matt, S. P., & Stassun, K. G. 2012, Mass Loss in Pre-main-sequence Stars via Coronal Mass Ejections and Implications for Angular Momentum Loss. *ApJ*, 760(1), 9.

Alvarado-Gómez, J. D., Drake, J. J., Cohen, O., Moschou, S. P., & Garraffo, C. 2018, Suppression of Coronal Mass Ejections in Active Stars by an Overlying Large-scale Magnetic Field: A Numerical Study. *ApJ*, 862(2), 93.

Alvarado-Gómez, J. D., Hussain, G. A. J., Cohen, O., Drake, J. J., Garraffo, C., Grunhut, J., & Gombosi, T. I. 2016, Simulating the environment around planet-hosting stars. I. Coronal structure. *A&A*, 588, A28.

† We emphasise that we consider both Alvarado-Gómez et al. (2018) and Veronig et al. (2021) to be correct in their respective works and conclusions. Indeed, our numerical modelling exemplify that a CME-caused coronal dimming process and magnetic confinement may be present in the same star.

Argiroffi, C., Reale, F., Drake, J. J., Ciaravella, A., Testa, P., Bonito, R., Miceli, M., Orlando, S., & Peres, G. 2019, A stellar flare-coronal mass ejection event revealed by X-ray plasma motions. *Nature Astronomy*, 3, 742–748.

Barenfeld, S. A., Bubar, E. J., Mamajek, E. E., & Young, P. A. 2013, A Kine-chemical Investigation of the AB Dor Moving Group "Stream". *ApJ*, 766(1), 6.

Cohen, O., Drake, J. J., Kashyap, V. L., Hussain, G. A. J., & Gombosi, T. I. 2010, The Coronal Structure of AB Doradus. *ApJ*, 721(1), 80–89.

Collier Cameron, A. & Unruh, Y. C. 1994, Doppler images of AB Doradus in 1992 January. *MNRAS*, 269, 814–836.

Donati, J. F., Collier Cameron, A., Hussain, G. A. J., & Semel, M. 1999, Magnetic topology and prominence patterns on AB Doradus. *MNRAS*, 302(3), 437–456.

Donati, J. F. & Landstreet, J. D. 2009, Magnetic Fields of Nondegenerate Stars. *ARA&A*, 47(1), 333–370.

Drake, J. J., Cohen, O., Garraffo, C., & Kashyap, V. Stellar flares and the dark energy of CMEs. In Kosovichev, A. G., Hawley, S. L., & Heinzel, P., editors, *Solar and Stellar Flares and their Effects on Planets* 2016, volume 320, pp. 196–201.

Drake, J. J., Cohen, O., Yashiro, S., & Gopalswamy, N. 2013, Implications of Mass and Energy Loss due to Coronal Mass Ejections on Magnetically Active Stars. *ApJ*, 764(2), 170.

Evensberget, D., Carter, B. D., Marsden, S. C., Brookshaw, L., & Folsom, C. P. 2021, The winds of young Solar-type stars in the Hyades. *MNRAS*, 506(2), 2309–2335.

Evensberget, D., Carter, B. D., Marsden, S. C., Brookshaw, L., Folsom, C. P., & Salmeron, R. 2022, The winds of young Solar-type stars in Coma Berenices and Hercules-Lyra. *MNRAS*, 510(4), 5226–5245.

Evensberget, D., Marsden, S. C., Carter, B. D., Salmeron, R., Vidotto, A. A., Folsom, C. P., Kavanagh, R. D., Pineda, J. S., Driessen, F. A., & Strickert, K. M. 2023, The winds of young Solar-type stars in the Pleiades, AB Doradus, Columba, and β Pictoris. *MNRAS*, 524(2), 2042–2063.

Hunter, J. D. 2007, Matplotlib: A 2D Graphics Environment. *Computing in Science and Engineering*, 9(3), 90–95.

Hussain, G. A. J., Jardine, M., Donati, J. F., Brickhouse, N. S., Dunstone, N. J., Wood, K., Dupree, A. K., Collier Cameron, A., & Favata, F. 2007, The coronal structure of AB Doradus determined from contemporaneous Doppler imaging and X-ray spectroscopy. *MNRAS*, 377(4), 1488–1502.

Jardine, M. & Collier Cameron, A. 2019, Slingshot prominences: nature's wind gauges. *MNRAS*, 482(3), 2853–2860.

Jin, M., Manchester, W. B., van der Holst, B., Oran, R., Sokolov, I., Toth, G., Liu, Y., Sun, X. D., & Gombosi, T. I. 2013, Numerical Simulations of Coronal Mass Ejection on 2011 March 7: One-temperature and Two-temperature Model Comparison. *ApJ*, 773, 50.

Kavanagh, R. D., Vidotto, A. A., Klein, B., Jardine, M. M., Donati, J.-F., & Ó Fionnagáin, D. 2021, Planet-induced radio emission from the coronae of M dwarfs: the case of Prox Cen and AU Mic. *MNRAS*, 504(1), 1511–1518.

Leitzinger, M., Odert, P., Greimel, R., Vida, K., Kriskovics, L., Guenther, E. W., Korhonen, H., Koller, F., Hanslmeier, A., Kővári, Z., & Lammer, H. 2020, A census of coronal mass ejections on solar-like stars. *MNRAS*, 493(3), 4570–4589.

Maehara, H., Shibayama, T., Notsu, S., Notsu, Y., Nagao, T., Kusaba, S., Honda, S., Nogami, D., & Shibata, K. 2012, Superflares on solar-type stars. *Nature*, 485(7399), 478–481.

Moschou, S.-P., Drake, J. J., Cohen, O., Alvarado-Gómez, J. D., Garraffo, C., & Fraschetti, F. 2019, The Stellar CME-Flare Relation: What Do Historic Observations Reveal? *ApJ*, 877(2), 105.

Namekata, K., Maehara, H., Honda, S., Notsu, Y., Nogami, D., & Shibata, K. 2022, Hunting for stellar coronal mass ejections. *arXiv e-prints*, arXiv:2211.05506.

Ó Fionnagáin, D., Vidotto, A. A., Petit, P., Neiner, C., Manchester, W., I., Folsom, C. P., & Hallinan, G. 2021, λ And: a post-main-sequence wind from a solar-mass star. *MNRAS*, 500(3), 3438–3453.

Osten, R. A. & Wolk, S. J. 2015, Connecting Flares and Transient Mass-loss Events in Magnetically Active Stars. *ApJ*, 809(1), 79.

Powell, K. G., Roe, P. L., Linde, T. J., Gombosi, T. I., & De Zeeuw, D. L. 1999, A Solution-Adaptive Upwind Scheme for Ideal Magnetohydrodynamics. *Journal of Computational Physics*, 154(2), 284–309.

Roussev, I. I., Forbes, T. G., Gombosi, T. I., Sokolov, I. V., DeZeeuw, D. L., & Birn, J. 2003, A Three-dimensional Flux Rope Model for Coronal Mass Ejections Based on a Loss of Equilibrium. *ApJ*, 588(1), L45–L48.

Semel, M. 1989, Zeeman-Doppler imaging of active stars. I - Basic principles. *A&A*, 225, 456–466.

Sokolov, I. V., van der Holst, B., Oran, R., Downs, C., Roussev, I. I., Jin, M., Manchester, Ward B., I., Evans, R. M., & Gombosi, T. I. 2013, Magnetohydrodynamic Waves and Coronal Heating: Unifying Empirical and MHD Turbulence Models. *ApJ*, 764(1), 23.

Strassmeier, K. G. 2009, Starspots. *A&Ar*, 17(3), 251–308.

Strickert, K. M., Evensberget, D., & Vidotto, A. A. 2024, High-latitude coronal mass ejections on the young solar-like star AB Dor. *MNRAS*, 533(1), 1156–1165.

Sun, X., Török, T., & DeRosa, M. L. 2022, Torus-stable zone above starspots. *MNRAS*, 509(4), 5075–5085.

Svestka, Z. & Cliver, E. W. History and Basic Characteristics of Eruptive Flares. In Svestka, Z., Jackson, B. V., & Machado, M. E., editors, *IAU Colloq. 133: Eruptive Solar Flares* 1992, volume 399, 1.

Titov, V. S. & Démoulin, P. 1999, Basic topology of twisted magnetic configurations in solar flares. *A&A*, 351, 707–720.

Tóth, G., van der Holst, B., Sokolov, I. V., De Zeeuw, D. L., Gombosi, T. I., Fang, F., Manchester, W. B., Meng, X., Najib, D., Powell, K. G., Stout, Q. F., Glocer, A., Ma, Y.-J., & Opher, M. 2012, Adaptive numerical algorithms in space weather modeling. *Journal of Computational Physics*, 231(3), 870–903.

van der Holst, B., Sokolov, I. V., Meng, X., Jin, M., Manchester, W. B., I., Tóth, G., & Gombosi, T. I. 2014, Alfvén Wave Solar Model (AWSoM): Coronal Heating. *The Astrophysical Journal*, 782(2), 81.

van der Walt, S., Colbert, S. C., & Varoquaux, G. 2011, The NumPy Array: A Structure for Efficient Numerical Computation. *Computing in Science and Engineering*, 13(2), 22–30.

Veronig, A. M., Odert, P., Leitzinger, M., Dissauer, K., Fleck, N. C., & Hudson, H. S. 2021, Indications of stellar coronal mass ejections through coronal dimmings. *Nature Astronomy*, 5, 697–706.

Vida, K., Leitzinger, M., Kriskovics, L., Seli, B., Odert, P., Kovács, O. E., Korhonen, H., & van Driel-Gesztelyi, L. 2019, The quest for stellar coronal mass ejections in late-type stars. I. Investigating Balmer-line asymmetries of single stars in Virtual Observatory data. *A&A*, 623, A49.

Vidotto, A. A., Bourrier, V., Fares, R., Bellotti, S., Donati, J. F., Petit, P., Hussain, G. A. J., & Morin, J. 2023, The space weather around the exoplanet GJ 436b. II. Stellar wind-exoplanet interactions. *A&A*, 678, A152.

Virtanen, P., Gommers, R., Oliphant, T. E., Haberland, M., Reddy, T., Cournapeau, D., Burovski, E., Peterson, P., Weckesser, W., Bright, J., van der Walt, S. J., Brett, M., Wilson, J., Jarrod Millman, K., Mayorov, N., Nelson, A. R. J., Jones, E., Kern, R., Larson, E., Carey, C., Polat, İ., Feng, Y., Moore, E. W., Vand erPlas, J., Laxalde, D., Perktold, J., Cimrman, R., Henriksen, I., Quintero, E. A., Harris, C. R., Archibald, A. M., Ribeiro, A. H., Pedregosa, F., van Mulbregt, P., & Contributors, S. 2020, SciPy 1.0: Fundamental Algorithms for Scientific Computing in Python. *Nature Methods*, 17, 261–272.

Solar and Stellar Coronal Mass Ejections
Proceedings IAU Symposium No. 388, 2026
N. Gopalswamy, O. Malandraki, A. Vidotto & W. Manchester, eds.
doi:10.1017/S1743921324001480

Searching for stellar CMEs in the Praesepe and Pleiades clusters

K. Vida[1,2], B. Seli[1,2,4], R. M. Roettenbacher[3], A. Görgei[1,2,4], L. Kriskovics[1,2], Zs. Kővári[1,2] and K. Oláh[1,2]

[1]Konkoly Thege Miklós Astronomical Institute, HUN-REN Research Centre for Astronomy and Earth Sciences
email: vidakris@konkoly.hu

[2]CSFK, MTA Centre of Excellence

[3]University of Michigan

[4]Eötvös Loránd University

Abstract. On the Sun, the energetic, erupting phenomena of flares and coronal mass ejections (CMEs) often occur together. While space-based photometry has revealed frequent white-light flares for vast numbers of stars, only a handful of coronal mass ejections have been detected. Space-based photometry reveals the timing and detailed structure of flares. To detect CME signatures, however, optical spectroscopy is essential, as the ejected plasma can be detected by Doppler-shifted emission bumps in the Balmer-regions. We present a dedicated ground-based multi-object spectroscopic observations of the young, nearby Praesepe (600 Myr) and Pleiades (135 Myr) clusters to detect CMEs and flares parallel with the observations of Praesepe by the TESS satellite. During the 10 days of overlapping observations, we did not find any obvious signs of CMEs or flares in the Hα region.

Keywords. Stars: activity, Stars: coronae, Stars: coronal mass ejections (CMEs), Stars: flare, Stars: late-type, Stars: magnetic field, open clusters and associations: individual:Pleiades, open clusters and associations: individual:Praesepe

---

1. Introduction

Flares and coronal mass ejections (CMEs) are powerful events that occur on the surface of stars, including the Sun. Flares are sudden, explosive releases of energy resulting from the twisting and reconnection of surface magnetic fields. This process releases bursts of radiation across various wavelengths, from radio waves to X-rays. On the Sun, most large flares are accompanied by massive eruptions of plasma known as CMEs. Assuming that solar and stellar activity signatures are driven by the same underlying physics, a similar relationship between flares and CMEs is expected on other stars. These events significantly impact the space environment around the star by producing energetic particles, accelerating charged particles to high speeds, and disrupting the interplanetary magnetic field. Large flares and accompanying CMEs may render planets uninhabitable (Vida et al. 2017; Roettenbacher & Kane 2017) or lead to the total loss of the planetary atmosphere (Khodachenko 2007; Lammer et al. 2011). Understanding these phenomena is important for studying the behavior of stars, as well as for predicting and mitigating their effects on space weather and potential habitability.

On the Sun, flares and CMEs are studied by dedicated monitoring, partly motivated by the immediate threat they pose to the Earth. On other stars, however, flares are much

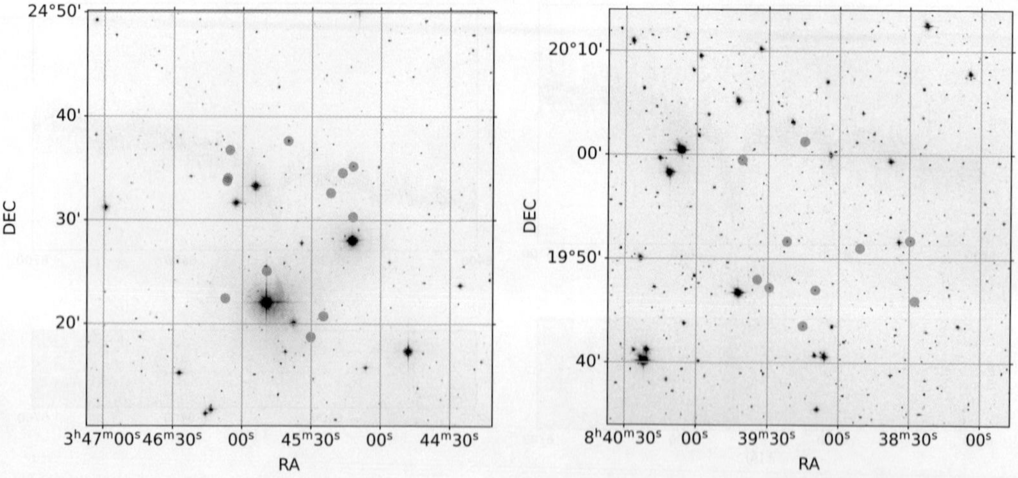

Figure 1. The field of view in the Pleiades (left) and Praesepe clusters (right) with the selected spectroscopic targets marked.

better characterized than CMEs, since the latter are challenging to observe. Most of the known stellar CMEs were detected serendipitously using optical spectroscopy when the Doppler-shifted signal of the ejecta caused extra emission in the blue wing of Balmer lines (see, e.g., Houdebine et al. 1990). Currently, almost all known CME events are detected on M-dwarfs (see Vida et al. 2019; Koller et al. 2021; Lu et al. 2022 and references therein). Even though there are numerous detections of superflares on solar-like stars (Maehara et al. 2012), the only one exhibiting signatures of a CME is the young solar analog EK Dra, where simultaneous spectroscopic and TESS photometry revealed evidence of a CME associated with a superflare (Namekata et al. 2021). Despite these advances, no comprehensive study of CME parameters and rates can be carried out. This poses a problem since CMEs have a large impact on the angular momentum evolution of young stars and can affect the atmospheres of exoplanets. Furthermore, current detections are based on either single-target observations, serendipitous detections, or archival data with a sub-optimal observation strategy for this purpose. A good solution could be observing open clusters, where several stars can be studied simultaneously – such multi-object observations have been performed by Guenther & Emerson (1997); Leitzinger et al. (2014); Korhonen et al. (2017). In this manuscript, we present an attempt to observe stellar CMEs in a sample of young stars in open clusters using contemporaneous ground-based spectroscopic and space-based photometric observations.

2. Target selection

We selected two young, nearby open clusters: Pleiades (M45) and Praesepe (M44), with ages of 135 and 600 Myr (Gossage et al. 2018), respectively. To maximize our chances of detecting a CME, we need to observe as many simultaneous targets as possible. Therefore, our best option is to pick a young open cluster with numerous magnetically active stars that are most likely to exhibit (super)flares and CMEs. To achieve a high S/N ratio, we need to select a nearby cluster. There are only three clusters that are suitable for our purposes: Hyades, Praesepe, and Pleiades. From these candidates, Praesepe and Pleiades were the best candidates for our purposes, as Hyades is the least dense cluster, and we could fit only half as many stars in a single FoV.

We analyzed photometric observations of the cluster members obtained earlier by the K2 mission and TESS to determine the chances of flaring in the possible fields. We

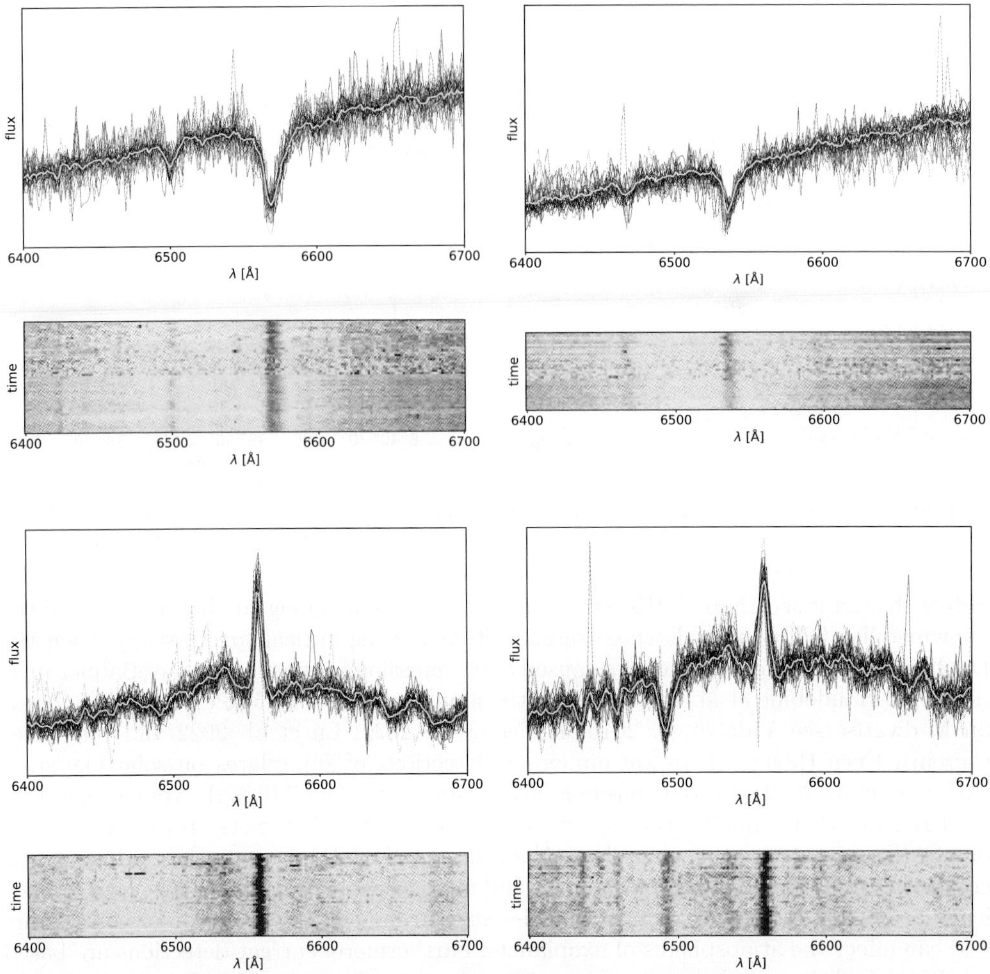

Figure 2. Example dynamic spectra and dynamic spectra from the Pleiades cluster showing the Hα region. The continuous thick line indicates the averaged spectra.

selected targets to observe from the catalog of Ilin et al. (2021), who searched for stellar flares in open clusters with Kepler K2 photometry. We selected flaring stars brighter than $T = 15^m$ (TESS magnitude) from Praesepe and Pleiades. We carried out a grid search in RA–Dec to find the optimal FoV with an 18' diameter, which contains the most flaring stars. With the best setup, we could include 10 stars from Praesepe and 12 from Pleiades (see Figure 1).

3. Early results

During the 10-day-long observing run in 2023 November, we used the OSMOS multi-object spectrograph on the MDM telescope (Arizona, US) with the moderate resolution VPH grism ($R \approx 1600$). This setup yields a velocity resolution of $\approx 180 \mathrm{km\,s^{-1}}$, enough to resolve larger coronal mass ejections (typical CMEs have velocities ranging from a few hundred to a few thousand $\mathrm{km\,s^{-1}}$) with good S/N while having decent temporal resolution. We had 12 hours of on-target time for the selected late-type stars in the Pleiades cluster and 15 hours of on-target time for the Praesepe cluster. During the

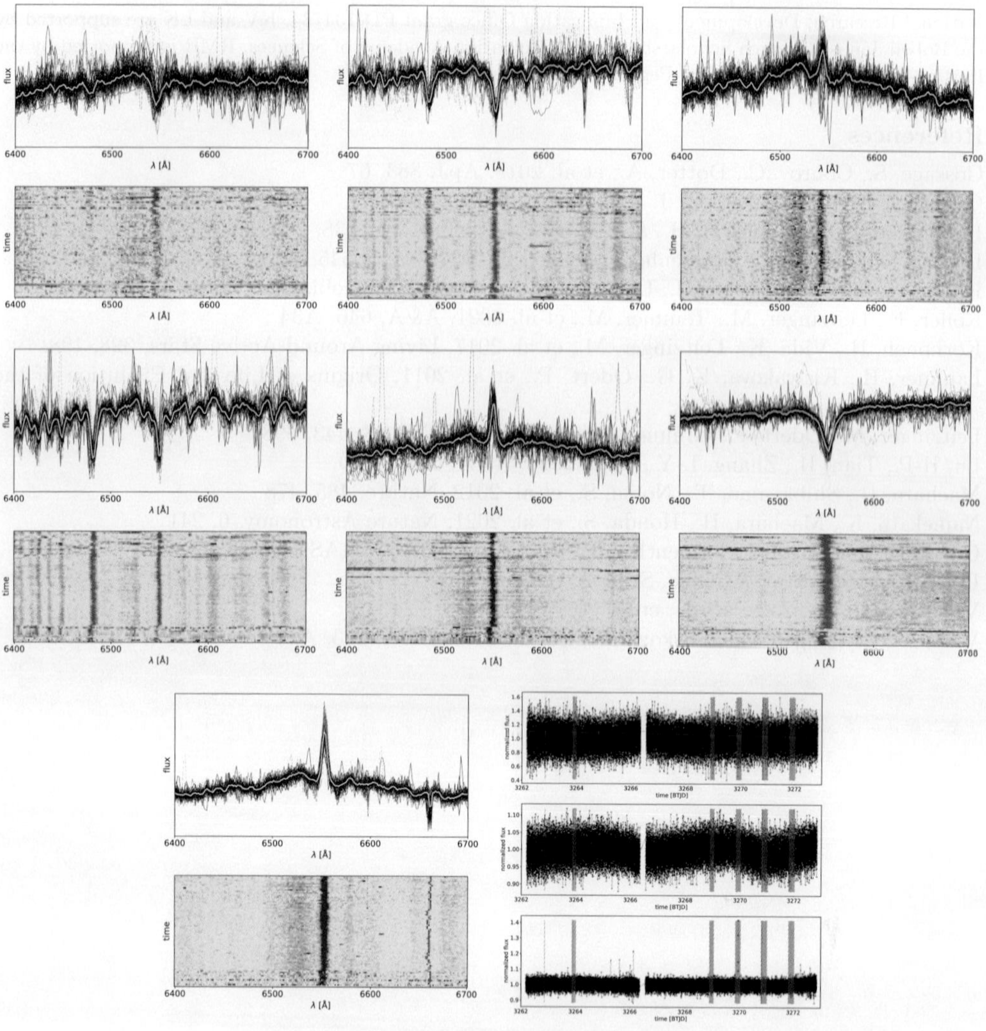

Figure 3. Example dynamic spectra and dynamic spectra from the Praesepe cluster showing the Hα region. The continuous thick line indicates the averaged spectra. The last plot shows a sample of TESS light curves with the times of the spectroscopic observations marked with red boxes.

MDM observing run the TESS also observed the cluster as part of the guest investigator program G06138. A sample of the resulting Hα spectra is shown in Figures 2 and 3.

We did not detect any obvious sign of coronal mass ejections or flares in the Hα region. The total on-source time for the targets in the two clusters was close to 300 hours – even if we discard the spectra with low signal-to-noise levels, based on the empirical estimations of Odert et al. (2020), we would expect at least one CME detection. It is possible that the relatively low spectral resolution, ionization of the ejecta (with the ionization of the plasma it cannot be detected in Hα), and projection effects could have an influence on the results. In the future we will use the observations to improve the existing statistics and to refine the upper limits of stellar coronal mass ejections.

Acknowledgements. This research was funded by the Hungarian National Research, Development and Innovation Office Élvonal grant KKP-143986. Authors acknowledge the financial support of the Austrian–Hungarian Action Foundation grants 112öu1. LK acknowledges the support of the Hungarian

National Research, Development and Innovation Office grant PD-134784. KV and LK are supported by the Bolyai János Research Scholarship of the Hungarian Academy of Sciences. RMR is supported by the Heising-Simons Foundation's 51 Pegasi b Fellowship.

References

Gossage, S., Conroy, C., Dotter, A., et al. 2018, ApJ, 863, 67
Guenther, E. W. & Emerson, J. P. 1997, A&A, 321, 803
Houdebine, E. R., Foing, B. H., & Rodono, M. 1990, A&A, 238, 249
Ilin, E., Schmidt, S. J., Poppenhäger, K., et al. 2021, A&A, 645, A42
Khodachenko, M. L., Ribas, I., Lammer, H., et al. 2007, Astrobiology, 7, 167
Koller, F., Leitzinger, M., Temmer, M., et al. 2021, A&A, 646, A34
Korhonen, H., Vida, K., Leitzinger, M., et al. 2017, Living Around Active Stars, 328, 198
Lammer, H., Kislyakova, K. G., Odert, P., et al. 2011, Origins of Life and Evolution of the Biosphere, 41, 503
Leitzinger, M., Odert, P., Greimel, R., et al. 2014, MNRAS, 443, 898
Lu, H-P., Tian, H., Zhang, L-Y., et al. 2022, A&A, 663, A140
Maehara, H., Shibayama, T., Notsu, S., et al. 2012, Nature, 485, 478
Namekata, K., Maehara, H., Honda, S., et al. 2021, Nature Astronomy, 6, 241
Odert, P., Leitzinger, M., Guenther, E. W., et al. 2020, MNRAS, 494, 3766
Roettenbacher, R. M. & Kane, S. R. 2017, ApJ, 851, 77
Vida, K., Kővári, Zs., Pál, A., et al. 2017, ApJ, 841, 124
Vida, K., Leitzinger, M., Kriskovics, L., et al. 2019, A&A, 623, A49

Solar and Stellar Coronal Mass Ejections
Proceedings IAU Symposium No. 388, 2026
N. Gopalswamy, O. Malandraki, A. Vidotto & W. Manchester, eds.
doi:10.1017/S174392132400142X

Spectroscopic investigation of the southern late-K/early-M dwarf binary CC Eri

M. Leitzinger[1], P. Odert[1], R. Greimel[2], P. Kabáth[3], J. Lipták[3,4], R. Karjalainen[3], P. Heinzel[3,5], J. Wollmann[3] and E.W. Guenther[6]

[1]Institute of Physics/AGP, University of Graz, Universitätsplatz 5, 8010 Graz, Austria

[2]RG Science, Schanzelgasse 17, 8010 Graz, Austria

[3]Astronomical Institute, The Czech Academy of Sciences, 25165 Ondřejov, Czech Republic

[4]Astronomical Institute of Charles University, V Holeóvićkćh 2, 180 00 Prague, Czech Republic

[5]University of Wrocław, Center of Scientific Excellence - Solar and Stellar Activity, Kopernika 11, 51-622 Wrocław, Poland

[6]Thüringer Landessternwarte, TLS, Sternwarte 5, 07778 Tautenburg, Germany

Abstract. In the recent past investigations of stellar Coronal Mass Ejections (CMEs) and flares/superflares gained special attention because of their relevance for planetary habitability and stellar evolution. With respect to that, we present an observing campaign at the ESO152 telescope on LaSilla operated by the PLATOSpec consortium of the late-K/early-M dwarf binary CC Eri. The ESO152 telescope is currently equipped with an Echelle spectrograph which enables the investigation of a broad spectral range, allowing the investigation of CMEs and flares/superflares in various spectral lines. Within the observing campaign CC Eri was not observed by TESS, but we have coordinated g'-band photometry to cross-check for white light flares. Preliminary results of the observing campaign are presented.

Keywords. stars: activity, stars: flare, stars: CMEs, stars: individual: CC Eri

1. Introduction

Superflares, i.e. very energetic flares defined by an energy threshold of 10^{33} erg (see e.g. Schaefer et al. 2000), have been statistically identified even on slow rotating solar-like stars a decade ago (Maehara et al. 2012). Spectroscopic observations of superflares are rare. Hawley and Pettersen (1991) present spectroscopic observations of a superflare on the dMe star AD Leo and only recently superflares have been observed in Hα on solar-like stars (Namekata et al. 2021; Namekata et al. 2024; Leitzinger et al. 2024). Here we present dedicated optical spectroscopic monitoring of the young and active southern binary CC Eri with the main goal to spectroscopically characterize superflares. CC Eri has been selected as a target because it shows one of the largest flare/superflare rates in the southern hemisphere (see Fig. 1), as obtained from TESS data. CC Eri is a dK7e+dM3e spectroscopic binary, located at a distance of 11.5 pc, with a logLx of 29.2 ... 29.8 erg s^{-1}, an orbital period of 1.56 d, an inclination of 40 ... 50°, and a mass ratio of 2 (see Amado et al. 2000; Karmakar et al. 2017). The expected flare rate of CC Eri is 10 ... 39 d^{-1} (E$_{flare}$ >10^{32} erg) for the XUV (estimated based on Audard et al. 2000), 0.6 ... 2 d^{-1} for Hα (estimated based on Leitzinger et al. 2020), and 4.3 d^{-1} derived from TESS data.

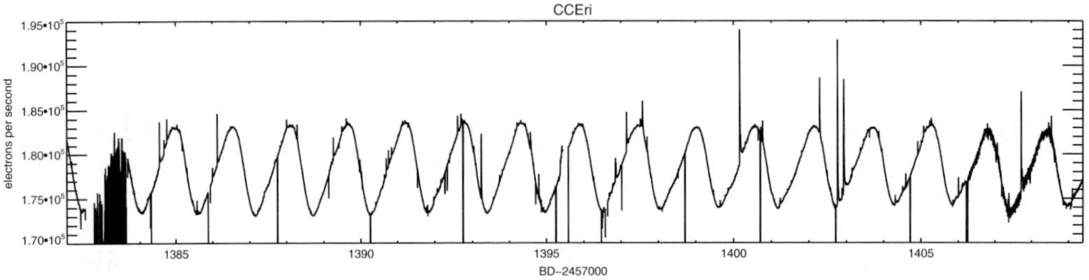

Figure 1. TESS light curve (sector 3) of CC Eri. One can already identify by eye few stronger and numerous smaller flares.

2. Observations

CC Eri was monitored from Sept. 2023 to Jan. 2024 with the ESO1.52m (ESO152) telescope on LaSilla, Chile hosted by the PLATOSpec consortium† consisting of the Astronomical Institute, Ondrejov of the Czech Academy of Sciences, Czech Republic, the Landessternwarte Tautenburg, Germany and the Universidad Catolica, Chile. PLATOSpec will perform spectroscopic follow-up observations (Echelle spectroscopy with R=70000) of PLATO targets. Before the installation of PLATOSpec at the ESO152, PUCHEROS+ (updated version of PUCHEROS; Vanzi et al. 2012), an Echelle spectrograph with R$\sim$18000 is installed. In addition, the ESO152 provides two finder telescopes each with 15cm apertures. OndCam, a CMOS camera with Sloan filters and GrazCam, also a CMOS camera with Sloan and Johnson filters as well as a transmission grating for low-resolution spectroscopy, are installed on the finder telescopes, for coordinated observations. Spectroscopic data of CC Eri were collected in more than 60 nights, or more than 2300 spectra, summing up to a total on-source time of close to 200 hours. For the majority of nights coordinated g'-band photometry is available, whereas only few nights are available with coordinated low-resolution spectroscopy. The spectra have been reduced and extracted with the CERES+ pipeline, an updated version of the CERES pipeline (Brahm et al. 2017). The photometric data have been reduced with a custom written pipeline (Fryda 2023) and light curves have been extracted using aperture photometry.

3. Results

To identify flares we have two possibilities, either we use the coordinated photometry or to derive light curves from the spectra. As we do not have coordinated photometry for every night and not every flare may have a detectable white light component, we identify flares from light curves derived from chromospheric spectral lines, such as the Balmer lines together with temperature sensitive lines such as the HeI (5876Å) line. We derive the equivalent width (EW) for every spectrum and line and plot then the EW time series to identify flares and superflares. In Fig. 2 we show the three strongest flares which we found in the spectroscopic time series data of CC Eri. In a first search we identified 18 flares. Only some flares have been fully captured, for some we have captured either the impulsive phase, or the gradual phase only. The three selected flares reveal Hα energies in the range of 2-8$\times 10^{32}$ erg, which is for the least energetic flare roughly one order of magnitude below the superflare threshold of 10^{33} erg. For one flare, namely for flare 2023-11-08 we estimate the g'-band energy which makes a much larger portion of the white-light energy especially if compared to spectral line energies, such as e.g. Hα. The g'-band energy for the flare is above the 10^{33} erg threshold. As CC Eri is a binary one needs to consider for the computation of flare energies the quiet flux level of the flaring component. However, for the calculation of the g'-band flare energies we

† `https://stel.asu.cas.cz/plato/`

Table 1. Flare characteristics of the flares shown in Fig. 2.

	$E_{g'-band}$ [erg]	$E_{H\alpha}$ [erg]	total duration [min]	duration impulsive phase [min]	duration gradual phase [min]
2023-10-27	-	2×10^{32}	>45	12	>33
2023-11-08	4×10^{33}	4×10^{32}	78	9	69
2023-11-14	-	8×10^{32}	>114	48	> 66

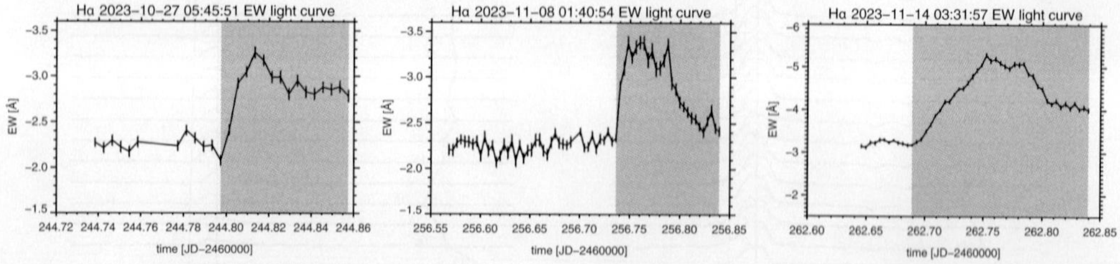

Figure 2. EW Hα light curves of the three strongest flares found in the spectroscopic monitoring data of CC Eri. The right most light curve shows the largest flare. The grey shaded area highlights the flare.

use the magnitude of the binary and not of the components as those are not available. Considering this, the g'-band energy is an upper limit. In Table 1 we list the flare energies (Hα, g'-band) and the flare durations. Two of the energetic flares/superflares reveal line wing asymmetries in Hα, those are 2023-10-27 and 2023-11-14. In Fig. 3 we show the spectral time series of both events (left column) together with their spectral residual Hα time series (right column) for better visibility of the line asymmetries. Flare 2023-10-27 reveals a blue extra emission in the impulsive flare phase, while flare 2023-11-14 reveals a red extra emission in the gradual flare phase.

4. Discussion and Outlook

We present preliminary results of an observing campaign aiming at the search for and the spectral characterization of superflares. CC Eri is an excellent target for this purpose as the binary shows numerous flares and also a few superflares. However, it is surprising that we do not see significant evidence of signatures of erupting filaments/prominences, although the target flares frequently and we have captured so far 18 flares spectroscopically. The analysis of asymmetries is hampered by the fact that the components need to be separated to analyze each of their line profiles. Up to now we only studied the wings of the lines opposite to the location of the secondary component. For asymmetries not overlapping with the secondary component we are able to perform an analysis of spectral line asymmetries and extra emissions/absorptions. This is the case for the two presented superflares in Fig. 3.

In the impulsive phase of flare 2023-10-27 starting at spectrum 14, a blue wing extra emission forms which lasts until spectrum 17 (one spectrum after the flare peak). Such blue wing asymmetries in the impulsive phase of flares are usually interpreted as chromospheric evaporation, if seen in hotter lines. During solar Hα flares gentle chromospheric evaporation is known to occur which is very slow and reaches few km s^{-1} (see e.g. Schmieder et al. 1987). Heinzel et al. (1994) discuss also blue asymmetries in the impulsive phase of solar flares and conclude that those are very short-lived (in the order of tens of seconds) and may stem from a small part of the flare. On the other hand blue extra emissions/absorptions in Balmer lines are also the signature of erupting prominences/filaments. But if the projected velocity of the signature is low, one can not unambiguously distinguish from flare related plasma motions (Houdebine et al. 1990;

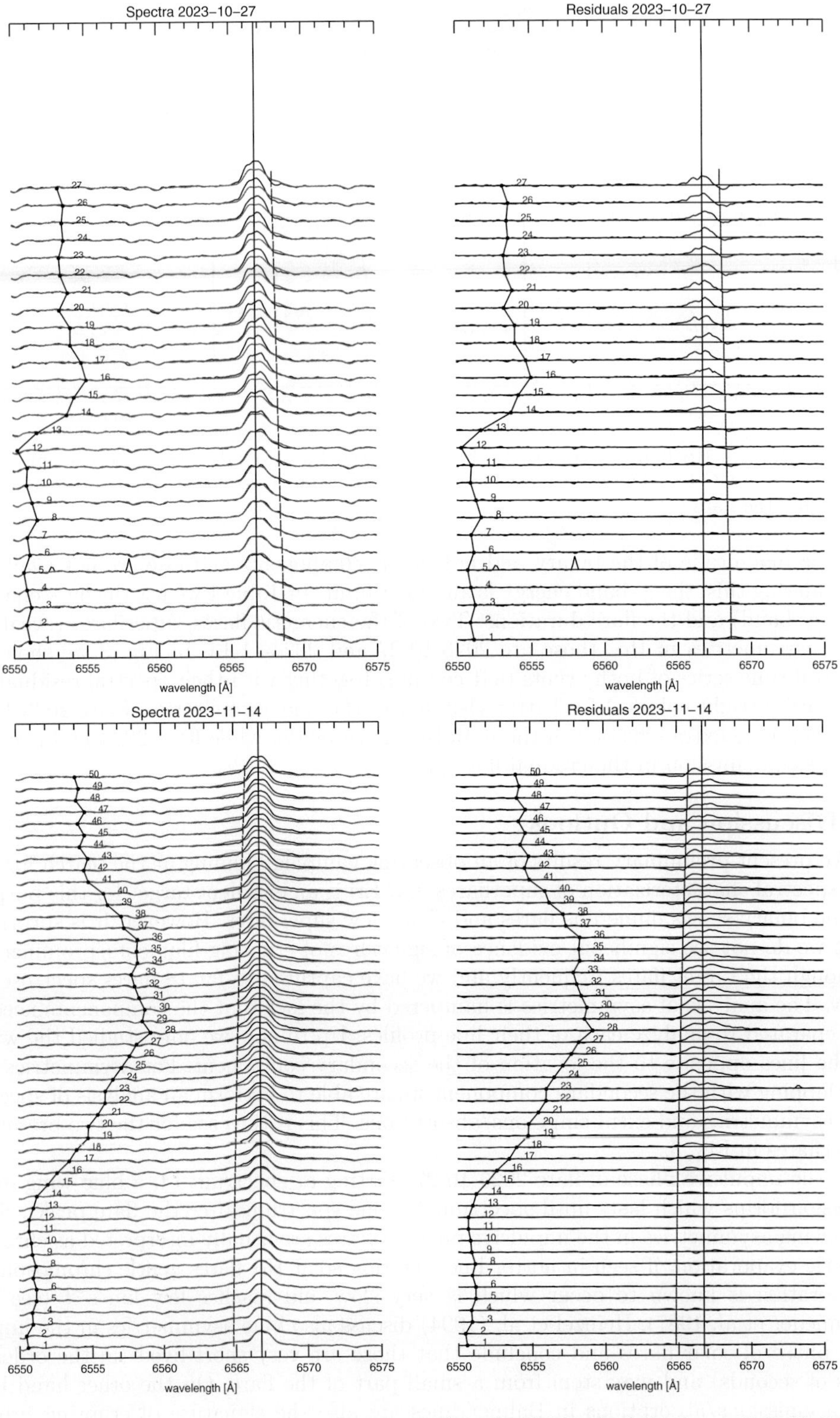

Figure 3. Left column: Hα spectral time series (black solid line) of the flares 2023-10-27 and 2023-11-14 overplotted with the corresponding pre-flare spectrum (red solid line). Overplotted is the EW Hα light curve (numbered data points) to identify impulsive and gradual flare phases. The blue vertical lines indicate the position of the Hα profile of the dMe component of the CC Eri system. The black solid vertical line denotes the Hα line core of the primary dKe component. Right column: Corresponding residual spectral Hα time series.

Gunn et al. 1994). In flare 2023-11-14 we see a long-lived red asymmetry in the gradual phase which arises in spectrum 32 and is visible until spectrum 41. In solar flares, red asymmetries are seen in the gradual flare phase as down-flowing plasma in post-flare loops also termed coronal rain. Only recently similar red asymmetries on the active dMe star AD Leo were presented by Wollmann et al. (2023) who interpreted those as coronal rain.

For stellar spectroscopic observations one has to keep two important things in mind. First, stellar signals are always disk-integrated, and second, line asymmetries in stellar spectra are always seen in projection. So even if, as reported in Heinzel et al. (1994), blue asymmetries stem from small parts of the flare, this may probably not or only hardly affect a disk integrated spectrum, even though on young and active stars, active regions and thus related flare areas may be larger than on the present-day Sun. Moreover, we usually do not know the orientation of the flare loops. Therefore, when interpreting stellar observations, this needs to be taken into account.

One main aim of this study is the spectral characterization of supeflares which is work in progress and has been so far only done on a simple comparative basis of quiet and flare peak spectrum for the event 2023-11-14. We see the rise of Balmer lines Hα, Hβ, Hγ, Hδ, the cores of the sodium lines NaID2+D1, the temperature sensitive line HeI (5876Å), the MgI triplet, several FeI, FeII, CaI, and CrI spectral lines, and few lines which still need to be identified. We did not detect the also temperature sensitive line HeII (4686 Å) which ususally appears in hot plasma (see e.g. Muheki et al. 2020). Moreover, a comparative spectral analysis of flares and superflares will be performed revealing possible differences between these two classes of flares.

Acknowledgements

The authors acknowledge the Austrian Science Fund (FWF): 10.55776/I5711 and the Czech Science Foundation (GACR): 22-30516K. PH was supported by the program 'Excellence Initiative - Research University' for years 2020–2026 at University of Wrocław, project No. BPIDUB.4610.96.2021.KG.

References

Amado, P. J., Doyle, J. G., Byrne, P. B.et al. 2000, *A&A*, 359, 159–167.
Audard, M., Güdel, M., Drake, J. J. et al. 2000, *ApJ*, 541, 396–409.
Brahm, R., Jordán, A., & Espinoza, N. 2017, *PASP*, 129(973), 034002.
Fryda, J.. 2023, Master Thesis, Charles University, Prague, Czech Republic
Gunn, A. G., Doyle, J. G., Mathioudakis, M. et al. 1994, *A&A*, 285, 489–496.
Hawley, S. L. & Pettersen, B. R. 1991, *ApJ*, 378, 725.
Heinzel, P., Karlicky, M., Kotrc, P. et al., 1994, *SolPhys*, 152, 393–408.
Houdebine, E. R., Foing, B. H., & Rodono, M. 1990, *A&A*, 238, 249–255.
Karmakar, S., Pandey, J. C., Airapetian, V. S. et al. 2017, *ApJ*, 840(2), 102.
Leitzinger, M., Odert, P., Greimel, R. et al. 2020, *MNRAS*, 493(3), 4570–4589.
Leitzinger, M., Odert, P., & Greimel, R. 2024, *MNRAS*, 532(2), 1486-1503.
Maehara, H., Shibayama, T., Notsu, S. et al. 2012, *Nature*, 485, 478–481.
Muheki, P., Guenther, E. W., Mutabazi, T. et al. 2020, *MNRAS*, 499(4), 5047–5058.
Namekata, K., Airapetian, V. S., Petit, P. et al. 2024, *ApJ*, 961(1), 23.
Namekata, K., Maehara, H., Honda, S. et al. 2021, *Nature Astronomy*, 6, 241–248.
Schaefer, B. E., King, J. R., & Deliyannis, C. P. 2000, *ApJ*, 529(2), 1026–1030.
Schmieder, B., Forbes, T. G., Malherbe, J. M. et al. 1987, *ApJ*, 317, 956–963.
Vanzi, L., Chacon, J., Helminiak, K. G. et al. 2012, *MNRAS*, 424(4), 2770–2777.
Wollmann, J., Heinzel, P., & Kabáth, P. 2023, *A&A*, 669, A118.

Solar and Stellar Coronal Mass Ejections
Proceedings IAU Symposium No. 388, 2026
N. Gopalswamy, O. Malandraki, A. Vidotto & W. Manchester, eds.
doi:10.1017/S1743921324001418

Spectroscopy of flares on AU Mic

P. Odert[1], M. Leitzinger[1], P. Kabáth[2], R. Greimel[3], J. Lipták[2,4], R. Karjalainen[2], P. Heinzel[2,5], J. Wollmann[2] and E.W. Guenther[6]

[1]Institute of Physics/AGP, University of Graz, Graz, Austria
[2]Astronomical Institute of the Czech Academy of Sciences, Ondřejov, Czech Republic
[3]RG Science, Graz, Austria
[4]Astronomical Institute of Charles University, Prague, Czech Republic
[5]Center of Scientific Excellence - Solar and Stellar Activity, University of Wrocław, Wrocław, Poland
[6]Thüringer Landessternwarte Tautenburg, Tautenburg, Germany

Abstract. Photometric space missions like Kepler and TESS have revealed unprecedented statistics of stellar flares, including the frequent detection of highly energetic superflares. Spectroscopic characterization of flares and especially superflares, however, is still not as commonly available. In contrast to broad-band photometry, spectroscopic data can give more insights into the physical parameters of flares, as well as allowing to identify possibly accompanying CMEs. In this context, we monitored the young planet-hosting flare star AU Mic for about 60 partial nights (>160 hours) between 2022 and 2023 spectroscopically with the ESO 1.52m telescope hosted by the PLATOSpec consortium. Here, we report on the search for signatures of flares and CMEs in these data, including the partial observation of a possible extreme superflare.

Keywords. stars: flare, stars: activity, stars: individual: AU Mic

1. Introduction

AU Mic is a young early M dwarf hosting several planets (Plavchan et al. 2020; Wittrock et al. 2023; Donati et al. 2023) and a debris disk (Kalas et al. 2004). Due to its young age (23 Myr; Mamajek and Bell 2014), the star is magnetically very active, possessing strong magnetic fields (Donati et al. 2023) and stellar spots (Ikuta et al. 2023). It exhibits frequent, strong flares and even superflares (Gilbert et al. 2022; Ikuta et al. 2023; Tristan et al. 2023). Superflares are flares with energies $> 10^{33}$ erg (Maehara et al. 2012), mainly detected and studied using broad-band photometry. Spectroscopic investigations of flares and superflares may provide more information on their physical properties and generation mechanisms than what can be extracted from broad-band photometry, but are far more time-consuming to obtain. Therefore, due to its high flare rate, AU Mic provides an ideal target to characterize its flares and superflares spectroscopically, which may also provide insights into the space weather environment of its planets.

2. Observations

Spectroscopic monitoring of AU Mic was performed with the ESO1.52m telescope (E152) located in La Silla, Chile, which is operated by the PLATOSpec consortium†.

† The PLATOSpec consortium is led by the Astronomical Institute of the Czech Academy of Sciences and consisting of Thüringer Landessternwarte Tautenburg and Universidad Catolica de Chile as major partners and of Universidad Adolfo Ibanez and Masaryk University as minor partners and University of Graz as collaborating partner (`https://stel.asu.cas.cz/plato/`).

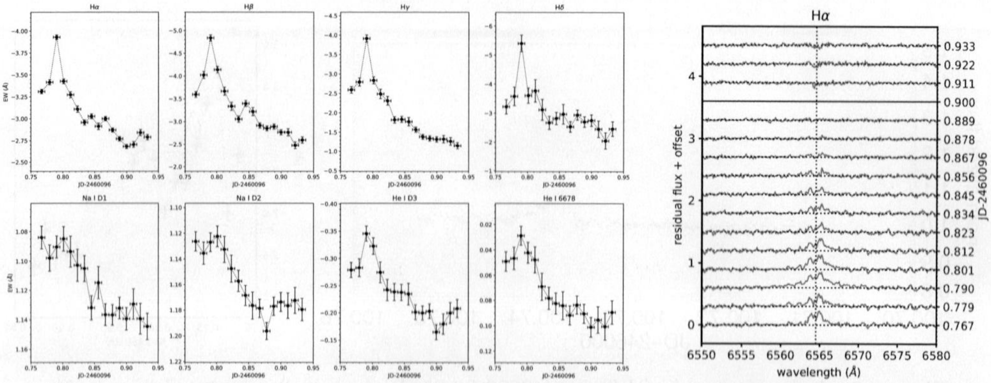

Figure 1. Left panels: Flare on 2023-06-01 in different chromospheric lines. Right panel: corresponding time series of residual spectra around the Hα line.

The telescope is currently equipped with the Echelle spectrograph PUCHEROS+, an upgraded version of PUCHEROS (Vanzi et al. 2012). Its resolving power is R≈18000 and it provides a wavelength coverage of about 400 to 700 nm. In addition, the E152 is equipped with two 15 cm finder telescopes, on which two cameras are installed. These can be used for simultaneous photometry and/or low-resolution slitless spectroscopy.

We observed AU Mic for 56 partial nights between 2022-10-31 and 2023-09-21 with PUCHEROS+. For 38 of these nights we obtained simultaneous photometry in the g'-band, and for two nights simultaneous low-resolution spectroscopy. Exposure times vary between 5, 10, and 15 min for the Echelle spectroscopy, and are typically a few seconds for the photometry. The PUCHEROS+ data are reduced with the CERES+ pipeline, which is an updated version of CERES (Brahm et al. 2017) and generates wavelength-calibrated, optimally extracted order-by-order spectra. More details on the instruments and the observing campaign can be found in Odert et al. (2025).

3. Flare detection

3.1. *Spectroscopy*

To detect flares in the spectroscopic data, we generate light curves of the equivalent widths (EWs) of several chromospheric lines (Hα, Hβ, Hγ, Hδ, Na I D1&D2, He I D3, He I 6678). We identify 24 flares in the light curves of these lines, an example from 2023-06-01 is shown in Fig. 1. One of the detected flares is a partial observation of an extreme event, which is described in more detail in Section 4. Excluding this event, the other 23 flares have Hα peak luminosities ranging from $\sim 10^{27}$ to $\sim 10^{28}$ erg s^{-1} and Hα energies between $\sim 3 \times 10^{30}$ and $\sim 5 \times 10^{31}$ erg.

A currently ongoing task is the search for filament/prominence eruptions as indicator for stellar CME activity. Therefore, we aim to identify spectral line asymmetries to find transient blue- or red-shifted line components. An example is shown in the right panel of Fig. 1, which shows the residual spectra around Hα for the flare event on 2023-06-01 displayed in the left panels. It reveals a small red asymmetry (excess emission) around the peak of the flare.

3.2. *Photometry*

From the accompanying photometric data we detect one g'-band flare (Fig. 2), associated with the most energetic Hα event for which we had simultaneous photometric observations of sufficient quality. The flare has an amplitude of ~27% and lasts for about

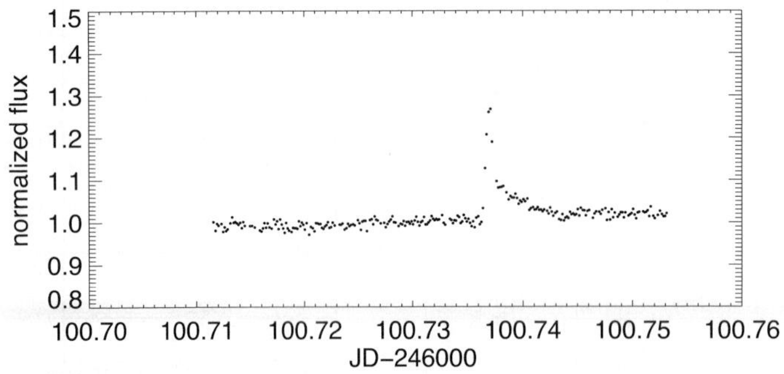

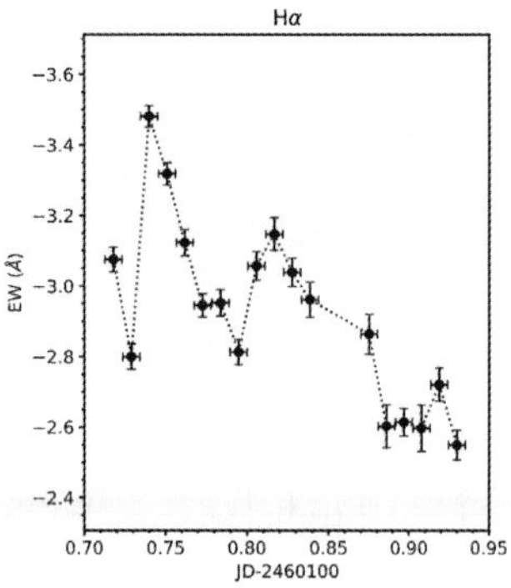

Figure 2. Left panel: g'-band flare on 2023-06-05. Right panel: associated flare in Hα.

11 min. Its estimated energy is $\sim 10^{32}$ erg in the g'-band, and we extrapolate a total white-light energy of $\sim 10^{33}$ erg following Shibayama et al. (2013), which makes it a superflare. The associated flare detected in the chromospheric lines lasts much longer, about 290 min. Its energy in Hα amounts to $\sim 4 \times 10^{31}$ erg, and in total to $\sim 10^{32}$ erg in all studied chromospheric spectral lines.

4. Extreme event

On 2023-09-16 we observed an extreme event with line fluxes increasing by factors of a few more than in any other night (see Fig. 3). Unfortunately, the observations covered only about 30 min, during which all line fluxes remained at a roughly similar level of enhancement. Thus, only a small part of a likely much longer event was observed. Upon close inspection of the EW light curves of the following night (2023-09-17), a decreasing trend is seen in the data. Possibly, this could resemble the decaying tail of the extreme event of 2023-09-16. If true, this would indicate a duration of this giant flare of more than 24 hours. We fit the combined EWs from the two nights for each spectral line with a simple flare light curve model and integrate over the fits to obtain an estimate of the energies. This results in superflare energies ($\gtrsim 10^{33}$ erg) already in Hα alone. To independently check this assumption, we utilize the relationship between flare peak luminosity and energy derived from the sample of AU Mic's other 23 flares we detected in our study (Odert et al. 2025). If we take the flare luminosity measured on 2023-09-16 as the peak luminosity, we obtain flare energies comparable to within a factor of two with our estimates from the light curve fits. This agreement substantiates our hypothesis of a rare, highly-energetic long duration superflare.

In addition to the prominent chromospheric lines presented here, we find a wealth of other spectral lines which were significantly enhanced or appeared during this extreme event. One example is the He II line at 4686 Å (see Fig. 3, lower right panel). A detailed analysis of flare-affected lines of this event will be the subject of a future study.

5. Conclusions and outlook

We present results from a spectroscopic monitoring campaign of the active young planet host star AU Mic. From the spectral line data, we identified 24 flares, including the partial observation of a rare extreme event. A detailed analysis of flare-affected spectral lines and line asymmetries, also comparatively between weaker flares and superflares, is currently ongoing.

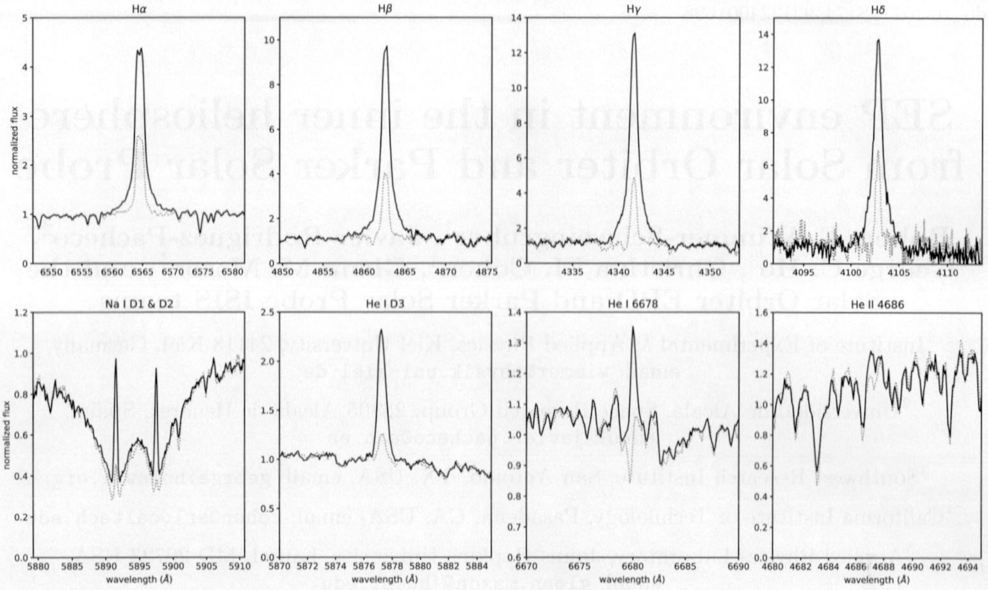

Figure 3. Comparison of spectral line profiles of the extreme event on 2023-09-16 (black solid line) compared with a low-activity spectrum from the following night 2023-09-17 (gray dotted line).

Acknowledgments

The authors acknowledge the Austrian Science Fund (FWF): 10.55776/I5711 and the Czech Science Foundation (GACR): 22-30516K. PH was supported by the program 'Excellence Initiative - Research University' for years 2020-2026 at University of Wrocław, project No. BPIDUB.4610.96.2021.KG.

Supplementary material

To view supplementary material for this article, please visit https://doi.org/10.1017/S1743921324001418

References

Brahm, R., Jordán, A., & Espinoza, N. 2017, *PASP*, 129, 034002
Donati, J. F., Cristofari, P. I., Finociety, B., *et al.* 2023, *MNRAS*, 525, 455
Gilbert, E. A., Barclay, T., Quintana, E. V., *et al.* 2022, *AJ*, 163, 147
Ikuta, K., Namekata, K., Notsu, Y., *et al.* 2023, *ApJ*, 948, 64
Kalas, P., Liu, M. C., & Matthews, B. C. 2004, *Science*, 303, 1990
Maehara, H., Shibayama, T., Notsu, S., *et al.* 2012, *Nature*, 485, 478
Mamajek, E. E. & Bell, C. P. M. 2014, *MNRAS*, 445, 2169
Odert, P., Leitzinger, M., Greimel, R., *et al.* 2025, *MNRAS*, 537, 537
Plavchan, P., Barclay, T., Gagné, J., *et al.* 2020, *Nature*, 582, 497
Shibayama, T., Maehara, H., Notsu, S., *et al.* 2013, *ApJS*, 209, 5
Tristan, I. I., Notsu, Y., Kowalski, A. F., *et al.* 2023, *ApJ*, 951, 33
Vanzi, L., Chacon, J., Helminiak, K. G., *et al.* 2012, *MNRAS*, 424, 2770
Wittrock, J. M., Plavchan, P. P., Cale, B. L., *et al.* 2023, *AJ*, 166, 232

Solar and Stellar Coronal Mass Ejections
Proceedings IAU Symposium No. 388, 2026
N. Gopalswamy, O. Malandraki, A. Vidotto & W. Manchester, eds.
doi:10.1017/S1743921324001698

SEP environment in the inner heliosphere from Solar Orbiter and Parker Solar Probe

Robert F. Wimmer-Schweingruber[1], Javier Rodriguez-Pacheco[2], George C. Ho[3], Christina M. Cohen[4], Glenn M. Mason[5], and the Solar Orbiter EPD and Parker Solar Probe ISIS teams

[1]Institute of Experimental & Applied Physics, Kiel University, 24118 Kiel, Germany, email: wimmer@physik.uni-kiel.de

[2]Universidad de Alcalá, Space Research Group, 28805 Alcalá de Henares, Spain, email: javier.pacheco@uah.es

[3]Southwest Research Institute, San Antonio, TX, USA, email: george.ho@swri.org

[4]California Institute of Technology, Pasadena, CA, USA, email: cohen@srl.caltech.edu

[5]Applied Physics Laboratory, Johns Hopkins University, Laurel, MD 20723 USA, email: glenn.mason@jhuapl.edu

Abstract. The Sun drives a supersonic wind which inflates a giant plasma bubble in our very local interstellar neighborhood, the heliosphere. It is bathed in an extremely variable background of energetic ions and electrons which originate from a number of sources. Solar energetic particles (SEPs) are accelerated in the vicinity of the Sun, whereas shocks driven by solar disturbances are observed to accelerate energetic storm particles (ESPs). Moreover, a dilute population with a distinct composition forms the anomalous cosmic rays (ACRs) which are of a mixed interstellar-heliospheric origin. Particles are also accelerated at planetary bow shocks. We will present recent observations of energetic particles by Solar Orbiter and Parker Solar Probe, as well as other spacecraft that allow us to study the acceleration and transport of energetic particles at multiple locations in the inner heliosphere.

Keywords. sun, heliosphere, energetic particles, Solar Orbiter, Parker Solar Probe

---

1. Introduction

The Integrated Science Investigation of the Sun (IS⊙IS, McComas et al. 2016) on Parker Solar Probe (PSP, Fox et al. 2016) and the Energetic Particle Detector (EPD, Rodríguez-Pacheco et al. 2020) on Solar Orbiter (Müller et al. 2020) basically have the same science goals which are worded slightly differently and are summarized here for convenience. McComas et al. (2016) states IS⊙IS' top level science goal as to determine the physical mechanisms that produce, accelerate, and transport energetic particles in the inner heliosphere. As usual, this is broken down into three lower-level science questions: a) What is the origin or seed population of solar energetic particles (SEPs)? b) How are these SEPs and other particle populations accelerated? c) What mechanisms are responsible for transporting the different particle populations into the heliosphere? On the other hand, EPD 's top-level science question is (see Rodríguez-Pacheco et al. 2020) to determine how solar eruptions produce energetic particle radiation that fills the heliosphere. This is also broken down into three more detailed science questions. a) How and where are energetic particles injected at the sources and, in particular, what are the seed populations for energetic particles? b) How and where are energetic particles accelerated at the Sun and in the interplanetary medium? c) How are energetic particles released from their sources and distributed in space and time?

Obviously there are three common science questions which can be broken down into three processes which are needed for energetic particles to be measured in the heliosphere: injection (sec. 2), acceleration (sec. 3), and transport (sec. 4). We have chosen to structure this paper according to these three processes and to provide examples for these processes that use data from Solar Orbiter, PSP, and other spacecraft that were recently reported. Because we have been tasked by the organizers of this workshop with providing an overview of results from PSP and Solar Orbiter, this paper does not report brand new results. Rather, it attempts to exemplify how the new instruments on these missions as well as the various constellations offered by them and other inner-heliosphere missions can elucidate these three important contributions to accelerating energetic particles that fill the heliosphere. As usual, we will provide a summary and discussion at the end of the paper.

2. Injection

In order to do their mischief in the heliosphere, energetic particles need to be injected into the acceleration process. If they are accelerated at a shock, their speed needs to exceed the thermal speed of the local particle population so that they can freely cross the acceleration region (e.g., a shock) multiple times. Just how they attain this elevated kinetic energy has been a highly debated topic in the past decades. Today we know that the solar wind does not serve as the sole or main reservoir for suprathermal particles but that there appears to be a population of such particles available for further acceleration. An important clue came from measurements of the composition of suprathermal particles. They are enriched in the rare ^{3}He isotope throughout the solar activity cycle (but do show a dependence on it, Wiedendeck and Mason (2013)). Desai and Giacalone (2016) provide an excellent overview of the measurements and their consequences for interpretation.

As is well known, the Sun is an active star with an activity cycle driven by the solar dynamo (Cliver et al. 2022). This is illustrated in the upper panel of Fig. 1 which shows sunspot number (Source: WDC-SILSO, Royal Observatory of Belgium, Brussels) which is a good indicator for solar activity as a function of time. The bottom panel shows oxygen differential flux spectra accumulated over the four Bartels rotations indicated by vertical lines in the upper panel and in the title of the Figure. The line styles in the lower panel correspond to those in the upper panel. Particle data were taken from three instruments on NASA's Advanced Composition Explorer (ACE, Stone et al. (1998c)), the Ultra-Low-Energy Isotope Spectrometer (ULEIS, Mason et al. (1998)) the Solar Isotope Spectrometer (SIS, Stone et al. (1998a)), and the Cosmic-Ray Isotope Spectrometer (CRIS, Stone et al. (1998b)).

Fig. 1 also shows the highest and lowest particle fluences observed during the course of the ACE mission and illustrates the enormous variability of particle fluxes which is bounded by the shaded area. It covers a dynamic range of about seven orders of magnitude even when averaged over entire Bartels rotations. Some further explanations are needed to interpret the figure. Fluxes at the highest energies covered by ULEIS during quiet times are affected by instrumental noise which is due to a malfunction of the second time-of-flight (ToF) measurement. The measured values (from the ACE Science Center) are plotted as a line plot with a "+" symbol. Because we know that the lower limits are not correct over the entire energy range, we have also plotted a logarithmic interpolation between the lowest-energy SIS data point and the ULEIS data point with the highest energy which is not affected by this background (around 1 MeV/nuc). Moreover, the CRIS instrument on ACE was not designed to measure high fluxes during solar particle events but to provide accurate measurements of the quiet-time galactic cosmic rays. It

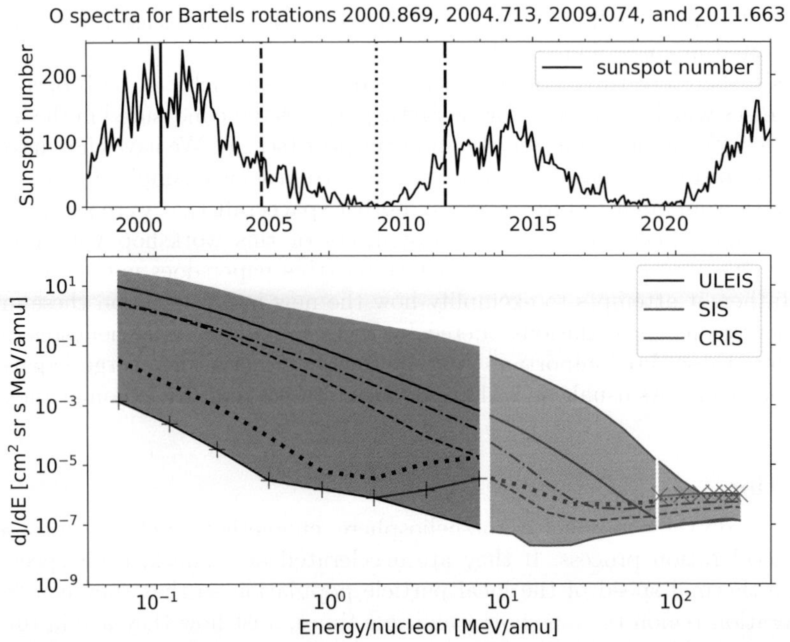

Figure 1. Upper panel: Sunspot numbers vs. time (from WDC-SILSO, Royal Observatory of Belgium, Brussels). Bottom panel: Oxygen differential flux spectra accumulated over the time periods indicated by vertical lines in the upper panel (decimal years 2000.869, 2004.787, 2009.074, and 2011.663). Line styles in the lower panel indicate the Bartels rotation of the observation as indicated in the upper panel. Colors indicate different instruments on the ACE spacecraft as discussed in the text. The shaded region shows the envelope of the variability of the Bartels-rotation-averaged O particle fluences from 1998 – 2024. See text for an explanation of the "x" and "+" symbols.

saturates during high-flux time periods† which is seen by the measured data points which are also shown as a line plot with an "x" symbol marking. To establish a likely envelope, shown as the solid line bounding the shaded region, we have used a fluence of 0.8 times the highest-energy SIS data point and 0.8 times the square root of that value multiplied by the second-lowest CRIS fluence value. This is also the reason for the "missing" CRIS data in Fig. 1. Figure 1 and its discussion above are updated versions of a similar figure shown in Wimmer-Schweingruber et al. (2023b).

Here we illustrate the injection stage with new measurments performed with the Suprathermal Ion Spectrograph (SIS Rodríguez-Pacheco et al. 2020) on Solar Orbiter. SIS measures total energy of and performs multiple time-of-flight measurements on suprathermal particles. This results in a very low background which allows measurements of the heliospheric quiet-time background of suprathermal particles (Mason et al. 2021, 2023). We will not consider those measurements here but rather investigate the origin of particles injected into and accelerated during two impulsive solar particle events observed by Solar Orbiter on the 9th of November 2022 and on the 8th of April 2023, as reported by Mason et al. (2023). The first of these events was very unusual in that it contained several hundreds of ultraheavy ions (with masses > 78 amu). The second event occurred during very low ambient activity which allowed those authors to observe particle spectra

† Dr. R. Leske, Caltech, private communication, 2022. See also the caveat at `https://izw1.caltech.edu/ACE/ASC/level2/cris_l2desc.html` which states that "Time periods during which solar activity is high are not covered by CRIS, since the instrument was not designed to operate during such conditions."

with only minimal pre-event background from the suprathermal seed population. These observations showed very low proton and ^{4}He intensities together with intensities of other ions that are typical of ^{3}He-rich events. Together these observations indicate that the enrichments in Fe/O, ^{3}He/^{4}He, etc. which have often been attributed to the injection process are more likely due to the acceleration process itself. In the model of Roth and Temerin (1997) and Temerin and Roth (1992) ions are accelerated by electromagnetic hydrogen cyclotron waves in a single-stage process. Ions are heated over a broad range of gyrofrequencies, but in the model presented by Mason et al. (2023) the spectrum of waves is strongly damped at the gyrofrequencies of the abundant H and ^{4}He. Thus the "unusual" abundance features such as enhanced Fe/O is not due to an enrichment of Fe but rather to a depletion of O because it is fully stripped and thus in resonance with the waves damped by ^{4}He. This process would also explain the enrichments of the neutron-rich isotopes of Ne, Mg, Si, etc. which have been reported in the past (Dwyer et al. 2001; Leske et al. 2001, 2007; Mason et al. 1994).

3. Acceleration

Acceleration of particles requires work to be done on them which in turn requires an electric field. Strong electric fields are difficult to maintain in a highly conductive plasma such as the corona or solar wind. Maxwell's equations allow a way out of this dilemma by inducing electric fields with a rapidly changing magnetic field. Rapid temporal variations can be achieved in solar flares as a consequence of reconnection. Rapid spatial variations such as shocks or other magnetic discontinuities translate to rapid temporal changes as they are convected into a plasma. Finally waves traveling through the plasma can also heat particles which are in resonance. Yang et al. (2023) and Trotta et al. (2023) have investigated the acceleration of suprathermal ions using data from the SupraThermal Electron Proton (STEP) sensor which is part of EPD on Solar Orbiter and offers unusually high energy and time resolution (Rodríguez-Pacheco et al. 2020; Wimmer-Schweingruber et al. 2021; Yang et al. 2023).

Yang et al. (2023) reconstructed the pitch-angle distributions (PADs) of suprathermal ions in the solar wind frame of reference and studied its temporal evolution, as well as the flux profile and velocity distribution function of these particles in the vicinity of a shock observed on the 3rd of November 2021. They found that suprathermal proton fluxes peaked ~ 12 to ~ 24s before the shock in the upstream region, and that close to the shock they exhibited anisotropic PADs that pointed away from the shock and tended to isotropize farther upstream. Downstream of the shock, the PADs showed anisotropies towards a pitch angle of 90°. Comparison of their measurements with theoretical models suggests that the acceleration of suprathermal protons at interplanetary shocks is highly dynamic on a timescale of ~ 10s, i.e., few proton gyroperiods. Furthermore, they conclude that shock-drift acceleration likely plays an important role in accelerating these suprathermal protons. They suggest that the strong variability of the suprathermal particle fluxes could be a result of time-varying acceleration efficiency at the shock or that it is a consequence of shock rippling (Johlander et al. 2016, 2018; Yang et al. 2018), i.e., a rapidly varying magnetic connection to different parts of the shock that would have different local θ_{Bn} and therefore different acceleration efficiencies.

Trotta et al. (2023) combined results from hybrid kinetic simulations with STEP measurements of suprathermal ions in the vicinity of the 30 October 2021 shock to show that the inherent variability of the injection process in both time and space must be considered to solve the problem of how suprathermal particle injection occurs in astrophysical systems. In other words, their results strongly support the second explanation given by Yang et al. (2023). Similar to that work, they found highly variable particle enhancements but in this case they showed clear velocity dispersion in the vicinity of the shock on time scales of 10 to 20 s. In their view, the particles are accelerated at different locations of the shock which are magnetically connected to the spacecraft at different times. These authors also used 2.5-dimensional kinetic simulations to model details of the shock and particle heating and acceleration. They found that suprathermal particles are not distributed uniformly in the vicinity of the shock. Their flux varies strongly along the shock front which indicates irregular injection into the acceleration process.

Together the results of Yang et al. (2023) and Trotta et al. (2023) underline the importance of detailed, high-resolution studies in the vicinity of interplanetary shocks to understand both the injection and acceleration of particles. They also drive home the point that injection (Sec. 2) and acceleration (Sec. 3) are not isolated processes but need to be considered together. We could just as well have presented these results in Sec. 2!

4. Transport

Particles can escape the injection/acceleration region after which we speak of "transport". A simple thought experiment starting with a point-like source may serve to illustrate this process. If we assume a point-like source then particles can propagate along the interplanetary magnetic field (IMF), they can experience drifts and diffusion perpendicular to the IMF, and further energy and momentum changes due to changes in the IMF at all temporal and spatial scales. Finally, flux tubes filled with energetic particles to different levels can be convected by an observer which results in interesting temporal behavior (see, e.g., Wimmer-Schweingruber et al. 2023a). To illustrate these effects we will discuss four publications by Trotta et al. (2024), Telloni et al. (2021), Kollhoff et al. (2021), and Mason et al. (2021).

How can we decide whether an observed feature in a particle event is due to injection/acceleration or due to transport? To answer that we need to know how the source, e.g., a shock evolves in the inner heliosphere where it accelerates particles. Trotta et al. (2024) observed a CME-driven interplanetary shock at 0.07 with Parker Solar Probe and at 0.7 with Solar Orbiter on the 5th of September 2022. This allowed them to study the shock properties at these two different heliocentric distances. Coincidentally, the 0.07 au or 15 $R_\odot$ are also the closest to the Sun that a shock has been observed to date. PSP (Solar Orbiter) was located at Carrington longitude 232° (252°) and latitude -1.8° (-3.6°). While this is not a perfect alignment the measured shock properties were nevertheless very similar at the two locations: θ_{Bn} was 53° (51°), shock magnetic compression ratio, r_B, was 2.3 (1.9), and the gas compression ratios was 1.6 (2.0). Shock speed decreased from 1520 km/s at PSP to 942 km/s at Solar Orbiter, but the fast magnetosonic and Alfvénic Mach numbers were very similar ($M_{fms} = 3.8$ (3.2) and $M_A = 3.9$ (3.8)). Despite these similarities, the shock at PSP was seen as a sharp transition without any signs of an upstream foreshock, but much more complex at Solar Orbiter. There a wavelet spectrum of the magnetic eld fluctuations shows enhanced power extending to small ($\sim$1 Hz) scales indicating upstream shocklet activity. At PSP the downstream level of fluctuations is enhanced by a factor $\sim$4 with respect to the upstream levels, whereas at Solar Orbiter this region is populated with strong compressive and non-compressive magnetic eld uctuations, which indicates that the shock propagated through a much more structured portion of the solar wind than at PSP. These authors conclude that the shock environment at Solar Orbiter is much more disturbed than at PSP.

Telloni et al. (2021) used the rst radial alignment between Parker Solar Probe and Solar Orbiter to investigate the evolution of solar wind turbulence in one and the same plasma parcel in the inner heliosphere. They allowed a tolerance of ±1.5 degrees in latitude and longitude to identify a 1.5 hour time period centered around 04:00 UT on the 27th of September 2020 at PSP (00:00 UT on the 2nd of October 2020 at Solar Orbiter) as corresponding to the same plasma parcel at 0.1 and 1 au. Using burst-mode magnetic field measurements at the two spacecraft, they could show that the two spacecraft indeed measured very similar characteristics of the solar wind stream. The stream originated from a quiescent region and showed exceptionally smooth magnetic field and a Parker-like field orientation without any switchbacks. They used standard diagnostics for spectral power, compressibility, and intermittency of the magnetic eld uctuations to show that at PSP turbulence was still highly Alfénic while it showed fully developed, intermittent turbulence at Solar Orbiter. They concluded that nonlinear interactions between Alfvén waves probably did not yet have time to fully develop at 0.1 au, where the solar wind sampled by PSP was still pristine and unaffected by stream-stream interactions. At Solar Orbiter (at 1 au) the same plasma has evolved considerably, the turbulent cascade has fully developed and Alfvénic uctuations have decreased.

Kollhoff et al. (2021) investigated the solar particle event observed by multiple spacecraft (ACE, Parker Solar Probe, SOHO, Solar Orbiter, and STEREO-A) across a longitudinal range of $> 230°$ on the 29th of November 2020. This was the first widespread event of solar cycle 25 and accelerated electrons to near-relativistic energies as well as ions to energies > 50 Mev. It was associated with an M4.4 class X-ray flare, a coronal mass ejection and an extreme ultraviolet (EUV) wave as well as a type II radio burst and multiple type III radio bursts. Kollhoff et al. (2021) used velocity dispersion and time-shift analysis techniques to identify the particle release times at the Sun to show that the delays in onset time increased with increasing angle between the flare foot point and the observer. They compared the thus determined onset times with the time the EUV wave intersected the estimated magnetic footpoints of each spacecraft and concluded that a simple scenario where the particle release is only determined or triggered by the EUV wave is unlikely. Anisotropic particle distributions were observed at SolO, Wind, and STEREO-A. Remarkably, they found that the largest anisotropy was observed by STEREO-A which was separated by $\sim 95°$ from the parent solar active region. Their observations suggest that particles are injected over a wide longitudinal range close to the Sun, in fact even crossing the heliospheric current sheet.

Mason et al. (2021) investigated the same event (29th of November 2021) as Kollhoff et al. (2021) in view of the heavy ion properties of energetic particles. SIS and ULEIS on ACE (which is a very similar instrument to SIS) saw only upper limits of $< 1\%$ and 0.03% to the enrichment of ^{3}He in this event. This could simply be a consequence of this being the first large solar particle event of solar cycle 25 and that the “reservoir” of suprathermal flare material (Wiedendeck and Mason 2013) had not yet been replenished enough to provide sufficient ^{3}He seed particles. Heavy ion abundances normalized to oxygen were observed to be about two time lower at PSP and ACE than long-term averages (Desai et al. 2006). PSP observed a nearly monotonic drop in Fe/O with increasing energy, consistent with measurements at STEREO and ACE. At low energies (below ~ 30 keV), however, PSP measured a significantly enhanced Fe/O ratio that increased with decreasing energy. These measurements extend below the energy range of most previous SEP events and may be a new feature.

5. Summary and Discussion

We have considered several new results from Solar Orbiter and Parker Solar Probe that illustrate the complexity of the question being tackled by the energetic particle instruments on these two spacecraft. The question “How do solar eruptions produce energetic particle radiation that fills the heliosphere?” sounds deceitfully simple but reveals a remarkable richness in phenomena once one digs into the details. We have broken it down into three stages which can bring an energetic particle to end up in one of the many instruments used to measure such radiation: injection, acceleration, and transport. We have seen that these different stages are strongly interlinked and are not easily separated.

The often reported enhancements of Fe/O were shown by Mason et al. (2023) to be due to a depletion of fully stripped O, at least for the two events discussed in that publication. They propose that the waves responsible for heating particles are strongly damped by protons and ^{4}He and thus there is less wave power available to ions with charge-to-mass ratios close to 0.5. This interpretation implies that not all variability in heavy-ion abundances (generally normalized to O) would be due to enrichments of those elements in the "seed population", but that the acceleration mechanism itself causes these apparent enrichments. Therefore the distinction between "injection" and "acceleration" becomes blurred and in a sense ill-defined.

The STEP sensor, part of Solar Orbiter's EPD suite of energetic particle instruments, is capable of measuring at an exceptionally high cadence of one second. Yang et al. (2023) and Trotta et al. (2023) used this 15-pixel pin-hole instrument to reconstruct high-resolution pitch-angle distributions in the vicinity of two shocks. Their high time-resolution results allowed them to catch shock-drift acceleration in the act (Yang et al. 2023) and revealed bursty, non-local acceleration close to IP shock (Trotta et al. 2023). STEP measures in a previously hardly accessible energy range close to the bulk energy of the solar wind. The remarkable dynamics in the vicinity of the shocks again demonstrate how blurred the distinction between "injection" and "acceleration" has become.

Once injected and accelerated, energetic particles leave their source region and experience "transport" effects. We may have been a little too inclusive here by also including the radial evolution of shocks in Sec. 4 (Transport) as discussed by Trotta et al. (2024), but this is an important effect. Shocks do evolve with distance from the Sun and the new constellations of spacecraft in the inner heliosphere allow us to measure this evolution. Similarly, Telloni et al. (2021) measured the radial evolution of turbulent fluctuations in a plasma parcel between 0.1 and 1 au. This went beyond the usual statistical sampling by identifying one and the same plasma parcel at PSP and Solar Orbiter. Kollhoff et al. (2021) and Mason et al. (2021) investigated the same event (29th of November 2021) at multiple spacecraft to elucidate especially the longitudinal spread of particles and the properties of heavy ions which serve as tracers for micro-physical processes in the injection/acceleration regions.

The constantly changing constellations of inner-heliosphere spacecraft (including spacecraft at one au) are allowing us to better understand radial and longitudinal gradients and spreads of particles and properties of the solar wind and interplanetary magnetic field through which they propagate. Solar Orbiter leaving the ecliptic in 2025 will add latitudinal transport to the equation.

References

Cliver, E. W., Schrijver, C. J., Shibata, K., & Usoskin, I. G. 2022, Extreme solar events. *Living Reviews in Solar Physics*, 19(1), 2.

Desai, M. & Giacalone, J. 2016, Large gradual solar energetic particle events. *Living Reviews in Solar Physics*, 13(1), 3.

Desai, M. I., Mason, G. M., Mazur, J. E., & Dwyer, J. R. 2006, The Seed Population for Energetic Particles Accelerated by CME-Driven Shocks. *Space Sci. Rev.*, 124, 261–275.

Dwyer, J. R., Mason, G. M., Mazur, J. E., Gold, R. E., Krimigis, S. M., Möbius, E., & Popecki, M. 2001, Isotopic Composition of Solar Energetic Particle Events Measured by Advanced Composition Explorer/ULEIS. *Astrophys. J.*, 563(1), 403–409.

Fox, N. J., Velli, M. C., Bale, S. D., Decker, R., Driesman, A., Howard, R. A., Kasper, J. C., Kinnison, J., Kusterer, M., Lario, D., Lockwood, M. K., McComas, D. J., Raouafi, N. E., & Szabo, A. 2016, The Solar Probe Plus Mission: Humanity's First Visit to Our Star. *Space Sci. Rev.*, 204(1-4), 7–48.

Johlander, A., Schwartz, S. J., Vaivads, A., Khotyaintsev, Y. V., Gingell, I., Peng, I. B., Markidis, S., Lindqvist, P. A., Ergun, R. E., Marklund, G. T., Plaschke, F., Magnes, W., Strangeway,

R. J., Russell, C. T., Wei, H., Torbert, R. B., Paterson, W. R., Gershman, D. J., Dorelli, J. C., Avanov, L. A., Lavraud, B., Saito, Y., Giles, B. L., Pollock, C. J., & Burch, J. L. 2016, Rippled Quasiperpendicular Shock Observed by the Magnetospheric Multiscale Spacecraft. *Phys. Rev. Lett.*, 117(16), 165101.

Johlander, A., Vaivads, A., Khotyaintsev, Y. V., Gingell, I., Schwartz, S. J., Giles, B. L., Torbert, R. B., & Russell, C. T. 2018, Shock ripples observed by the MMS spacecraft: ion reflection and dispersive properties. *Plasma Physics and Controlled Fusion*, 60(12), 125006.

Kollhoff, A., Kouloumvakos, A., Lario, D., Dresing, N., Gómez-Herrero, R., Rodríguez-García, L., Malandraki, O. E., Richardson, I. G., Posner, A., Klein, K. L., Pacheco, D., Klassen, A., Heber, B., Cohen, C. M. S., Laitinen, T., Cernuda, I., Dalla, S., Espinosa Lara, F., Vainio, R., Köberle, M., Kühl, R., Xu, Z. G., Berger, L., Eldrum, S., Brüdern, M., Laurenza, M., Kilpua, E. J., Aran, A., Rouillard, A. P., Bučík, R., Wijsen, N., Pomoell, J., Wimmer-Schweingruber, R. F., Martin, C., Böttcher, S. I., Freiherr von Forstner, J. L., Terasa, J. C., Boden, S., Kulkarni, S. R., Ravanbakhsh, A., Yedla, M., Janitzek, N., Rodríguez-Pacheco, J., Prieto Mateo, M., Sánchez Prieto, S., Parra Espada, P., Rodríguez Polo, O., Martínez Hellín, A., Carcaboso, F., Mason, G. M., Ho, G. C., Allen, R. C., Bruce Andrews, G., Schlemm, C. E., Seifert, H., Tyagi, K., Lees, W. J., Hayes, J., Bale, S. D., Krupar, V., Horbury, T. S., Angelini, V., Evans, V., O'Brien, H., Maksimovic, M., Khotyaintsev, Y. V., Vecchio, A., Steinvall, K., & Asvestari, E. 2021, The first widespread solar energetic particle event observed by Solar Orbiter on 2020 November 29. *Astron. & Astrophys.*, 656, A20.

Leske, R. A., Mewaldt, R. A., Cohen, C. M. S., Christian, E. R., Cummings, A. C., Slocum, P. L., Stone, E. C., von Rosenvinge, T. T., & Wiedenbeck, M. E. Isotopic abundances in the solar corona as inferred from ACE measurements of solar energetic particles. In Wimmer-Schweingruber, R. F., editor, *Joint SOHO/ACE workshop "Solar and Galactic Composition"* 2001, volume 598 of *American Institute of Physics Conference Series*, pp. 127–132. AIP.

Leske, R. A., Mewaldt, R. A., Cohen, C. M. S., Cummings, A. C., Stone, E. C., Wiedenbeck, M. E., & von Rosenvinge, T. T. 2007, Solar Isotopic Composition as Determined Using Solar Energetic Particles. *Space Sci. Rev.*, 130(1-4), 195–205.

Mason, G. M., Cohen, C. M. S., Ho, G. C., Mitchell, D. G., Allen, R. C., Hill, M. E., Andrews, G. B., Berger, L., Boden, S., Böttcher, S., Cernuda, I., Christian, E. R., Cummings, A. C., Davis, A. J., Desai, M. I., de Nolfo, G. A., Eldrum, S., Elftmann, R., Kollhoff, A., Giacalone, J., Gómez-Herrero, R., Hayes, J., Janitzek, N. P., Joyce, C. J., Korth, A., Kühl, P., Kulkarni, S. R., Labrador, A. W., Espinosa Lara, F., Lees, W. J., Leske, R. A., Mall, U., Martin, C., Martínez Hellín, A., Matthaeus, W. H., McComas, D. J., McNutt, R. L., Mewaldt, R. A., Mitchell, J. G., Pacheco, D., Parra Espada, P., Prieto, M., Rankin, J. S., Ravanbakhsh, A., Rodríguez-Pacheco, J., Rodríguez Polo, O., Roelof, E. C., Sánchez-Prieto, S., Schlemm, C. E., Schwadron, N. A., Seifert, H., Stone, E. C., Szalay, J. R., Terasa, J. C., Tyagi, K., Freiherr von Forstner, J. L., Wiedenbeck, M. E., Wimmer-Schweingruber, R. F., Xu, Z. G., & Yedla, M. 2021, Solar energetic particle heavy ion properties in the widespread event of 2020 November 29. *Astron.& Astrophys.*, 656, L12.

Mason, G. M., Gold, R. E., Krimigis, S. M., Mazur, J. E., Andrews, G. B., Daley, K. A., Dwyer, J. R., Heuerman, K. F., James, T. L., Kennedy, M. J., Lefevere, T., Malcolm, H., Tossman, B., & Walpole, P. H. 1998, The Ultra-Low-Energy Isotope Spectrometer (ULEIS) for the ACE spacecraft. *Space Sci. Rev.*, 86, 409–448.

Mason, G. M., Ho, G. C., Allen, R. C., Kouloumvakos, A., Wimmer-Schweingruber, R. F., & Rodriguez-Pacheco, J. 2023, Quiet-time suprathermal ions in the inner heliosphere during the rising phase of solar cycle 25. *Astron.& Astrophys.*, 673, L12.

Mason, G. M., Mazur, J. E., & Halmilton, D. C. 1994, Heavy-Ion Isotopic Anomalies in 3He-rich Solar Particle Events. *Astrophys. J.*, 425, 843.

McComas, D. J., Alexander, N., Angold, N., Bale, S., Beebe, C., Birdwell, B., Boyle, M., Burgum, J. M., Burnham, J. A., Christian, E. R., Cook, W. R., Cooper, S. A., Cummings, A. C., Davis, A. J., Desai, M. I., Dickinson, J., Dirks, G., Do, D. H., Fox, N., Giacalone, J., Gold,

R. E., Gurnee, R. S., Hayes, J. R., Hill, M. E., Kasper, J. C., Kecman, B., Klemic, J., Krimigis, S. M., Labrador, A. W., Layman, R. S., Leske, R. A., Livi, S., Matthaeus, W. H., McNutt, R. L., Mewaldt, R. A., Mitchell, D. G., Nelson, K. S., Parker, C., Rankin, J. S., Roelof, E. C., Schwadron, N. A., Seifert, H., Shuman, S., Stokes, M. R., Stone, E. C., Vandegriff, J. D., Velli, M., von Rosenvinge, T. T., Weidner, S. E., Wiedenbeck, M. E., & Wilson, P. 2016, Integrated Science Investigation of the Sun (ISIS): Design of the Energetic Particle Investigation. *Space Sci. Rev.*, 204(1-4), 187–256.

Müller, D., St. Cyr, O. C., Zouganelis, I., Gilbert, H. R., Marsden, R., Nieves-Chinchilla, T., Antonucci, E., Auchère, F., Berghmans, D., Horbury, T. S., Howard, R. A., Krucker, S., Maksimovic, M., Owen, C. J., Rochus, P., Rodriguez-Pacheco, J., Romoli, M., Solanki, S. K., Bruno, R., Carlsson, M., Fludra, A., Harra, L., Hassler, D. M., Livi, S., Louarn, P., Peter, H., Schühle, U., Teriaca, L., del Toro Iniesta, J. C., Wimmer-Schweingruber, R. F., Marsch, E., Velli, M., De Groof, A., Walsh, A., & Williams, D. 2020, The Solar Orbiter mission. Science overview. *Astron. & Astrophys.*, 642, A1.

Rodríguez-Pacheco, J., Wimmer-Schweingruber, R. F., Mason, G. M., Ho, G. C., Sánchez-Prieto, S., Prieto, M., Martín, C., Seifert, H., Andrews, G. B., Kulkarni, S. R., Panitzsch, L., Boden, S., Böttcher, S. I., Cernuda, I., Elftmann, R., Espinosa Lara, F., Gómez-Herrero, R., Terasa, C., Almena, J., Begley, S., Böhm, E., Blanco, J. J., Boogaerts, W., Carrasco, A., Castillo, R., da Silva Fariña, A., de Manuel González, V., Drews, C., Dupont, A. R., Eldrum, S., Gordillo, C., Gutiérrez, O., Haggerty, D. K., Hayes, J. R., Heber, B., Hill, M. E., Jüngling, M., Kerem, S., Knierim, V., Köhler, J., Kolbe, S., Kulemzin, A., Lario, D., Lees, W. J., Liang, S., Martínez Hellín, A., Meziat, D., Montalvo, A., Nelson, K. S., Parra, P., Paspirgilis, R., Ravanbakhsh, A., Richards, M., Rodríguez-Polo, O., Russu, A., Sánchez, I., Schlemm, C. E., Schuster, B., Seimetz, L., Steinhagen, J., Tammen, J., Tyagi, K., Varela, T., Yedla, M., Yu, J., Agueda, N., Aran, A., Horbury, T. S., Klecker, B., Klein, K. L., Kontar, E., Krucker, S., Maksimovic, M., Malandraki, O., Owen, C. J., Pacheco, D., Sanahuja, B., Vainio, R., Connell, J. J., Dalla, S., Dröge, W., Gevin, O., Gopalswamy, N., Kartavykh, Y. Y., Kudela, K., Limousin, O., Makela, P., Mann, G., Önel, H., Posner, A., Ryan, J. M., Soucek, J., Hofmeister, S., Vilmer, N., Walsh, A. P., Wang, L., Wiedenbeck, M. E., Wirth, K., & Zong, Q. 2020, The Energetic Particle Detector. Energetic particle instrument suite for the Solar Orbiter mission. *Astron. & Astrophys.*, 642, A7.

Roth, I. & Temerin, M. 1997, Enrichment of ^{3}He and Heavy Ions in Impulsive Solar Flares. *Astrophys. J.*, 477(2), 940–957.

Stone, E. C., Cohen, C. M. S., Cook, W. R., Cummings, A. C., Gauld, B., Kecman, B., Leske, R. A., Mewaldt, R. A., Thayer, M. R., Dougherty, B. L., Grumm, R. L., Milliken, B. D., Radocinski, R. G., Wiedenbeck, M. E., Christian, E. R., Shuman, S., Trexel, H., von Rosenvinge, T. T., Binns, W. R., Crary, D. J., Dowkontt, P., Epstein, J., Hink, P. L., Klarmann, J., Lijowski, M., & Olevitch, M. A. 1998,b The Cosmic-Ray Isotope Spectrometer for the Advanced Composition Explorer. *Space Sci. Rev.*, 86b, 285–356.

Stone, E. C., Cohen, C. M. S., Cook, W. R., Cummings, A. C., Gauld, B., Kecman, B., Leske, R. A., Mewaldt, R. A., Thayer, M. R., Dougherty, B. L., Grumm, R. L., Milliken, B. D., Radocinski, R. G., Wiedenbeck, M. E., Christian, E. R., Shuman, S., & von Rosenvinge, T. T. 1998,a The Solar Isotope Spectrometer for the Advanced Composition Explorer. *Space Sci. Rev.*, 86a, 357–408.

Stone, E. C., Frandsen, A. M., Mewaldt, R. A., Christian, E. R., Margolies, D., Ormes, J. F., & Snow, F. 1998,c The Advanced Composition Explorer. *Space Sci. Rev.*, 86c, 1–22.

Telloni, D., Sorriso-Valvo, L., Woodham, L. D., Panasenco, O., Velli, M., Carbone, F., Zank, G. P., Bruno, R., Perrone, D., Nakanotani, M., Shi, C., D'Amicis, R., De Marco, R., Jagarlamudi, V. K., Steinvall, K., Marino, R., Adhikari, L., Zhao, L., Liang, H., Tenerani, A., Laker, R., Horbury, T. S., Bale, S. D., Pulupa, M., Malaspina, D. M., MacDowall, R. J., Goetz, K., de Wit, T. D., Harvey, P. R., Kasper, J. C., Korreck, K. E., Larson, D., Case, A. W., Stevens, M. L., Whittlesey, P., Livi, R., Owen, C. J., Livi, S., Louarn, P., Antonucci, E., Romoli, M., O'Brien, H., Evans, V., & Angelini, V. 2021, Evolution of Solar Wind Turbulence from 0.1 to 1 au during the First Parker Solar Probe-Solar Orbiter Radial Alignment. *Astrophys. J. Lett.*, 912(2), L21.

Temerin, M. & Roth, I. 1992, The production of ^{3}He and heavy ion enrichments in ^{3}He-rich flares by electromagnetic hydrogen cyclotron waves. *Astrophys. J.*, 391, L105 – L108.

Trotta, D., Horbury, T. S., Lario, D., Vainio, R., Dresing, N., Dimmock, A., Giacalone, J., Hietala, H., Wimmer-Schweingruber, R. F., Berger, L., & Yang, L. 2023, Irregular Proton Injection to High Energies at Interplanetary Shocks. *Astrophys. J. Lett.*, 957(2), L13.

Trotta, D., Larosa, A., Nicolaou, G., Horbury, T. S., Matteini, L., Hietala, H., Blanco-Cano, X., Franci, L., Chen, C. H. K., Zhao, L., Zank, G. P., Cohen, C. M. S., Bale, S. D., Laker, R., Fargette, N., Valentini, F., Khotyaintsev, Y., Kieokaew, R., Raouafi, N., Davies, E., Vainio, R., Dresing, N., Kilpua, E., Karlsson, T., Owen, C. J., & Wimmer-Schweingruber, R. F. 2024, Properties of an Interplanetary Shock Observed at 0.07 and 0.7 au by Parker Solar Probe and Solar Orbiter. *Astrophys. J.*, 962(2), 147.

Wiedendeck, M. E. & Mason, G. M. The solar-cycle variation of 3 He from solar energetic particle events. In *Proceedings of the 33rd international cosmic ray conference, CBPF* 2013,.

Wimmer-Schweingruber, R. F., Berger, L., Kollhoff, A., Kühl, P., Heber, B., Yang, L., Heidrich-Meisner, V., Klassen, A., Gomez-Herrero, R., Rodriguez-Pacheco, J., Ho, G. C., Mason, G. M., Janitzek, N. P., Kouloumvakos, A., Wang, L., Warmuth, A., Lario, D., Carcaboso, F., Owen, C. J., Bučík, R., Pacheco, D., Malandraki, O., Allen, R. C., Rodriguez, L., Shukhobodskaia, D., Espinosa Lara, F., Cernuda, I., Böttcher, S. I., Eldrum, S., Fleth, S., & Xu, Z. 2023,a Unusually long path length for a nearly scatter-free solar particle event observed by Solar Orbiter at 0.43 au. *Astron.& Astrophys.*, 678a, A98.

Wimmer-Schweingruber, R. F., Janitzek, N. P., Pacheco, D., Cernuda, I., Espinosa Lara, F., Gómez-Herrero, R., Mason, G. M., Allen, R. C., Xu, Z. G., Carcaboso, F., Kollhoff, A., Kühl, P., Freiherr von Forstner, J. L., Berger, L., Rodriguez-Pacheco, J., Ho, G. C., Andrews, G. B., Angelini, V., Aran, A., Boden, S., Böttcher, S. I., Carrasco, A., Dresing, N., Eldrum, S., Elftmann, R., Evans, V., Gevin, O., Hayes, J., Heber, B., Horbury, T. S., Kulkarni, S. R., Lario, D., Lees, W. J., Limousin, O., Malandraki, O. E., Martín, C., O'Brien, H., Prieto Mateo, M., Ravanbakhsh, A., Rodriguez-Polo, O., Sánchez Prieto, S., Schlemm, C. E., Seifert, H., Terasa, J. C., Tyagi, K., Vainio, R., Walsh, A., & Yedla, M. K. 2021, First year of energetic particle measurements in the inner heliosphere with Solar Orbiter's Energetic Particle Detector. *Astron.& Astrophys.*, 656, A22.

Wimmer-Schweingruber, R. F., Kollhoff, A., Pacheco, D., Yang, L., Rodríguez-Pacheco, J., Gómez-Herrero, R., Mason, G. M., Berger, L., Kühl, P., Böttcher, S. I., Heber, B., Xu, Z., Allen, R., Aran, A., Bucik, R., Wang, L., Espinosa Lara, F., Cernuda, I., Prieto, M., Ho, G. C., Eldrum, S., & Fleth, S. Advances in energetic particle physics with Solar Orbiter. In *27th European Cosmic Ray Symposium* 2023,b, 9.

Yang, L., Heidrich-Meisner, V., Berger, L., Wimmer-Schweingruber, R. F., Wang, L., He, J., Zhu, X., Duan, D., Kollhoff, A., Pacheco, D., Kühl, P., Xu, Z., Keilbach, D., Rodríguez-Pacheco, J., & Ho, G. C. 2023, Acceleration of suprathermal protons near an interplanetary shock. *Astron.& Astrophys.*, 673, A73.

Yang, L., Wang, L., Li, G., Wimmer-Schweingruber, R. F., He, J., Tu, C., Tian, H., & Bale, S. D. 2018, The Strongest Acceleration of >40 keV Electrons by ICME-driven Shocks at 1 au. *Astrophys. J.*, 853(1), 89.

Solar and Stellar Coronal Mass Ejections
Proceedings IAU Symposium No. 388, 2026
N. Gopalswamy, O. Malandraki, A. Vidotto & W. Manchester, eds.
doi:10.1017/S1743921324001790

High Time Resolution Heavy Ion Sensor Observations across a Collisionless Shock

B. L. Alterman[1,2], S. A. Livi[2,3], S. T. Lepri[3], J. M. Raines[3], R. M. Dewey[3], D. Trotta[5], S. A. Spitzer[4], C. M. Bert[2,3], L. M. Kistler[7], A. Galvin[7], G. Nicolaou[8], T. S. Horbury[5], Y. J. Rivera[6], C. J. Owen[8], D. Verscharen[8], J. Coburn[8], G. Lewis[8], R. A. D'Amicis[9], D. T. Carpenter[3], E. P. Shimoun[3], I. Gershkovich[3], K. Ploof[3], T. J. Eddy[3], A. Fedorov[10], D. Lario[2], H. Kucharek[7], K. Ogasawara[1], F. Allegrini[1], L. Ellis[7], P. Louran[10], T. Stubbs[2]

[1]NASA Goddard Space Flight Center
8800 Greenbelt Road
Greenbelt, MD 20771
USA

[2]Space Science
Southwest Research Institute
6220 Culebra Rd.
San Antonio, TX, 78238
USA

[3]Department of Climate and Space Sciences & Engineering
University of Michigan
Climate and Space Research Building
2455 Hayward St.
Ann Arbor, MI, 48109
USA

[4]Weizmann Institute of Science
234 Herzl Street
Rehovot 7610001
Israel

[5]The Blackett Laboratory
Department of Physics
Imperial College London
London SW7 2AZ
UK

[6]Center for Astrophysics | Harvard & Smithsonian
60 Garden Street
Cambridge, MA 02138
USA

[7]Space Science Center
University of New Hampshire
8 College Rd
Durham, NH 03824
USA

[8]Department of Space and Climate Physics
Mullard Space Science Laboratory
University College London
Holmbury St. Mary
Dorking
Surrey RH5 6NT
UK

[9]INAF-IFSI
Via del Fosso del Cavaliere 100
00133 Rome
Italy
[10]Institut de Recherche en Astrophysique et Planétologie
9 Avenue du Colonel ROCHE
BP 4346, 31028 Toulouse Cedex 4
France

Abstract. Charged particle acceleration and heating in the heliosphere requires detailed observations of the velocity distribution functions (VDFs) of these particles across a range of mass and charge states. Such particle energization is often observed concurrent with transient phenomena. Proper characterization of such energization processes and their impact on charged particles requires careful consideration of instrument observation timing and the timing at which a transient phenomenon is observed. Solar Orbiter's Heavy Ion Sensor (HIS) observes the composition of the solar wind along with suprathermal and pickup ions. As such, it provides the observations necessary to resolve the mass and charge states appropriate for properly characterizing charged particle energization. In this work, we describe the current status of data aggregation at variable time intervals and how that enables proper characterization of charged particle energization concurrent with transient phenomena.

Keywords. Interplanetary shocks (829), Solar coronal mass ejections (310)

---

1. Introduction

Transient phenomena such as shocks, coronal mass ejections (CMEs), and stream interaction regions (SIRs) are key drivers of heliospheric variability. In particular, they are known to heat and accelerate charged particles. Two of Solar Orbiter's four science objectives aim to understand transient phenomena in the heliosphere and the origin of energetic particles (Müller et al. 2013, 2020). Fully characterizing these phenomena requires high time resolution, statistically significant observations of charged particles across a wide range of energies from thermal solar wind energies up to energetic particles (Müller et al. 2013, 2020). The ability to resolve ion mass (M) and charge state (Q) is critical to such investigations because various theories of charged particle energization depend on these quantities and/or their charge-to-mass ratio (Q/M). Solar Orbiter's Heavy Ion Sensor (HIS) is a time-of-flight (ToF) mass spectrometer that provides the mass- and charge-resolved composition observations necessary to fully characterize transient phenomena, as listed in Solar Orbiter's science objectives (Müller et al. 2013, 2020; Owen et al. 2020; Livi et al. 2023).

In this paper, we provide a brief description of HIS, including its data products. We also demonstrate that HIS's high resolution observations can be aggregated in custom intervals that account for transient events. We then overview an interplanetary coronal mass ejection (ICME) driven shock observed by Solar Orbiter on March 11, 2022 and illustrate custom aggregation of HIS observations in a manner accounting for shock timing.

2. The Heavy Ion Sensor (HIS)

HIS is one of three instruments in the Solar Wind Analyzer (SWA) suite. HIS measures solar wind heavy, suprathermal, and pickup ions (PUI) over the energy-per-charge (E/Q) range $0.5 - 75\,\mathrm{keV/e}$. Owen et al. (2020); Livi et al. (2023) provide a detailed description of the instrument, its operational capabilities, and the first results. In addition to HIS,

SWA is composed of the Electron Analyzer System (EAS) and the Proton-Alpha Sensor (PAS). Together, these three instruments make *in situ* observations of solar wind ions and electrons. Owen et al. (2020) describe the instrument suite along with all three instruments in detail.

HIS derives heritage from the Advanced Composition Explore (ACE) (Stone et al. 1998) Solar Wind Ion Composition Spectrometer (SWICS) (Gloeckler et al. 1998). Owen et al. (2020, Fig. 10) is a photo of the delivered instrument. Livi et al. (2023, Fig. 1) and Owen et al. (2020, Fig. 11) are schematics illustrating HIS's principle of operation.

HIS' Level 2 (L2) data include individual ion measurements, called pulse height analyzed (PHA) words. PHA words consist of the observed E/Q, residual energy, and time-of-flight (ToF) through the instrument of an ion observed at a given azimuth (ϕ) and elevation (θ) along with additional operationally determined and/or required quantities (Owen et al. 2020; Livi et al. 2023). Level 3 (L3) data is "science ready" and includes various products. The products of interest in this paper include ion speeds and temperatures. The method of converting PHAs to physical quantities for HIS data is comparable to that applied to ACE/SWICS (Livi et al. 2023; Shearer et al. 2014), which primarily focuses on data aggregation, ion species separation, and conversion to physical quantities. Livi et al. (2023) find a robust comparison between HIS composition data and corresponding ACE/SWICS observations. L2 and L3 HIS data are publicly available on the Solar Orbiter Archive (SOAR, `soar.esac.esa.int`)

2.1. *HIS Data*

HIS observes the solar wind from $-33°$ to $+66°$ in azimuthal (ϕ) and $\pm 17°$ in elevation (θ) bins (Owen et al. 2020). A single HIS scan consists of observing all relevant ϕ and θ across the full energy-per-charge range from 0.5 to 75 keV/e. HIS covers this energy-per-charge range by stepping its electrostatic analyzer from its 75 keV/e maximum down. While HIS is capable of observing solar wind H^{1+} at its lowest energy steps, HIS often avoids the energy range associated with H^{1+} to protect and prolong the lifetime of key hardware components. This is colloquially referred to as "proton avoidance".

Solar Orbiter observed a shock on March 11, 2022 at 19:52 UT (Dresing et al. 2024; Trotta et al. 2024). The shock is described in Section 3. Figure 1 is a plot of L2 HIS observations of E/Q as a function of ToF upstream and downstream of the shock. Both intervals are accumulated over 30 minutes to illustrate the changes across the shock. The purple region separated by a black dashed line indicates observations corresponding to spurious or accidental ToF signals. The plotted threshold corresponds to a speed $v < 0.7 \times v_{\mathrm{sw}}$, where v_{sw} is the solar wind speed, and we use it for illustrative purposes here. Livi et al. (2023) utilize $v_{\mathrm{sw}} - 2w$, where v_{sw} and w (the thermal speed) are both derived from O^{6+} observations. (E/Q,ToF) tracks are plotted for H^{1+}, He^{2+}, C^{6+}, and Fe^{8+} to Fe^{12+}. Numbers on the top axis indicate these ions charge states; only even Fe charge states are labeled for visual clarity. The centers and widths of these tracks are derived from ground calibration and comparison with in-flight data.

Figure 1 shows several key features of the HIS observations. In the downstream region, the shaded region corresponding to $v < 0.7 \times v_{\mathrm{sw}}$ covers a larger range of E/Q steps, corresponding to an increase in v_{sw} of $\sim 200\,\mathrm{km\,s^{-1}}$ across the shock (c.f. Figure 2). HIS steps from high to low E/Q. To protect the lifetime of certain instrument components, HIS implements a "proton avoidance", which does not step to certain low E/Q values during a given scan when the flux is high. The minimum E/Q reached varies with each scan. Here, we provide a brief overview of the effect. In upstream and downstream region, HIS's proton avoidance cut off portions of at least the H^{1+} and He^{2+} peaks. In the upstream region, this is less pronounced as the overall solar wind speed and flux are

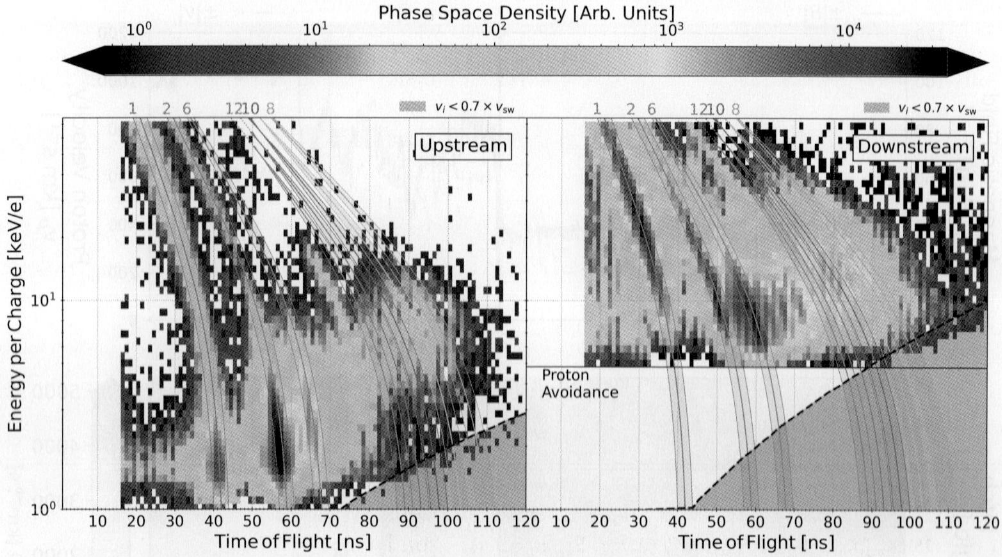

Figure 1. Energy-per-Charge (E/Q) vs Time-of-Flight (ToF) plots aggregated for 30 minutes upstream and downstream of the shock. (E/Q,ToF) tracks are plotted for H^{1+}, He^{2+}, C^{6+}, and Fe^{8+} to Fe^{12+}. Numbers on the top axis indicate these charge states; only even numbered Fe charge states are labeled for clarity. The region in the lower right corner of each plot covered by a purple region indicate where ToF accidentals occur, as determined by an ion having an incident speed $< 0.7 \times v_H$.

smaller. In the downstream region, the flux is markedly higher and the C^{6+} peak is also cut off because HIS does not step to a larger range of voltages. We indicate this threshold in the downstream region as E/Q steps below the horizontal line at 4.82 keV/e. Livi et al. (2023) demonstrate proton avoidance in greater detail, including how it varies with each scan. As the bulk of Fe^{8+} to Fe^{12+} enter the instrument at E/Q steps above proton avoidance, their peaks are observed in both the upstream and downstream regions. The center of these Fe peaks is at higher E/Q in the downstream region and the extent of these distributions in (E/Q,ToF) space is also larger in the downstream region.

Similar to ACE/SWICS, HIS data processing utilizes a maximum likelihood estimator (MLE) to attribute count rates to specific ions based on the distribution in (E,ToF) space (Livi et al. 2023; Shearer et al. 2014). Livi et al. (2023, Fig. 2) illustrates the relationship between observations in (E/Q,ToF) space and (E,ToF) space and discusses the relevant aspects of the method. Although computationally expensive, this method separates ions with comparatively smaller count rates or that have overlapping distributions in measurement space Figure 1 (Shearer et al. 2014). Level 3 (L3) is considered "science" data, i.e. physical quantities useful for immediate data analysis. L3 quantities for each ion are derived by taking moments of the velocity distribution functions (VDFs) after they have been assigned to a given ion by MLE. **Such observations are typically plotted as the phase space density (PSD) as a function of E/Q.** HIS has two operational modes: normal and burst mode (Owen et al. 2020; Livi et al. 2023). L2 burst mode data is collected at 4s cadence and discussed in Owen et al. (2020); Livi et al. (2023). We use "normal mode" L2 data, which are collected at 30s cadence. Version 01 of the L3 data is distributed at 10min resolution due to instrumental considerations (Livi et al. 2023, Tbl. 2). As of this paper's writing, data in the SOAR covers the time period January 2022 to April 2023.

We have developed a preliminary version of the data processing software that enables variable time intervals to be selected for aggregating L2 observations. Figure 1 is

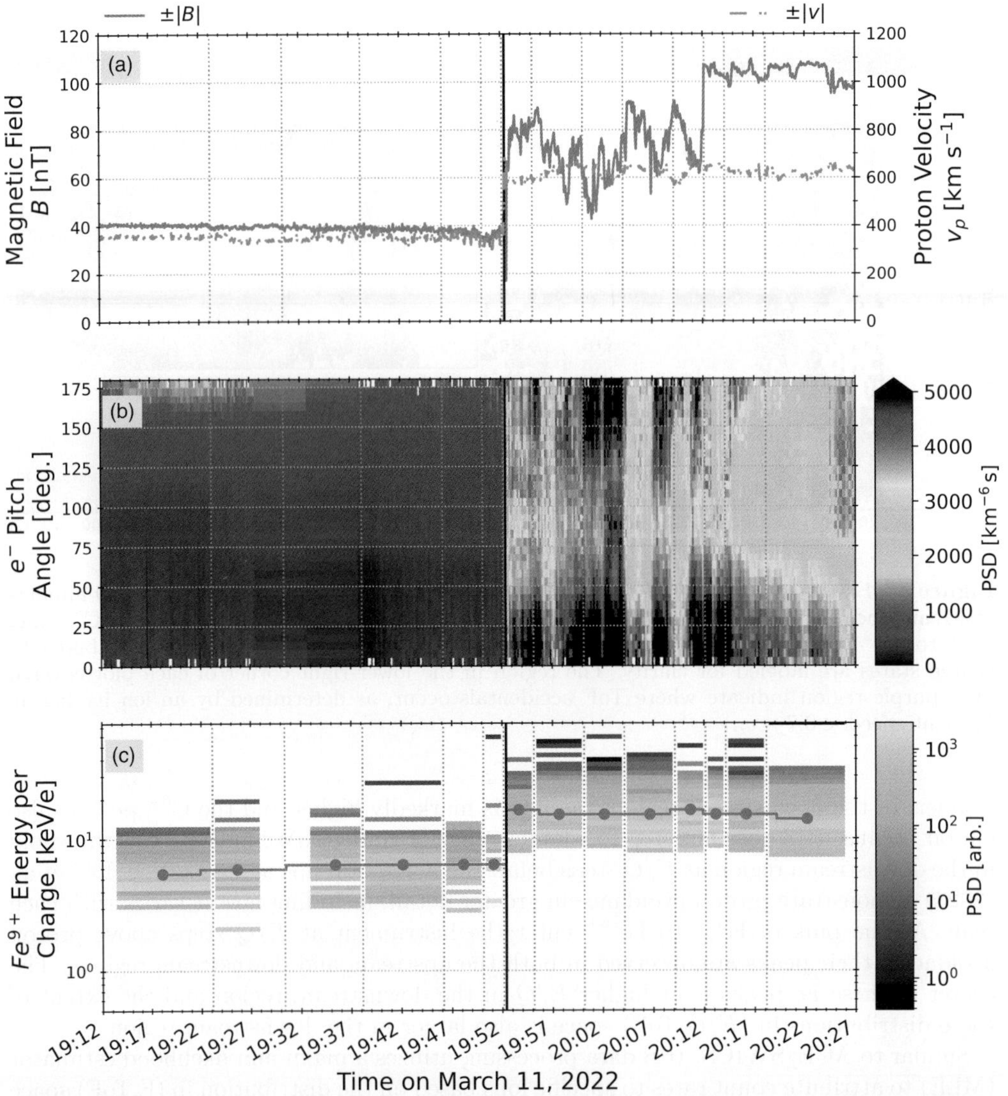

Figure 2. Overview of the event observed by HIS on March 11, 2022 at ∼19:52. Panels are **(a, left)** magnetic field magnitude from MAG, **(a, right)** proton speed from PAS, **(b)** electron pitch angle distribution from EAS, and **(c)** a spectrogram of Fe^{9+} aggregated to the indicated intervals. Gaps in panel (c) and vertical dotted lines indicate the variable width intervals in which HIS observations are aggregated. The vertical solid line at ∼19:52 indicates when SolO observed the shock.

aggregated into 30min intervals so that count rates are high and features in the observations are visually recognizable. In the following section, we show that observations can also be aggregated into intervals several minutes in length to enable studies of transient events that do not align with the 10min cadence of current L3 data.

3. March 11, 2022 Shock

On March 11, 2022 Solar Orbiter observed a shock at 0.44 AU while inbound towards the 1st science perihelion. This fast-forward shock was embedded in a interplanetary CME (ICME). The shock parameters are derived following Trotta et al. (2022). Briefly,

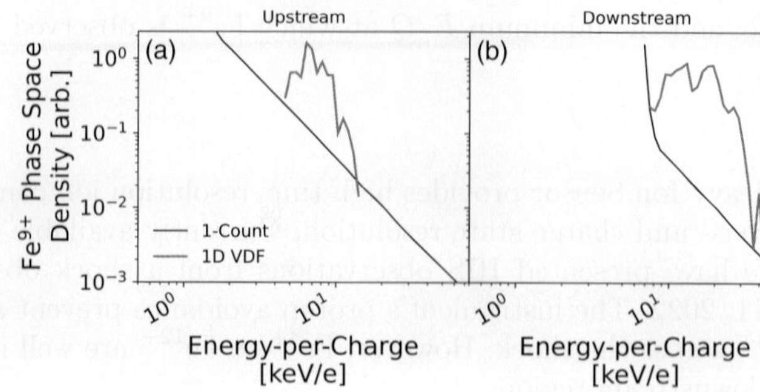

Figure 3. Fe^{9+} VDFs **plotted as the phase space density (PSD)** from the **(a)** upstream and **(b)** downstream region as a function of energy-per-charge. The upstream VDF is observed at 19:42. The downstream VDF is observed at 20:22. The 1-count levels are shown as black lines.

the shock normal angle is $\theta_{Bn} = 27°$; the Mach number is $M_A = 3.2$; and the shock speed is $737\,\mathrm{km\,s^{-1}}$ in the spacecraft frame (Dresing et al. 2024; Trotta et al. 2024).

Figure 2 provides a timeseries overview of the shock. Panel (a) plots (left) the magnetic field magnitude observed by MAG (Horbury et al. 2020) and (right) the proton speed observed by PAS (Owen et al. 2020). Panel (b) is an electron pitch angle distribution (PAD) from EAS (Owen et al. 2020). Panel (c) is a 1D spectrogram for Fe^{9+}, which means that PHA words are accumulated over both azimuth and elevation angles to improve overall statistics. Both the HIS and EAS spectrograms are plotted as a phase space density (PSD). The typical unit for a VDF is $\mathrm{km}^{-6}\,\mathrm{s}^3$. For this, deriving such units requires cross calibration with PAS, which is still being refined. As such, this Fe^{9+} spectrogram units are arbitrary. The spectrogram peaks are plotted with dots and connected to aid the eye. The time intervals are variable in length. The spectrogram gaps indicate 30s separation between adjacent accumulations. In contrast to the 30min intervals utilized in Figure 1 and the 10min cadence intervals used by Livi et al. (2023), these intervals range from ~ 2 to ~ 8 minutes. The gaps and vertical dotted lines indicate 30s intervals where PHA words are not utilized because either the start or end time did not fall into the specified accumulation interval. A vertical black line indicates the shock time at 19:52 UT.

There are multiple changes across the shock that can be observed in the data. Panel (a, left) shows that $\boldsymbol{B}$ is approximately constant and $|\boldsymbol{B}| \approx 40\,\mathrm{nT}$ in the upstream region. In the downstream region, $|\boldsymbol{B}| \approx 80\,\mathrm{nT}$ and is highly variable. In Panel (a, right), the proton velocity increases across the shock from ~ 350 to $\sim 580\,\mathrm{km\,s^{-1}}$. In Panel (b), the electron PAD shows periods that are bi-directional and uni-directional with PSD spread that is wide in pitch-angle similar to shock sheath of ICMEs where it is likely the shock efficiently scatters electrons in pitch-angle (Carcaboso et al. 2020). As expected from Figure 1, the E/Q corresponding to the peak of the Fe spectrogram in Panel (c) increases markedly across the shock. The distribution also broadens in the downstream region.

Figure 3 is an example of Fe^{9+} VDFs from the (a) upstream and (b) downstream regions observed at 19:42 UT and 20:22 UT, respectively. Both are 7.5min in length. The upstream interval is chosen to be close to the shock. The downstream period is chosen to be far downstream. The units are arbitrary for the same reason that the HIS PSD in Figure 2 are. The black lines indicate the 1-count level in the upstream region. Although the lowest E/Q observation in the downstream region is close to the 1-count level, all higher E/Q observations are significantly above this level. Consistent with Figure 2, the upstream distribution peaks at $\sim 6.5\,\mathrm{keV/e}$; the downstream distribution peaks at $\sim 18\,\mathrm{keV/e}$; and the minimum E/Q at which Fe^{9+} is observed increases in the downstream region.

4. Results

Solar Orbiter's Heavy Ion Sensor provides high time resolution ion composition observations with both mass and charge state resolution. Currently available data has 10min time resolution. We have presented HIS observations from a shock observed by Solar Orbiter on March 11, 2022. The instrument's proton avoidance prevent a comparison of H^{1+}, He^{2+}, and O^{6+} across the shock. However, Fe^{8+} to Fe^{12+} are well resolved in both the upstream and downstream regions.

We have aggregated L2 PHA words collected in HIS' normal mode into variably defined intervals that range from ~ 2 to ~ 8 minutes. To improve statistics, PHAs have also been accumulated over azimuth and elevation angles. We show that the peak E/Q for Fe^{9+} increases across the shock. The VDFs derived from these spectra are above the 1-count level. Future, detailed analysis of multiple charge states is necessary to characterize the interaction between the observed heavy ions and the shock.

Further developments for HIS data are ongoing. Owen et al. (2020) describes the full complement of possible products. Several products under active development include expanded Fe charge states, high confidence derived physical quantities for an expanded set of elements and charge states, VDFs derived from onboard classification, and angular calibrations. Future cross-calibration with PAS will provide an absolute density calibration and intercalibration with the Suprathermal Ion Spectrograph (SIS) in the Energetic Particle Detector suite (EPD, Rodríguez-Pacheco et al. 2020; Wimmer-Schweingruber et al. 2021) will also provide observations of suprathermals across the energy ranges necessary to characterize multiple aspects of charged particle energization.

5. Acknowledgements

This work is funded by NASA contract NNG10EK25C.

References

Carcaboso, F., Gómez-Herrero, R., Espinosa Lara, F., et al. 2020, Astronomy & Astrophysics, 635, A79, doi: 10.1051/0004-6361/201936601

Dresing, N., Yli-Laurila, A., Valkila, S., et al. 2024, Astronomy & Astrophysics, 687, A72, doi: 10.1051/0004-6361/202449831

Gloeckler, G., Cain, J., Ipavich, F. M., et al. 1998, Space Sci. Rev., 86, 497, doi: 10.1023/A:1005036131689

Horbury, T. S., O'Brien, H., Carrasco Blazquez, I., et al. 2020, Astronomy & Astrophysics, 642, A9, doi: 10.1051/0004-6361/201937257

Livi, S., Lepri, S. T., Raines, J. M., et al. 2023, Astronomy & Astrophysics, 676, A36, doi: 10.1051/0004-6361/202346304

Müller, D., Marsden, R. G., St. Cyr, O. C., & Gilbert, H. R. 2013, Solar Physics, 285, 25, doi: 10.1007/s11207-012-0085-7

Müller, D., St. Cyr, O. C., Zouganelis, I., et al. 2020, Astronomy & Astrophysics, 642, A1, doi: 10.1051/0004-6361/202038467

Owen, C. J., Bruno, R., Livi, S. A., et al. 2020, Astronomy & Astrophysics, 642, A16, doi: 10.1051/0004-6361/201937259

Rodríguez-Pacheco, J., Wimmer-Schweingruber, R. F., Mason, G., et al. 2020, Astronomy & Astrophysics, 642, A7, doi: 10.1051/0004-6361/201935287

Shearer, P., von Steiger, R., Raines, J. M., et al. 2014, The Astrophysical Journal, 789, 60, doi: 10.1088/0004-637X/789/1/60

Stone, E. C., Frandsen, A. M., Mewaldt, R. A., et al. 1998, Space Science Reviews, 86, 1, doi: 10.1023/A:1005082526237

Trotta, D., Vuorinen, L., Hietala, H., et al. 2022, Frontiers in Astronomy and Space Sciences, 9, 1005672, doi: 10.3389/fspas.2022.1005672

Trotta, D., Hietala, H., Dresing, N., et al. 2024, Solar Orbiter Cycle 25 Interplanetary Shock List, Zenodo, doi: 10.5281/ZENODO.12518015

Wimmer-Schweingruber, R. F., Janitzek, N. P., Pacheco, D., et al. 2021, Astronomy & Astrophysics, 656, A22, doi: 10.1051/0004-6361/202140940

Solar and Stellar Coronal Mass Ejections
Proceedings IAU Symposium No. 388, 2026
N. Gopalswamy, O. Malandraki, A. Vidotto & W. Manchester, eds.
doi:10.1017/S174392132400139X

Origin of ^{3}He abundance enhancements in gradual solar energetic particle events

Radoslav Bučík[1], Samuel T. Hart[2,1], Maher A. Dayeh[1,2], Mihir I. Desai[1,2], Glenn M. Mason[3] and Mark E. Wiedenbeck[4]

[1]Southwest Research Institute, San Antonio, TX, USA

[2]University of Texas at San Antonio, San Antonio, TX, USA

[3]Johns Hopkins University, Applied Physics Laboratory, Laurel, MD, USA

[4]Jet Propulsion Laboratory, California Institute of Technology, Pasadena, CA, USA

Abstract. We examined the origin of ^{3}He abundance enhancement in 23 high-energy (25–50 MeV) solar proton events that coincide with ^{3}He-rich periods detected by ACE ULEIS in 1997–2021. In seven events, ^{3}He enhancement was due to ^{3}He leftover from preceding events or independent ^{3}He events occurring during proton events. One event is the most likely impulsive (^{3}He-rich), and another is unclear. Reaccelerated remnant flare material was the most probable cause of ^{3}He enhancements in the remaining 14 proton events. Imaging observations showed coronal jets in the parent active regions in six of these 14 events. Remarkably, the highest ^{3}He/^{4}He occurred in events with jets, implying their contribution to ^{3}He enhancement.

Keywords. Elemental composition, Particle acceleration, CME

1. Introduction

Coronal mass ejection (CME)-shock-related (or gradual) solar energetic particle (SEP) events show extensive event-to-event compositional variations whose origin is largely unexplored. The events often exhibit high abundance enhancement of ^{3}He and heavy ions, typical of flare-related (or impulsive) SEP events. Mason et al. (1999) examined 12 gradual SEP (GSEP) events and found in eight events the 0.5–2.0 MeV/nuc ^{3}He/^{4}He between 5 and 135 times the average slow solar wind (SSW) ratio (Gloeckler and Geiss 1998). Desai et al. (2016) found in 27 out of 46 GSEP events the 0.5–2.0 MeV/nuc ^{3}He/^{4}He between 2 and 194 times the average SSW ratio. Two competing ideas have been suggested: 1) reacceleration of remnant flare suprathermal ions by CME-driven shocks in corona and interplanetary (IP) space (e.g., Mason et al. 1999) and 2) simultaneous flare (jet)-related acceleration in parent active regions (ARs) or other sites in the corona (e.g., Cane et al. 2003).

2. Results

2.1. *Event selection*

We browsed Hart et al. (2022) catalog of ^{3}He-rich periods, covering 1997 Sep 29 – 2021 Mar 1, for coincidences with GSEP events measured by SOHO ERNE (Torsti et al. 1995). During the SDO era (after May 2010), our interval of interest, 174 ACE ULEIS (Mason et al. 1998) ^{3}He-rich periods were reported in the catalog. We selected proton events that are sufficiently intense at high energies. Specifically, we require that

Table 1. 25–50 MeV proton event characteristics

#	#	Proton event start day	GOES X-ray flare		CME		Jet	^{3}He/^{4}He ($\times 10^2$)	Ref.
			Start	Class	Speed	Width			
1	470	2010-Jun-12[b]	00:54:00	M2.0	486	119	N	1.54±0.18	i, j, k, l
2	473	2010-Aug-31[a,b]	...	...	1304	360	Y	0.68±0.39	i, k, l
3	494	2011-Jan-28[b]	00:44:00	M1.3	606	119	Y	4.08±0.93	i
4	508	2011-Apr-21[a,b]	...	...	475	111	Y	43.57±19.83	i
5	524	2011-Aug-02[b]	05:19:00	M1.4	712	268	N	...	i
6	525	2011-Aug-08	18:00:00	M3.5	1343	237	N	1.05±0.09	i
7	539	2011-Nov-26[b]	06:09:00	C1.2	933	360	N	0.32±0.03	i, m
8	554	2012-Jan-19[b]	13:44:00	M3.2	1120	360	N	0.99±0.08	i
9	566	2012-May-17[b]	01:25:00	M5.1	1582	360	N	0.35±0.04	i, m
10	571	2012-Jun-14[b]	12:52:00	M1.9	987	360	N	0.29±0.06	i
11	604	2013-Feb-26[a,b]	...	...	987	360	N	0.59±0.19	i
12	611	2013-May-02	04:58:00	M1.1	671	99	Y	20.05±2.57	i, j, n
13	641	2013-Sep-30[b]	21:43:00	C1.2	1179	360	N	0.36±0.04	i, j
14	647	2013-Oct-25[b]	07:53:00	X1.7	587	360	Y	2.45±0.33	i
			14:51:00	X2.1	1081	360	N		i
15	665	2013-Dec-26[a,b]	...	...	1011	>171	N	2.28±0.56	i
16	668	2014-Jan-04	18:47:00	M4.0	977	360	Y	0.29±0.08	m
17	682	2014-Feb-25[b]	00:39:00	X4.9	2147	360	N	0.30±0.03	
18	692	2014-Apr-18	12:31:00	M7.3	1202	360	Y	0.69±0.06	j
19	700	2014-May-07	16:07:00	M1.2	923	360	Y	1.96±0.57	
20	715	2014-Jun-12	21:34:00	M3.1	684	186	Y	4.20±0.30	
21	781	2015-Jul-19	09:22:00	C2.1	782	194	N	0.35±0.27	
22	789	2015-Sep-20	17:32:00	M2.1	1239	360	N	0.49±0.07	
23	790	2015-Sep-30	16:54:00	C2.4	...	...	Y	0.58±0.06	j

Notes. [a]Events with backside source. [b]Events measured at least on one of the two STEREO spacecraft.
References. (i) Richardson et al. (2014), (j) Cliver et al. (2019), (k) Wiedenbeck et al. (2013), (l) Bučík et al. (2016), (m) Desai et al. (2016), (n) Nitta et al. (2015)

the ratio between 25–50 MeV proton intensity peak and 24-hr background before the event onset is greater than two. In addition to ^{3}He periods that occurred during GSEP events (i.e., coincidences), we found cases when ^{3}He periods closely ended before the events. Since these periods would contain suprathermal seed particles for further re-acceleration by CME-driven shocks, we included them in this study. Namely, we adopted ^{3}He periods ending within one day before proton event onset. We note we did not find in the catalog a ^{3}He-period ending within two days before proton event onset. 41 coincidences fulfill the above criteria. Ten events at low (<1 MeV/nuc) ULEIS energies did not show ^{4}He (and often O) intensity increase above the previous background and were not included. Furthermore, seven ion enhancements were measured during corotating interaction regions (CIRs) and one during an IP shock. These events were also excluded. The CIRs were identified based on typical solar wind plasma and IP magnetic field behavior. The IP shock is reported in the database by Oliveira (2023). Thus, we have 23 events left to explore.

Table 1 shows the solar characteristics of proton events. Column 1 marks the event number. Column 2 shows the corresponding ^{3}He-rich period number (Hart et al. 2022). Column 3 indicates the start day of the proton event at 25–50 MeV. Columns 4 and 5 show GOES X-ray start time (with day in column 3) and class as obtained from the NOAA Space Weather Prediction Center (SWPC) Edited Events list. Events #2, 4, 11, and 15 have a backside source. A flare for event #13 started on an earlier day than the proton event start day. Note that event #14 was associated with multiple flares. Columns 6 and 7 indicate CME speed (km/s) and width (°) from the LASCO manually identified CME catalog† (Yashiro et al. 2004). Event #23 has not reported CME. Column 8 indicates with

† `https://cdaw.gsfc.nasa.gov/CME_list/`

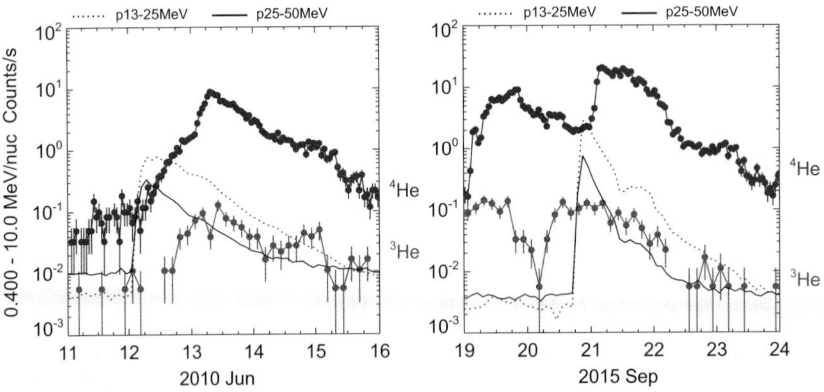

Figure 1. ACE ULEIS 3-hr ^{3}He (2.8–3.1 AMU) and 1-hr ^{4}He (3.5–6.0 AMU) count rates in events #1 (left) and #22 (right). Overplotted are SOHO ERNE proton fluxes (in arbitrary units) in two energy ranges.

Y a jet in the parent AR as observed by SDO AIA (Lemen et al. 2012) or STEREO EUVI (Howard et al. 2008). Column 9 shows 0.5–2.0 MeV/nuc ^{3}He/^{4}He. Note that independent ISEP events that occurred during some proton events were excluded from the ^{3}He/^{4}He calculation (details in Section 2.2). The last column marks references to earlier studies of these events. All events except #21 and #23 were associated with type II radio bursts. The radio bursts were reported in the SWPC list or identified using STEREO-A, -B (Bougeret et al. 2008), or Wind (Bougeret et al. 1995) radio spectrograms. 14 events were measured at least on one of the two STEREO spacecraft by HET (von Rosenvinge et al. 2008) at the proton energy range 24–41 MeV. The minimum longitudinal separation between Earth and STEREO was 70°, implying that these events were widespread. Five events (#7, 9, 13, 17, and 18) are on the NOAA list of Solar Proton Events Affecting the Earth's Environment‡.

2.2. *Observations*

It has been argued that similar time profiles of ^{3}He with other ion species suggest the common acceleration and transport origin (Mason et al. 1999; Wiedenbeck et al. 2000). Four events (#15, 18, 21, 22) show distinct time profiles of ^{3}He and ^{4}He, caused by ^{3}He left over from preceding impulsive (^{3}He-rich) events. Six events (#1, 3, 4, 5, 9, 14) from the remaining 19 events contain independent ^{3}He-rich SEP enhancement, which also leads to distinct time profiles. The periods of independent ^{3}He enhancements were excluded from four events (#1, 5, 9, 14) by adjusting the integration interval. It was not possible in two events (#3, 4) where the independent ^{3}He-rich period spanned nearly the whole proton event. Thus, we have 17 events to explore. Figure 1 (left) shows event #1 (included in the final list) with preceding ^{3}He ions, similar time profiles of ^{3}He and ^{4}He during proton flux decay until the beginning of 2010 Jun 14. Later, the increase in ^{3}He count rates is due to an independent ISEP event accompanied by solar energetic electrons detected by ACE EPAM (Gold et al. 1998). Figure 1 (right) shows event #22 (excluded from the final list) with ^{3}He left over from the preceding ISEP event. During a sudden rise in ^{4}He count rates on 2015 Sep 21, associated with the GSEP event, ^{3}He count rates continued to decrease.

‡ `https://umbra.nascom.nasa.gov/SEP/`

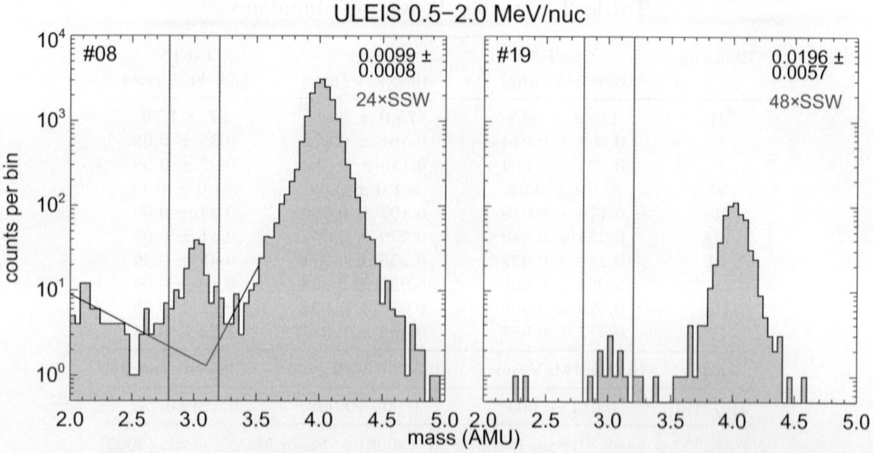

Figure 2. He mass histograms for events #8 (left) and #19 (right). The vertical red lines mark the ^{3}He mass range. The ^{3}He/^{4}He ratios are shown in the upper right corner. Left: The two tilted lines are the least square fits of the background and ^{4}He spillover (black) and the extrapolations to the ^{3}He range (blue).

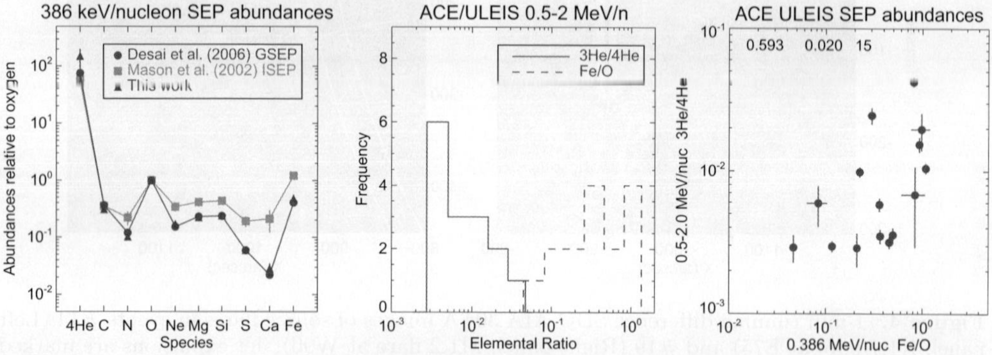

Figure 3. Left: Survey-averaged abundances relative to O in this work (blue triangles), GSEP (red circles), and ISEP (orange squares) reference abundances. Middle: Distribution of 0.5–2.0 MeV/nuc ^{3}He/^{4}He (red solid line) and 386 keV/nuc Fe/O (blue dashed line). Right: ^{3}He/^{4}He vs. Fe/O. The three numbers indicate the Spearman correlation coefficient, p-value, and sample size.

The ^{3}He/^{4}He ratio was determined using helium (2–5 AMU) mass histograms. The ratio is calculated in the broad energy range of 0.5–2.0 MeV/nuc, ensuring enough ^{3}He counts. We fitted the background at 2.00–2.64 AMU and spillover from ^{4}He at 3.28–3.52 AMU with two $\log(y) = a + bx$ functions, where a and b are parameters, and x corresponds to mass and y to ion counts. Both fits were extrapolated to the ^{3}He range (2.8–3.2 AMU), and the number of counts under the fit was subtracted from the total number of ^{3}He counts. Net ^{4}He counts were obtained from the 3.5–5.0 AMU range. Figure 2 shows event #08, where background and spillover were removed, and event #19, with no background and spillover. After adjusting integration intervals to eliminate independent ^{3}He events, the ^{3}He counts in #5 dropped to zero. In the remaining events, the ^{3}He peaks are finite - the number of net ^{3}He counts is 2σ or more above zero (same as in Desai et al. 2006).

Table 2. Average heavy ion abundances

Element	GSEP[a] (386 keV/nuc)	GSEP[b] (385 keV/nuc)	ISEP[c] (385 keV/nuc)
^{4}He	149.5 ± 50.7	75.0 ± 23.6	57 ± 7.79
C	0.308 ± 0.014	0.361 ± 0.012	0.35 ± 0.08
N	0.102 ± 0.004	0.119 ± 0.003	0.22 ± 0.06
O	≡1.0 ± 0.02	≡1.0 ± 0.02	≡1.0 ± 0.14
Ne	0.170 ± 0.015	0.152 ± 0.005	0.34 ± 0.06
Mg	0.223 ± 0.018	0.229 ± 0.007	0.41 ± 0.07
Si	0.240 ± 0.022	0.235 ± 0.011	0.43 ± 0.06
S	0.059 ± 0.005	0.059 ± 0.004	0.19 ± 0.04
Ca	0.029 ± 0.006	0.022 ± 0.002	0.21 ± 0.09
Fe	0.472 ± 0.087	0.404 ± 0.047	1.21 ± 0.14
Ratio	(0.5–2.0 MeV/nuc)	(0.5–2.0 MeV/nuc)	(385 keV/nuc)
^{3}He/^{4}He	0.011±0.003	0.006±0.002	0.354±0.137

Notes. [a]This work. [b]From Desai et al. (2006) . [c]From Mason et al. (2002).

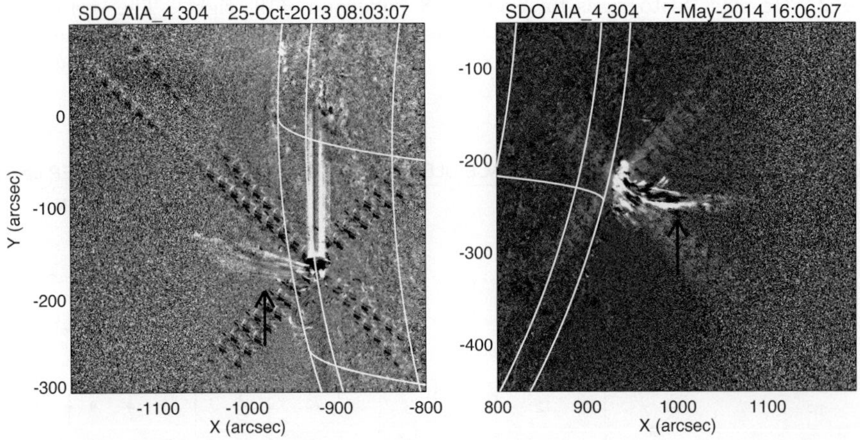

Figure 4. 1-min running difference SDO AIA 304 Å images of source flares in events #14 (Left panel; X1.7 flare at E75) and #19 (Right panel; M1.2 flare at W90). Jet expulsions are marked by arrows. Vertical bright stripes in the left panel are saturation, and diagonal patterns in both panels are diffraction fringes due to a strong EUV signal.

Figure 3 (left) displays the survey-averaged abundances using 16 events in Table 1 (#1, 2, 5–11, 13, 14, 16, 17, 19, 20, 23). Event #12 was not included in the averaging. This event was on the ISEP event list by Nitta et al. (2015). Indeed, the event 0.5–2.0 MeV/nuc ^{3}He/^{4}He = 0.20±0.03 (492×SSW value) is exceptionally high. Furthermore, the abundances of Si, S, and Fe are ISEP-like, and Ne and Mg are enhanced compared to GSEP values. Figure 3 (left) shows that the relative abundances in this work are consistent with reference GSEP abundances. Table 2 lists the average heavy-ion abundances of this survey along with GSEP (Desai et al. 2006) and ISEP (Mason et al. 2002) average abundances. Figure 3 (middle) indicates that ^{3}He/^{4}He peaks at the lowest values while Fe/O shows a more uniform distribution. Some high Fe/O can be due to rigidity-dependent transport effects. This will be explored in the next study. Minimum, maximum, mean, and median ^{3}He/^{4}He expressed as multipliers of the SSW value $(4.08\pm0.25)\times10^{-4}$ are 7, 103, 26±7, and 15, respectively. Figure 3 (right) shows a moderate positive correlation (r =0.593) between ^{3}He/^{4}He and Fe/O with a low probability of 2% that ratios are a random sample. Note that because ^{3}He dropped to zero in #5, both the middle and right panels involve 15 events. Desai et al. (2006) reported r =0.66 (for 29 events with a finite ^{3}He peak), but after excluding two outliers, r dropped to 0.026. The authors concluded that Fe/O is independent of ^{3}He/^{4}He in GSEP events. Similar observations have been reported in ISEP events (Mason et al. 1986; Reames et al. 1994).

Figure 4 shows the source flare for two events. Along with the bright flares that left some artifacts in the EUV images, ejections of jets from the sources were observed. Jets, a signature of magnetic reconnection involving field lines open to IP space, have been associated with the source of ^{3}He-rich SEPs (e.g., Bučík 2020; Nitta et al. 2023, and references therein).

Note that events #1 and #2 were included in the list of ISEP events by Wiedenbeck et al. (2013) and Bučík et al. (2016). The events had ^{3}He/^{4}He of 38×SSW and 17×SSW ratios, respectively. These values are typical for ^{3}He-enriched GSEP events. The heavy ions showed mixed composition (N, S with GSEP and Ne, Fe with ISEP abundances) in event #1 and GSEP composition (except Fe) in event #2. In addition, these events were widespread and associated with type II radio bursts.

3. Conclusion

We explored the ^{3}He origin in 23 high-energy (25–50 MeV) proton events. Except for independent ISEP events starting just before or during proton events, the ^{3}He origin in 15 events has no simple cause (presumed ISEP event #12 is not counted). 14 out of 15 events in this study showed preceding remnant ^{3}He. All these 14 events showed similar ^{3}He & ^{4}He intensity-time profiles, suggesting that reacceleration of remnant material by the associated CME shock was the most probable cause of ^{3}He enhancement. The origin of ^{3}He ions in one event (#10) remains unclear. In 6 events (those with preceding remnant ^{3}He ions), there were observed simultaneous jets in parent ARs. Interestingly, three events (#19, 14, 20) with the highest ^{3}He/^{4}He were associated with jets. This signifies the role of jets as a simultaneous source of energetic ^{3}He in GSEP events.

R.B. acknowledges support by NASA grants 80NSSC21K1316 and 80NSSC22K0757. The LASCO C2 CME catalog is generated and maintained at the CDAW Data Center by NASA and The Catholic University of America in cooperation with the Naval Research Laboratory. SOHO is a project of international cooperation between ESA and NASA.

References

Bougeret, J. L., Goetz, K., Kaiser, M. L., *et al.* 2008, *Space Sci. Revs*, 136(1-4), 487–528.
Bougeret, J. L., Kaiser, M. L., Kellogg, P. J., *et al.* 1995, *Space Sci. Revs*, 71(1-4), 231–263.
Bučík, R. 2020, *Space Sci. Revs*, 216(2), 24.
Bučík, R., Innes, D. E., Mason, G. M., & Wiedenbeck, M. E. 2016, *ApJ*, 833(1), 63.
Cane, H. V., von Rosenvinge, T. T., Cohen, C. M. S., & Mewaldt, R. A. 2003, *Geophys. Res. Lett.*, 30(12), 8017.
Cliver, E. W., Kahler, S. W., Kazachenko, M., & Shimojo, M. 2019, *ApJ*, 877(1), 11.
Desai, M. I., Mason, G. M., Dayeh, M. A., *et al.* 2016, *ApJ*, 816(2), 68.
Desai, M. I., Mason, G. M., Gold, R. E., *et al.* 2006, *ApJ*, 649(1), 470–489.
Gloeckler, G. & Geiss, J. 1998, *Space Sci. Revs*, 84, 275–284.
Gold, R. E., Krimigis, S. M., Hawkins, S. E., I., *et al.* 1998, *Space Sci. Revs*, 86, 541–562.
Hart, S. T., Dayeh, M. A., Bučík, R., *et al.* 2022, *ApJS*, 263(2), 22.
Howard, R. A., Moses, J. D., Vourlidas, A., *et al.* 2008, *Space Sci. Revs*, 136(1-4), 67–115.
Lemen, J. R., Title, A. M., Akin, D. J., *et al.* 2012, *Solar Phys.*, 275(1-2), 17–40.
Mason, G. M., Gold, R. E., Krimigis, S. M., *et al.* 1998, *Space Sci. Revs*, 86, 409–448.
Mason, G. M., Mazur, J. E., & Dwyer, J. R. 1999, *ApJ Letters*, 525(2), L133–L136.
Mason, G. M., Reames, D. V., Klecker, B., *et al.* 1986, *ApJ*, 303, 849.
Mason, G. M., Wiedenbeck, M. E., Miller, J. A., *et al.* 2002, *ApJ*, 574(2), 1039–1058.

Nitta, N. V., Bučík, R., Mason, G. M., *et al.* 2023, *Front. Astron. Space Sci.*, 10, 50.
Nitta, N. V., Mason, G. M., Wang, L., *et al.* 2015, *ApJ*, 806(2), 235.
Oliveira, D. M. 2023, *Front. Astron. Space Sci.*, 10, 1240323.
Reames, D. V., Meyer, J. P., & von Rosenvinge, T. T. 1994, *ApJS*, 90, 649.
Richardson, I. G., von Rosenvinge, T. T., Cane, H. V., *et al.* 2014, *Solar Phys.*, 289(8), 3059–3107.
Torsti, J., Valtonen, E., Lumme, M., *et al.* 1995, *Solar Phys.*, 162(1-2), 505–531.
von Rosenvinge, T. T., Reames, D. V., Baker, R., *et al.* 2008, *Space Sci. Revs*, 136(1-4), 391–435.
Wiedenbeck, M. E., Christian, E. R., Cohen, C. M. S., *et al.* In Mewaldt, R. A., Jokipii, J. R., Lee, M. A., Möbius, E., & Zurbuchen, T. H., editors, *Acceleration and Transport of Energetic Particles Observed in the Heliosphere* 2000, volume 528 of *American Institute of Physics Conference Series*, pp. 107–110.
Wiedenbeck, M. E., Mason, G. M., Cohen, C. M. S., *et al.* 2013, *ApJ*, 762(1), 54.
Yashiro, S., Gopalswamy, N., Michalek, G., *et al.* 2004, *J. Geophys. Res. Space Phys.*, 109(A7), A07105.

Solar and Stellar Coronal Mass Ejections
Proceedings IAU Symposium No. 388, 2026
N. Gopalswamy, O. Malandraki, A. Vidotto & W. Manchester, eds.
doi:10.1017/S1743921324001522

Energetic Particle Transport during Solar Gamma Ray Flares

Frederic Effenberger[1,3], Julien Dörner[1], Horst Fichtner[1], Meng Jin[2], Wei Liu[2,3], Gang Li[4], Vahé Petrosian[5] and Julia Becker Tjus[1]

[1]Theoretical Physics IV, Faculty for Physics and Astronomy and Ruhr Astropaticle and Plasma Physics Center (RAPP Center), Ruhr University Bochum, 44780 Bochum, Germany

[2]Lockheed Martin Solar and Astrophysics Laboratory, 3251 Hanover Street, Palo Alto, CA 94304, USA

[3]Bay Area Environmental Research Institute, NASA Research Park, Moffett Field, CA 94035, USA

[4]General Linear Space Plasma Lab LLC, Foster City, CA 94404, USA

[5]Department of Physics, Stanford University, Stanford, CA 94305, USA

Abstract. Gamma-ray emission during solar eruptive events is thought to be caused mainly by high-energy (MeV-TeV) protons propagating to the lower layers of the solar atmosphere, i.e. the chromosphere and photosphere, and the subsequent pion production and decay. A potential source of these high-energy particles are coronal mass ejection (CME)-driven shocks, followed by particle back-precipitation to the Sun. Alternatively, a scenario where particles propagate from a compact acceleration site along extended coronal loops towards the emission region is also a possibility. In both cases, the solar magnetic field configuration is a key ingredient to understand the origins of the gamma-ray emission as observed, e.g. by the *Fermi* Large Area Telescope (LAT). Here, we investigate the propagation of energetic protons and their subsequent gamma-ray production in a comprehensive modeling framework based on the CRPropa code, which is a well-established tool in high-energy astrophysics. The advantage of this approach is that the proton transport and gamma ray production can be studied with a single, unified code by following individual particle trajectories. We present first results of particle dynamics and the resulting gamma-ray distributions based on simplified field structures in a potential field (PFSS) model. We discuss further prospects of our approach that can include outputs from detailed MHD simulations of CME evolution and extensions to other energy regimes.

Keywords. Solar Gamma Ray Flares, Coronal Mass Ejections, Energetic Particles, Particle Transport

---

1. Introduction

Gamma-ray emission from eruptive events on the Sun has, by now, been observed for decades (e.g. Forrest et al. 1986; Seckel et al. 1991), while observations with the Large Area Telescope (LAT) on board the *Fermi Gamma-ray Space Telescope* (*Fermi:* Atwood et al. 2009) have increased the number of flares with observed gamma ray emission by more than an order of magnitude (Ajello et al. 2021). Important and puzzling events are especially the long-duration events that show gamma-ray emission that can last for hours at a time (Ryan 2000), and the so-called behind-the-limb events (Pesce-Rollins et al. 2015; Ackermann et al. 2017), where the primary flare site is behind the solar limb and cannot be observed directly from Earth, while still LAT observations clearly register gamma ray emissions from the Earth-facing part of the Sun. Additionally, the centroid position of the gamma ray emission in such events is observed to be moving in time (e.g. Ajello et al. 2014), which could point to a time-dependent injection and highlights the importance of magnetic connectivities. *Fermi*-LAT also confirmed the existence of a gamma ray component from the quiet Sun (Giglietto et al. 2012), as indicated earlier in the analysis of the Energetic Gamma Ray Experiment Telescope (EGRET) data (Orlando & Strong 2008).

In general, both electrons and protons can produce high-energy gamma ray photons by either relativistic bremsstrahlung and inverse compton scattering of electrons or decay of neutral pions produced by the interaction of high-energy protons with the background ions. The need for sufficient densities and column depths for pion production require the transport of protons to the lower layers of the solar atmosphere, i.e. the denser chromosphere and photosphere. At least two scenarios are discussed as explanations for the long-duration gamma-ray flares and the differences in location between the optical/X-ray flare location and the observed gamma-ray centroids in some of the events. The first scenario proposes coronal mass ejection (CME) shock acceleration as the primary source of high-energy protons. Their subsequent propagation back to the Sun could cause long-duration gamma-ray emission which, depending on the magnetic field configuration, can also be far away from the original flaring active region and CME release site. This CME scenario has first been described in Cliver et al. (1993) and more recent studies have investigated different aspects of this idea (e.g. Plotnikov et al. 2017; Jin et al. 2018; Gopalswamy et al. 2019; Mäkelä et al. 2023). Additional connections of particle acceleration and gamma-ray emission to EUV waves are also discussed (e.g. Kouloumvakos et al. 2020; Pesce-Rollins et al. 2022). An open issue in this scenario is the magnetic mirror effect in the strongly converging solar magnetic field. Depending on the magnetic field ratio experienced by the particles, scattering into the loss-cone by local magnetic turbulence might be necessary to enable the back-propagation of a sufficient number of particles from the evolving CME shock to allow for the production of observable gamma rays (Effenberger & Petrosian 2018; Hutchinson et al. 2022). Observational connections of in-situ solar energetic particles and the observed gamma-ray signatures may also provide additional information (e.g. Bruno et al. 2023). The second scenario proposes the acceleration of high-energy protons in the flaring region and the subsequent storage and potential continuous re-acceleration of particles in large-scale coronal loop structures (e.g., Ryan & Lee 1991). For shorter-duration gamma ray events, this has for example been discussed in Pesce-Rollins et al. (2024), while the consistent explanation of long-duration gamma-ray events and moving sources remains more challenging in this scenario. Figure 1 illustrates both scenarios with the field configuration from an MHD simulation (Jin et al. 2018). One possible observation which may differentiate these two scenarios is through energetic neutral atoms (ENAs) (Li et al. 2023). In all of such situations, detailed particle transport studies including the interactions with the background plasma can improve our understanding of the relevant processes and lead to more quantitative explanations of the observed emissions.

In this proceedings paper, we present first steps towards a quantitative model of particle transport and gamma-ray emission during solar events, based on CRPropa 3.2.1 (Alves Batista et al. 2022; Aerdker et al. 2023). A more detailed, full paper on the model, numerical tests and applications is in preparation.

2. Transport model and pion-decay gamma ray model with CRPropa

Our new model for particle transport in solar magnetic fields and associated gamma-ray emission is based on the publicly available CRPropa framework†, originally designed

† `https://github.com/CRPropa/CRPropa3`

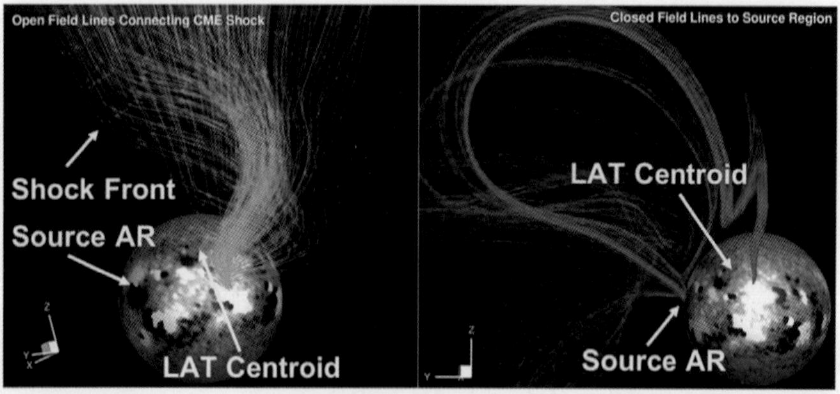

Figure 1. Illustration of the two scenarios for solar gamma ray emission during eruptive events. Shown is the 3D field configuration from a MHD simulation of the September 8, 2014 event at t=30 minutes. The open (left) and closed (right) field lines based on the same field but with different footpoints are given with important features marked. For details on the model and the original figures, see Jin et al. (2018).

for ultra-high-energy astrophysics studies. With its general modular structure, it provides two propagation modes for particles, namely the full-orbit test particle integration and a stochastic formulation based on stochastic differential equations (Strauss & Effenberger 2017) that assumes a diffusive behavior of particle transport. In the current setup, we utilize only the full-orbit mode, since it allows for a more straightforward implementation of complex field structures and the associated magnetic mirror effect. The code uses a Boris push algorithm to solve the Newton-Lorentz equations for particle transport in a prescribed magnetic field.

For the solar magnetic field, we use a potential field model (PFSS) based on the freely available python PFSS model by A. Yeates†. The model allows to use magnetogram maps from GONG or SDO/HMI as inputs for all available Carrington rotations (CR). We use CR2154 and CR2157 as example cases in the following.

In Figure 2, two different types of trajectories for charged particles are shown. All particles have an energy of $E = 1\,\mathrm{GeV}$, an initial pitch angle cosine of $\mu = 0.999999$ and start at a distance of $2.5R_{\odot}$. The cyan line on the right side corresponds to particles which are mirrored in the magnetic field, and leave the region along the same field line again. The green trajectory on the left side also follows an open field line, but it allows the particles to reach the chromosphere and to produce gamma rays.

The hadronic interaction of high energy particles uses a Monte-Carlo description, similar to other interactions implemented in CRPropa. The basic idea is to decide in every propagation step Δs, if an interaction is occurring, by comparing the interaction probability $p = n(\vec{r})\,\sigma\,\Delta s$ with a random number. Here, $n(\vec{r})$ is the density at the position of the particle and σ is the total inelastic cross-section as described in Kafexhiu et al. (2014). The energy of the secondary gamma-rays is randomly drawn from the parametrisation of the inclusive cross section by Orusa et al. (2023), which includes the production of π^0 and other more rare particles. The details of the implementation of the hadronic interactions into CRPropa are given in Dörner et al. (2024).

The quantitative calculation of gamma-ray emission requires a solar background density model for the chromosphere. We utilize the density model of Avrett & Loeser (2008) as a first approach, even though some modifications to such a model are to be expected in a dynamic situation during a solar eruption, especially within the flaring active region.

† `https://github.com/antyeates1983/pfss`

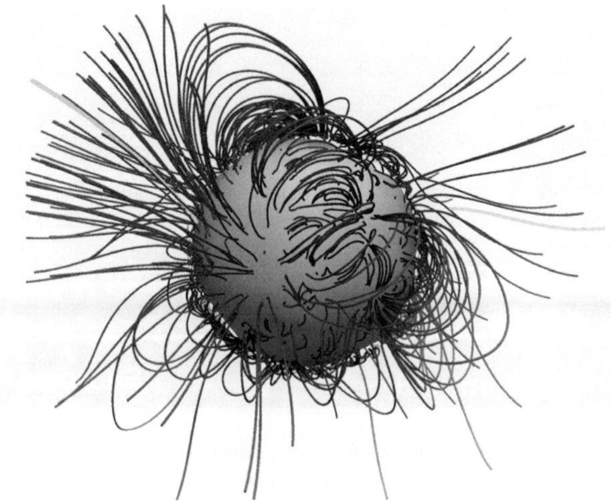

Figure 2. Two particle trajectories starting in different regions around the Sun. The particle following the cyan trajectory (right side) is mirrored early and propagates back, while the particle following the green trajectory (left side) is reaching lower layers of the solar atmosphere. Additionally, field lines of the PFSS model for CR 2157 are shown.

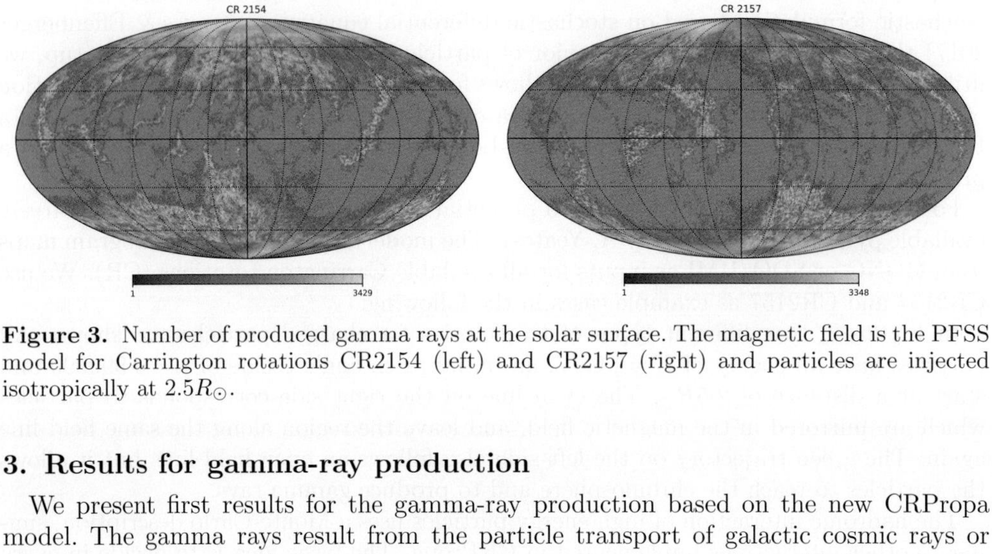

Figure 3. Number of produced gamma rays at the solar surface. The magnetic field is the PFSS model for Carrington rotations CR2154 (left) and CR2157 (right) and particles are injected isotropically at $2.5R_{\odot}$.

3. Results for gamma-ray production

We present first results for the gamma-ray production based on the new CRPropa model. The gamma rays result from the particle transport of galactic cosmic rays or solar energetic ions and their subsequent chromospheric interactions. These are meant to illustrate the capabilities of the model and do not represent detailed case studies as of yet. In particular, particles are injected homogeneously as well as isotropically at $2.5R_{\odot}$ and all produced gamma rays are accumulated statistically. In the future, the model will be adapted to address more specific questions related to the above described scenarios for solar transient events and long-term, high-energy cosmic ray interactions with the solar atmosphere.

In Figure 3 the production position of gamma rays projected back onto the solar surface is shown. The figure shows all gamma rays independent of their direction. Most of the gamma rays will have a direction pointing into the Sun, and will not be observable at Earth. For relativistic particles, the emission will be directed in a very narrow cone around the direction of travel of the particle. We can see that most of the gamma rays are produced at the poles, where the magnetic field has more open field lines connected to the starting surface. Closer to the equator, the gamma-ray density becomes more clustered, as the cosmic rays are concentrated by open field lines to specific regions.

Some regions do not show any gamma ray production. Here two effects prevent the cosmic rays from reaching the surface and producing gamma-ray emission: (1) the magnetic mirroring of cosmic rays along the field lines, which can be seen in Figure 2 and (2) the absence of open field lines reaching $2.5R_{\odot}$. Without open field lines, the particles are deflected away or reach the solar surface at different positions, which leads to the clustering of the emission patterns.

4. Discussion

The structured appearance of the positions of gamma-ray production for different magnetic field configurations shows that gamma ray observation contain information about the solar magnetic field structure. The spatial and spectral properties of the emission will thus also depend on solar activity and the solar cycle (Linden et al. 2018). While a majority of open field lines in the PFSS fields is often concentrated at the solar poles, the input magnetogram maps based on Earth position observations in the ecliptic are not well resolved at high latitudes. It will therefor be of interest to improve the models at the poles, potentially with novel upcoming observations of Solar Orbiter (Müller et al. 2020) while it starts to move to higher latitudes.

A number of improvements to the simple PFSS approach presented here are possible. This includes a sensitivity study on the outer boundary distance of $2.5R_{\odot}$ which might arbitrarily force closed field to be open at the boundary. Elliptical boundaries have also been considered (Kruse et al. 2021) as an alternative, see also Zhao & Hoeksema (2010). It should be noted that the fields during a CME eruption will be non-potential in order to provide the free energy needed in the eruption process. Thus, the use of PFSS fields has limitations that should be addressed in future studies. Other field extrapolation methods such as non-linear force free approaches and the direct usage of MHD simulation outputs should further clarify the influence of different magnetic field structures on the resulting gamma ray production.

Scattering of particles from small-scale turbulent features in the magnetic field may play an important role in particle dynamics. We started to implement a first scattering model based on isotropic scattering by assigning a new particle velocity direction randomly following a prescribed mean free path (Kocharov et al. 1998). We could successfully reproduce the results of Hutchinson et al. (2022) for scattering and mirroring in a Parker magnetic field with our code setup and will study the effects for fields closer to the Sun in the future. Depending on the properties of particle scattering, the pitch-angle changes may improve the chances of particle populations to overcome the magnetic mirror effect, especially if the CME particle injection site has moved further away from the Sun.

Other energy regimes and emission mechanisms, including leptonic processes, can in principle also be addressed with our setup. For example, the High-Altitude Water Cherenkov Gamma-Ray Observatory (HAWC) observations of 6 years of solar TeV gamma rays (Albert et al. 2023) indicate an anticorrelation with the solar cycle, which may be consistent with our findings of a magnetic field influence and the importance of open field line structures. The differences of halo and disk components of gamma-ray emission resulting from inverse compton processes have also been studied in the past (e.g. Orlando & Strong 2008) and offer further constraints on modelling efforts.

5. Conclusions and Outlook

We have presented a novel model for the transport and subsequent gamma-ray production of energetic particles in the solar atmosphere and magnetic field. With the first application to the so-called quiet Sun disk emission we have established a proof of concept.

With forthcoming detailed event studies and extensions to a broad range of energies and physical processes involved in solar gamma-ray emissions, and with further constraints from supplemental observations of solar magnetic fields and atmospheric properties, we should be able to improve our understanding of the relevant physical processes that govern high-energy solar emissions.

Acknowledgements

Helpful discussions at the IAU meeting on the topic of this paper are gratefully acknowledged. We thank A. Yeates for providing the freely available python PFSS solver to the community. We acknowledge support from NASA LWS grant 80NSSC21K1327 and DFG grants EF 98/4-1, FI 706/27-1, and TJ 62/7-1 as well as via the Collaborative Research Center (Sonderforschungsbereich, SFB) 1491. Additional support from the DFG under grant HE 8392/2-1 (project 508335258) was helpful to initialize this international collaboration.

References

Ackermann, M., Allafort, A., Baldini, L., *et al.* 2017, Fermi-LAT Observations of High-energy Behind-the-limb Solar Flares. *ApJ*, 835(2), 219.

Aerdker, S., Alves Batista, R., Becker Tjus, J., Dörner, J., Dundovic, A., Eichmann, B., Frie, A., Heiter, C., Hoerbe, M., Kampert, K.-H., Merten, L., Müller, G., Reichherzer, P., Rossoni, S., Saveliev, A., Schlegel, L., Sigl, G., van Vliet, A., & Winchen, T. 2023, CRPropa 3.2: a public framework for high-energy astroparticle simulations. *arXiv e-prints*, arXiv:2308.09532.

Ajello, M., Albert, A., Allafort, A., *et al.* 2014, Impulsive and Long Duration High-energy Gamma-Ray Emission from the Very Bright 2012 March 7 Solar Flares. *ApJ*, 789(1), 20.

Ajello, M., Baldini, L., Bastieri, D., *et al.* 2021, First Fermi-LAT Solar Flare Catalog. *ApJS*, 252(2), 13.

Albert, A., Alfaro, R., Alvarez, C., *et al.* 2023, Discovery of Gamma Rays from the Quiescent Sun with HAWC. *Phys. Rev. Lett.*, 131(5), 051201.

Alves Batista, R., Becker Tjus, J., Dörner, J., Dundovic, A., Eichmann, B., Frie, A., Heiter, C., Hoerbe, M. R., Kampert, K.-H., Merten, L., Müller, G., Reichherzer, P., Saveliev, A., Schlegel, L., Sigl, G., van Vliet, A., & Winchen, T. 2022, CRPropa 3.2 - an advanced framework for high-energy particle propagation in extragalactic and galactic spaces. *JCAP*, 2022(9), 035.

Atwood, W. B., Abdo, A. A., Ackermann, M., *et al.* 2009, *ApJ*, 697(2), 1071–1102.

Avrett, E. H. & Loeser, R. 2008, Models of the Solar Chromosphere and Transition Region from SUMER and HRTS Observations: Formation of the Extreme-Ultraviolet Spectrum of Hydrogen, Carbon, and Oxygen. *ApJS*, 175(1), 229–276.

Bruno, A., de Nolfo, G. A., Ryan, J. M., Richardson, I. G., & Dalla, S. 2023, Statistical Relationship between Long-duration High-energy Gamma-Ray Emission and Solar Energetic Particles. *ApJ*, 953(2), 187.

Cliver, E. W., Kahler, S. W., & Vestrand, W. T. 1993, On the Origin of Gamma-Ray Emission from the Behind-the-Limb Flare on 29 September 1989. *23rd International Cosmic Ray Conference (ICRC23), Volume 3*, 3, 91.

Dörner, J., Morejon, L., Becker Tjus, J., & Kampert, K.-H. 2024,. Hadronic interaction in CRPropa, in preperation.

Effenberger, F. & Petrosian, V. 2018, The Relation between Escape and Scattering Times of Energetic Particles in a Turbulent Magnetized Plasma: Application to Solar Flares. *ApJ*, 868(2), L28.

Forrest, D. J., Vestrand, W. T., Chupp, E. L., Rieger, E., Cooper, J., & Share, G. H. 1986, Very energetic gamma-rays from the 3 June 1982 solar flare. *Adv. Sp. Res.*, 6(6), 115–118.

Giglietto, N., Allafort, A., Brigida, M., Longo, F., Omodei, N., Takahashi, H., Tanaka, Y., & Fermi-LAT Collaboration 2012, Fermi LAT observation of quiet gamma-ray emission from the Sun and first solar flares detection. *Nucl. Instrum. Methods Phys. Res. A*, 692, 262–264.

Gopalswamy, N., Mäkelä, P., Yashiro, S., Lara, A., Akiyama, S., & Xie, H. 2019, On the Shock Source of Sustained Gamma-Ray Emission from the Sun. *Journal of Physics Conference Series*, 1332, 012004.

Hutchinson, A., Dalla, S., Laitinen, T., de Nolfo, G. A., Bruno, A., Ryan, J. M., & Waterfall, C. O. G. 2022, Energetic proton back-precipitation onto the solar atmosphere in relation to long-duration gamma-ray flares. *A&A*, 658, A23.

Jin, M., Petrosian, V., Liu, W., Nitta, N. V., Omodei, N., Rubio da Costa, F., Effenberger, F., Li, G., Pesce-Rollins, M., Allafort, A., & Manchester, Ward, I. 2018, Probing the Puzzle of Behind-the-limb γ-Ray Flares: Data-driven Simulations of Magnetic Connectivity and CME-driven Shock Evolution. *ApJ*, 867(2), 122.

Kafexhiu, E., Aharonian, F., Taylor, A. M., & Vila, G. S. 2014, Parametrization of gamma-ray production cross sections for p p interactions in a broad proton energy range from the kinematic threshold to PeV energies. *Phys. Rev. D*, 90(12), 123014.

Kocharov, L., Vainio, R., Kovaltsov, G. A., & Torsti, J. 1998, Adiabatic Deceleration of Solar Energetic Particles as Deduced from Monte Carlo Simulations of Interplanetary Transport. *Sol. Phys.*, 182(1), 195–215.

Kouloumvakos, A., Rouillard, A. P., Share, G. H., Plotnikov, I., Murphy, R., Papaioannou, A., & Wu, Y. 2020, Evidence for a Coronal Shock Wave Origin for Relativistic Protons Producing Solar Gamma-Rays and Observed by Neutron Monitors at Earth. *ApJ*, 893(1), 76.

Kruse, M., Heidrich-Meisner, V., & Wimmer-Schweingruber, R. F. 2021, Evaluation of a potential field source surface model with elliptical source surfaces via ballistic back mapping of in situ spacecraft data. *A&A*, 645, A83.

Li, G., Shih, A. Y., Allen, R. C., Ho, G. C., Cohen, C. M. S., Desai, M., Dayeh, M. A., & Mason, G. M. 2023, Modeling Solar Energetic Neutral Atoms from Solar Flares and CME-driven Shocks. *ApJ*, 944(2), 196.

Linden, T., Zhou, B., Beacom, J. F., Peter, A. H. G., Ng, K. C. Y., & Tang, Q.-W. 2018, Evidence for a New Component of High-Energy Solar Gamma-Ray Production. *Phys. Rev. Lett.*, 121(13), 131103.

Mäkelä, P., Gopalswamy, N., Akiyama, S., Xie, H., & Yashiro, S. 2023, Speed and Acceleration of Coronal Mass Ejections Associated with Sustained Gamma-Ray Emission Events Observed by Fermi/LAT. *ApJ*, 954(1), 79.

Müller, D., St. Cyr, O. C., Zouganelis, I., Gilbert, H. R., Marsden, R., Nieves-Chinchilla, T., Antonucci, E., Auchère, F., Berghmans, D., Horbury, T. S., Howard, R. A., Krucker, S., Maksimovic, M., Owen, C. J., Rochus, P., Rodriguez-Pacheco, J., Romoli, M., Solanki, S. K., Bruno, R., Carlsson, M., Fludra, A., Harra, L., Hassler, D. M., Livi, S., Louarn, P., Peter, H., Schühle, U., Teriaca, L., del Toro Iniesta, J. C., Wimmer-Schweingruber, R. F., Marsch, E., Velli, M., De Groof, A., Walsh, A., & Williams, D. 2020, The Solar Orbiter mission. Science overview. *A&A*, 642, A1.

Orlando, E. & Strong, A. W. 2008, Gamma-ray emission from the solar halo and disk: a study with EGRET data. *A&A*, 480(3), 847–857.

Orusa, L., Di Mauro, M., Donato, F., & Korsmeier, M. 2023, New determination of the production cross section for γ rays in the Galaxy. *Phys. Rev. D*, 107(8), 083031.

Pesce-Rollins, M., Klein, K.-L., Krucker, S., Warmuth, A., Veronig, A. M., Omodei, N., & Monstein, C. 2024, Evidence for flare-accelerated particles in large scale loops in the behind-the-limb gamma-ray solar flare of September 29, 2022. *A&A*, 683, A208.

Pesce-Rollins, M., Omodei, N., Krucker, S., Di Lalla, N., Wang, W., Battaglia, A. F., Warmuth, A., Veronig, A. M., & Baldini, L. 2022, The Coupling of an EUV Coronal Wave and Ion Acceleration in a Fermi-LAT Behind-the-Limb Solar Flare. *ApJ*, 929(2), 172.

Pesce-Rollins, M., Omodei, N., Petrosian, V., Liu, W., Rubio da Costa, F., Allafort, A., & Chen, Q. 2015, First Detection of >100 MeV Gamma Rays Associated with a Behind-the-limb Solar Flare. *ApJ*, 805(2), L15.

Plotnikov, I., Rouillard, A. P., & Share, G. H. 2017, The magnetic connectivity of coronal shocks from behind-the-limb flares to the visible solar surface during γ-ray events. *A&A*, 608, A43.

Ryan, J. M. 2000, Long-Duration Solar Gamma-Ray Flares. *Space Sci. Rev*, 93, 581–610.

Ryan, J. M. & Lee, M. A. 1991, On the Transport and Acceleration of Solar Flare Particles in a Coronal Loop. *ApJ*, 368, 316.

Seckel, D., Stanev, T., & Gaisser, T. K. 1991, Signatures of Cosmic-Ray Interactions on the Solar Surface. *ApJ*, 382, 652.

Strauss, R. D. T. & Effenberger, F. 2017, A Hitch-hiker's Guide to Stochastic Differential Equations. Solution Methods for Energetic Particle Transport in Space Physics and Astrophysics. *Space Sci. Rev*, 212(1-2), 151–192.

Zhao, X. P. & Hoeksema, J. T. 2010, The Magnetic Field at the Inner Boundary of the Heliosphere Around Solar Minimum. *Sol. Phys.*, 266(2), 379–390.

Solar and Stellar Coronal Mass Ejections
Proceedings IAU Symposium No. 388, 2026
N. Gopalswamy, O. Malandraki, A. Vidotto & W. Manchester, eds.
doi:10.1017/S1743921325000031

Co-occurrence of SGRE and SEP events

Pertti Mäkelä[1,2], Nat Gopalswamy[2], Sachiko Akiyama[1,2], Seiji Yashiro[1,2] and Hong Xie[1,2]

[1]The Catholic University of America, Washington, DC 20064, USA
email: pertti.makela@nasa.gov

[2]NASA Goddard Space Flight Center, Greenbelt, MD 20771, USA

Abstract. Coronal mass ejections (CMEs) have been suggested to be the source of high-energy protons producing sustained gamma-ray emission (SGRE) events at the Sun, where the >100 MeV gamma-ray emission observed by Fermi Large Area Telescope (LAT) lasts beyond the end of the associated X-ray flare. Large SEP events observed at 1 au are produced by shocks driven by fast and wide (FW) CMEs. However, there are more large SEP events than SGRE events and more FW CMEs than the large SEP events. Magnetic connectivity of the observer to the shock nose affects the detectability of SEP events at 1 au. Mäkelä et al. (2023) found that the SGRE-producing halo CMEs associated with type II radio bursts and/or a large SEP events have higher space speeds and initial accelerations than those without an SGRE event. In this paper, we consider characteristics of CME eruptions associated with co-occurring SEP and SGRE events.

Keywords. solar energetic particles, gamma-ray emission, coronal mass ejections

1. Introduction

In sustained gamma-ray emission (SGRE) events the >100 MeV emission continues beyond the flare impulsive phase, sometimes up to almost a day (Forrest et al. 1985). Although the Large Area Telescope (LAT, Atwood et al. 2009) on board the Fermi satellite does not observe the Sun continuously, Fermi-LAT observations have shown that SGRE events are relatively common (e.g., Share et al. 2018). Recently, Ajello et al. (2021) published the first Fermi-LAT solar flare catalog, where they listed a total of 45 solar gamma-ray flares in the 30 MeV–10 GeV energy range. Delayed emission beyond the impulsive, prompt component was detected during 37 flares. Most of the observed SGRE events lasted more than two hours. Previously in the pre-Fermi era, SGRE events were know as long-duration gamma-ray flares, and they were detected less frequently (Ryan 2000).

The >100 MeV gamma-ray emission is produce when >300 MeV protons propagate down into the dense solar atmosphere and interact with the ambient plasma producing neutral pions decaying into gamma-rays (e.g., Gopalswamy 2020). The acceleration region of the >300 MeV protons is still debated. The suggested acceleration sites are solar flares and shock fronts driven by fast coronal mass ejections (CMEs). Solar flares are known to produce impulsive gamma-ray emission. However, SGRE events require coronal trapping of high-energy protons for long periods well after the flare ended, and the required prolonged coronal trapping is difficult to achieve. Detection of gamma-ray emission during behind-the-limb flares requires that gamma-ray source must have wide area, suggesting the wide CME shock as a possible source of the >300 MeV protons (e.g., Pesce-Rollins et al. 2015; Plotnikov et al. 2017). Recently, Gopalswamy et al.

(2018, 2019) found a correlation between the SGRE durations and the durations and end frequencies of decameter-hectometric type II radio bursts, providing additional support for the CME-shock scenario.

We know that SGRE events are associated with ultrafast (deprojected speeds $\geq$2000 km s^{-1}) halo CMEs similar to CMEs producing ground level enhancements (GLEs; Gopalswamy et al. 2016). As mentioned, SGRE events are also associated with interplanetary type II radio bursts in the decameter-to-kilometer wavelength domains, mirroring again a similar type II association observed in GLEs Gopalswamy et al. (2005). Previously, Mäkelä et al. (2023) have shown that CMEs with metric type II radio bursts and having an associated major solar energetic particle (SEP) and SGRE event have high space speed and initial acceleration. However, in their study they did not consider the possibility of missed SGRE detection due to a gap in the Fermi-LAT solar observations. In this study, we analyze only non-SGRE events where LAT observed the Sun during the flare associated with the event.

2. Data analysis

The starting point of our analysis is a list of GOES soft X-ray flares with a peak magnitude $\geq$C5 (5×10^{-6} W m^{-2}) and a known on-disk location. From the list, we select flares that have LAT solar observation periods starting $\leq$20 minutes after the flare peak time and the flare peak time must occur $\geq$10 minutes before the end time of the LAT solar observation period. In addition, the duration of the LAT solar observation period must be $\geq$20 minutes. These time limits for the LAT solar observations are arbitrary but we expect them to reasonably guarantee that, if the flare has an associated SGRE event, Fermi-LAT should be able to detect it. Therefore, we will not be misclassify a flare with an SGRE event to be a flare without an SGRE event due to a gap in LAT observations. The flare peak is selected as a reference time because SGRE emission is assumed to start after the flare impulsive phase.

The list of SGRE events observed by LAT include the on-disk events with CMEs observed 2008–2023 selected from the list published by Ajello et al. (2021) with a few more recent events not included in their list. For the major SEP events we use the list of major SEP events that have the GOES >10 MeV or STEREO-A/B GOES-equivalent peak flux $\geq$10 cm^{-2} s^{-1} sr^{-1} (particle flux unit). Because STEREO particle instruments do not observe particles above 100 MeV, the GOES-equivalent flux is calculated by fitting a power law to the energy spectra of the STEREO High Energy Telescope (von Rosenvinge et al. 2008), and the >100 MeV flux is calculated integrating the fitted power law over the energy range of 10–150 MeV (see Gopalswamy et al. 2016).

The space speed (V_{space}) of the halo CMEs are estimated using the cone model by Xie et al. (2004). The space speed of non-halo CMEs are estimated by dividing the observed sky-plane speed by $\cos(90° - l)$, where l is the longitude of the CME solar source. We include non-halo CMEs if the source longitude of the non-halo CME is $\geq$ 30°. The initial acceleration (a_{ini}) of the CME is calculated by assuming that CME acceleration occurs during the impulsive phase of the associated flare (Zhang and Dere 2006): $a_{ini} = V_{space}/(t_{peak} - t_{onset})$, where t_{onset} is the onset time and t_{peak} the peak time of the associated soft X-ray flare.

In Table 1, we have listed the number of events and the estimated mean, median and range of the CME space speed and initial acceleration for on-disk SGRE events that had an associated major SEP event observed at 1 au.

In Table 2, we have listed the number of events and the mean, median and range of the CME space speed and initial acceleration for those CMEs that has an observed major SEP event at 1 au without an SGRE event and Fermi-LAT was observing the Sun during the associated on-disk flare. The number of major SEP events in Table 2

Table 1. CMEs with a major SEP event and an on-disk SGRE.

	Count	Mean	Median	Range
Space speed V_{space} (km s^{-1})	28	1890	1816	599-3136
Initial acceleration a_{ini} (km s^{-2})	28	1.80	1.54	0.71-4.54

Table 2. CMEs with a major SEP event but without an SGRE while Fermi/LAT was observing the Sun.

	Count	Mean	Median	Range
Space speed V_{space} (km s^{-1})	8	1414	1366	960-1905
Initial acceleration a_{ini} (km s^{-2})	8	0.77	0.53	0.01-2.88

is significantly smaller than in Table 1 because Fermi/LAT does not observe the Sun continuously. Comparison of Tables 1 and 2 shows that CMEs with an major SEP and SGRE event have higher space speed and initial acceleration than CMEs with an major SEP event but without an SGRE event.

3. Discussion and Conclusions

Gopalswamy et al. (2016) found a hierarchy between the CME initial acceleration and the power-law index of the fluence spectrum: CMEs associated with GLEs accelerate very fast and produce major SEP events with the hardest fluence spectrum. CMEs associated with filament eruption accelerate slowly and produce major SEP events with very soft spectrum. Regular major SEP events fall between the two extreme groups of CMEs. The hierarchy also applies to less-intense regular SEP events (Mäkelä et al. 2019).

In their study, Mäkelä et al. (2023) required that CMEs have an observed metric type II radio bursts (a shock). They found that for SEP events without an SGRE event the CME mean space speed is similar ($\approx$1500 km s^{-1}) but mean initial acceleration higher ($\approx$1.15 km s^{-2}) than in Table 2, i.e., formation of a detectable metric type II burst selects CMEs with slightly higher initial acceleration as expected. Based on our results and the previous studies by Gopalswamy et al. (2016) and Mäkelä et al. (2023), we conclude that CMEs with a major SEP event and SGRE event have very high space speeds and initial acceleration, hence the fluence spectrum of the accelerated protons is very hard, i.e., the shock is very likely to be able to accelerate proton above 300 MeV. Therefore, our results support the idea that CME-shock is producing the >300 MeV protons that are producing the SGRE events.

Ajello et al. (2021) listed two SGRE events that do not have an associated fast CME: the 2012 Oct 23 X1.8-class flare with an estimated SGRE duration of 1.9$\pm$0.5 hours. Although, a possible slow CME with a speed of 224 km s^{-1} was observed at 04:46 UT by The Large Angle Spectroscopic Coronagraph (LASCO; Brueckner et al. 1995) on SOHO. The other M1.6-class flare without any clear mass motion seen in the LASCO images had a SGRE duration of 0.166 $\pm$ 0.025 hours. Recently, Pesce-Rollins et al. (2024) have also suggested that the behind-the-limb SGRE event on September 29, 2022, cannot be due to a CME-driven shock but is due to flare-accelerated particles in large scale loops. Eruption had a CME (416 km s^{-1}) and type II radio burst starting late compared to the gamma-ray emission. However, all these three SGRE events have a short duration. The average SGRE event duration is about 4 hours, but SGRE durations vary significantly and the longest SGRE event observed so far on 2012 Mar 7 lasted about 20 hours. Therefore, in a few short-duration SGRE events without a fast CME, the source of the >300 MeV protons might be a flare.

Our conclusions are that most likely SEPs accelerated in shocks driven by fast accelerating CMEs are the source of SGRE. In a few events that have a very short >100 MeV SGRE duration, the flare-accelerated protons could be the SGRE source.

Acknowledgments. We thank the Fermi/LAT, GOES and SOHO/LASCO teams for providing the data. P.M. and S.A. were partially supported by NSF grant AGS-2043131. N.G. was supported by NASA's STEREO project and the Living With a Star program. H.X. was partially supported by NSF grant AGS-2228967. P.M. likes to thank NASA's Astrophysics Division and Planetary Science Division for providing a travel grant 80NSSC24K0732 supporting attendance of U.S. scientists the symposium.

References

Ajello, M., Baldini, L., Bastieri, D., Bellazzini, R., Berretta, A., Bissaldi, E., Blandford, R. D., Bonino, R., Bruel, P., Buson, S., Cameron, R. A., Caputo, R., Cavazzuti, E., Cheung, C. C., Chiaro, G., Costantin, D., Cutini, S., D'Ammando, F., de Palma, F., Desiante, R., Di Lalla, N., Di Venere, L., Dirirsa, F. F., Fegan, S. J., Fukazawa, Y., Funk, S., Fusco, P., Gargano, F., Gasparrini, D., Giordano, F., Giroletti, M., Green, D., Guiriec, S., Hays, E., Hewitt, J. W., Horan, D., Jóhannesson, G., Kovac'evic', M., Kuss, M., Larsson, S., Latronico, L., Li, J., Longo, F., Lovellette, M. N., Lubrano, P., Maldera, S., Manfreda, A., Martí-Devesa, G., Mazziotta, M. N., Mereu, I., Michelson, P. F., Mizuno, T., Monzani, M. E., Morselli, A., Moskalenko, I. V., Negro, M., Omodei, N., Orienti, M., Orlando, E., Paneque, D., Pei, Z., Persic, M., Pesce-Rollins, M., Petrosian, V., Piron, F., Porter, T. A., Principe, G., Racusin, J. L., Rainò, S., Rando, R., Rani, B., Razzano, M., Razzaque, S., Reimer, A., Reimer, O., Serini, D., Sgrò, C., Siskind, E. J., Spandre, G., Spinelli, P., Tak, D., Troja, E., Valverde, J., Wood, K., & Zaharijas, G. 2021, First Fermi-LAT Solar Flare Catalog. *ApJS*, 252(2), 13.

Atwood, W. B., Abdo, A. A., Ackermann, M., Althouse, W., Anderson, B., Axelsson, M., Baldini, L., Ballet, J., Band, D. L., Barbiellini, G., Bartelt, J., Bastieri, D., Baughman, B. M., Bechtol, K., Bédérède, D., Bellardi, F., Bellazzini, R., Berenji, B., Bignami, G. F., Bisello, D., Bissaldi, E., Blandford, R. D., Bloom, E. D., Bogart, J. R., Bonamente, E., Bonnell, J., Borgland, A. W., Bouvier, A., Bregeon, J., Brez, A., Brigida, M., Bruel, P., Burnett, T. H., Busetto, G., Caliandro, G. A., Cameron, R. A., Caraveo, P. A., Carius, S., Carlson, P., Casandjian, J. M., Cavazzuti, E., Ceccanti, M., Cecchi, C., Charles, E., Chekhtman, A., Cheung, C. C., Chiang, J., Chipaux, R., Cillis, A. N., Ciprini, S., Claus, R., Cohen-Tanugi, J., Condamoor, S., Conrad, J., Corbet, R., Corucci, L., Costamante, L., Cutini, S., Davis, D. S., Decotigny, D., DeKlotz, M., Dermer, C. D., de Angelis, A., Digel, S. W., do Couto e Silva, E., Drell, P. S., Dubois, R., Dumora, D., Edmonds, Y., Fabiani, D., Farnier, C., Favuzzi, C., Flath, D. L., Fleury, P., Focke, W. B., Funk, S., Fusco, P., Gargano, F., Gasparrini, D., Gehrels, N., Gentit, F. X., Germani, S., Giebels, B., Giglietto, N., Giommi, P., Giordano, F., Glanzman, T., Godfrey, G., Grenier, I. A., Grondin, M. H., Grove, J. E., Guillemot, L., Guiriec, S., Haller, G., Harding, A. K., Hart, P. A., Hays, E., Healey, S. E., Hirayama, M., Hjalmarsdotter, L., Horn, R., Hughes, R. E., Jóhannesson, G., Johansson, G., Johnson, A. S., Johnson, R. P., Johnson, T. J., Johnson, W. N., Kamae, T., Katagiri, H., Kataoka, J., Kavelaars, A., Kawai, N., Kelly, H., Kerr, M., Klamra, W., Knödlseder, J., Kocian, M. L., Komin, N., Kuehn, F., Kuss, M., Landriu, D., Latronico, L., Lee, B., Lee, S. H., Lemoine-Goumard, M., Lionetto, A. M., Longo, F., Loparco, F., Lott, B., Lovellette, M. N., Lubrano, P., Madejski, G. M., Makeev, A., Marangelli, B., Massai, M. M., Mazziotta, M. N., McEnery, J. E., Menon, N., Meurer, C., Michelson, P. F., Minuti, M., Mirizzi, N., Mitthumsiri, W., Mizuno, T., Moiseev, A. A., Monte, C., Monzani, M. E., Moretti, E., Morselli, A., Moskalenko, I. V., Murgia, S., Nakamori, T., Nishino, S., Nolan, P. L., Norris, J. P., Nuss, E., Ohno, M., Ohsugi, T., Omodei, N., Orlando, E., Ormes, J. F., Paccagnella, A., Paneque, D., Panetta, J. H., Parent, D., Pearce, M., Pepe, M., Perazzo, A., Pesce-Rollins, M., Picozza, P., Pieri, L., Pinchera, M., Piron, F., Porter, T. A., Poupard, L., Rainò, S., Rando, R., Rapposelli, E., Razzano, M., Reimer, A., Reimer, O., Reposeur, T., Reyes, L. C., Ritz, S., Rochester, L. S., Rodriguez, A. Y., Romani, R. W., Roth, M., Russell, J. J., Ryde, F., Sabatini, S., Sadrozinski, H. F. W., Sanchez, D., Sander, A., Sapozhnikov, L., Parkinson, P. M. S., Scargle, J. D., Schalk, T. L., Scolieri, G., Sgrò, C., Share, G. H., Shaw, M., Shimokawabe, T., Shrader, C., Sierpowska-Bartosik, A., Siskind, E. J., Smith,

D. A., Smith, P. D., Spandre, G., Spinelli, P., Starck, J. L., Stephens, T. E., Strickman, M. S., Strong, A. W., Suson, D. J., Tajima, H., Takahashi, H., Takahashi, T., Tanaka, T., Tenze, A., Tether, S., Thayer, J. B., Thayer, J. G., Thompson, D. J., Tibaldo, L., Tibolla, O., Torres, D. F., Tosti, G., Tramacere, A., Turri, M., Usher, T. L., Vilchez, N., Vitale, V., Wang, P., Watters, K., Winer, B. L., Wood, K. S., Ylinen, T., & Ziegler, M. 2009, The Large Area Telescope on the Fermi Gamma-Ray Space Telescope Mission. *ApJ*, 697(2), 1071–1102.

Brueckner, G. E., Howard, R. A., Koomen, M. J., Korendyke, C. M., Michels, D. J., Moses, J. D., Socker, D. G., Dere, K. P., Lamy, P. L., Llebaria, A., Bout, M. V., Schwenn, R., Simnett, G. M., Bedford, D. K., & Eyles, C. J. 1995, The Large Angle Spectroscopic Coronagraph (LASCO). *SoPh*, 162(1-2), 357–402.

Forrest, D. J., Vestrand, W. T., Chupp, E. L., Rieger, E., Cooper, J. F., & Share, G. H. Neutral Pion Production in Solar Flares. In *19th International Cosmic Ray Conference (ICRC19), Volume 4* 1985, volume 4 of *International Cosmic Ray Conference*, 146.

Gopalswamy, N. 2020, Positron Processes in the Sun. *Atoms*, 8(2), 14.

Gopalswamy, N., Mäkelä, P., Yashiro, S., Lara, A., Akiyama, S., & Xie, H. On the Shock Source of Sustained Gamma-Ray Emission from the Sun. In *Journal of Physics Conference Series* 2019, volume 1332 of *Journal of Physics Conference Series*, 012004.

Gopalswamy, N., Mäkelä, P., Yashiro, S., Lara, A., Xie, H., Akiyama, S., & MacDowall, R. J. 2018, Interplanetary Type II Radio Bursts from Wind/WAVES and Sustained Gamma-Ray Emission from Fermi/LAT: Evidence for Shock Source. *ApJL*, 868(2), L19.

Gopalswamy, N., Xie, H., Yashiro, S., & Usoskin, I. Coronal Mass Ejections and Ground Level Enhancements. In *29th International Cosmic Ray Conference (ICRC29), Volume 1* 2005, volume 1 of *International Cosmic Ray Conference*, 169.

Gopalswamy, N., Yashiro, S., Thakur, N., Mäkelä, P., Xie, H., & Akiyama, S. 2016, The 2012 July 23 Backside Eruption: An Extreme Energetic Particle Event? *ApJ*, 833(2), 216.

Mäkelä, P., Gopalswamy, N., Akiyama, S., Xie, H., & Yashiro, S. 2023, Speed and Acceleration of Coronal Mass Ejections Associated with Sustained Gamma-Ray Emission Events Observed by Fermi/LAT. *ApJ*, 954(1), 79.

Mäkelä, P., Gopalswamy, N., Xie, H., Akiyama, S., Yashiro, S., & Thakur, N. 2019, On the Properties of Solar Energetic Particle Events Associated with Metric Type II Radio Bursts. *Sun and Geosphere*, 14, 123–129.

Pesce-Rollins, M., Klein, K.-L., Krucker, S., Warmuth, A., Veronig, A. M., Omodei, N., & Monstein, C. 2024, Evidence for flare-accelerated particles in large scale loops in the behind-the-limb gamma-ray solar flare of September 29, 2022. *A&A*, 683, A208.

Pesce-Rollins, M., Omodei, N., Petrosian, V., Liu, W., Rubio da Costa, F., Allafort, A., & Chen, Q. 2015, First Detection of &gt100 MeV Gamma Rays Associated with a Behind-the-limb Solar Flare. *ApJL*, 805(2), L15.

Plotnikov, I., Rouillard, A. P., & Share, G. H. 2017, The magnetic connectivity of coronal shocks from behind-the-limb flares to the visible solar surface during γ-ray events. *A&A*, 608, A43.

Ryan, J. M. 2000, Long-Duration Solar Gamma-Ray Flares. *SSRv*, 93, 581–610.

Share, G. H., Murphy, R. J., White, S. M., Tolbert, A. K., Dennis, B. R., Schwartz, R. A., Smart, D. F., & Shea, M. A. 2018, Characteristics of Late-phase ¿100 MeV Gamma-Ray Emission in Solar Eruptive Events. *ApJ*, 869(2), 182.

von Rosenvinge, T. T., Reames, D. V., Baker, R., Hawk, J., Nolan, J. T., Ryan, L., Shuman, S., Wortman, K. A., Mewaldt, R. A., Cummings, A. C., Cook, W. R., Labrador, A. W., Leske, R. A., & Wiedenbeck, M. E. 2008, The High Energy Telescope for STEREO. *SSRv*, 136(1-4), 391–435.

Xie, H., Ofman, L., & Lawrence, G. 2004, Cone model for halo CMEs: Application to space weather forecasting. *JGR*, 109(A3), A03109.

Zhang, J. & Dere, K. P. 2006, A Statistical Study of Main and Residual Accelerations of Coronal Mass Ejections. *ApJ*, 649(2), 1100–1109.

Solar and Stellar Coronal Mass Ejections
Proceedings IAU Symposium No. 388, 2026
N. Gopalswamy, O. Malandraki, A. Vidotto & W. Manchester, eds.
doi:10.1017/S1743921324001182

Stellar energetic particle and cosmic ray effects in exoplanetary atmospheres

D. Rodgers-Lee

Dublin Institute for Advanced Studies

Abstract. Energetic particles, in the form of stellar energetic particles and cosmic rays, can lead to disequilibrium chemical effects in exoplanetary atmospheres. In Earth-like atmospheres, energetic particles can drive the formation of prebiotic molecules, the building blocks of life. Here instead, I study the transport of energetic particles through a hydrogen-dominated exoplanet atmosphere and calculate the resulting ionisation rate of molecular hydrogen using a Monte Carlo energetic particle transport model. I focus on a GJ436 b-like atmosphere at orbital distances between 0.01-0.2 au which includes the orbital distance of the exoplanet GJ436 b (0.028 au). I found that stellar energetic particles lead to high ionisation rates in a GJ436 b-like atmosphere between 0.01-0.2 au. These results motivate the use of chemical models of gas giant atmospheres including energetic particle ionisation to ultimately produce synthetic James Webb Space Telescope (JWST) and Ariel transmission spectra in the future.

Keywords. Stellar energetic particles, Galactic cosmic rays, exoplanet atmospheres

1. Introduction

There are two types of energetic particles that are important for exoplanet atmospheres. These are stellar energetic particles from the coronal mass ejections and flares of low-mass stars like the Sun and Galactic cosmic rays that enter an astrosphere† from the interstellar medium (ISM). These two types of energetic particles then reach the location of an exoplanet. Here, they encounter any existing exoplanetary magnetosphere and interact with the atmosphere of the exoplanet.

Energetic particles are of interest in the context of exoplanet atmospheres because they can cause a number of interesting effects. They can drive the formation of prebiotic molecules in Earth-like atmospheres (Airapetian *et al.* 2016; Dong *et al.* 2019) which are the molecules thought to be important for the origin of life on Earth. On the other hand, they can also cause fake biosignatures in Earth-like atmospheres (Grenfell *et al.* 2012). A biosignature is a chemical signature in an exoplanet atmosphere thought to be indicative of life. Thus, since energetic particles can also contribute to fake biosignatures it is important to constrain their effect so that it can be removed when searching for real biosignatures. Last, to isolate the chemical effect of energetic particles, Helling & Rimmer (2019) and Barth *et al.* (2021) found that energetic particles can lead to the formation of exotic molecules in hydrogen-dominated atmospheres that are not expected to form otherwise which are known as fingerprint ions.

More broadly, studying the effect of energetic particles is important because high fluxes of energetic particles affect life-forms by damaging DNA (Herbst *et al.* 2019; Atri 2020). It has even been suggested that energetic particles (in this case Galactic cosmic rays

† An astrosphere is the stellar equivalent of the Sun's heliosphere.

but the same logic applies to stellar energetic particles) indirectly left an imprint on the helicity of DNA (Globus & Blandford 2020).

2. Methodology

Here, I will focus on the impact of energetic particles in the GJ436 system. GJ436 is a well-studied M dwarf star with a known close-in planet, GJ436 b orbiting at 0.028au (Butler *et al.* 2004). GJ436 b is a warm mini-Neptune (with a calculated effective temperature of $T = 880$K, Rodgers-Lee *et al.* 2023) and is an interesting exoplanet because it has a cometary-like outflow detected via a Ly-α transit with the Hubble Space Telescope (Ehrenreich *et al.* 2015). While these hydrogen-dominated gas giant atmospheres are less likely to be hospitable to any life-as-we-know-it in comparison to an Earth-like atmosphere, they do have an advantage: gas giant atmospheres constitute the majority of JWST and Ariel exoplanet targets. Their low mean molecular weight results in a large scale height, making them easier to observe with transmission spectroscopy than an Earth-like atmosphere with a higher mean molecular weight.

In order to model the transport of energetic particles in the GJ436 system a stellar wind model and a model for the exoplanetary atmosphere are required. In Rodgers-Lee *et al.* (2023) we use a stellar wind model from Mesquita *et al.* (2020). The stellar wind properties necessary to model Galactic cosmic ray transport from the ISM into an astrosphere are the stellar wind velocity, magnetic field strength and mass loss rate. The level of turbulence present in the magnetic field, rather than the field strength itself, dictates the energetic particle transport through the magnetised stellar wind. The turbulence properties for winds other than the solar wind are unknown and we simply adopt similar properties as for the solar wind. See Rodgers-Lee *et al.* (2023) for more details.

2.1. *Energetic particle spectra at the top of the exoplanet atmosphere*

The spectrum for Galactic cosmic rays in the local ISM (known as the Local Interstellar Spectrum), outside of the solar system, has been measured by *Voyager 1 and 2* (e.g. Cummings *et al.* 2016). The Galactic cosmic ray spectrum is then suppressed by the solar wind, mainly at cosmic ray energies below $\sim$30 GeV so that a somewhat reduced spectrum is detected at Earth with instruments such as *PAMELA* (e.g. Potgieter & Vos 2017). Thus, we calibrated our 1D energetic particle transport model (Rodgers-Lee *et al.* 2020) using the *Voyager* and *PAMELA* detections. The LIS is used as the input spectrum outside the astrosphere of another stellar system. The energetic particle transport model from Rodgers-Lee *et al.* (2020), based on the diffusive-advection transport equation from Parker (1965), is used in combination with the stellar wind properties (Case 'A' presented in Mesquita *et al.* 2020) to calculate the Galactic cosmic ray fluxes at different orbital distances within the GJ436 b system. These fluxes at the different orbital distances are the values we adopt as the values at the top of the hypothetical exoplanet atmospheres.

The Galactic cosmic ray fluxes are suppressed as they travel through the magnetised stellar wind because they are charged particles. The solar wind, and therefore we assume stellar winds from other low-mass stars, are turbulent. This means that Galactic cosmic rays diffuse into the solar, or stellar, system as they are deflected and bounce off perturbations in the magnetic field rather than travelling ballistically. The stellar wind is expanding, relating to its velocity, which advects the Galactic cosmic rays out of the solar system. The Galactic cosmic rays suffer adiabatic losses due to the divergence of the stellar wind's velocity field, which also can be thought of as advection in momentum space. The overall level of suppression that the Galactic cosmic rays experience in any given stellar system is a balance between these diffusive and advective processes.

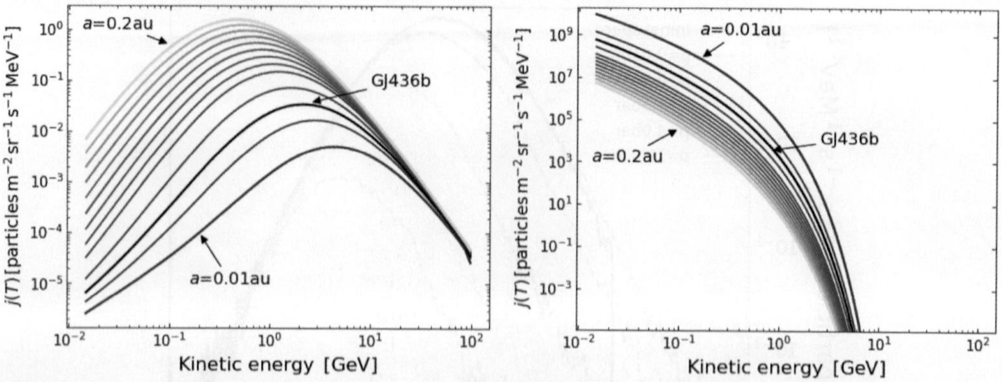

Figure 1. Differential intensities of *(left)* Galactic cosmic rays and *(right)* stellar energetic particles as a function of cosmic ray kinetic energy at orbital distances between 0.01 - 0.2 au for the GJ436 system (from Rodgers-Lee *et al.* 2023).

For GJ436b, the resulting Galactic cosmic ray differential intensities at orbital distances between 0.01 - 0.2 au are shown in the left panel of Fig. 1. The Galactic cosmic ray fluxes increase with increasing orbital distance. The differential intensity at the orbital distance of GJ436 b, shown by the black line, is lower than the differential intensities observed at Earth.

The right panel of Fig. 1 shows the stellar energetic particle differential intensities for the same orbital distances. Again, the black line represents the differential intensities at the orbital distance of GJ436 b. It is important to note that the scale on the left and right panels is significantly different. An interesting feature of these spectra is that the Galactic cosmic ray spectra peak at ~GeV energies. In comparison for the stellar energetic particles, the intensities are much higher generally but they begin to decrease rapidly at ~Gev energies. The stellar energetic particle spectra are produced based on 3 assumptions. The first is that the acceleration mechanism produces a power-law index of -2 for the low-energy part of the spectrum (indicative of diffusive shock acceleration). There is a maximum momentum that the star can accelerate stellar energetic particles to, beyond this momentum the differential intensities decrease exponentially. The final assumption is that the stellar energetic particles are constant in time.

The exoplanet atmosphere properties (e.g. T, p) in Rodgers-Lee *et al.* (2023) were obtained using the HELIOS model (Malik *et al.* 2017, 2019). The temperature-pressure profiles are shown in Fig. 3 of Rodgers-Lee *et al.* (2023). The density is the most important atmospheric property to model energetic particle transport in exoplanet atmospheres which was calculated using the ideal gas law. In Rodgers-Lee *et al.* (2023), for the energetic particle transport we assume a hydrogen-dominated atmosphere.

3. Analysis and results

Here, I show our results for the penetration of cosmic rays in a GJ436 b-like atmosphere. Each of the spectra shown in Fig. 1 is propagated down through an exoplanet atmosphere using a Monte Carlo cosmic ray transport code (Rimmer & Helling 2013). For example, the solid black line in Fig. 2 shows the initial Galactic cosmic ray spectrum at the top of the atmosphere for $a = 0.2$au. The dashed coloured lines show the subsequent spectra for increasing pressures in the atmosphere as the Galactic cosmic rays lose energy by interacting with the atmosphere. The resulting spectra for the Galactic cosmic rays and the stellar energetic particles is then used to calculate the ionisation rate of molecular hydrogen at each height in the atmosphere.

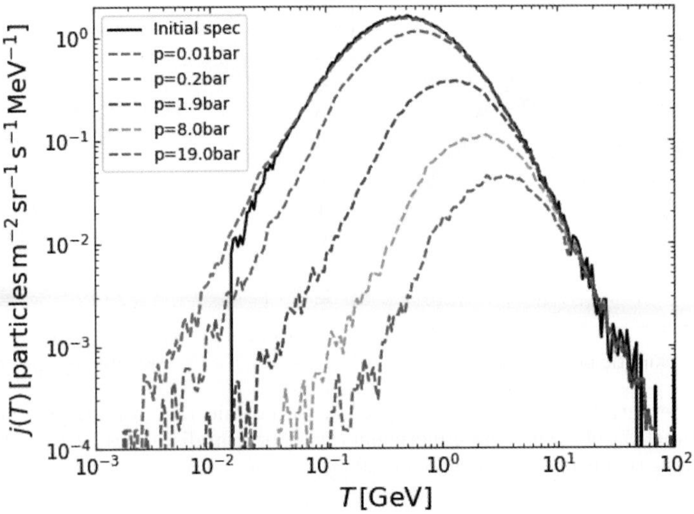

Figure 2. Differential intensities of Galactic cosmic rays as a function of cosmic ray kinetic energy at different heights in an exoplanet atmosphere for a GJ436 b-like exoplanet at $a = 0.2$ au in the GJ436 system (from Rodgers-Lee *et al.* 2023).

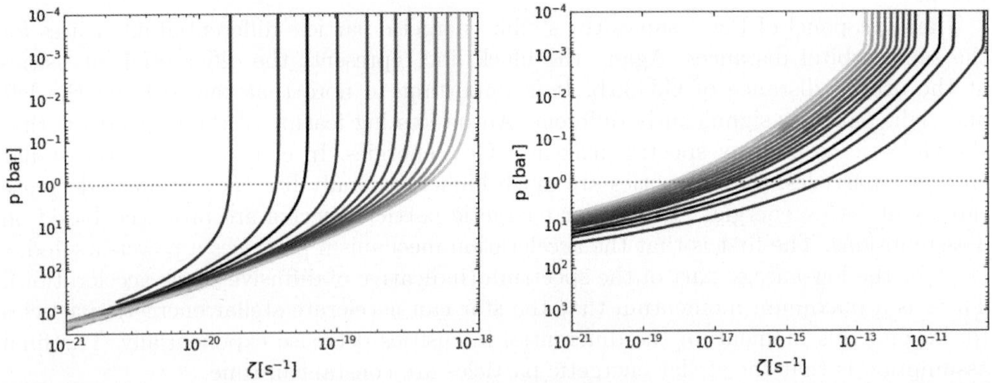

Figure 3. The ionisation rate of molecular hydrogen due to *(left)* Galactic cosmic rays and *(right)* stellar energetic particles as a function of atmospheric pressure for a GJ436 b-like planet at orbital distances between 0.01 - 0.2 au for the GJ436 system (adapted from Rodgers-Lee *et al.* 2023).

3.1. *Ionisation rate of molecular hydrogen*

The ionisation rate of molecular hydrogen, ζ, due to energetic particles gives an indication of how important, and where, they might be at driving chemistry in the exoplanet atmosphere. The H_2 ionisation rate is calculated from the differential intensities and the H_2 ionisation cross-section (Eq.4 from Rodgers-Lee *et al.* 2023).

Fig. 3 shows the ionisation rate of molecular hydrogen, ζ, as a function of pressure for a GJ436 b-like planet for $a = 0.01 - 0.2$ au (adapted from Rodgers-Lee *et al.* 2023). The left and right panels plot ζ as a result of ionisation due to Galactic cosmic rays and stellar energetic particles, respectively. The black solid line represents ζ in an exoplanet atmosphere at 0.028 au, the orbital distance of GJ436 b. The dotted grey line denotes $P = 1$ bar for reference. For comparison, the LIS results in $\zeta \sim 10^{-17} s^{-1}$.

Additionally, for both Galactic cosmic rays and stellar energetic particles, the value of ζ remains constant with increasing pressure up to a certain point. In this region of the atmosphere the energetic particles are not losing a significant amount of their energy. For Galactic cosmic rays, ζ begins to decrease between $p \sim 10^1 - 10^{-1}$bar for $a = 0.01 - 0.2$au, respectively. The stellar energetic particles are absorbed at much higher heights in the atmosphere, corresponding to $p \sim 10^{-2}$ bar. Thus, stellar energetic particles dominate as a source of ionisation over Galactic cosmic rays in the upper atmosphere. At high pressures, $p \gtrsim 1$bar, although overall the ionisation rate is low Galactic cosmic rays dominate instead. Transmission spectroscopy with JWST and Ariel will probe between $10^{-4} - 10^{-1}$bar in hydrogen dominated atmospheres (Welbanks & Madhusudhan 2019). Thus, transmission spectroscopy with JWST and Ariel is more likely to be sensitive to chemistry driven by stellar energetic particles for the atmospheres and orbital distances studied here.

4. Discussion and Summary

I presented our results of the ionising effect of stellar energetic particle and Galactic cosmic rays on a hydrogen-dominated exoplanet atmosphere for $a = 0.01 - 0.2$au in the M dwarf system, GJ436. This includes the orbital distance of the known mini-Neptune, GJ436 b. I discussed how the stellar wind properties, along with the composition and density of an exoplanet atmosphere determine the energetic particle fluxes in the exoplanet atmosphere. Stellar energetic particles lead to high ionisation rates in a GJ436 b-like atmosphere at the orbital distances discussed here, namely between 0.01-0.2 au. The Galactic cosmic ray ionisation rates for a hypothetical exoplanet at these orbital distances were generally much lower but dominate over stellar energetic particles deep in the exoplanet atmosphere.

The results presented here did not include the effect of a planetary magnetic field. A planetary magnetic field would shield the atmosphere by deflecting both stellar energetic particles and Galactic cosmic rays, up to a certain energy, towards the poles (see e.g. Herbst *et al.* 2019). Investigating the effect of a planetary magnetic is of great interest. As mentioned above, the stellar energetic particle fluxes are assumed to be constant here. In reality, stellar flares and CMEs are unlikely to produce constant fluxes of associated stellar energetic particles that impact an exoplanet between $a = 0.01 - 0.2$au for this system, given that it is not an active star. It is not clear what impact this would have on chemistry in the atmosphere at this point. Chemical modelling of an exoplanet atmosphere including energetic particle ionisation are needed to address this time dependent effect.

Using chemical models, Helling & Rimmer (2019) identified H_3^+ and H_3O^+ as ions indicative of Galactic cosmic ray ionisation in a brown dwarf/gas giant atmosphere. The important pathway for creating these ions is:

$$H_2 + CR \rightarrow H_2^+ + e^- + CR \tag{1}$$

$$H_2^+ + H_2 \rightarrow H_3^+ + H \tag{2}$$

$$H_3^+ + H_2O \rightarrow H_3O^+ + H_2 \tag{3}$$

where 'CR' is short for cosmic ray. Stellar energetic particles will produce similar effects (e.g. Barth *et al.* 2021). Note, the presence of water is important. Scaling the abundances of H_3^+ and H_3O^+ from Helling & Rimmer (2019), Bourgalais *et al.* (2020) produced synthetic JWST and Ariel transmission spectra for a GJ1214 b-like atmosphere, showing that absorption features from H_3^+ and H_3O^+ in the near-infrared would be strong enough to be observed with Ariel and NIRSpec on JWST (see their Fig. 6). This remains to date the only synthetic JWST and Ariel transmission spectrum for a hydrogen-dominated atmosphere that includes H_3^+ and H_3O^+, formed as a result of energetic particle ionisation.

The stellar energetic particle and Galactic cosmic ray spectra shown here can be used in the future with chemical and radiative transfer models to produce a suite of synthetic transmission spectra, building on the results presented in Bourgalais *et al.* (2020). This would identify the energetic particle ionisation rate needed to result in sufficient abundances of H_3^+ and H_3O^+ to be detected with transmission spectroscopy. In turn this would help us identify the types of exoplanets to target, in terms of orbital distance and stellar activity, to detect these features. Alongside this more direct connection with observations, more sophisticated 2D/3D energetic particle transport models can be employed in the future (see e.g. Engelbrecht *et al.* 2024).

References

Airapetian V. S., Glocer A., Gronoff G., *et al.* 2016, Nature Geoscience, 9, 452
Atri D., 2020, *Scientific Reports*, 1011646
Barth P., Helling C. Stüeken E. E., *et al.* 2021, *MNRAS*, 502, 6201
Bourgalais J., et al., 2020, *ApJ*, 895, 77
Butler R. P., Vogt S. S., Marcy G. W., *et al.* 2004, *ApJ*, 617, 580
Cummings A. C. *et al.* 2016, *ApJ*, 831, 18
Dong, C. F., Lingam, M., Fang, X. H. *et al.* 2019, *The First Billion Years: Habitability*, Conference Proceedings, 2134, 1046
Engelbrecht, N. Eugene, Herbst, K., Strauss, R. Du Toit *et al.* 2024, *ApJ*, 964, 89
Ehrenreich, D., Bourrier, V., Wheatley, P. J. *et al.* 2015, *Nature*, 522, 459
Globus N., Blandford R. D., 2020, *ApJ*, 895, L11
Grenfell, J. L., Grießmeier, J.-M., von Paris, P., et al. 2012, *Astrobiology*, 12, 1109
Helling C., Rimmer P. B., 2019, *Philosophical Transactions of the Royal Society of London Series A*, 377, 20180398
Herbst, K., Grenfell, J. L., Sinnhuber, M. *et al.* 2019, *A&A*, 631, 101
Malik M., Grosheintz L., Mendonça, J. M., *et al.* 2017, *AJ*, 153, 56
Malik M., Kitzmann D., Mendonça, J. M., *et al.* 2019, *AJ*, 157, 170
Mesquita A. L., Rodgers-Lee D., Vidotto A. A., 2021, *MNRAS*, 505, 1817
Parker E. N., 1965, *Planetary and Space Science*, 13, 9
Potgieter, M. S., Vos, E. E., *A&A*, 601, 23
Rimmer P. B., Helling C., 2013, *ApJ*, 774, 108
Rodgers-Lee D., Vidotto A. A., Taylor A. M., *et al.* 2020, *MNRAS*, 499, 2124
Rodgers-Lee, D., Rimmer, P. B., Vidotto, A. A., *et al.* 2023, *MNRAS*, 521, 4
Welbanks L., Madhusudhan N., 2019, *AJ*, 157, 206

Solar and Stellar Coronal Mass Ejections
Proceedings IAU Symposium No. 388, 2026
N. Gopalswamy, O. Malandraki, A. Vidotto & W. Manchester, eds.
doi:10.1017/S1743921324001169

Understanding Coronal Mass Ejections in Magnetically-Active Stars

Julián D. Alvarado-Gómez

Leibniz Institute for Astrophysics Potsdam, An der Sternwarte 16, 14483, Potsdam, Germany
julian.alvarado-gomez@aip.de

Abstract. While Coronal Mass Ejections (CMEs) are known to be the most energetic manifestation of magnetic activity on the Sun, their stellar counterparts have remained, for the most part, elusive to our detection attempts. This situation, which is in stark contrast with the significantly enhanced observed level of flaring activity of stars, questions our understanding of the extension of the solar flare-CME connection to the stellar domain. In this presentation, I have emphasized the crucial role played by the stellar magnetic field in shaping the properties of CMEs in active stars. With the aid of state-of-the-art numerical simulations, we have explored the process of magnetic confinement of CMEs by a strong large-scale stellar magnetic field. This exploration includes multiple consequences ranging from the existence of the eruption itself, to different observational signatures emerging as this process takes place in the stellar atmosphere. These results are placed in context with our current observational knowledge of stellar CMEs and avenues are given to continue improving our understanding of these transient events on other stars

Keywords. Magnetohydrodynamics (MHD), stars: activity, stars: coronae, stars: flare, stars: winds, outflows, Sun: coronal mass ejections (CMEs), Sun: flares

1. Introduction

Flares and coronal mass ejections (CMEs) are the solar system's most energetic phenomena. Flares manifest as sudden radiation bursts emerging from the solar atmosphere across the electromagnetic spectrum, while CMEs entail a localized discharge of magnetized plasma into interplanetary space. Their generation typically involves a sequence of processes intricately linked to coronal magnetic field reorganization and reconnection (Chen 2011; Shibata & Magara 2011; Webb & Howard 2012; Benz 2017; Fan 2021). Recent studies have consistently shown that the most intense solar flares, particularly those classified as X-class in the GOES scale, are almost always accompanied by a CME (Andrews 2003; Yashiro & Gopalswamy 2009; Compagnino et al. 2017), and that CME mass and kinetic energy are positively correlated with the X-ray fluence of the flare (Lamy et al. 2019). Furthermore, it is now evident that CMEs carry a significantly larger portion of the available energy compared to X-ray/EUV flares (Emslie et al. 2005, 2012). Nonetheless, when accounting for thermal and non-thermal emissions across various wavelengths such as white light/UV continuum, the combined energies of flares and CMEs seem to be comparable (Kretzschmar et al. 2010; Kretzschmar 2011; Emslie et al. 2012; Aschwanden et al. 2017).

Continuous solar monitoring has spurred substantial advancements in the field, as has been clearly demonstrated by the talks during this conference. In contrast, observations suggest that the flare-CME relationship in stellar cases may deviate from its solar counterpart. Evidence of this comes in different forms. On one hand, the coronae of very active stars (particularly M-dwarfs) exhibit a seemingly perpetual flaring activity (see Kashyap et al. 2002, Huenemoerder et al. 2010; Kowalski 2024), often described by a superposition of flares in their light curves (e.g., Audard et al. 2000; Caramazza et al. 2007; Tristan et al. 2023). This heightened flare activity potentially implies a corresponding prevalence of stellar CMEs, a notion explored by various authors in the context of stellar evolution and environmental conditions surrounding stellar systems (e.g., Lammer et al. 2007; Khodachenko 2007; Cohen et al. 2011; Aarnio et al. 2012; Drake 2013; Schnepf et al. 2015; Takahashi et al. 2016; Cranmer 2017, Kay et al. 2019; Cherenkov et al. 2017). Notably, flares in highly active stars can be orders of magnitude more energetic than solar flares and occur with much greater frequency (e.g., Osten et al. 2007; Kowalski et al. 2010; Shibayama et al. 2013; Osten 2016; Davenport et al. 2019; Kowalski 2024).

On the other hand, as extensively discussed by Drake 2013, extrapolating the solar flare-CME relationships to active stars leads to problematic outcomes. One significant issue is the prediction of a substantial contribution from CMEs to the stellar mass loss rate ($\dot{M}$), resulting in $\dot{M}$ values that are up to four orders of magnitude higher than the accepted solar value (attributed to the solar wind) and approximately two orders of magnitude above the highest $\dot{M}$ estimate derived from astrospheric Lyman-α absorption (Wood et al. 2002, 2005, 2014, 2021). (Osten & Wolk 2015) achieved comparably large $\dot{M}$ predictions by extending flare-CME relationships to cover the various energy bandpasses and flare frequencies observed in solar-type and low-mass stars. Cranmer (2017), using a slightly different method, considered a solar-based relationship between surface-averaged magnetic fluxes and the contribution of CMEs to the total stellar mass loss. According to this model, CMEs would significantly dominate the mass loss budget in young stars (a few Myr old), by factors of 10 to 100, surpassing the contribution from the steady stellar wind even in later stages (younger than 1 Gyr). Odert (2017) introduced an empirical model (also incorporating solar extrapolations), which shows relatively good agreement on the $\dot{M}$ for moderately active stars (i.e., surface X-ray fluxes $F_{\rm X} < 10^6$ erg cm^{-2} s^{-1}), but still predicts large values at the high-activity end. Still, all of these approaches would eventually lead to CME-driven mass loss rates that are too large to be reconciled with the required radio transparency of the wind (see the discussion in Lim & White 1996). Note also that the recent study of Kislyakova et al. (2024) reports relatively small mass loss rate values for 3 cool stellar systems (2 binaries, one single star) using an independent methodology based on the stellar wind-ISM charge-exchange X-ray signature. As such, it must be concluded that our current knowledge on stellar winds from cool main-sequence stars does not support the expectation of high contributions from stellar CMEs based on solar extrapolations.

Drake (2013) also pointed out that maintaining solar-like flare-driven CME activity would require implausibly high energy requirements, reaching up to ten percent of the stellar bolometric luminosity for active stars in the saturated regime. They concluded that either the relationships between CME mass/speed and flare energy must flatten for X-ray energies $\gtrsim 10^{31}$ erg, or the flare-CME association rate must significantly decrease below 1 for more energetic events (contrary to the solar case).

To exacerbate these issues and contrary to the "high CME activity expectation", solid evidence for stellar CMEs remains scarce and challenging to obtain. A few studies have noted blue-wing enhancements in the Balmer lines of M dwarfs, interpreted as eruptive stellar filaments (e.g., Houdebine et al. 1990; Gunn et al. 1994; Vida et al. 2016). Guenther & Emerson (1997) identified a distinct Hα blue-wing enhancement in the spectrum of the weak T Tauri star DZ Cha (RX J1149.8–7850), attributing it to a CME and inferring an ejected mass in the range of 10^{18} – 10^{19} g (larger than the most massive solar CMEs). By characterizing the column density absorption profile evolution after a large flare on Algol, Moschou et al. (2017) deduced an even larger ejected mass in the range of $10^{21} - 10^{22}$ g (see also Favata & Schmitt 1999). Still, several additional attempts to detect stellar CMEs using radio and optical data have yielded no significant signatures, even in dedicated observations (e.g., Leitzinger et al. 2014; Crosley et al. 2016; Villadsen 2017; Crosley and Osten 2018, Muheki et al. 2020a, 2020b).

In a recent development, Veronig (2021) documented a series of 21 coronal dimmings across 13 active late-type stars, interpreting them as indirect signatures of CMEs. The phenomenon of coronal dimmings associated with CMEs has been a consistent observation on the Sun for over two decades, particularly at EUV and soft X-ray wavelengths (Hudson et al. 1996; Sterling & Hudson 1997; Zarro et al. 1999). These dimmings are observable across various coronal temperatures but are most conspicuous in the $1 - 2$ million Kelvin plasma of the quiet corona (Vanninathan et al. 2018; Dissauer et al. 2018). They typically manifest at the bipolar ends of closed magnetic field lines, which become stretched or temporarily opened during an eruption, resulting in the depletion of coronal plasma due to expansion and mass loss caused by the CME (Thompson et al. 1998; Harra et al. 2016). While the X-ray dimming signatures reported by Veronig (2021) dramatically exceeded the nominal values observed on the Sun (by more than one order of magnitude), this study did not provide estimates for the physical properties of these potentially extreme events. Based on the same dimming phenomenology, Loyd et al. (2022) recently derived a upper limits of mass and kinetic energy of a few 10^{15} g and 10^{31} erg respectively for a candidate CME event observed after a flare in the K2V star ϵ Eridani.

To date, the most compelling stellar CME candidates come from the research conducted by Argiroffi et al. (2019) and Namekata et al. (2021). In the former, blue-shifted cool coronal emission following a flare on the giant star HR 9024 (G1III) was reported. The estimated properties for this event ($M_{\mathrm{CME}} \sim 10^{21}$ g, $K_{\mathrm{CME}} \sim 10^{34}$ erg) surpass the typical parameter space observed in solar eruptions. In this case, the CME developed under magnetic and coronal conditions comparable to those expected for active Sun-like stars (Borisova et al. 2016) but it was hosted by a star with significantly reduced surface gravity. In the latter study, a blue-shifted hydrogen absorption component with a high velocity of -510 km/s was observed shortly after a large flare on the young Sun-like star EK Dra (G5V). The temporal changes in the spectra closely resembled those of solar filament eruptions, strongly suggesting a stellar CME event. The erupted filament mass of $\sim 10^{18}$ g was notably larger than those of the largest solar coronal mass ejections.

In a prior study, Moschou (2019) reexamined the most promising stellar flare-CME candidates documented in the literature, obtaining constraints on CME parameters such as mass and kinetic energy (see Fig. 1). Their investigation encompassed events identified through sustained X-ray absorption subsequent to a stellar flare (e.g., Favata & Schmitt 1999; Moschou et al. 2017), as well as potential candidates inferred from post-flare Doppler-shift signatures observed in optical or UV spectral lines (e.g., Houdebine et al. 1990; Leitzinger et al. 2011). The compilation presented by Moschou (2019) offered crucial insights into the behavior of the flare-CME relationship within the stellar context: While the estimated masses of the CMEs generally aligned with a solar extrapolation based on the observed X-ray energy of the flares, the kinetic energy associated with each eruption fell notably below expectations. As depicted in Fig. 1, the characteristics of the CME events deduced by Argiroffi et al. (2019) and Namekata et al. (2021) likewise adhere to this overarching trend.

2. Suppression of CMEs: The role of the large-scale magnetic field

One mechanism that may decrease the association between solar flares and CMEs in active stars, thus circumventing the energy/mass loss issues highlighted by Drake

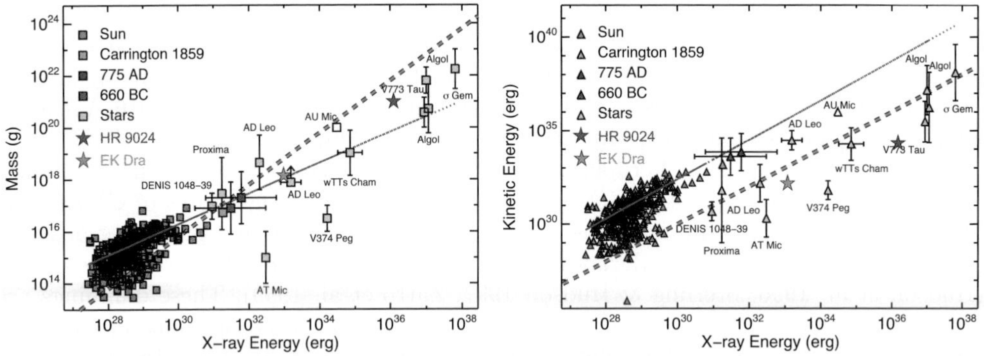

Figure 1. Current observational constraints on the flare-CME connection in cool stars. Derived CME masses (left) and kinetic energies (right) are plotted against the X-ray radiated energy of the associated flare. Recent (gray) and historical (yellow, blue, purple) solar events are shown alongside stellar candidate eruptions (green). The star symbols correspond to the individual events currently regarded as stellar CME detections, observed in the stars HR 9024 (G1III, red) and EK Dra (G1.5V, cyan). Red lines represent fits and extrapolations of the solar data. Gray dashed lines indicate a constant ratio of CME mass to flare X-ray energy (left) and parity between flare X-ray and CME kinetic energies (right). Note that while the stellar CME masses appear roughly aligned with the solar extrapolation, the kinetic energy of those events is significantly lower than the expectation (by several orders of magnitude). **Adapted from:** Moschou (2019), Argiroffi et al. (2019) and Namekata et al. (2021).

(2013), involves the suppression of plasma ejection by a robust overlying magnetic field (Drake et al. 2016, also discussed by Odert 2017). This theory has been proposed to elucidate the behavior of specific active regions on the Sun that are rich in flares but devoid of CMEs (e.g., Thalmann et al. 2015, 2017; Sun et al. 2015; Liu et al. 2016). In the stellar regime, spectropolarimetry and Zeeman-Doppler Imaging (ZDI) have uncovered the widespread presence of magnetic fields in cool main-sequence stars (Donati & Landstreet 2009; Reiners 2012; Kochukhov 2021). ZDI, which is sensitive to large-scale magnetic components, reveals that active stars possess surface field strengths significantly greater than those on the Sun (e.g., Morgenthaler et al. 2012; Jeffers et al. 2017; See et al. 2019). These variations correspond to photospheric large-scale magnetic fields ranging from 10 G to several 10^2 G in young Sun-like stars (e.g., Alvarado-Gómez et al. 2015; Hussain et al. 2016; Folsom et al. 2016, 2018), and reaching up to a few 10^3 G in M-dwarfs (e.g., Morin et al. 2008; Morin et al. 2010; Lavail et al. 2018). Similar average unsigned magnetic field strengths have also been reported from Zeeman Broadening/Intensification techniques applied to Sun-like stars (Kochukhov et al. 2020) and M-dwarfs (Reiners et al. 2022).

The concept of magnetic suppression of stellar CMEs was briefly explored in a magnetohydrodynamic (MHD) simulation by Drake et al. (2016), where a large solar CME failed to erupt when subjected to the large-scale magnetic field configuration akin to that of a much more active star (AB Doradus; Hussain et al. 2002). Our recent work, which integrates information on stellar magnetism with numerical simulations, has been pivotal in understanding the disparity between flare and CME occurrence in stars and interpreting the limited available data. Employing realistic 3D MHD models, we demonstrated that the stellar large-scale magnetic field can create a suppression threshold that prevents CMEs of certain energies from escaping (Fig. 2; see also Alvarado-Gómez 2018; Alvarado-Gómez et al. 2022). Our analysis also showed that the overlying magnetic field significantly reduces the overall speeds of CMEs (and consequently their associated kinetic energies), which contrasts with extrapolations from solar data. Conversely, the

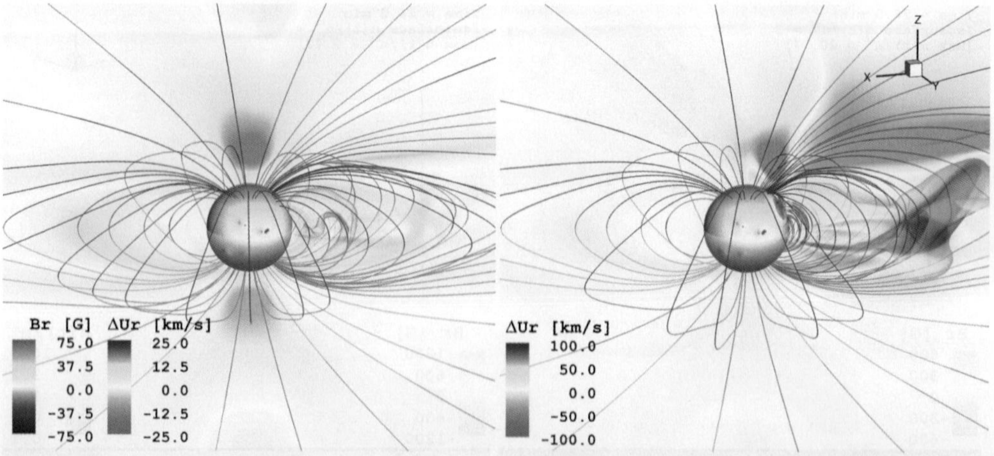

Figure 2. Examples of a confined (left) and escaping (right) CME evolving within a solar magnetic field configuration modified to host a stronger large-scale dipolar field (75 G). The escaping CME is driven by a flux-rope eruption containing ~ 15 times more magnetic energy compared to the confined case. The central sphere represents the stellar surface, colored by the magnetic field that drives the model. A secondary color scale indicates a proxy (ΔUr) for the coronal Doppler shift velocity relative to pre-CME conditions. The detected eruption is shown as a translucent yellow iso-surface ($\times 3$ density enhancement). Selected magnetic field lines related to the large-scale field (gray) and surrounding the eruptive active region (magenta) are shown. The field of view in both panels spans 12 $R_{\star}$. **Adapted from:** Alvarado-Gómez (2018).

total mass perturbed in the simulated events approximately followed the solar flare-CME relationship extended to the stellar context. Notably, these two predictions are consistent with the properties of the stellar CME candidates observed to date (see Fig. 1).

3. Exploring the stellar CME confinement and its observational signatures: The case of M-dwarf stars

As previously noted, M-dwarfs are characterized by frequent and energetic flaring (e.g., Howard et al. 2018; Günther et al. 2020), persisting even at relatively old ages in their evolutionary stages (France et al. 2020), and occurring even when they appear "quiet" in optical data (Loyd et al. 2020). Their surface magnetic fields, reaching up to a few kG, are the strongest observed among cool main-sequence stars (Reiners 2012; Kochukhov 2021). Importantly, these stars are primary targets for the atmospheric characterization of many current and future exoplanet missions. Due to their lower luminosity, the habitable zone —where an exoplanet could maintain surface liquid water— is situated very close to the star, sometimes up to ten or more times closer than the distance between the Sun and Earth (Kopparapu et al. 2014; Shields et al. 2016). These close-in orbits have short periods, ranging from a few days to a couple of weeks, which facilitates the detection and follow-up of these systems.

Numerous factors then underscore the importance of numerically investigating CMEs in M-dwarfs, including their potential confinement, which holds significant relevance for both stellar and exoplanetary studies. Such simulations can offer valuable insights and constraints for ongoing and future observational efforts targeting these energetic phenomena. In our work (Alvarado-Gómez et al. 2019, 2020), we conducted the first realistic modelling of CME events in M dwarf stars. Our findings revealed marked disparities in how the corona and stellar wind respond to these eruptions under magnetic confinement, contrasting with their solar counterparts (see Fig. 3). Depending on the CME's energy

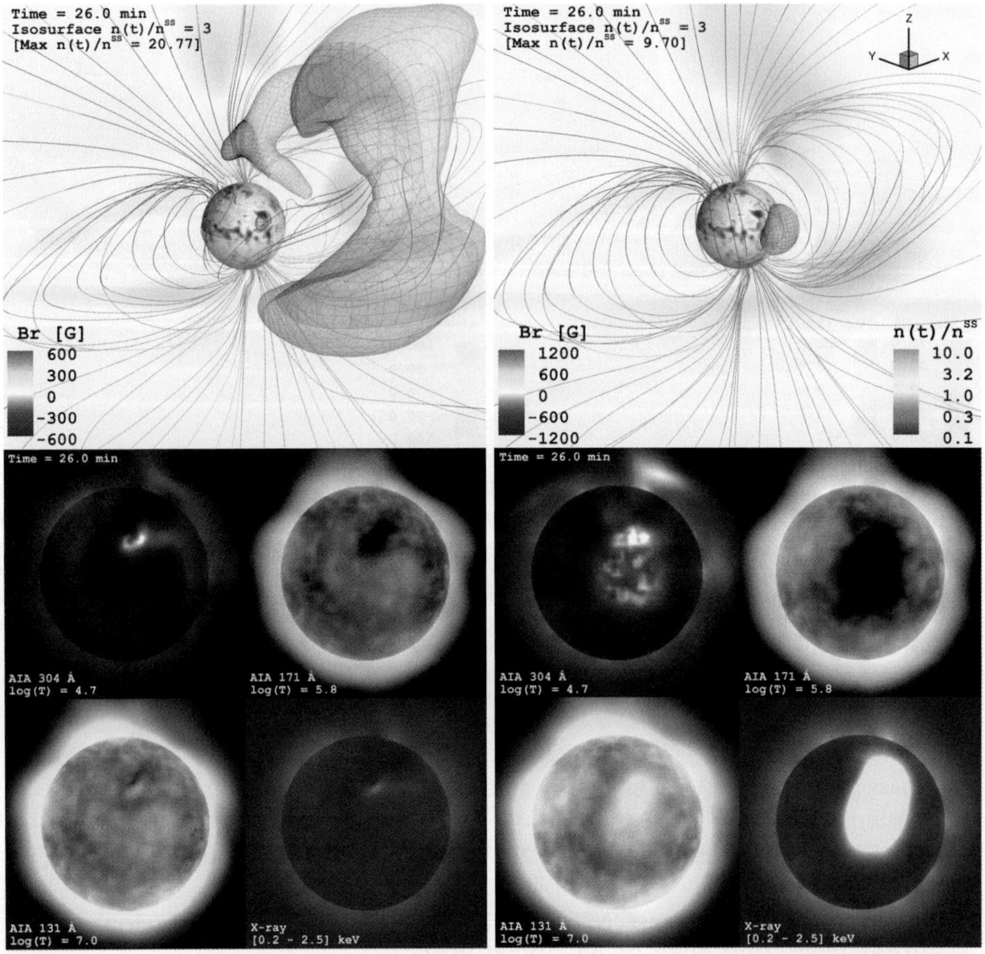

Figure 3. Snapshots from two time-dependent numerical simulations comparing the corona and stellar wind response during escaping (left) and confined (right) CME events on the flare-star Proxima Centauri (M5.5V, 4.85 Gyr). The geometry of the magnetic field driving the models is provided by the 3D dynamo simulation performed by Yadav et al. (2016). Each case employs a different scaling (*Left:* $B_{\rm r}^{\rm max} = \pm\, 600\,{\rm G}$, *Right:* $B_{\rm r}^{\rm max} = \pm\, 1200\,{\rm G}$) adjusted to reach the values observed in low- and moderately-active M dwarfs (Reiners et al. 2022). While an identical CME is initiated in both cases ($M_{\rm CME} = 4 \times 10^{17}$ g, $E_{\rm CME}^{\rm K} = 10^{35}$ erg), the eruption is only able to escape in the weaker magnetic field configuration, as indicated by the density ratio visualizations in the top row. Naturally, the coronal response is completely different in both cases, as can be seen from the simulated full-disk images in the bottom panels, synthesized in X-ray and EUV wavelengths (observer towards the $-y$ axis). **Adapted from:** Alvarado-Gómez et al. (2019).

and the characteristics of the stellar magnetic field (i.e. the confinement properties), notable signatures include:

- In CME events with weak suppression, a sudden increase–by factors ranging from approximately 5 to more than 10–in integrated X-ray coronal emissions (0.2 – 2.5 keV), lasting from tens of minutes to an hour is obtained. These brightenings resemble typical stellar flare events but are fueled not by magnetic reconnection but by rapid compression of coronal material.

- The evolution of these flare-like transients show a swift evolution –under one hour– of Doppler shift signals (from red to blue), with velocities reaching up to 200 km s^{-1}, transitioning from hotter ($\log(T) \gtrsim 6.8$) to cooler ($\log(T) \lesssim 6.0$) coronal lines. Slower velocities (< 100 km s^{-1}) and longer durations (> 1 h) are anticipated for more strongly suppressed eruptions (either due to stronger stellar fields or weaker eruptions).
- Fully confined CME events would lead to a gradual enhancement of soft X-ray coronal brightness (by factors of approximately 2-3) over several hours. This emission would exhibit redshifts (< -50 km s^{-1}), indicative of in-falling material (referred to as coronal rain clouds).
- In all the simulated events (weakly and strongly suppressed) a decrease in the emission at mid-coronal temperatures is obtained. These decreases in emission resemble coronal dimming events on the Sun which display a very high association rate with CMEs (Harra et al. 2016; Dissauer et al. 2018). However, in our simulations, these also occur during fully confined eruptions where the material might be removed from the eruption site but launched to higher latitudes to fall back onto the star (so that no mass is ejected at the end). We also noticed important changes in thermodynamic structure of the corona due to plasma motions and compression which might also contribute to the reduction of flux at a particular temperature without the removal of coronal mass.
- In radio wavelengths, Type II bursts signal the presence of an MHD shock in the corona or inner heliosphere and are strongly linked to CMEs and Solar Energetic Particle events (SEPs). These shocks and their associated particle acceleration occur close to the solar surface (see Gopalswamy et al. 2009; Zucca et al. 2014). The frequency of a Type II radio burst is proportional to the ambient density ($\propto \sqrt{n}$), and their frequency drift can reveal properties of the shock. However, as noted in Sect. 2 and demonstrated in Alvarado-Gómez et al. (2020), due to magnetic suppression stellar CMEs are significantly slowed down becoming super-Alfvénic (and thus inducing shocks) much farther from the star. This effect is heightened in more active stars, which have higher Alfvén speeds in their corona due to stronger surface magnetic fields, creating less favorable conditions for shocks near the star (Villadsen and Hallinan 2019). Consequently, the associated Type II radio bursts occur at much lower frequencies, often near or below the ionospheric cutoff at 10 MHz. Therefore, the magnetic suppression of stellar CMEs significantly impedes their detection via Type II radio bursts from the ground.

4. Concluding Remarks

The stellar eruptive behavior, crucial for high-energy stellar astrophysics and exoplanetary science, remains largely unexplored beyond the Sun. While solar CME knowledge offers a fundamental basis, blindly applying its paradigm to other stars poses theoretical and observational challenges, especially concerning more active stars. Detailed numerical simulations are essential to study and characterize these energetic phenomena in cool stars under these constraints.

Our numerical experiments underscore the pivotal role of stellar magnetic fields in shaping CME properties. Advanced models of CME propagation in active stars demonstrate that these events can be partially or completely suppressed by the star's large-scale magnetic field. This magnetic suppression mechanism effectively reduces the flare-CME association rate as well as lowers the CME speed and kinetic energy, aligning with limited observational data. This process significantly affects expected CME signatures across different wavelengths, potentially limiting detection following solar analogies, such as Type II radio bursts.

This suppression mechanism extends to stronger and more complex magnetic field regimes, such as those found in M-dwarfs. Contrary to what might appear at first glance, simulations indicate both escaping and suppressed CME events could pose significant risks to exoplanets. Direct impacts on exoplanet atmospheres and ionizing radiation from associated coronal responses (like compression flaring and up-flows/down-flows) may be detectable by upcoming high-energy astrophysics instruments.

Finally, understanding CMEs and associated phenomena in exoplanetary atmospheric studies requires realistic stellar-specific CME assessments, avoiding direct solar extrapolations. Accurate characterization of stellar CME properties and occurrence rates demands integrated multi-wavelength observations and sophisticated numerical modeling approaches.

Acknowledgments

J.D.A.G. would like to thank the organizers of the IAU Symposium 388 for the invitation to present this work and for the financial support received to attend the conference.

Questions/Discussion

A. Veronig: You demonstrated in your simulations that failed CMEs may form strong density enhancements due to compression which may produce flare-like light curves. Are the temperatures that can be achieved by this process similarly high as in flares?

J. D. Alvarado-Gómez: According to our models yes, some of the temperatures can be very high. Perhaps they will not reach the highest temperatures seen in stellar flares but definitely they reach X-ray and EUV emitting temperatures (so a few MK). Note that we do see these 'flare-like' signatures in all cases: when the CME escapes and when it is confined. The degree of confinement affects their duration (the less confined ones generate shorter, more impulsive 'flare-like' profiles whereas for more confined events the induced emission will last a bit longer, evolving in a more gradual way). While we have not estimated properly their energies, to a first order, the profiles emerging from fully-confined CME events will radiate more energy in XUV wavelengths compared to the ones generated for escaping CMEs.

K. Dissauer: From an observational point of view, what do you think would be required to confirm magnetic confinement?

J. D. Alvarado-Gómez: Interesting question. I believe that there are multiple observational hints already that magnetic confinement is taking place for the stellar regime. As I indicated in my talk, the fact that our searches for stellar CMEs have returned empty handed, contrary to our expectations based on solar behaviour, is probably one of them. In addition, as nicely presented by Tom Ayres in his 2015 paper on EK Dra (Ayres 2015), HST observations of this star reveal the persistent presence of 'downflows' visible in transition region/chromospheric lines. This behaviour is consistent with the picture of material trying to get ejected after reconnection events (i.e. flares) but not managing to escape and falling back onto the star. This chain of events is also very similar to what we obtain with our numerical models of suppressed CME events. While these observations might not completely 'confirm' the magnetic confinement of CMEs, they strongly support this scenario for active stars.

References

Aarnio, A. N., Matt, S. P., & Stassun, K. G. 2012, ApJ, 760, 9, doi: 10.1088/0004-637X/760/1/9

Alvarado-Gómez, J. D., Drake, J. J., Cohen, O., et al. 2022, Astronomische Nachrichten, 343, e10100, doi: 10.1002/asna.20210100

Alvarado-Gómez, J. D., Drake, J. J., Cohen, O., Moschou, S. P., & Garraffo, C. 2018, ApJ, 862, 93, doi: 10.3847/1538-4357/aacb7f
Alvarado-Gómez, J. D., Drake, J. J., Moschou, S. P., et al. 2019, ApJ, 884, L13, doi: 10.3847/2041-8213/ab44d0
Alvarado-Gómez, J. D., Hussain, G. A. J., Grunhut, J., et al. 2015, A&A, 582, A38, doi: 10.1051/0004-6361/201525771
Alvarado-Gómez, J. D., Drake, J. J., Fraschetti, F., et al. 2020, ApJ, 895, 47, doi: 10.3847/1538-4357/ab88a3
Andrews, M. D. 2003, Sol. Phys., 218, 261, doi: 10.1023/B:SOLA.0000013039.69550.bf
Argiroffi, C., Reale, F., Drake, J. J., et al. 2019, Nature Astronomy, 3, 742, doi: 10.1038/s41550-019-0781-4
Aschwanden, M. J., Caspi, A., Cohen, C. M. S., et al. 2017, ApJ, 836, 17, doi: 10.3847/1538-4357/836/1/17
Audard, M., Güdel, M., Drake, J. J., & Kashyap, V. L. 2000, ApJ, 541, 396, doi: 10.1086/309426
Ayres, T. R. 2015, AJ, 150, 7, doi: 10.1088/0004-6256/150/1/7
Benz, A. O. 2017, Living Reviews in Solar Physics, 14, 2, doi: 10.1007/s41116-016-0004-3
Borisova, A., Aurière, M., Petit, P., et al. 2016, A&A, 591, A57, doi: 10.1051/0004-6361/201526726
Caramazza, M., Flaccomio, E., Micela, G., et al. 2007, A&A, 471, 645, doi: 10.1051/0004-6361:20077195
Chen, P. F. 2011, Living Reviews in Solar Physics, 8, doi: 10.12942/lrsp-2011-1
Cherenkov, A., Bisikalo, D., Fossati, L., & Möstl, C. 2017, ApJ, 846, doi: 10.3847/1538-4357/aa82b2
Cohen, O., Kashyap, V. L., Drake, J. J., Sokolov, I. V., & Gombosi, T. I. 2011, ApJ, 738, 166, doi: 10.1088/0004-637X/738/2/166
Compagnino, A., Romano, P., & Zuccarello, F. 2017, Sol. Phys., 292, 5, doi: 10.1007/s11207-016-1029-4
Cranmer, S. R. 2017, ApJ, 840, 114, doi: 10.3847/1538-4357/aa6f0e
Crosley, M. K., & Osten, R. A. 2018, ApJ, 856, doi: 10.3847/1538-4357/aaaec2
Crosley, M. K., Osten, R. A., Broderick, J. W., et al. 2016, ApJ, 830, 24, doi: 10.3847/0004-637X/830/1/24
Davenport, J. R. A., Covey, K. R., Clarke, R. W., et al. 2019, ApJ, 871, 241, doi: 10.3847/1538-4357/aafb76
Dissauer, K., Veronig, A. M., Temmer, M., Podladchikova, T., & Vanninathan, K. 2018, ApJ, 855, 137, doi: 10.3847/1538-4357/aaadb5
Donati, J.-F., & Landstreet, J. D. 2009, ARA&A, 47, 333, doi: 10.1146/annurev-astro-082708-101833
Drake, J. J., Cohen, O., Garraffo, C., & Kashyap, V. 2016, in IAU Symposium, Vol. 320, Solar and Stellar Flares and their Effects on Planets, ed. A. G. Kosovichev, S. L. Hawley, & P. Heinzel, 196–201, doi: 10.1017/S1743921316000260
Drake, J. J., Cohen, O., Yashiro, S., & Gopalswamy, N. 2013, ApJ, 764, 170, doi: 10.1088/0004-637X/764/2/170
Emslie, A. G., Dennis, B. R., Holman, G. D., & Hudson, H. S. 2005, Journal of Geophysical Research (Space Physics), 110, A11103, doi: 10.1029/2005JA011305
Emslie, A. G., Dennis, B. R., Shih, A. Y., et al. 2012, ApJ, 759, 71, doi: 10.1088/0004-637X/759/1/71
Fan, Y. 2021, Living Reviews in Solar Physics, 18, 5, doi: 10.1007/s41116-021-00031-2
Favata, F., & Schmitt, J. H. M. M. 1999, A&A, 350, 900
Folsom, C. P., Petit, P., Bouvier, J., et al. 2016, MNRAS, 457, 580, doi: 10.1093/mnras/stv2924
Folsom, C. P., Bouvier, J., Petit, P., et al. 2018, MNRAS, 474, 4956, doi: 10.1093/mnras/stx3021
France, K., Duvvuri, G., Egan, H., et al. 2020, AJ, 160, 237, doi: 10.3847/1538-3881/abb465
Gopalswamy, N., Thompson, W. T., Davila, J. M., et al. 2009, Sol. Phys., 259, 227, doi: 10.1007/s11207-009-9382-1
Guenther, E. W., & Emerson, J. P. 1997, A&A, 321, 803

Gunn, A. G., Doyle, J. G., Mathioudakis, M., Houdebine, E. R., & Avgoloupis, S. 1994, A&A, 285, 489
Günther, M. N., Zhan, Z., Seager, S., et al. 2020, AJ, 159, 60, doi: 10.3847/1538-3881/ab5d3a
Harra, L. K., Schrijver, C. J., Janvier, M., et al. 2016, Sol. Phys., 291, 1761, doi: 10.1007/s11207-016-0923-0
Houdebine, E. R., Foing, B. H., & Rodono, M. 1990, A&A, 238, 249
Howard, W. S., Tilley, M. A., Corbett, H., et al. 2018, ApJ, 860, L30, doi: 10.3847/2041-8213/aacaf3
Hudson, H. S., Acton, L. W., & Freeland, S. L. 1996, ApJ, 470, 629, doi: 10.1086/177894
Huenemoerder, D. P., Schulz, N. S., Testa, P., et al. 2010, ApJ, 723, 1558, doi: 10.1088/0004-637X/723/2/1558
Hussain, G. A. J., van Ballegooijen, A. A., Jardine, M., & Collier Cameron, A. 2002, ApJ, 575, 1078, doi: 10.1086/341429
Hussain, G. A. J., Alvarado-Gómez, J. D., Grunhut, J., et al. 2016, A&A, 585, A77, doi: 10.1051/0004-6361/201526595
Jeffers, S. V., Boro Saikia, S., Barnes, J. R., et al. 2017, MNRAS, 471, L96, doi: 10.1093/mnrasl/slx097
Kashyap, V. L., Drake, J. J., Güdel, M., & Audard, M. 2002, ApJ, 580, 1118, doi: 10.1086/343869
Kay, C., Airapetian, V. S., Lüftinger, T., & Kochukhov, O. 2019, ApJ, 886, L37, doi: 10.3847/2041-8213/ab551f
Khodachenko, M. L., Ribas, I., Lammer, H., et al. 2007, Astrobiology, 7, 167, doi: 10.1089/ast.2006.0127
Kislyakova, K. G., Güdel, M., Koutroumpa, D., et al. 2024, Nature Astronomy, 8, 596, doi: 10.1038/s41550-024-02222-x
Kochukhov, O. 2021, A&A Rev., 29, 1, doi: 10.1007/s00159-020-00130-3
Kochukhov, O., Hackman, T., Lehtinen, J. J., & Wehrhahn, A. 2020, A&A, 635, A142, doi: 10.1051/0004-6361/201937185
Kopparapu, R. K., Ramirez, R. M., SchottelKotte, J., et al. 2014, ApJ, 787, L29, doi: 10.1088/2041-8205/787/2/L29
Kowalski, A. F. 2024, Living Reviews in Solar Physics, 21, 1, doi: 10.1007/s41116-024-00039-4
Kowalski, A. F., Hawley, S. L., Holtzman, J. A., Wisniewski, J. P., & Hilton, E. J. 2010, ApJ, 714, L98, doi: 10.1088/2041-8205/714/1/L98
Kretzschmar, M. 2011, A&A, 530, A84, doi: 10.1051/0004-6361/201015930
Kretzschmar, M., de Wit, T. D., Schmutz, W., et al. 2010, Nature Physics, 6, 690, doi: 10.1038/nphys1741
Lammer, H., Lichtenegger, H. I. M., Kulikov, Y. N., et al. 2007, Astrobiology, 7, 185, doi: 10.1089/ast.2006.0128
Lamy, P. L., Floyd, O., Boclet, B., et al. 2019, Space Sci. Rev., 215, 39, doi: 10.1007/s11214-019-0605-y
Lavail, A., Kochukhov, O., & Wade, G. A. 2018, MNRAS, 479, 4836, doi: 10.1093/mnras/sty1825
Leitzinger, M., Odert, P., Greimel, R., et al. 2014, MNRAS, 443, 898, doi: 10.1093/mnras/stu1161
Leitzinger, M., Odert, P., Ribas, I., et al. 2011, A&A, 536, A62, doi: 10.1051/0004-6361/201015985
Lim, J., & White, S. M. 1996, ApJ, 462, L91, doi: 10.1086/310038
Liu, L., Wang, Y., Wang, J., et al. 2016, ApJ, 826, 119, doi: 10.3847/0004-637X/826/2/119
Loyd, R. O. P., Shkolnik, E. L., France, K., Wood, B. E., & Youngblood, A. 2020, Research Notes of the American Astronomical Society, 4, 119, doi: 10.3847/2515-5172/aba94a
Loyd, R. O. P., Mason, J. P., Jin, M., et al. 2022, ApJ, 936, 170, doi: 10.3847/1538-4357/ac80c1
Morgenthaler, A., Petit, P., Saar, S., et al. 2012, A&A, 540, A138, doi: 10.1051/0004-6361/201118139
Morin, J., Donati, J.-F., Petit, P., et al. 2010, MNRAS, 407, 2269, doi: 10.1111/j.1365-2966.2010.17101.x

—. 2008, MNRAS, 390, 567, doi: 10.1111/j.1365-2966.2008.13809.x
Moschou, S.-P., Drake, J. J., Cohen, O., Alvarado-Gomez, J. D., & Garraffo, C. 2017, ApJ, 850, 191, doi: 10.3847/1538-4357/aa9520
Moschou, S.-P., Drake, J. J., Cohen, O., et al. 2019, ApJ, 877, 105, doi: 10.3847/1538-4357/ab1b37
Muheki, P., Guenther, E. W., Mutabazi, T., & Jurua, E. 2020a, A&A, 637, A13, doi: 10.1051/0004-6361/201936904
—. 2020b, MNRAS, 499, 5047, doi: 10.1093/mnras/staa3152
Namekata, K., Maehara, H., Honda, S., et al. 2021, Nature Astronomy, 6, 241, doi: 10.1038/s41550-021-01532-8
Odert, P., Leitzinger, M., Hanslmeier, A., & Lammer, H. 2017, MNRAS, 472, 876, doi: 10.1093/mnras/stx1969
Osten, R. 2016, Solar explosive activity throughout the evolution of the solar system, ed. C. J. Schrijver, F. Bagenal, & J. J. Sojka, 23–55, doi: 10.1017/CBO9781316106778.003
Osten, R. A., Drake, S., Tueller, J., et al. 2007, ApJ, 654, 1052, doi: 10.1086/509252
Osten, R. A., & Wolk, S. J. 2015, ApJ, 809, 79, doi: 10.1088/0004-637X/809/1/79
Reiners, A. 2012, Living Reviews in Solar Physics, 9, 1, doi: 10.12942/lrsp-2012-1
Reiners, A., Shulyak, D., Käpylä, P. J., et al. 2022, A&A, 662, A41, doi: 10.1051/0004-6361/202243251
Schnepf, N. R., Lovelace, R. V. E., Romanova, M. M., & Airapetian, V. S. 2015, MNRAS, 448, 1628, doi: 10.1093/mnras/stv056
See, V., Matt, S. P., Folsom, C. P., et al. 2019, ApJ, 876, 118, doi: 10.3847/1538-4357/ab1096
Shibata, K., & Magara, T. 2011, Living Reviews in Solar Physics, 8, 6, doi: 10.12942/lrsp-2011-6
Shibayama, T., Maehara, H., Notsu, S., et al. 2013, The Astrophysical Journal Supplement Series, 209, doi: 10.1088/0067-0049/209/1/5
Shields, A. L., Ballard, S., & Johnson, J. A. 2016, Physics Reports., 663, 1, doi: 10.1016/j.physrep.2016.10.003
Sterling, A. C., & Hudson, H. S. 1997, ApJ, 491, L55, doi: 10.1086/311043
Sun, X., Bobra, M. G., Hoeksema, J. T., et al. 2015, ApJ, 804, L28, doi: 10.1088/2041-8205/804/2/L28
Takahashi, T., Mizuno, Y., & Shibata, K. 2016, ApJ, 833, L8, doi: 10.3847/2041-8205/833/1/L8
Thalmann, J. K., Su, Y., Temmer, M., & Veronig, A. M. 2015, ApJ, 801, L23, doi: 10.1088/2041-8205/801/2/L23
—. 2017, ApJ, 844, L27, doi: 10.3847/2041-8213/aa7f6f
Thompson, B. J., Plunkett, S. P., Gurman, J. B., et al. 1998, Geo. Phys. Rev. Lett., 25, 2465, doi: 10.1029/98GL50429
Tristan, I. I., Notsu, Y., Kowalski, A. F., et al. 2023, ApJ, 951, 33, doi: 10.3847/1538-4357/acc94f
Vanninathan, K., Veronig, A. M., Dissauer, K., & Temmer, M. 2018, ApJ, 857, 62, doi: 10.3847/1538-4357/aab09a
Veronig, A. M., Odert, P., Leitzinger, M., et al. 2021, Nature Astronomy, 5, 697, doi: 10.1038/s41550-021-01345-9
Vida, K., Kriskovics, L., Oláh, K., et al. 2016, A&A, 590, doi: 10.1051/0004-6361/201527925
Villadsen, J., & Hallinan, G. 2019, ApJ, 871, 214, doi: 10.3847/1538-4357/aaf88e
Villadsen, J. R. 2017, PhD thesis, California Institute of Technology
Webb, D. F., & Howard, T. A. 2012, Living Reviews in Solar Physics, 9, doi: 10.12942/lrsp-2012-3
Wood, B. E., Müller, H.-R., Redfield, S., & Edelman, E. 2014, ApJ, 781, L33, doi: 10.1088/2041-8205/781/2/L33
Wood, B. E., Müller, H.-R., Zank, G. P., & Linsky, J. L. 2002, ApJ, 574, 412, doi: 10.1086/340797
Wood, B. E., Müller, H.-R., Zank, G. P., Linsky, J. L., & Redfield, S. 2005, ApJ, 628, L143, doi: 10.1086/432716

Wood, B. E., Müller, H.-R., Redfield, S., et al. 2021, ApJ, 915, 37, doi: 10.3847/1538-4357/abfda5

Yadav, R. K., Christensen, U. R., Wolk, S. J., & Poppenhaeger, K. 2016, ApJ, 833, L28, doi: 10.3847/2041-8213/833/2/L28

Yashiro, S., & Gopalswamy, N. 2009, in IAU Symposium, Vol. 257, Universal Heliophysical Processes, ed. N. Gopalswamy & D. F. Webb, 233–243, doi: 10.1017/S1743921309029342

Zarro, D. M., Sterling, A. C., Thompson, B. J., Hudson, H. S., & Nitta, N. 1999, ApJ, 520, L139, doi: 10.1086/312150

Zucca, P., Carley, E. P., Bloomfield, D. S., & Gallagher, P. T. 2014, A&A, 564, A47, doi: 10.1051/0004-6361/201322650

Solar and Stellar Coronal Mass Ejections
Proceedings IAU Symposium No. 388, 2026
N. Gopalswamy, O. Malandraki, A. Vidotto & W. Manchester, eds.
doi:10.1017/S1743921325000547

Implications of the abundance of halo coronal mass ejections for the strength of solar cycle 25

Nat Gopalswamy[1], Grzegorz Michałek[2], Seiji Yashiro[3,1], Pertti Mäkelä[3,1], Sachiko Akiyama[3,1] and Hong Xie[3,1]

[1]NASA Goddard Space Flight Center, Greenbelt, MD, USA
email: nat.gopalswamy@nasa.gov

[2]Jagiellonian University, Kraków, Poland

[3]The Catholic University of America, Washington, DC, USA

Abstract. We assess the relative strength of solar cycle (SC) 25 with respect to SCs 23 and 24 based on the abundance of halo coronal mass ejections (CMEs). We make use of the halo CME database (https://cdaw.gsfc.nasa.gov/CME_list/halo/halo.html) to compare the halo CME abundance during the first four years in each of SCs 23–25. The main result is that in several aspects such as the abundance, occurrence rate, source locations, and halo heights, halo CMEs are similar between SCs 24 and 25 but different from SC 23. This result follows from the fact that weaker cycles have low heliospheric total pressure, whose backreaction on CMEs allows them to expand more and hence enhancing the chance of becoming a halo. The solar cycle variation of halo CME properties is consistent with the precursor-based cycle prediction methods that indicate SC 25 is similar to or only slightly stronger than SC 24.

Keywords. coronal mass ejections, solar cycle, heliospheric state

1. Introduction

A coronal mass ejections (CME) is referred to a halo CME when its apparent width in the sky plane is 360°. The occurrence rate of halo CMEs is correlated with the sunspot number (SSN), so it is likely that this correlation has a bearing on the prediction of cycle strengths. During the maximum phase of solar cycle (SC) 24 it was recognized that the halo CME abundance had a discordant behavior with SSN: the halo CME abundance normalized to SSN was higher in the weaker cycle 24 than in SC 23 (Gopalswamy et al. 2015; Dagnew et al. 2022). This was explained to be a consequence of the anomalous expansion of CMEs due to the weak heliospheric state (Gopalswamy et al. 2014). Comparing occurrence rates of halo CMEs during the rise phases of SCs 23–25, it was recently reported that the halo CME abundance is higher in SC 25 as well. Furthermore, the halo CME abundance indicates that the strength of SC 25 is intermediate between SCs 23 and 24, but closer to SC 24 (Gopalswamy et al. 2023). Thus, the halo CME abundance provides clues to the relative strength of solar cycles. Given the widely distributed predictions of the strength of SC 25, the halo CME abundance is able to discriminate among the predictions that vary from less than half to almost twice the strength of SC 24 (Nandy 2021). In this paper, we extend the results of Gopalswamy et al. (2023) to the maximum phase thereby including a bigger sample of CMEs over the first four years of cycles 23–25 to assess their relative strengths.

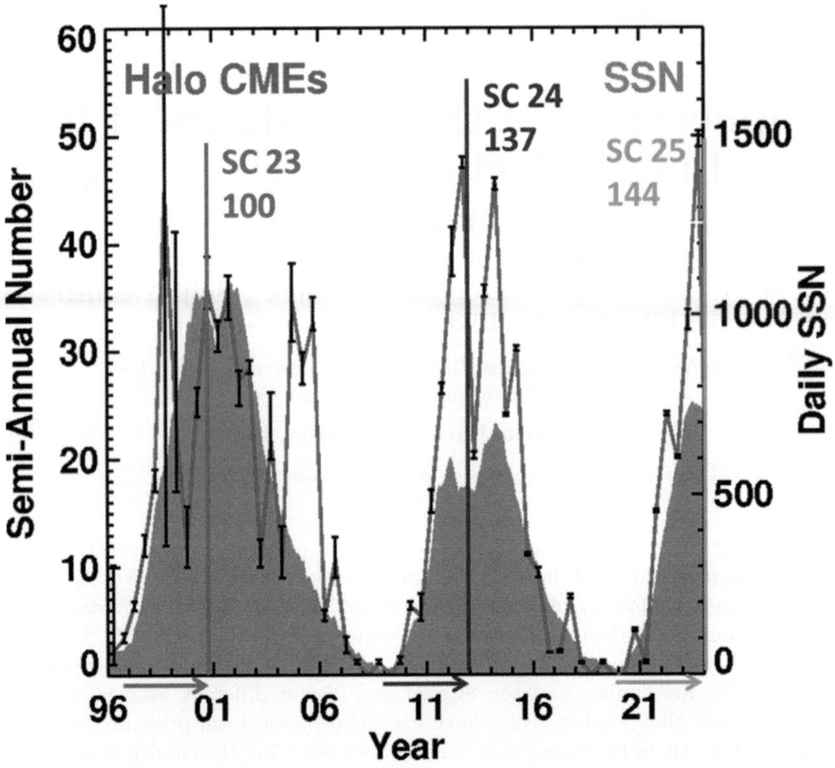

Figure 1. Semiannual variation of the number of halo CMEs. During the first 49 months (indicated by arrows), 100, 137, and 144 halos were observed in SCs 23, 24 and 25, respectively. SC 23: 1996 May 10 to 2000 June 9; SC 24: 2008 December 1 to 2012 December 31; SC 25: 2019 December 1 to 2023 December 31. The error bars are based on the data gaps in the LASCO data. SSN (Source: WDC-SILSO, Royal Observatory of Belgium, Brussels) is shown for reference.

2. Observations

We use the halo CME catalog (`https://cdaw.gsfc.nasa.gov/CME_list/halo/halo.html`, Gopalswamy et al. 2010) based on the observations from the Large Angle and Spectrometric Coronagraph (LASCO; Brueckner et al. 1995) on board the Solar and Heliospheric Observatory (SOHO). The catalog includes the solar sources of the halo CMEs (heliographic coordinates). We use the v2 SSN from the Sunspot Index and Long-term Solar Observations (SILSO, World Data Center, `https://sidc.be/SILSO/datafiles`). Figure 1 shows the semiannual number of halo CMEs as a function of time from 1996 to the end of 2023. We see that the number of halos during the first 49 months of SC 23 is only 100 compared to 137 in SC 24 and 144 in SC 25. Thus, the halo CME abundance increased in SCs 24 and 25 although the SSN decreased significantly.

3. Analysis and Results

3.1. *Halo CME abundance*

Table 1 compares the number of halos during the first 49 months of each cycle and the relative changes with respect to cycle 23 as absolute values and relative to the average SSN. In cycle 23, there was a 4-month data gap. If we assume that halo CMEs occurred during the data gaps at the same rate as the average rate over the whole interval, the estimated number is 109 CMEs instead of 100. We use the corrected value (109) in comparisons. In terms of absolute numbers, we see that the number of halos increased

Table 1. Comparison of the number of halos in SCs 23–25

	SC 23	SC 24	SC 25
Total # Halos	100 (109)	137	144
Change		+26%	+32%
Avg SSN	82.7	47.8	60.5
Change		-42%	-17%
#Halo/<SSN>	1.32	2.87	2.38
Change		+117%	+80%

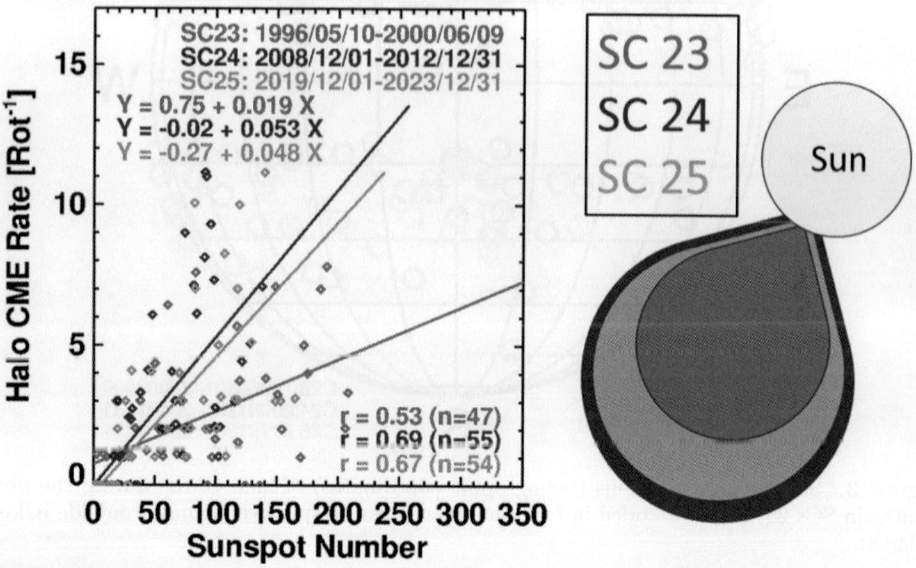

Figure 2. (left) Scatter plot between SSN and the occurrence rate of halo CMEs per Carrington rotation (CR) period. The number of data points during the first 49 months is not the same in the three cycles due to data gaps. The beginning and end dates of the compared intervals, the correlation coefficients and equations of the regression lines are shown on the plot. (right) Schematic showing the size CMEs in each cycle for a given CME speed.

by 26% and 32% in SCs 24 and 25, respectively. On the other hand, the average SSN declined by 42% in SC 24 and 27% in SC 25. When we normalize the halo number to SSN, the relative numbers are 1.32, 2.87, and 2.38 per SSN. Clearly, the normalized halo CME abundance approximately doubled in SCs 24 and 25. Furthermore, the SC 24 abundance is the highest. The SC 25 abundance is much closer to that of SC 24.

3.2. *Correlation between SSN and halo CME occurrence rate*

Figure 2 shows the scatter plot between SSN and the number of halo CMEs in each Carrington rotation (CR) period. The SSN used in the plot is an average over CR periods. We see that the halo CME rate is positively correlated with SSN in each of the three SCs. All correlations are significant because the Pearson's critical coefficients for 47, 55, and 54 samples are 0.372, 0.363, and 0.360, respectively for a p value of 1% (the probability that the obtained correlation is by chance). The regression lines have similar slopes in SCs 24 and 25 but significantly higher than that in SC 23. These plots indicate that there are more halo CMEs for a given SSN in SCs 24 and 25, confirming the result shown in Table 1 for the whole interval. The effect of the weak heliosphere on the CME size is shown schematically in Fig. 2. The size of the CMEs is inversely related to the total

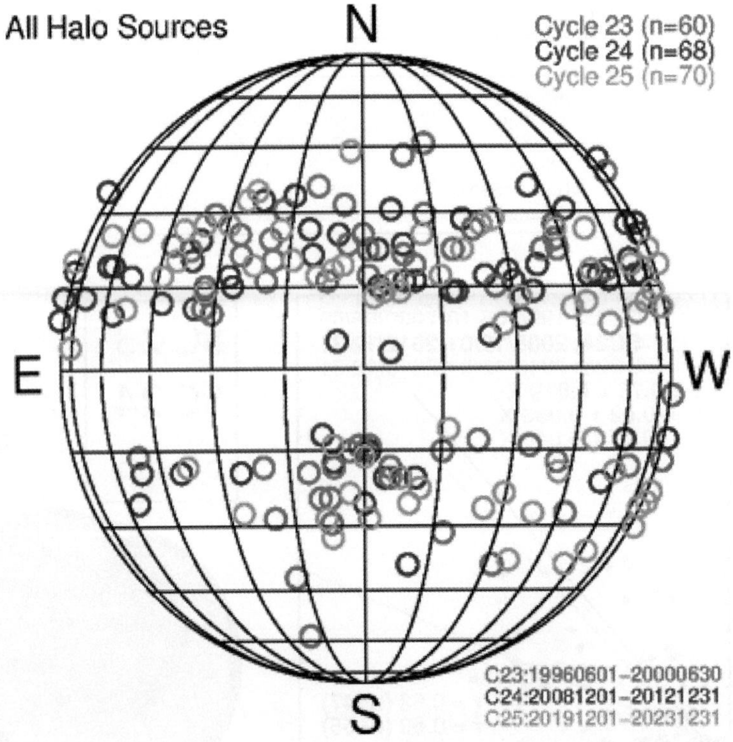

Figure 3. Solar source locations (heliographic coordinates) of halo CMEs during the first 49 months in SCs 23–25, color coded in red, blue, and green, respectively. Only frontside halos are included.

pressure in the heliosphere. For a given speed, CMEs expand more in SCs 24 and 25 than in SC 23 resulting in the higher abundance of halo CMEs in SCs 24 and 25.

3.3. *Distribution of solar source locations*

The distribution of halo CME sources on the solar disk is shown in Fig. 3. The source locations are generally concentrated in the latitude range 15°–30°, which corresponds to the active region belt. Most halo CMEs originate from active regions, where large amount of free energy can be stored and released. Most of the source locations of SC 23 CMEs are concentrated near the disk center (within the longitude range of ±45°). On the other hand, the halos from SCs 24 and 25 originate from a broader range of longitudes. The mean absolute longitudes are 23°.4 (SC 23), 37°.0 (SC 24), and 36°.2 (SC 25) with standard deviations of 20°.7, 26°.9, and 26°.1, respectively. In SC 23, only one halo source is located beyond a central meridian distance (CMD) of 45° in the eastern hemisphere and 6 in the western hemisphere out of the 60 frontside halos (or ∼12%). On the other hand, there are 25 (out of 68) and 24 (out of 70) halos beyond a CMD of 45° in SCs 24 and 25, respectively amounting to 37% and 34% of the frontside halos in these cycles. Clearly, the large-CMD halos in SCs 24 and 25 outnumber those in SC 23 by a factor of ∼3. Once again, we see that the longitudinal distribution of halo CME sources in SCs 24 and 25 are close to each other, but much wider than that in SC 23. Halos originating beyond a CMD of 60° have a similar pattern: 4 out of 60 (or 7%) in SC 23 compared to 17 out of 68 (or 25%) in SC 24 and 15 out of 70 (or 21%) in SC 25. We also note that the large-CMD halo sources in SC 25 are intermediate between SCs 23 and 24 but closer to SC 24.

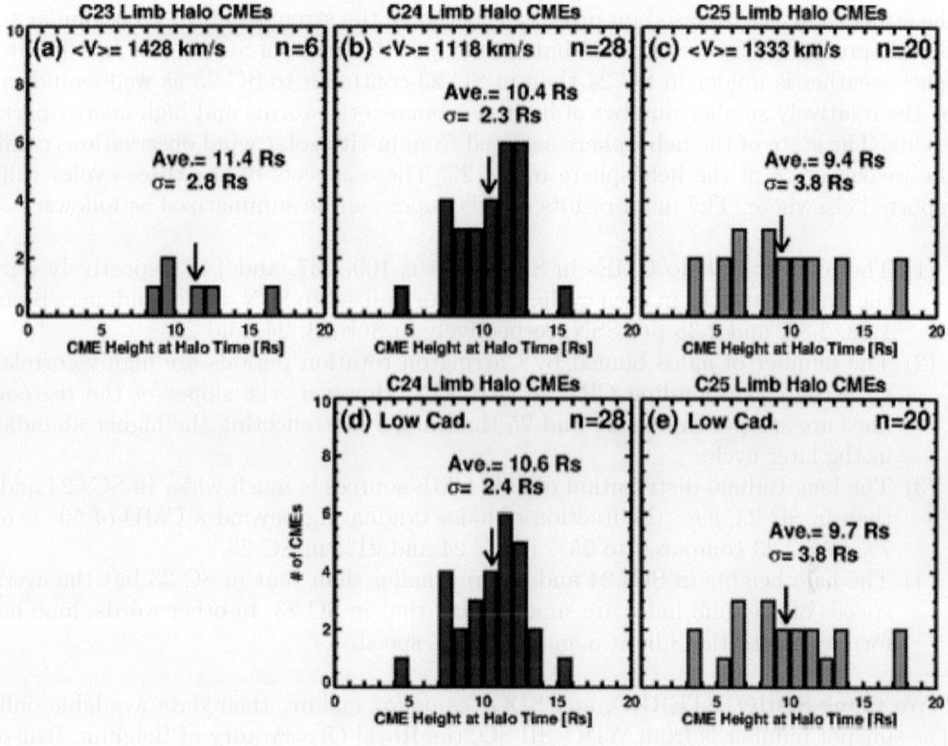

Figure 4. (top) Halo heights of limb CMEs in SCs 23–25 in panels a–c. The average of the distributions and their standard deviations are shown on the plots. The number of limb halos is higher than in Fig. 3 because we have included cases in which the sources are behind the limb but within 30°. (bottom) Halo heights when the LASCO image cadence in SCs 24 and 25 reduced to match that in SC 23. Panels a–c display the average speeds of limb halos in three cycles. The speeds are obtained from `https://cdaw.gsfc.nasa.gov/CME_list/halo/halo.html`.

3.4. *Halo heights of limb halo CMEs*

Gopalswamy et al. (2020) introduced a new parameter known as "halo height" defined as the height of the leading edge of a limb CME when it appears as a halo for the first time in the LASCO FOV. In such limb halos, a CME originating close to one of the solar limbs ends up with features appearing on the opposite limb. Figure 4 shows the distributions of halo heights in the three cycles. The halo heights in SCs 24 and 25 are smaller than that in SC 23. Furthermore, the average speeds of the limb halos are lower in SCs 24 and 25. In other words, limb halos formed earlier in SCs 24 and 25 and at lower speeds. The limb halo sample size is small, so the results are not very firm. However, the trend is similar to what was seen between SCs 23 and 24 when the full-cycle data were used (Gopalswamy et al. 2020).

4. Discussion and Summary

We investigated the properties of halo CMEs in SCs 23–25 that occurred during the first 49 months of each cycle, focusing on the abundance, occurrence rate, source locations, and halo heights. The results are consistent with the expected backreaction of the weaker heliospheric states in SCs 24 and 25 than that in SC 23. In all the aspects studied, we find that the halo CME properties are similar between SCs 24 and 25 and different from SC 23. Furthermore, the halo CME properties in SC 25 are intermediate between those in SCs 23 and 24, but closer to SC 24. These results strongly support the predictions of the strength of SC 25 based on precursor methods: the strength of SC 25 is similar to or slightly greater than that of SC 24 but much lower than that of SC 23. The trend that the space weather is milder in SC 24 than in SC 23 continues to SC 25 as well as indicated by the relatively smaller number of intense geomagnetic storms and high-energy particle events. The state of the heliosphere assessed from in-situ solar wind observations confirm the weaker state of the heliosphere in SC 25. These aspects of the three cycles will be reported elsewhere. The main results of this paper can be summarized as follows:

(1) The number of halo CMEs in SCs 23–24 is 109, 137, and 144 respectively during the first 49 months in each cycle. When normalized to SSN, the abundances become 1.32, 2.87, and 2.38 per SSN, respectively in SCs 23, 24, and 25.
(2) The number of halos binned by Carrington rotation periods are highly correlated with the corresponding CR-averaged SSN. However, the slopes of the regression lines are steeper in SCs 24 and 25 than in SC 23, reflecting the higher abundance in the later cycles.
(3) The longitudinal distribution of halo CME sources is much wider in SCs 24 and 25 than in SC 23, e.g., the fraction of halos originating beyond a CMD of 60° is only 7% in SC 23 compared to 25% in SC 24 and 21% in SC 25.
(4) The halo heights in SCs 24 and 25 are smaller than that in SC 23 but the average speeds of the limb halos are smaller than that in SC 23. In other words, limb halos form closer to the Sun at a smaller CME speed.

We thank SOHO, STEREO, and SDO teams for making their data available online. The sunspot number is from WDC–SILSO, the Royal Observatory of Belgium, Brussels. SOHO is a joint project of ESA and NASA. STEREO is a mission in NASA's Solar Terrestrial Probes program. Work supported by NASA's LWS and STEREO programs. H.X. was partially supported by NSF grant AGS-2228967. P.M. was partially supported by NSF grant AGS-2043131.

References

Brueckner, G. E., Howard, R. A., Koomen, M. J., et al. 1995, *SoPh*, 162, 357, doi: 10.1007/BF00733434

Dagnew, F. K., Gopalswamy, N., Tessema, S. B., Akiyama, S., & Yashiro, S. 2022, *ApJ*, 936, 122, doi: 10.3847/1538-4357/ac8744

Gopalswamy, N., Akiyama, S., & Yashiro, S. 2020, *ApJL*, 897, L1, doi: 10.3847/2041-8213/ab9b7b

Gopalswamy, N., Akiyama, S., Yashiro, S., et al. 2014, *GeoRL*, 41, 2673, doi: 10.1002/2014GL059858

Gopalswamy, N., Michalek, G., Yashiro, S., et al. 2023, *ApJL*, 952, L13, doi: 10.3847/2041-8213/acdde2

Gopalswamy, N., Xie, H., Akiyama, S., et al. 2015, *ApJL*, 804, L23, doi: 10.1088/2041-8205/804/1/L23

Gopalswamy, N., Yashiro, S., Michalek, G., et al. 2010, *SunGe*, 5, 7

Nandy, D. 2021, *SoPh*, 296, 54, doi: 10.1007/s11207-021-01797-2

Solar and Stellar Coronal Mass Ejections
Proceedings IAU Symposium No. 388, 2026
N. Gopalswamy, O. Malandraki, A. Vidotto & W. Manchester, eds.
doi:10.1017/S1743921324001388

Extreme solar events: An update

Ilya Usoskin
University of Oulu, Finland

Abstract. Before a dozen years ago, we believed that the Sun is a mildly active star unable to produce disastrous events. However, as discovered in 2012, extreme events can be produced by the Sun rarely on the millennial timescales. These events include extreme solar particle events with enormous fluxes of solar energetic particles as known from cosmogenic-isotope proxies recorded in independently dateable natural archives like tree rings and polar ice cores; and super-flares as known from an analysis of thousands of sun-like stars observed recently by the space-borne telescope *Kepler*. While both types of events imply that the Sun can rarely produce eruptive events several orders of magnitude stronger than those we observed during the last decades, the statistics of the two types of events mismatch quantitatively. Here, we briefly summarize the start-of-the-art knowledge of the extreme events updating more detailed recent reviews and listing the open questions still remaining to be answered.

Keywords. Solar flares, solar energetic particles, extreme events, Space Climate

1. Introduction

Just recently, before 2012, we thought that the Sun is a mildly active star which can occasionally produce strong eruptive events which attract strong academic interest but do not pose serious hazards to human beings. This was based on the practical knowledge collected during the past decades started with the onsets of geophysical (since the middle of the 19th century) and space (since the middle of the 20th century) exploration epochs. However, this century-long interval of scientific exploration of the Earth's environment appears too short to encompass the full range of solar variability. The situation changed when several discoveries made in 2012 pushed the boundaries of what we can expect from the Sun much further.

One discovery was related to a very short and strong spike of high-resolution radiocarbon Δ^{14}C measurements made in a Japanese cedar tree that corresponded to the year 775 CE (Miyake et al. 2012). This was independently identified also by Usoskin and Kovaltsov (2012) in low-resolution decadal data from Intcal09 (Reimer et al. 2009) and ^{10}Be data in polar ice cores. The spike was so strong that Miyake et al. (2012) initially ascribed it to a nearby supernova, but it was soon proven by Usoskin et al. (2013) that the event was caused by an extreme solar particle event (ESPE) whose intensity was several orders of magnitude stronger than that of any solar particle storm we have experienced during the modern era. Several other ESPEs have been found since then throughout the Holocene and even beyond, as discussed in Section 2.

Another important discovery also made in 2012 was related to the occurrence of super-flares on sun-like stars (Maehara et al. 2012) suggesting that the Sun can rarely produce powerful flares one – two orders of magnitude stronger than those directly observed solar flares. Although the fact that the Sun can potentially produce super-flares is generally accepted, their occurrence frequency and strength are still debated as discussed in greater detail in Section 3.

After the discovery of the extreme events, big efforts have been made by the research community to systematise, analyse and understand the extreme events on the Sun. The first dedicated workshop on extreme solar events was organised in Nagoya in October 2018 (`https://cicr.isee.nagoya-u.ac.jp/site1/info_e/workshop/2018/1002.html`) and led to the first textbook where all available at that time information was collected (Miyake et al. 2020). The community-driven effort continued as an international team *Solar Extreme Events: Setting Up a Paradigm* (SEESUP – `https://www.issibern.ch/teams/solextremevent`) at the International Space Science Institute in Bern in 2020 – 2022. As a result, a summary paper was wrapped up (Usoskin et al. 2023) summarising recent developments. Another recent review about the state of the art in extreme solar events was published by Cliver et al. (2022) with a broader view on the field of extreme events.

Thus, the paradigm of extreme solar events is set now but many questions are still open. Here, we briefly review the current status of research in this field with an emphasis on the most recent developments. The reader is directed to the reviews mentioned above for more detailed information.

2. Extreme solar particle events (ESPEs)

Earth is continuously bombarded by energetic particles of Galactic origin. This forms galactic cosmic rays (GCRs) with an omnipresent and slightly variable flux of energetic (>1 GeV/nucleon) particles whose composition ranges from protons ($\approx 90\%$) to iron and beyond. The GCR flux is slightly modulated by solar/heliospheric activity and geomagnetic field (Potgieter 2013). GCRs initiate nucleonic-muon-electromagnetic cascades (showers) in the Earth's atmosphere and produce trace amounts of radioactive isotopes such as ^{14}C (known as radiocarbon), ^{10}Be, ^{36}Cl, etc. (Beer et al. 2012). Since this forms the only (main) source of such isotopes in the terrestrial system, they are called cosmogenic isotopes which can be stored in natural dateable archives and provide a very useful tool to reconstruct solar activity in the past, mostly for the last twelve millennia of the Holocene (e.g., Usoskin et al. 2003; Solanki et al. 2004). In addition to the variable omnipresent GCR flux, sporadic solar energetic particle (SEP) events can be produced by the Sun, when intense fluxes of energetic particles (in the energy range from tens MeV to several GeV) are accelerated by solar flares or coronal mass ejections and dominate the near-earth radiation field for hours – days (e.g., Desai and Giacalone 2016). SEPs can potentially also produce cosmogenic isotopes (Lingenfelter and Ramaty 1970) but it was believed that such events are undetectable because of too low yield of the isotopes (Usoskin et al. 2006) for the known SEP events.

Accordingly, when the first spike corresponding to the year 775 CE was discovered by Miyake et al. (2012), it was initially ascribed to exotic sources such as a nearby supernova, gamma-ray burst or even a cometary impact on Earth. However, it was soon demonstrated by Usoskin et al. (2013) that the only feasible explanation for that event was an extreme solar particle event with the SEP fluence being several orders of magnitude stronger than that for known SEP events. At the time of this paper's writing, five ESPEs were known and confirmed and four candidates are still waiting for confirmation with independent data sources, as summarized in Table 1.

Productions of different cosmogenic isotopes have different effective energies of SEPs, ranging from 30 to 300 MeV (Koldobskiy et al. 2022). Accordingly, the use of different isotopes makes it possible to estimate the energy/rigidity spectrum of the primary particles (Mekhaldi et al. 2015; O'Hare et al. 2019). A recent reconstruction of the spectral fluences of several strong ESPEs (Koldobskiy et al. 2023) is shown in Figure 1 along with those for typical (soft-, moderate- and hard-spectrum) directly observed strong SEP

Table 1. List of known extreme solar particle events including their date, estimated F_{200} fluence and key references. For comparison, the strongest directly observed SEP event of 23-Feb-1956 is also shown.

Date	F_{200} (in 10^9 cm^{-2})	References
774 CE	9.6 ± 4.1	Miyake et al. (2012); Usoskin et al. (2013) Büntgen et al. (2017)
7176 BCE	8.1 ± 3.6	Brehm et al. (2021); Paleari et al. (2022)
660 BCE	7.4 ± 3.4	O'Hare et al. (2019); Sakurai et al. (2020)
5259 BCE	7.0 ± 3.1	Brehm et al. (2022)
993 CE	5.2 ± 2.3	Miyake et al. (2013); Büntgen et al. (2017)
5410 BCE †	$\approx 4.5^*$	Miyake et al. (2021)
1279 CE †	$\approx 4.0^*$	Brehm et al. (2021); Miyahara et al. (2022)
1052 CE †	$\approx 3.2^*$	Brehm et al. (2021); Terrasi et al. (2020)
≈12300 BCE †	n/a	Bard et al. (2023)
23-Feb-1956	0.137	Usoskin et al. (2020); Hayakawa et al. (2024)

† awaits for independent confirmation. * preliminary estimate.

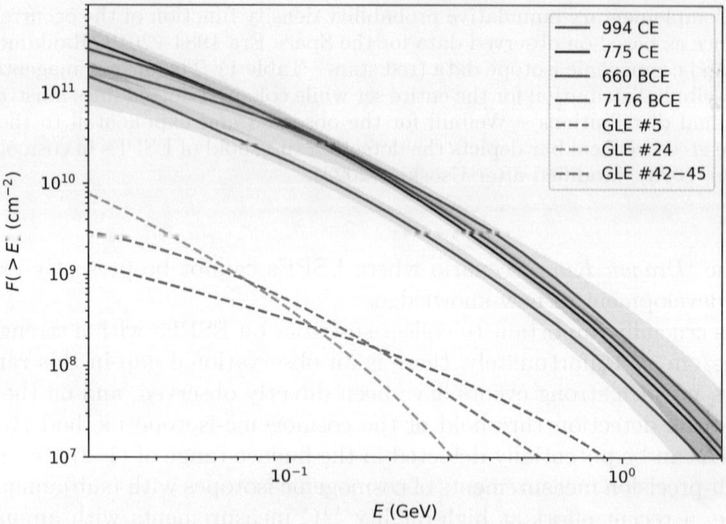

Figure 1. Spectral fluence estimates for four ESPEs (solid lines with shaded uncertainties, as indicated in the legend) in comparison with typical observed strong SEP events: hard-spectrum (GLE #5, 23-Feb-1956), soft-spectrum (GLE #24, 04-Aug-1972), and a moderate spectrum (series of GLEs #42–24, 20-Sep through 24-Oct-1989) ones. After Koldobskiy et al. (2023).

events. As seen, the ESPE spectra are quite similar in shape to those of the SEP events but two orders of magnitude higher.

The probability distribution of SEP event occurrence is shown in Figure 2 jointly for data directly obtained for the Space exploration era (blue dots) and for ESPEs reconstructed from cosmogenic-isotope proxies (red stars). One can observe a large gap between the directly measured and proxy-based datasets raising a question of whether the two sets are manifestations of the same phenomenon (solar eruptive events) and belong to the same distribution or form two different distributions. The former case is illustrated by the jointly fitted distribution (magenta curve in Figure 2) and would imply the so-called '*Black Swan*' scenario where the extreme events can be understood in the framework of the existing knowledge. The latter case is indicated by the two different distributions (blue and red dotted curves) implying that the two phenomena are different in nature

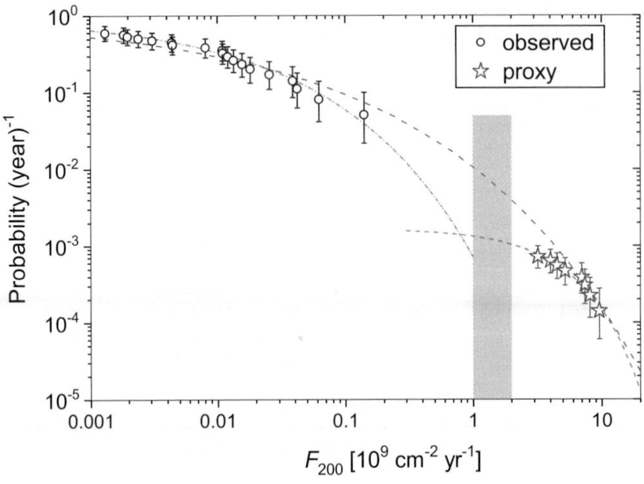

Figure 2. Complementary cumulative probability density function of the occurrence of annual SEP F_{200} uence as based on observed data for the Space Era 1984 – 2019 (Raukunen et al. 2022, – blue dots) and cosmogenic isotope data (red stars – Table 1). The dashed magenta line denotes the best-fit Weibull distribution for the entire set while coloured dotted lines illustrate the formal fits to individual distributions – Weibull for the observed and exponential to the proxy-based datasets. The green vertical bar depicts the detection threshold of ESPEs in cosmogenic isotopes (Usoskin et al. 2020). Modified after Usoskin (2023).

and form the '*Dragon King*' scenario where ESPEs cannot be presently explained and require the development of new knowledge.

Thus, it is crucially important to collect statistics on ESPEs with a strength of $F_{200} \approx 10^9$ particles/cm^{-2}. Unfortunately, there is an observational gap in this range because, on one hand, no such strong events have been directly observed, and on the other hand, it is close to the detection threshold of the cosmogenic-isotope method (Usoskin et al. 2020). ESPEs can be potentially detected in the fluence range of (1–2)$\cdot 10^9$ cm^{-2} but this requires high-precision measurements of cosmogenic isotopes with (sub)annual resolution. For example, a recent effort at high-quality ^{14}C measurements with annual resolution for the past millennium not only confirmed the ESPE of 993 CE but also led to the discovery of two other ESPE event candidates in 1052 CE and 1279 CE (see Table 1). At present, about 2000 years (about 16%) of the Holocene are covered by high-quality annual resolution of ^{14}C (Brehm et al. 2022). Thus, while it is unlikely that the largest ESPEs ($F_{200} \geq 10^{10}$ cm^{-2}) would be still unknown for the Holocene, many milder events ($F_{200} < 5 \cdot 10^9$ cm^{-2}) may still be expected to be discovered but this requires time.

An analysis of the known ESPEs implies that they are more likely to occur during periods of moderate solar activity, viz. neither grand minima nor grand maxima of activity (Usoskin et al. 2023). The association of the ESPE to the solar cycle phase is difficult to assess since such events greatly distort the 11-year solar cycle visible in cosmogenic isotopes (Usoskin et al. 2021). No periodicities have been found in the ESPE occurrence throughout the Holocene. Some clustering of the events can be observed during the 10th – 13th centuries (three events between 993 – 1279 CE) but this might be accidental. The question of the possible existence of the maximum strength of ESPEs is still open – although the event of 774 CE is the strongest one over the Holocene, the newly discovered event of ca. 12350 BCE (Bard et al. 2023) has not yet been analysed and may be found even stronger. We expect to get more statistics on ESPEs in the near future to be able to answer the main question of the main mechanism producing such events.

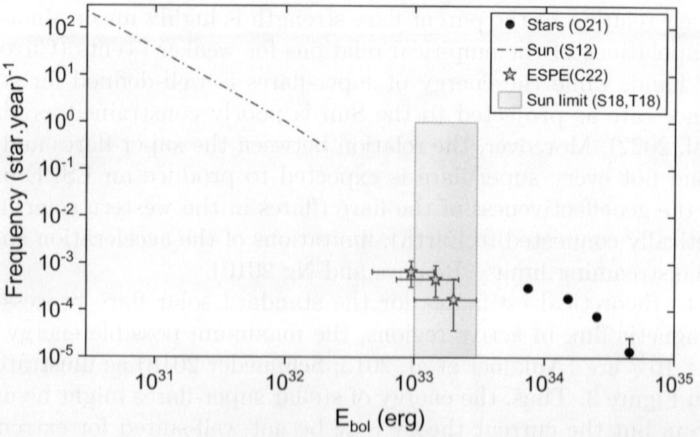

Figure 3. Complementary cumulative probability density function of the occurrence of flares with the bolometric energy exceeding E_{bol}. The red dashed line denotes an approximate power-law of solar flare distribution (S12 – Schrijver et al. 2012), blue dots super-flares on sun-like stars (O21 – Okamoto et al. 2021), red stars projection of ESPEs (C22 – Cliver et al. 2022). The vertical grey bar denotes a theoretical upper limit of the solar-flare energy (Schmieder 2018; Tschernitz et al. 2018).

3. Super-flares

The most energetic directly observed solar flare was recorded on 04 November 2003 with the bolometric energy of $\approx 4.3 \cdot 10^{32}$ erg (Emslie et al. 2012), which is comparable to the estimated strength of the Carrington flare (e.g., Cliver and Dietrich 2013). Can stronger super-flares (often defined as flares with the bolometric energy E_{bol} exceeding 10^{33} erg) be unleashed by the Sun? This question is difficult to answer since we do not know any detectable signatures solar flares can leave in natural archives. Indirect projections are highly uncertain. For example, the estimates of the strength of the flare corresponding to the 774 ESPE range from $2 \cdot 10^{33}$ to $3 \cdot 10^{34}$ erg (see Sect. 7.8 in Cliver et al. 2022).

Another way to estimate the super-flares is to look not at the Sun through millennia but at sun-like stars directly observed nowadays by space-based telescopes. Although the period of observations is short, only a few years, the population of sun-like stars is large, including several thousand samples, making the statistics comparable to that of the Sun (tens of thousands of star*years). This method, based on the stellar observations made by the *Kepler* mission during roughly four years, led Maehara et al. (2012) to discover super-flares on sun-like stars. The energy of these super-flares ranged from 10^{34} to 10^{36} erg. However, a direct projection of the stellar data on the Sun is not straightforward since such parameters of the star as age, temperature, rotation period, depth of the convective zone and the differential rotation may affect the flaring activity of the star. Also, the data may be subjected to observational biases since flares on more active stars are typically better observed, and their rotational period is better defined. This makes the stellar projection to the Sun unclear (Reinhold et al. 2020). Several updates and revisions to the analysis of stellar super-flares have been performed (Maehara et al. 2015; Okamoto et al. 2021; Vasilyev et al. 2022), but the results are still not well defined. In particular, the occurrence rate and/or strength of super-flares reduced to the solar conditions are much higher, by a factor of 10 – 30, than those estimated for the ESPEs, as seen in Figure 3. There may be different factors related to this discrepancy. On the one hand, the occurrence rate of ESPEs is defined robustly, especially for the strongest events, but their relation to the parent flare strength is highly uncertain as based on far-reaching extrapolations of the empirical relations for weaker events (Cliver et al. 2022). On the other hand, while the energy of super-flares is well-defined for stellar samples, their occurrence rate as projected to the Sun is poorly constrained as discussed above (Vasilyev et al. 2022). Moreover, the relation between the super-flares and ESPEs is not one-to-one, and not every super-flare is expected to produce an ESPE on Earth. This is defined by the geoeffectiveness of the flare (flares in the western solar hemisphere are better magnetically connected to Earth), limitations of the acceleration and transport of SEPs (e.g., the streaming limit – Reames and Ng 2010).

According to theoretical estimates for the standard solar flare process based on the amount of magnetic flux in active regions, the maximum possible energy of solar flares is limited to $< 10^{34}$ erg (Aulanier et al. 2013; Schmieder 2018) as illustrated by the grey vertical bar in Figure 3. Thus, the energy of stellar super-flares might be unreachable for the modern Sun but the current theory may be not well-suited for extreme cases when the energy partition between various processes may differ from that for regular cases.

4. Discussion and open questions

The topic of extreme solar events was opened a dozen years ago and has significantly matured since then with essential progress being reached. The current paradigm, created as a result of the hard work of many research groups internationally, acknowledges that the modern Sun can rarely produce extreme eruptive events several orders of magnitude stronger than 'typical' solar storms experienced during the last decades of the Space exploration era. While such extreme events might have passed unnoticed by the earlier society, they would have led to dramatic consequences should they appear now. This would include, for example, damage or destruction of most of the spacecraft and satellites, leading to long-lasting navigation and communication blackouts, malfunction of various control and operational systems, and damage to landlines (e.g., Miyake et al. 2020).

An interesting application of the ESPEs is found in archaeology and glaciology. The signatures of these events are so pronounced that allow absolute dating of floating chronologies of dendrochronology and ice-core analyses. For example, the analysis of the ESPE of 774 CE led to a correction of the ice-core dating by 5 – 7 years (e.g., Sukhodolov et al. 2017), while the radiocarbon spikes corresponding to the ESPE of 993 CE allowed precise dating of the Viking's settlements in Greenland and America (Kuitems et al. 2022; Philippsen et al. 2022).

Accordingly, understanding and modelling of such events is crucial not only for academic research including Solar and Stellar Physics, Space Physics, and Geosciences, but also for stakeholders beyond Academia. Although the very existence of solar extreme events and their severity are beyond doubt, many questions are still open and need to be answered in the future.

One important question is related to the nature of the extreme events – Are they *Black Swans* (far tail of the known event distribution) or *Dragon Kings* (yet unknown type of events). To answer this question, the gap in the observed/reconstructed event distribution in the F_{200} range of $(1-3)\cdot 10^9$ cm^{-2} (see Figure 2) needs to be filled. This will require extensive work on high-precision measurements of large amounts of tree-ring and ice-core samples (e.g., Brehm et al. 2021) to cover a major fraction of the Holocene.

Another important question is – What is the strongest solar event? So far, the event of 774 CE is the strongest confirmed event over the Holocene. The newly discovered event of ca. 12300 BCE is the strongest in the Δ^{14}C increase (Bard et al. 2023) but, since it took place during the ice age, it is not yet analysed in the sense of the ESPE strength.

The relation between the statistics of the projection of super-flares ($E_{\rm bol} > 10^{33}$ erg) and ESPEs reconstructed from cosmogenic isotopes (Figure 3) needs to be resolved. The two statistics mismatch by one – two orders of magnitude, and it is presently unknown whether this is caused by statistical, viz. the missing ESPE events over the Holocene or incorrect projection of the stellar data to the Sun, or physical reasons, viz. only a small fraction of super-flares can produce ESPEs.

The latter is also linked to a theoretical question of what the upper limit for the solar eruptive events and the related SEP events is. Extrapolations of the existing models suggest that super-flares observed for cool starts are unlikely to occur on the Sun but these extrapolations are not validated for extreme cases. For the ESPEs, at present there is no theoretical model able to provide a self-consistent consideration of energetic particle acceleration and transport.

In conclusion, the paradigm of a new, previously unknown type of energetic eruptive events, which can strongly affect the Earth's environment, is now set, and the events are being studied. However, the time of fast discoveries is likely over and more work is needed to deepen and broaden our knowledge about ESPEs.

Acknowledgements

The author is grateful to the organisers of the IAU Symposium #388 for the opportunity to present this review motivated by the work supported by the International Space Science Institution (ISSI, Bern) via the International Team #510 (Solar Extreme Events: Setting Up a Paradigm – SEESUP, 2020 – 2022) and a visiting fellowship in 2024.

References

Aulanier, G., Démoulin, P., Schrijver, C. J., Janvier, M., Pariat, E., & Schmieder, B. 2013, The standard flare model in three dimensions. II. Upper limit on solar flare energy. *Astron. Astrophys.*, 549, A66.

Bard, E., Miramont, C., Capano, M., Guibal, F., Marschal, C., Rostek, F., Tuna, T., Fagault, Y., & Heaton, T. 2023, A radiocarbon spike at 14300 cal yr BP in subfossil trees provides the impulse response function of the global carbon cycle during the Late Glacial. *Phil. Trans. R. Soc. A*, 381, 20220206.

Beer, J., McCracken, K., & von Steiger, R. 2012,. *Cosmogenic Radionuclides: Theory and Applications in the Terrestrial and Space Environments.* Springer, Berlin.

Brehm, N., Bayliss, A., Christl, M., Synal, H., Adolphi, F., Beer, J., Kromer, B., Muscheler, R., Solanki, S., Usoskin, I., Bleicher, N., Bollhalder, S., Tyers, C., & Wacker, L. 2021, Eleven-year solar cycles over the last millennium revealed by radiocarbon in tree rings. *Nature Geosci.*, 14, 10–15.

Brehm, N., Christl, M., Knowles, T. D. J., Casanova, E., Evershed, R., Adolphi, F., Muscheler, R., Synal, H.-A., Mekhaldi, F., Paleari, C., Leuschner, H.-H., Bayliss, A., Nicolussi, K., Pichler, T., Schlüchter, C., Pearson, C., Salzer, M., Fonti, P., Nievergelt, D., Hantemirov, R., Brown, D., Usoskin, I., & Wacker, L. 2022, Tree-rings reveal two strong solar proton events in 7176 and 5259 BCE. *Nature Comm.*, 13, 1196.

Büntgen, U., Eggertsson, Ó., Wacker, L., Sigl, M., Ljungqvist, F. C., Di Cosmo, N., Plunkett, G., Krusic, P. J., Newfield, T. P., Esper, J., Lane, C., Reinig, F., & Oppenheimer, C. 2017, Multi-proxy dating of Iceland's major pre-settlement Katla eruption to 822-823 CE. *Geology*, 45, 783–786.

Cliver, E. W. & Dietrich, W. F. 2013, The 1859 space weather event revisited: limits of extreme activity. *J. Space Weath. Space Clim.*, 3(26), A31.

Cliver, E. W., Schrijver, C. J., Shibata, K., & Usoskin, I. G. 2022, Extreme solar events. *Liv. Rev. Solar Phys.*, 19(1), 2.

Desai, M. & Giacalone, J. 2016, Large gradual solar energetic particle events. *Liv. Rev. Solar Phys.*, 13, 3.

Emslie, A. G., Dennis, B. R., Shih, A. Y., Chamberlin, P. C., Mewaldt, R. A., Moore, C. S., Share, G. H., Vourlidas, A., & Welsch, B. T. 2012, Global Energetics of Thirty-eight Large Solar Eruptive Events. *Astrophys. J.*, 759, 71.

Hayakawa, H., Koldobskiy, S., Mishev, A., Poluianov, S., Gil, A., Usoskina, I., & Usoskin, I. 2024, Revision of the strongest solar energetic particle event of 23 February 1956 (GLE #5) based on the rediscovered original records. *Astron. Astrophys.*, 684, A46.

Koldobskiy, S., Mekhaldi, F., Kovaltsov, G., & Usoskin, I. 2023, Multiproxy Reconstructions of Integral Energy Spectra for Extreme Solar Particle Events of 7176 BCE, 660 BCE, 775 CE, and 994 CE. *J. Geophys. Res. (Space Phys.)*, 128(3), e2022JA031186.

Koldobskiy, S. A., Usoskin, I., & Kovaltsov, G. 2022, Effective energy of cosmogenic isotope (^{10}Be, ^{14}C and ^{36}Cl) production by solar energetic particles and galactic cosmic rays. *J. Geophys. Res. Space Phys.*, 127, e2021JA029919.

Kuitems, M., Wallace, B., Lindsay, C., Scifo, A., Doeve, P., Jenkins, K., Lindauer, S., Erdil, P., Ledger, P., Forbes, V., Vermeeren, C., Friedrich, R., & Dee, M. 2022, Evidence for European presence in the Americas in AD 1021. *Nature*, 601(7893), 388–391.

Lingenfelter, R. & Ramaty, R. Astrophysical and geophysical variations in c-14 production. In Olsson, I., editor, *Proc. 12th Nobel symposium, Radiocarbon variations and absolute chronology* 1970, pp. 513–537. John Wiley & Sons, NY.

Maehara, H., Shibayama, T., Notsu, S., Notsu, Y., Nagao, T., Kusaba, S., Honda, S., Nogami, D., & Shibata, K. 2012, Superflares on solar-type stars. *Nature*, 485, 478–481.

Maehara, H., Shibayama, T., Notsu, Y., Notsu, S., Honda, S., Nogami, D., & Shibata, K. 2015, Statistical properties of superflares on solar-type stars based on 1-min cadence data. *Earth, Planets and Space*, 67, 59.

Mekhaldi, F., Muscheler, R., Adolphi, F., Aldahan, A., Beer, J., McConnell, J., Possnert, G., Sigl, M., Svensson, A., Synal, H., Welten, K., & Woodruff, T. 2015, Multiradionuclide evidence for the solar origin of the cosmic-ray events of AD 774/5 and 993/4. *Nature Comm.*, 6, 8611.

Miyahara, H., Tokanai, F., Moriya, T., Takeyama, M., Sakurai, H., Ohyama, M., Horiuchi, K., & Hotta, H. 2022, Recurrent Large-Scale Solar Proton Events Before the Onset of the Wolf Grand Solar Minimum. *Geophys. Res. Lett.*, 49(5), e2021GL097201.

Miyake, F., Masuda, K., & Nakamura, T. 2013, Another rapid event in the carbon-14 content of tree rings. *Nature Comm.*, 4, 1748.

Miyake, F., Nagaya, K., Masuda, K., & Nakamura, T. 2012, A signature of cosmic-ray increase in ad 774-775 from tree rings in Japan. *Nature*, 486, 240–242.

Miyake, F., Panyushkina, I. P., Jull, A. J. T., Adolphi, F., Brehm, N., Helama, S., Kanzawa, K., Moriya, T., Muscheler, R., Nicolussi, K., Oinonen, M., Salzer, M., Takeyama, M., Tokanai, F., & Wacker, L. 2021, A single year cosmic ray event at 5410 BCE registered in 14C of tree rings. *Geophys. Res. Lett.*, 48, e2021GL093419.

Miyake, F., Usoskin, I., & Poluianov, S., editors 2020,. *Extreme Solar Particle Storms: The Hostile Sun.* IOP Publishing, Bristol, UK.

O'Hare, P., Mekhaldi, F., Adolphi, F., Raisbeck, G., Aldahan, A., Anderberg, E., Beer, J., Christl, M., Fahrni, S., Synal, H.-A., Park, J., Possnert, G., Southon, J., Bard, E., Aster Team, & Muscheler, R. 2019, Multiradionuclide evidence for an extreme solar proton event around 2,610 B.P. (660 BC). *Proc. Nat. Acad. Sci.*, 116(13), 5961–5966.

Okamoto, S., Notsu, Y., Maehara, H., Namekata, K., Honda, S., Ikuta, K., Nogami, D., & Shibata, K. 2021, Statistical Properties of Superflares on Solar-type Stars: Results Using All of the Kepler Primary Mission Data. *Astrophys. J.*, 906(2), 72.

Paleari, C., Mekhaldi, F., Adolphi, F., M., C., Vockenhuber, C., Gautschi, P., Beer, J., Brehm, N., Erhardt, T., Synal, H., Wacker, L., F., W., & Muscheler, R. 2022, Cosmogenic radionuclides reveal an extreme solar particle storm near a solar minimum 9125 years bp. *Nature Comm.*, 13, 214.

Philippsen, B., Feveile, C., Olsen, J., & Sindbæk, S. 2022, Single-year radiocarbon dating anchors Viking Age trade cycles in time. *Nature*, 601(7893), 392–396.

Potgieter, M. 2013, Solar Modulation of Cosmic Rays. *Living Rev. Solar Phys.*, 10, 3.

Raukunen, O., Usoskin, I., Koldobskiy, S., Kovaltsov, G., & Vainio, R. 2022, Annual integral solar proton fluences for 1984–2019. *Astron. Astrophys.*, 660, A65.

Reames, D. V. & Ng, C. K. 2010, Streaming-limited Intensities of Solar Energetic Particles on the Intensity Plateau. *Astrophys. J.*, 723, 1286–1293.

Reimer, P. J., Baillie, M. G. L., Bard, E., Bayliss, A., Beck, J. W., Blackwell, P. G., Ramsey, C. B., Buck, C. E., Burr, G. S., Edwards, R. L., Friedrich, M., Grootes, P. M., Guilderson, T. P., Hajdas, I., Heaton, T. J., Hogg, A. G., Hughen, K. A., Kaiser, K. F., Kromer, B., McCormac, F. G., Manning, S. W., Reimer, R. W., Richards, D. A., Southon, J. R., Talamo, S., Turney, C. S. M., van der Plicht, J., & Weyhenmeye, C. E. 2009, Intcal09 and marine09 radiocarbon age calibration curves, 0-50000 years cal bp. *Radiocarbon*, 51(4), 1111–1150.

Reinhold, T., Shapiro, A. I., Solanki, S. K., Montet, B., Krivova, N. A., Cameron, R. H., & Amazo-Gómez, E. M. 2020, The Sun is less active than other solar-like stars. *Science*, 368(6490), 518–521.

Sakurai, H., Tokanai, F., Miyake, F., Horiuchi, K., Masuda, K., Miyahara, H., Ohyama, M., Sakamoto, M., Mitsutani, T., & Moriya, T. 2020, Prolonged production of ^{14}C during the 660 BCE solar proton event from Japanese tree rings. *Sci. Rep.*, 10, 660.

Schmieder, B. 2018, Extreme solar storms based on solar magnetic field. *J. Atmos. Solar-Terr. Phys.*, 180, 46–51.

Schrijver, C. J., Beer, J., Baltensperger, U., Cliver, E., Güdel, M., Hudson, H., McCracken, K., Osten, R., Peter, T., Soderblom, D., Usoskin, I., & Wolff, E. 2012, Estimating the frequency of extremely energetic solar events, based on solar, stellar, lunar, and terrestrial records. *J. Geophys. Res.*, 117, A08103.

Solanki, S. K., Usoskin, I. G., Kromer, B., Schüssler, M., & Beer, J. 2004, Unusual activity of the sun during recent decades compared to the previous 11,000 years. *Nature*, 431, 1084–1087.

Sukhodolov, T., Usoskin, I., Rozanov, E., Asvestari, E., Ball, W., Curran, M., Fischer, H., Kovaltsov, G., Miyake, F., Peter, T., Plummer, C., Schmutz, W., Severi, M., & Traversi, R. 2017, Atmospheric impacts of the strongest known solar particle storm of 775 AD. *Sci. Rep.*, 7, 45257.

Terrasi, F., Marzaioli, F., Buompane, R., Passariello, I., Porzio, G., Capano, M., Helama, S., Oinonen, M., Njd, P., Uusitalo, J., Jull, A., Panyushkina, I., Baisan, C., M., M., Varga, T., Kovaltsov, G., Poluianov, S., & Usoskin, I. 2020, Can the 14C production in 1055 CE be affected by SN1054? *Radiocarbon*, 62(5), 1403–1418.

Tschernitz, J., Veronig, A., Thalmann, J., Hinterreiter, J., & Pötzi, W. 2018, Reconnection Fluxes in Eruptive and Confined Flares and Implications for Superflares on the Sun. *Astrophys. J.*, 853(1), 41.

Usoskin, I., Miyake, F., Baroni, M., Brehm, N., Dalla, S., Hayakawa, H., Hudson, H., Jull, A. J., Knipp, D., Koldobskiy, S., Maehara, H., Mekhaldi, F., Notsu, Y., Poluianov, S., Rozanov, E., Shapiro, A., Spiegl, T., Sukhodolov, T., Uusitalo, J., & Wacker, L. 2023, Extreme Solar Events: Setting up a Paradigm. *Space Sci. Rev.*, 219(8), 73.

Usoskin, I., Solanki, S., Kovaltsov, G., Beer, J., & Kromer, B. 2006, Solar proton events in cosmogenic isotope data. *Geophys. Res. Lett.*, 33, L08107.

Usoskin, I. G. 2023, A History of Solar Activity over Millennia. *Living Rev. Solar Phys.*, 20, 2.

Usoskin, I. G., Koldobskiy, S. A., Kovaltsov, G. A., Rozanov, E. V., Sukhodolov, T. V., Mishev, A. L., & Mironova, I. A. 2020, Revisited Reference Solar Proton Event of 23 February 1956: Assessment of the Cosmogenic-Isotope Method Sensitivity to Extreme Solar Events. *J. Geophys. Res. (Space Phys.)*, 125(6), e27921.

Usoskin, I. G. & Kovaltsov, G. A. 2012, Occurrence of Extreme Solar Particle Events: Assessment from Historical Proxy Data. *Astrophys. J.*, 757, 92.

Usoskin, I. G., Kromer, B., Ludlow, F., Beer, J., Friedrich, M., Kovaltsov, G. A., Solanki, S. K., & Wacker, L. 2013, The AD775 cosmic event revisited: the Sun is to blame. *Astron. Astrophys.*, 552, L3.

Usoskin, I. G., Solanki, S. K., Krivova, N., Hofer, B., Kovaltsov, G. A., Wacker, L., Brehm, N., & Kromer, B. 2021, Solar cyclic activity over the last millennium reconstructed from annual ^{14}C data. *Astron. Astrophys.*, 649, A141.

Usoskin, I. G., Solanki, S. K., Schüssler, M., Mursula, K., & Alanko, K. 2003, Millennium-scale sunspot number reconstruction: Evidence for an unusually active sun since the 1940s. *Phys. Rev. Lett.*, 91, 211101.

Vasilyev, V., Reinhold, T., Shapiro, A. I., Krivova, N. A., Usoskin, I., Montet, B. T., Solanki, S. K., & Gizon, L. 2022, Superflares on solar-like stars. A new method for identifying the true flare sources in photometric surveys. *Astron. Astrophys.*, 668, A167.

Solar and Stellar Coronal Mass Ejections
Proceedings IAU Symposium No. 388, 2026
N. Gopalswamy, O. Malandraki, A. Vidotto & W. Manchester, eds.
doi:10.1017/S1743921324001170

Radio Observations of Stellar CMEs

Rachel A. Osten

Space Telescope Science Institute
Johns Hopkins University
osten@stsci.edu

Abstract. Radio observations have the potential advantage to provide not only direct confirmation about the occurrence of stellar CMEs, but are also a potential tool to probe the systematic properties of stellar CMEs. I will describe the rationale behind using radio observations to probe the incidence and properties of stellar CMEs. I'll start with a brief review of solar radio emissions related to CMEs, then switch to observational attempts to identify stellar type II bursts, which so far have resulted in no confident detections. I will step through the several assumptions made regarding stellar type II bursts and examine their astrophysical significance. I'll then pan out to consider other possibilities in the radio domain such as type IV bursts and negative radio bursts, and end with some perspective on future opportunities for making progress in studying stellar CMEs using radio observations.

Keywords. stellar radio, stellar flares, stellar CMEs, solar-stellar connection

1. Introduction

1.1. *Solar vs stellar, flares and CMEs*

The study of space weather really began in 1859 with the Carrington event on the Sun (Carrington 1859), and has made vast advances since then. The Carrington event is acknowledged to be the first detection of a solar flare, which is widely understood to occur as a result of rearrangement of magnetic fields on the surface of the Sun and consequent liberation of magnetic energy from reconnection, which goes into particle acceleration, plasma heating, and mass motions (Benz and Güdel 2010). In the solar context these energy releases have a maximum of about 10^{32} ergs. Routine solar CME detection began in 1971 (Gosling et al. 1974), and advances in this field have yielded an appreciation that CMEs are the primary driver of space weather, as the most geoeffective component of a solar eruptive event. CMEs have masses ranging from 10^{16}-10^{17}g and maximum energies around 10^{33} erg. There are numerous connections between solar flares and CMEs based on the large number of joint events.

In contrast, stellar flares were first detected in 1924 (Hertzsprung 1924). These events tend to be much more energetic, with the largest values going up to 10^{36} and higher. Stellar flares are studied across the electromagnetic spectrum, and there have been several studies connecting solar and stellar flares, such as those relating flare peak coronal temperature and emission measure (Shibata and Yokoyama 1999), and white-light flare studies (Namekata et al. 2017). Astronomers are still in the process of coming up with detection strategies for stellar CMEs, and the ultimate goal is to connect stellar flares with stellar CMEs, as well as with stellar CMEs and solar CMEs, to form an understanding of the relation of mass ejection to magnetic reconnection across different types of stars and stellar environments.

Table 1. Table of New and/or Upgraded Radio Facilities

Facility	Frequency Range
LOFAR	10-240 MHz
MWA	80-300 MHA
uGMRT	50-1450 MHz
ASKAP	700-1800 MHz
MeerKAT	540 MHz-3.5 GHz
FAST	70 MHz - 3 GHz
JVLA	58 MHz - 50 GHz
ALMA	30 -950 GHz

1.2. *What do radio observations bring to the stellar case?*

The term "radio" spans a wide range of frequency/wavelength, from meter-wavelengths to decimeter even into the mm. In the current age where nations are exploring the lunar surface and the possibility of radio telescopes on the far side of the moon, this term can be expanded even further still to consider km-wavelengths. One advantage of radio-wavelength observations is the possibility of routine access to all four Stokes parameters, which enable diagnostics of density and/or magnetic field, given assumptions about the emission mechanism. Current radio wavelength facilities capture emission as a function of frequency and time, enabling the construction of dynamic spectra, useful for characterizing and analyzing the emitted radiation and aiding in diagnosis of emission mechanism. On the down side, there can be a wide range of possible emission mechanisms contributing at these wavelengths. These emission mechanisms are generally divided into two broad categories – incoherent and coherent. For incoherent mechanisms (processes like bremsstrahlung, gyoresonance, gyrosynchrotron, synchrotron), the particles do not act in phase, and the total intensity can be predicted from the number density of particles doing the emitting. On the other hand, coherent processes are often quite variable and point to dynamic situations in the stellar atmosphere. Coherent processes are often the result of collective processes with particles operating in phase. Various wave-wave interactions can occur, and once there is a coupling to transverse waves there is a possibility to observe electromagnetic radiation. There is no dependence of the intensity on the number of particles so the expected intensity is more difficult to predict. The two main processes are closely linked to fundamental plasma parameters, namely the electron density (plasma frequency), and magnetic field (cyclotron maser emission). These can often be distinguished in the stellar case by the extreme levels of circular polarization which can approach 100%. In the last decade there have been a plethora of new and/or upgraded facilities that offer increased sensitivity and bandwidths. Table 13 listing facilities, frequency range, and web links to follow for interested readers.

2. On the potential for type II bursts

Type II bursts provide a unique radio signature from super-Alfvénic shocks as CMEs propagate through the stellar atmosphere. This results in a predictable shape in frequency and time and offers diagnostics of the propagation speed. The frequency drift rate, $d\nu/dt$, is given by the observing frequency ν, the speed at which the exciter is travelling through the atmosphere (v_B), the angle between this velocity and the line of sight θ, and the plasma scale height of the atmosphere H_n, as in

$$\frac{d\nu}{dt} = \frac{\nu \cos\theta v_B}{2H_n}. \tag{1}$$

Due to the relation between the observing frequency and the local electron density, the observed frequencies enable constraints on the ambient electron density through which the shock is passing, as well as the height in the atmosphere (with a suitable model for the density scale height). The phenomenon is expected to occur at "low" frequencies – type II bursts on the Sun can start as high as several hundred MHz, and go down to kHz frequencies as the event transitions to interplanetary space. A new generation of low frequency radio telescopes (LOFAR, JVLA, MWA) combines increased sensitivity and frequency coverage to make this search feasible. A strong point in favor of this signature is that it originates from the CME itself, and not flares, and can potentially explore the systematic behavior of stellar CMEs if observed on a range of stellar types, without the need to invoke a strange property of a flare or to argue whether the signature arises from the flare as opposed to the CME.

A detection of a stellar radio type II burst would naturally lead to inferences of the properties of the CME from that of the shock producing the type II burst. Crosley et al. (2017) used a selection of Green Bank Solar Radio Burst Spectrometer full-disk observations of solar type II spectra, coupled with GOES 1-8 Å sun-as-a-star X-ray light curves, to examine how well such observations would do in reproducing key properties of CMEs compared with measurements made from spatially resolved solar coronagraphic observations. The velocity can be derived directly from the dynamic spectrum; Crosley et al. determined that these coronagraphy-derived CME velocities and those derived from a dynamic spectrum analysis were generally in agreement to within 50%. Several papers have presented relationships between CME properties and flare properties – Osten and Wolk (2015) motivated the use of an equipartition between CME kinetic energy and flare bolometric energy,

$$\frac{1}{2} M_{\rm CME} v_{\rm CME}^2 = \frac{E_{\rm rad}}{\epsilon f_{\rm rad}} \tag{2}$$

where $M_{\rm CME}$ is the CME mass, $v_{\rm CME}$ the velocity of the CME, $E_{\rm rad}$ the radiated flare energy in a particular passband, $f_{\rm rad}$ the fraction of the total flare bolometric energy which appears in this passband, taking into account the multi-wavelength nature of flares, which involve multiple stellar atmospheric layers, and ϵ the factor relating the CME kinetic energy to the flare bolometric energy. Aarnio et al. (2012); Drake (2013) individually established a statistical correlation between solar CME mass and solar flare X-ray energy, characterized by a power-law of the form

$$M_{\rm CME} \propto E_{\rm flare}^{\gamma} \tag{3}$$

where $E_{\rm flare}$ is the solar X-ray flare energy, and γ is the exponent of the relationship between CME mass and flare energy as explored by Aarnio et al. and Drake. Crosley et al. (2017) used the GOES X-ray light curve information which accompanied the GBSRBS data to derive CME mass and kinetic energy estimates, which could then be compared with catalogged values. The CME mass estimates were within an order of magnitude, while kinetic energies spanned three orders of magnitude.

A series of papers (Crosley et al. 2016; Crosley and Osten 2018a,b) and Villadsen and Hallinan (2019) provided the first large-scale systematic attempts to detect stellar CMEs via searches for type II bursts. The expectations follow from solar flare-CME relationships: the most energetic solar flares have a 100% association with CMEs (Yashiro et al. 2006). CMEs with the highest velocities (above 2000 km/s) have a 100% association with producing type II bursts (Gopalswamy et al. 2008). The central question being addressed by these studies is whether a high flaring rate gives rise to a high rate of CMEs. Crosley and Osten (2018a) reported dynamic spectra of EQ Peg at P band. EQ Peg is a binary system, unresolved in the P-band measurements, with a high flaring rate. Over 20 hours of overlapping radio and optical data, there were several moderate optical flares but no signals over **XX** MHz span from 300-450 MHz. Crosley and Osten (2018b) pursued another 44 hours of JVLA only measurements of the star, and while two low-frequency bursts were detected, the features of these bursts (bandwidth, drift rate, duration) were not consistent with expectations for a type II burst. No type II bursts were observed in 64 hours of monitoring EQ Peg. We expected 1.2 flares/hr above the flare energy where all solar flares have an associated CME. Using a more conservative interpretation and using a large-scale model corona to identify likely Alfvén speed profiles from coronal base measurements, we would expect 1 flare every 27 hours to drive an observable shock. Villadsen and Hallinan (2019) also explored 5 active M dwarfs at radio wavelengths, monitoring at P band with the VLA and higher frequencies (1-2 GHz and 2.4-4 GHZ). Over a total of 58 hours they found almost 2 dozen coherent bursts, but nothing able to be interpreted as a type II burst.

This large number of null results is cause to re-examine the underlying assumptions, addressed separately here.

(1) *Could this be due to a mismatch between the type II parameters and the observing sensitivity?* The radio searches described above utilized particular wavelength ranges and specific radio telescopes. While there is no agreed-upon flux scaling for solar type II bursts with frequency with which to apply to stellar type II burst searches, simple geometric arguments predict the expected flux density for a source of a given solid angle and brightness temperature. Generally speaking detection requires large angular sizes (relative to the stellar disk) and high brightness temperatures (see Figure 5 in Crosley et al. 2016), and it is entirely possible that both conditions do not present at the same time. Alvarado-Gómez et al. (2020) performed 3D simulations of weakly confined and partially confined CME event in M dwarf coronae, finding that these CMEs can generate shocks, but they occur much farther away from the stellar surface than solar events – the transition from sub-Alfvenic to super-Alfvenic (necessary to create the shock that is the type II burst) happens at 10-20 $R_\star$, and this will produce emission at frequencies below the ionospheric cutoff.

(2) *Is it possible that CMEs are occurring but they are not able to exceed the local Alven speed and drive a type II burst?* The existence of a super-Alfvénic shock is the necessary precondition to form a type II burst. Observations on the Sun (Gopalswamy et al. 2001) show that often times type II bursts are not observed until further away from the solar active region, and this can be deduced as a consequence of the enhanced magnetic field which causes a subsequent increase in the local Alfvén speed. Gopalswamy et al. described the importance of magnetic structures in forming shocks observable as type II bursts on the Sun: weaker disturbances can drive shocks in quiet regions, while stronger disturbances have trouble exceeding the local Alfvén speed. Additionally, the lifetime of a shock depends on its initial speed, the lifetime of the driver, and the origin on the stellar surface as well as location of propagation.

(3) *Is it possible that we have a situation where flares are occurring but eruptions are not?* Active stars have large magnetic field strengths on their surfaces. Large overlying fields above an active region can prevent breakout or eruption. This has been seen in the Sun where solar active region 12192 produced many X-ray flares but few CMEs (Sun et al. 2015). Magnetic suppression of CMEs has been modelled in the solar case with helical kink instabilities of a twisted magnetic flux rope (Török and Kliem 2005).

(4) *Do the large scale fields on M dwarfs prevent breakout?* The simulations in Alvarado-Gómez et al. (2020) do show that it is possible for large scale overlying fields to confine eruptions.

3. The intrigue of type IV signatures

Type IV bursts can be used as signposts of CMEs and space weather. In the solar case, there are two classes of type IV bursts: moving type IV bursts are often preceded by a type II burst, and produced by gyrosynchrotron or plasma emission from fast-moving electrons trapped in the magnetic fields of an erupting CME. Stationary type IV bursts are often associated with large long-duration flares which provide energetic particles, and tend not to be associated with CMEs. Type IV emission in both cases arises from the presence of accelerated particles, but by itself does not provide strong proof that an eruption has taken place.

Zic et al. (2020) reported on dynamic spectra from Proxima where they interferometrically detected a coherent burst temporally coincident with an optical flare on Proxima, using ASKAP at a frequency of 888 GHz. The event had a ratio of circularly polarized intensity to total intensity of more than 50%, along with narrow-band and sharp cutoff spectral features evident in the dynamic spectrum. Given the association with an optical flare, Zic et al. suggested this was driven by large-scale magnetic field reconfiguration associated with a post-eruptive loop arcade. They noted this was "highly suggestive" of a CME.

More recently, Mohan et al. (2024) observed coherent burst in AD Leo with uGMRT which they identified as a type IV burst based on the dynamic spectrum characteristics and circular polarization values. This observation was at a similar frequency range as reported by Zic – 550 to 800 MHz, although with no accompanying observations. Similarly to the conclusion drawn by Zic, these authors note the event is "consistent with a type IV radio burst from flare-accelerated particles trapped in magnetic loops, which *could* be a CME signature".

4. A speculative look at "negative radio bursts" in stellar radio flares

Elsewhere in this symposium the subject of absorption dimmings has been raised as a signature of mass ejection. In addition to the possibility of seeing these dimmings at optical, EUV or X-ray wavelengths, "negative radio bursts" are an intriguing possibility to consider in the stellar context. Grechnev et al. (2008) detailed analysis of a filament eruption on the Sun which filled one-fourth of the visible solar surface, and this provides a good case study for consideration and extrapolation. There were numerous indications of a CME: coronagraph images, a type II burst, long-lived dimmings seen with SOHO/EIT at 304 Å and 195 Å, as well as in Hα. On top of this, the event was observed at multiple radio frequencies from 1-9.4 GHz. In all of these light curves (see Figure 1) there was a prominent dip in the decay phase of the radio flare lasting about an hour, whose depth varied as a function of frequency from a maximum of $\sim$10% at 1 GHz to $\sim$5% at 5 GHz and no apparent decrease at the highest frequency. Grechnev et al. modelled this as free-free absorption due to overlying material, and was able to fit the frequency-dependence of the dimming depths with a model involving chromospheric material, and an absorbing layer inserted in the corona. They could constrain the temperature of the absorbing material, and mass estimates were about 2×10^{14} g. Notably, one part of the ejecta appeared to escape as a CME, while the other part (seen in absorption at radio wavelengths) fell back onto the Sun. These had apparently equal mass.

This is relevant to the search for stellar CMEs. Not only does it provide a testable hypothesis for the presence of a CME, the frequency dependence required for the free-free absorption provides additional credence to any interpretation. Gyrosynchrotron emission from stellar flares tends to show a predictable exponential decay at microwave frequencies. The poster contribution by Tristan et al. (2024) in this conference proceedings provides excellent examples of the behavior. Previous multi-frequency observations of some stellar

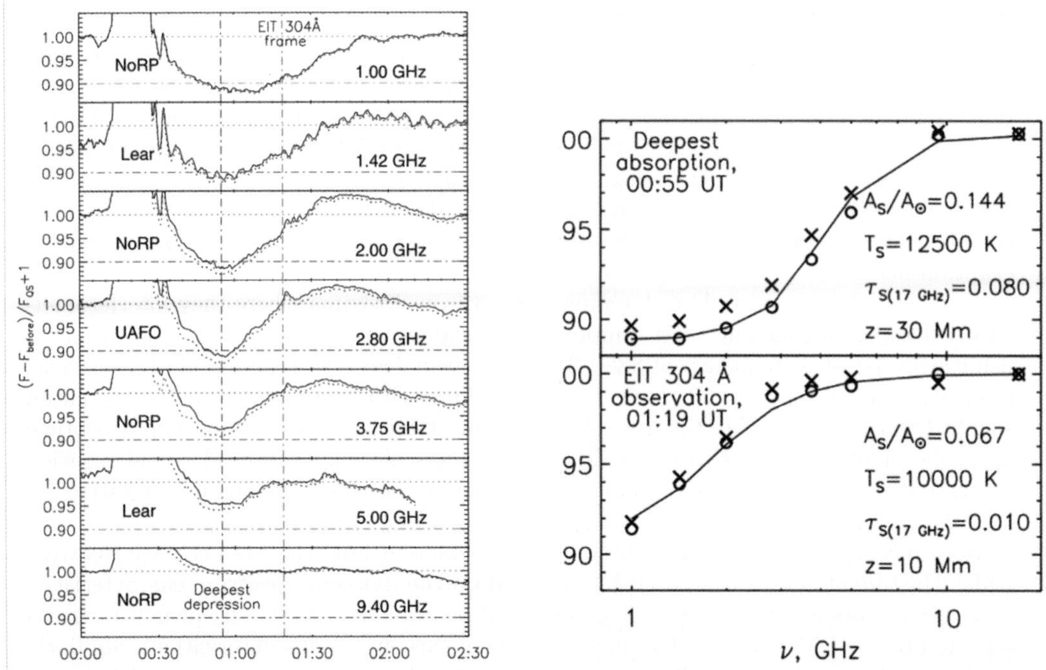

Figure 1. Example "negative radio burst" from Grechnev et al. (2008). *(left panel)* Light curves at different radio frequencies, normalized to the pre-flare intensity, showing the appearance of a frequency-dependent "dip" during the decay phase. The vertical lines indicate the times of the deepest depression as well as the time when an EIT image of the event was obtained. The *right panel* shows the absorption depth as a function of frequency for these two times, along with best-fit model parameters.

microwave flares show a frequency dependence in the initial stages of the radio flare, where the lower frequency emission is temporarily suppressed (e.g. Osten et al. 2005, Osten et al. in prep.). While there could be several interpretations for this frequency-dependent change in flux density, such as optical depth effects in the flaring loop, absorption from overlying eruptive material is an interesting addition to ponder and will be the subject of additional work.

5. Conclusion

Radio observations provide versatility when it comes to the study of stellar bursts and eruptions. Recent advancements in detector technology and array size have brought with them advances in the sensitivity of the telescopes and they have become increasingly relevant to the study of stellar transient mass loss. The potential for multiple radio diagnostics to probe the incidence of stellar eruptions is particularly exciting. The type II radio burst is still the most promising signature to pursue, but its rarity suggests the need to explore alternate avenues. Ideally, a signature of transient stellar mass loss would have multiple observational constraints which are individually reproducible and falsifiable, to add weight to this emerging sub-topic.

References

Aarnio, A. N., Matt, S. P., & Stassun, K. G. 2012, *ApJ*, 760, 9

Alvarado-Gómez, J. D., Drake, J. J., Fraschetti, F., Garraffo, C., Cohen, O., Vocks, C., Poppenhäger, K., Moschou, S. P., Yadav, R. K., & Manchester, Ward B., I. 2020, *ApJ*, 895, 47

Benz, A. O. & Güdel, M. 2010, *ARA&A*, 48, 241
Carrington, R. C. 1859, *MNRAS*, 20, 13.
Crosley, M. K. & Osten, R. A. 2018a, *ApJ*, 856, 39
—. 2018b, *ApJ*, 862, 113.
Crosley, M. K., Osten, R. A., Broderick, J. W., Corbel, S., Eislöffel, J., Grießmeier, J. M., van Leeuwen, J., Rowlinson, A., Zarka, P., & Norman, C. 2016, *ApJ*, 830, 24.
Crosley, M. K., Osten, R. A., & Norman, C. 2017, *ApJ*, 845, 67
Drake, J. J., Cohen, O., Yashiro, S., & Gopalswamy, N. 2013, *ApJ*, 764, 170
Gopalswamy, N., Lara, A., Kaiser, M. L., & Bougeret, J. L. 2001, *JGR*, 106, 25261
Gopalswamy, N., Yashiro, S., Akiyama, S., Mäkelä, P., Xie, H., Kaiser, M. L., Howard, R. A., & Bougeret, J. L. 2008, *Annales Geophysicae*, 26, 3033
Gosling, J. T., Hildner, E., MacQueen, R. M., Munro, R. H., Poland, A. I., & Ross, C. L. 1974, *JGR*, 79, 4581.
Grechnev, V. V., Uralov, A. M., Slemzin, V. A., Chertok, I. M., Kuzmenko, I. V., & Shibasaki, K. 2008, *Sol. Phys.*, 253, 263
Hertzsprung, E. 1924, *Bull. Astron. Inst. Netherlands*, 2, 87.
Mohan, A., Mondal, S., Wedemeyer, S., & Gopalswamy, N. 2024, *A&A*, 686, A51
Namekata, K., Sakaue, T., Watanabe, K., Asai, A., Maehara, H., Notsu, Y., Notsu, S., Honda, S., Ishii, T. T., Ikuta, K., Nogami, D., & Shibata, K. 2017, *ApJ*, 851, 91
Osten, R. A., Hawley, S. L., Allred, J. C., Johns-Krull, C. M., & Roark, C. 2005, *ApJ*, 621, 398
Osten, R. A. & Wolk, S. J. 2015, *ApJ*, 809, 79
Shibata, K. & Yokoyama, T. 1999, *ApJ*, 526, L49
Sun, X., Bobra, M. G., Hoeksema, J. T., Liu, Y., Li, Y., Shen, C., Couvidat, S., Norton, A. A., & Fisher, G. H. 2015, *ApJ*, 804, L28
Török, T. & Kliem, B. 2005, *ApJ*, 630, L97
Villadsen, J. & Hallinan, G. 2019, *ApJ*, 871, 214
Yashiro, S., Akiyama, S., Gopalswamy, N., & Howard, R. A. 2006, *ApJ*, 650, L143
Zic, A., Murphy, T., Lynch, C., Heald, G., Lenc, E., Kaplan, D. L., Cairns, I. H., Coward, D., Gendre, B., Johnston, H., MacGregor, M., Price, D. C., & Wheatland, M. S. 2020, *ApJ*, 905, 23

Solar and Stellar Coronal Mass Ejections
Proceedings IAU Symposium No. 388, 2026
N. Gopalswamy, O. Malandraki, A. Vidotto & W. Manchester, eds.
doi:10.1017/S1743921325000845

DH type II radio bursts during solar cycles 23-25: Origin and association with solar eruptive events

Bhuwan Joshi[1], Binal D. Patel[1], Kyung-Suk Cho[2] and Rok-Soon Kim[2]

[1]Udaipur Solar Observatory, Physical Research Laboratory, Udaipur 313001, India
email: `bhuwan@prl.res.in`

[2]Space Science Division, Korea Astronomy and Space Science Institute, Daejeon 34055, Republic of Korea

Abstract. We analyses occurrence of DH type II solar radio bursts spanning over solar cycles 23-25 during which a total of 590 DH type II bursts are reported with confirmed 568 and 462 cases of associated CME and flares, respectively. We find short-term yet important differences in DH type II activity when the data is examined in terms of event counts and their durations, e.g., temporal shift in the peak activity during cycle 24 and variation in the growth rate of the activity level during cycle 25. For an in-depth exploration, DH type II bursts are classified in 3 categories based on their end-frequencies: Low-, Medium-, and High- Frequency Groups (LFG, MFG, and HFG, respectively). The HFG category is the most populous (≈47%) while the LFG category occupy about a quarter of the events (≈24%). The LFG events show a clear inclination toward fastest CMEs and X-class flares with a quarter of events exhibiting end frequency below 50 MHz.

Keywords. Coronal mass ejections, solar radio bursts, type II bursts, solar cycle

1. Introduction

It is well recognized that the powerful CMEs generate shocks in the coronal and interplanetary medium. In radio dynamic spectrum, these shocks are identified as type II bursts. Depending on the energetics of CMEs, the type II bursts are observed in metric (m; 30 MHz $\leq$ f $\leq$ 300 MHz), decametric-hectometric (DH; 300 kHz $\leq$ f $\leq$ 30 MHz), and kilometric (km; 30 kHz $\leq$ f $\leq$ 300 kHz) wavelength domains. The extension or origin of type II radio bursts in the DH domain implies the cases of stronger MHD shocks propagating from the inner corona and entering the interplanetary (IP) medium (Gopalswamy et al. 2001, 2005). Hence the study of shocks in DH domain, together with their associated CME-flare events, becomes extremely important to infer not only the propagation characteristics of CMEs but also to develop their forecasting tools (Reiner et al. 2007; Joshi et al. 2018; Syed Ibrahim et al. 2019). Contextually, the CME associated with type II radio bursts in metric and/or DH wavelength domain are termed as radio loud CMEs (Michalek et al. 2007; Gopalswamy et al. 2008).

In this paper, we present a statistical study of DH type II bursts occurred during solar cycles 23, 24, and 25. For the present analysis, we have obtained data from the following sources: (1) Wind/WAVES Type-II Burst Catalogue (`https://cdaw.gsfc.nasa.gov/CME_list/radio/waves_type2.html`), (2) Solar and Heliospheric Observatory (SOHO) Large Angle and Spectroscopic Coronagraph (LASCO) CME Catalogue(`https://cdaw.`

Table 1. Occurrence of DH type II radio bursts from January 1996 to June 2023, covering solar cycles 23-25, along with the identified solar events, i.e., CMEs and flares, occurred in association with the reported bursts.

Solar Cycle	No. of Events DH type II (% out of total)	Identified solar events CME	Flare
23	335 (57%)	320	275
24	180 (30%)	173	127
25	75 (13%)	75	60
23+24+25	590	568	462

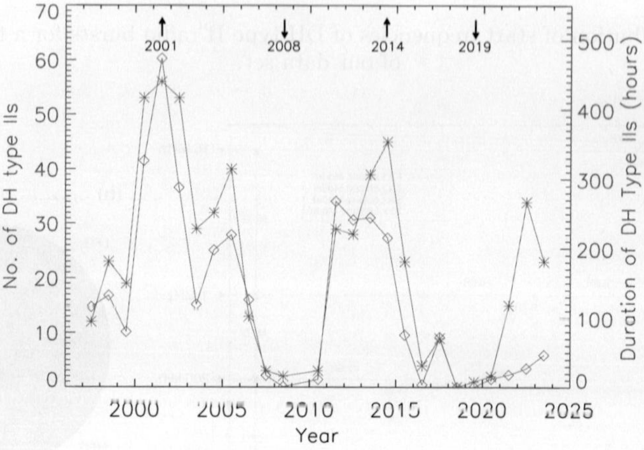

Figure 1. Annual occurrence of DH type II radio bursts from January 1996 to June 2023, covering solar cycle 21, 22, and rising phase of cycle 23 in terms of counts and cumulative durations over a year. Up and down arrows represent the maximum and minimum years of solar cycles.

`gsfc.nasa.gov/CME_list/`). This work highlights the importance of analysing DH type II bursts in terms of their end frequencies (Patel et al. 2021, 2022). Notably, the end frequencies of these IP bursts directly relate to the heliocentric distance up to which a shock can survive and, therefore, have implications in exploring the energetics and propagation characteristics of CMEs.

2. Occurrence and frequency-dependent characteristics

2.1. *DH type II bursts during cycles 23-25*

Table 1 and Figure 1 present inter-cycle variations of DH type II activity during solar cycles 23 (January 1996–December 2008), 24 (January 2009–Dec 2017, and 25 (January 2018–June 2023). We note a significant reduction of 46% in the occurrence of DH type II bursts for cycle 24 in comparison to the previous cycle, consistent with the work of Gopalswamy et al. (2019) who authored the Wind/WAVES type II burst catalogue. The solar cycle 25 is still running, however, in the beginning phase of ≈5 years, considered here, it has exhibited a deficit in DH type II activity compared to the previous two cycles along with a slow rise. To understand the activity level of DH type II radio bursts with the evolution of solar cycles, we characterize the type II activity in terms of their total yearly counts and cumulative durations (Figure 1). Both indicators reflect an obvious cyclic behaviour of ≈11 years with some noteworthy differences: the maximum activity period for the two parameters differs for cycle 24, and type II event counts show a steep growth during 2021–2023 while the cumulative durations indicates a slow and gradual

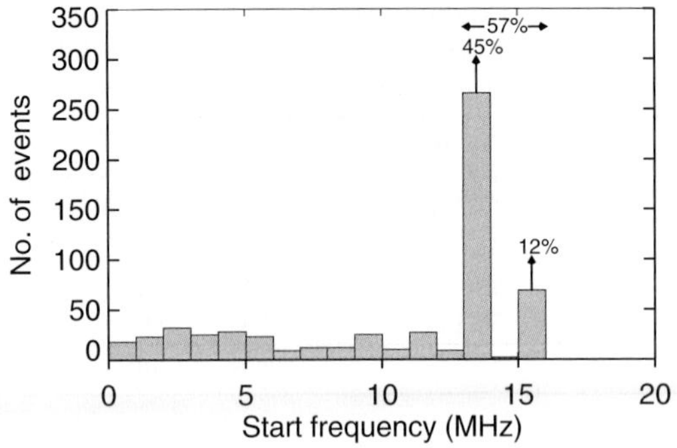

Figure 2. Distribution of start frequencies of DH type II radio bursts for a total of 590 events of our data set.

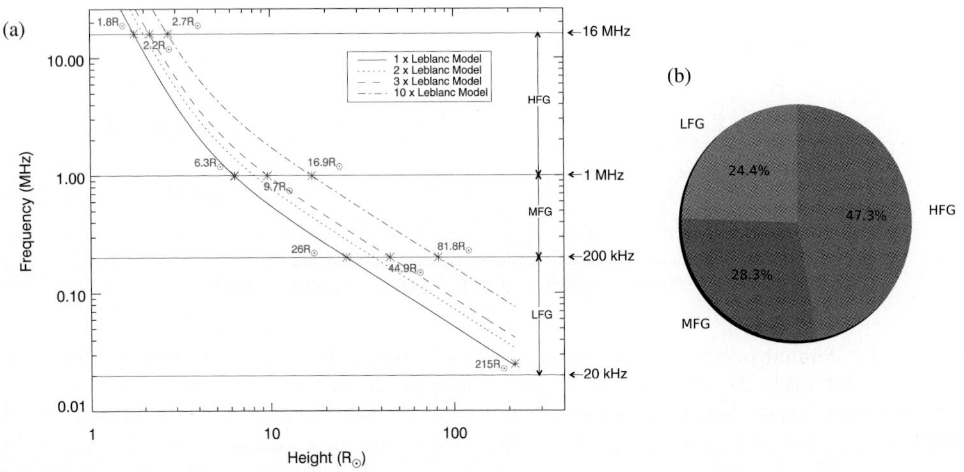

Figure 3. (a) Variation of heliocentric burst height ($R_\odot$) with plasma frequency estimated from Leblanc electron density model. (b) Pie chart showing the fraction of LFG (red), MFG (blue), and HFG (green) DH type II radio bursts occurred during cycles 23-25.

rise. To have a clear understanding of the observed DH type II events in terms of the frequency at which the bursts originated or first detected, we present a histogram of the starting frequencies in Figure 2. Here we clearly notice that the histogram bars representing the limiting frequency range of Wind/WAVES (13–14 MHz) and SWAVES (15–16 MHz) populate 45% and 12% events, respectively. From this statistics, it is clear that a sizeable population of DH type II events exist as an extension of type II emission starting at metric frequencies.

2.2. *Frequency-dependent categories and their characteristics*

The characteristics feature of type II burst is slow frequency drift toward lower frequencies which is physically attributed to the burst driver moving outward from the solar atmosphere to larger heliocentric distances with ever decreasing electron densities. This frequency-height relationship of the type II burst is illustrated using the atmospheric density model of Leblanc et al. (1998) in Figure 3. Contextually, we note that the height estimations from the atmospheric density model provide a "coarse" estimation only as

Table 2. Occurrence of DH type II events during Solar Cycles 23, 24, and 25 for LHG, MFG, and HFG categories.

Solar Cycle	Number of events			
	LFG $f \leq 200$ kHz	MFG 200 kHz $< f \leq 1$ MHz	HFG 1 MHz $< f \leq 16$ MHz	Total
23	81(24%)	96 (29%)	158 (47%)	335 (57%)
24	57 (32%)	55 (30%)	68 (38%)	180 (31%)
25	6 (8%)	16 (21%)	53 (71%)	75 (13%)
23+24+25	144 (24.4%)	167 (28.3%)	279 (47.3%)	590

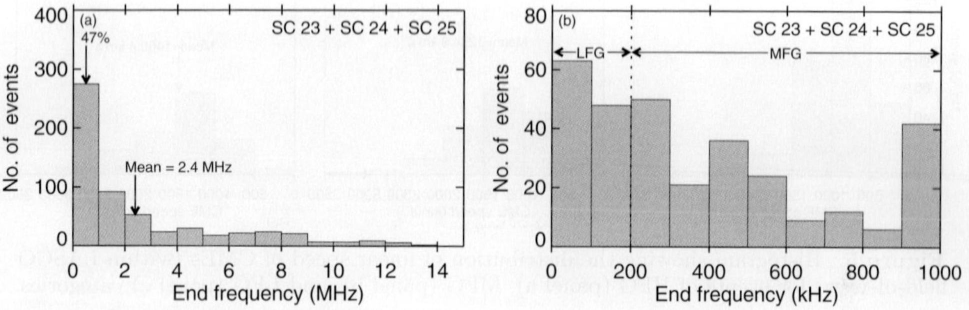

Figure 4. (a) Distribution of DH type II radio bursts in terms of their end frequencies. (b) Distribution of end frequencies exclusively for LFG and MFG events.

the density of the medium, through which shock is propagating, evolves with the spatially and temporally varying coronal and interplanetary conditions. Thus, rationally, we consider a multiplier (or suitable "fold") to the basic atmospheric density model.

In Figure 3a, we draw frequency-height relation for 1-, 2-, 3-, and 10-fold Leblanc models to incorporate density variations of upto an order. The frequency-height relationship suggests that the observing window of 16 MHz-20 kHz essentially represents a large heliocentric distance from $\approx 2R_{\odot}$ to 1 AU. Therefore, to explore the characteristic of DH type II radio bursts and associated solar eruptive events, we divide this vast frequency range into three groups: Low-Frequency Group (LFG; 20 kHz $\leq$ f $\leq$ 200 kHz), Medium-Frequency Group (MFG; 200 kHz $\leq$ f $\leq$ 1 MHz), and High Frequency Group (HFG; 1 MHz $\leq$ f $\leq$ 16 MHz. Figure 3a readily shows that our end-frequency classification essentially provides a quantitative estimation on the distance up to which a shock can survive. We further note that HFG and MFG events represent shocks terminating within the lower and upper coronal heights ($9.7R_{\odot}$ and $44.9R_{\odot}$, as per three-fold Leblanc model), respectively. On the other hand, the events under LFG group represent shock travelling in the interplanetary medium (beyond $\approx 45R_{\odot}$).

In Table 2, we list the counts for LFG, MFG, and HFG events for solar cycle 23, 24, and 25. The same has been pictorially shown in a pie chart in Figure 3b. In Figure 4a, we present the histogram of ending frequency for DH type II events which reveal 49% of events to lie below 1 MHz frequency bin. We further show the distribution of ending frequencies from 1 MHz down to 20 KHz in Figure 4b to explore the occurrence of LFG and MFG events exclusively.

In Table 3, we specifically give number of type II bursts that end at or below 50 KHz as this frequency roughly corresponds to heliocentric distance of 0.5 AU (see Figure 3a); We find cycle 24 to be highly deficient in comparison to cycle 23 (12% versus 37%) while rise phase of cycle 25 completely lacks such events. Here, we note that cycle 25 is still running and the present analysis considers DH type II radio bursts occurred during a period of 66 months since the onset of the cycle (January 2018–June 2023). Therefore,

Table 3. Occurrence of DH type II events having end frequency $\leq$50 kHz.

Solar Cycle	No. of Events (total events)	%
23	30 (81)	37%
24	7 (57)	12%
25	0 (6)	0
23+24+25	37 (144)	26

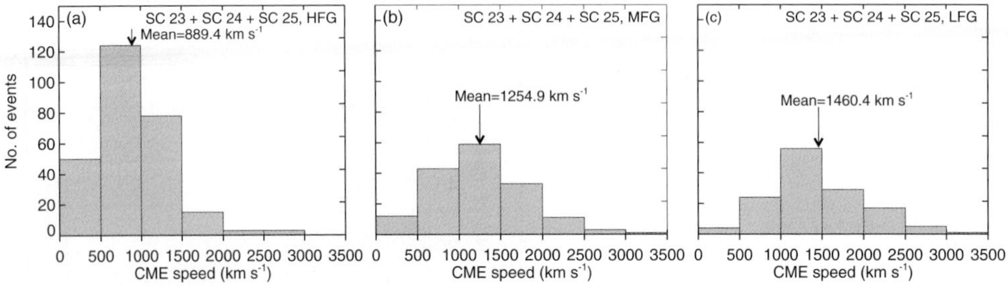

Figure 5. Histogram showing the distribution of linear speed of CMEs (within LASCO field-of-view) for events of HFG (panel a), MFG (panel b), and LFG (panel c) categories.

for an appropriate inter-cycle comparison, we further consider the occurrence of DH type II events during the same interval for all the cycles. We find that during the rise phase of 66 months, cycles 23 and 24 witnessed a total of 9 and 7 events, respectively, with end frequency $\leq$50 KHz.

3. Association with CMEs and flares

In Figure 5, we present the distribution of linear speed of CMEs within LASCO field-of-view associated with DH type II radio bursts for HFG, MFG, and LFG categories. The histograms clearly reveal a larger difference of $\approx$570 km s^{-1} between the mean CME speeds for two limiting cases i.e., LFG and HFG categories. Further, as expected, the mean speed for MFG events lies between the LFG and HFG classes. To investigate whether the CME speeds for HFG, MFG, and LFG categories belong to the same distribution or not (i.e., if the difference in mean speeds of CMEs for the three successive categories is statistically significant), we perform the two-sample Kolmogorov-Smirnov test (K-S test; Press et al. 1992) and the results of the test is presented in Table 4. The two-sample K-S test suggests that the difference in the mean speed of CMEs between the events of three categories of DH type II radio bursts is statistically highly significant.

In Figure 6, we present the histogram showing the association of flares of different GOES classes (viz B, C, M, and X) with DH type II radio bursts. We find that the highest number of events are associated with the M class flares ($\approx$55%) for all the categories. We further find that the fraction of occurrence of X-class flares becomes almost double as we go from HFG to LFG categories (12, 23, and 36%, respectively). Notably, this is due to the causal chain of the flares, CMEs, and type IIs, i.e., X-class flares are more likely to be associated with fast CMEs which produce DH type II bursts of LFG category.

4. Conclusions

The present study make a detailed survey of DH type II radio bursts occurred during solar cycles 23-25 involving a total of 590 DH type II bursts from January 1996 to June 2023, with confirmed 568 and 462 cases of associated CME and flares, respectively. Our approach is to classify radio loud CMEs based on the end-frequencies of the corresponding

Table 4. The results of the two-sample K-S test assessing the observed difference between mean CME speeds of HFG, MFG, and LFG categories. N_1 and N_2 represent the number of events for respective categories. The K-S test statistics and probability are denoted by D and P.

Category	ΔV (km s^{-1})	Sample Size N_1	N_2	D	P
$\Delta V_{MFG-HFG}$	365	162	273	0.33	2.5×10^{-10}
$\Delta V_{LFG-MFG}$	206	136	162	0.205	0.0032
$\Delta V_{LFG-HFG}$	571	136	273	0.515	5.7×10^{-22}

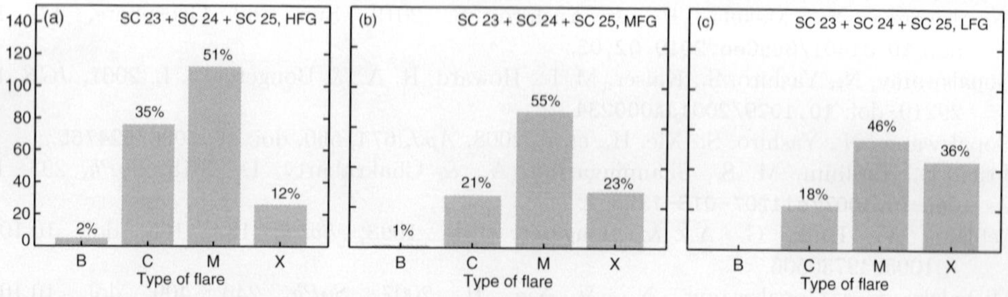

Figure 6. Histograms showing the distribution of flare counts of GOES classes (B, C, M, and X) associated with DH-type II radio bursts of HFG (panel a), MFG (panel b), and LFG (panel c) categories.

DH type II radio bursts. Accordingly, three categories of CMEs have been considered viz. Low-Frequency Group (LFG; 20 kHz $\leq$ f $\leq$ 200 kHz), Medium-Frequency Group (MFG; 200 kHz $\leq$ f $\leq$ 1 MHz), and High Frequency Group (HFG; 1 MHz $\leq$ f $\leq$ 16 MHz. This end-frequency classification plausibly describes the heliocentric distance up to which a shock can survive. In the following, we summarize the main results of the paper:

(1) The occurrence rate of DH type bursts follow a periodic behaviour with the phases of solar cycle. However, subtle but important differences are seen when the data is examined in terms of event counts and event durations. This result has important implication in understanding the origin of solar activity and its coronal and heliospheric consequence.

(2) The HFG category is the most populous, occupying almost half of the events of the entire sample ($\approx$47%). The LFG category (f $\leq$ 200 kHz), most important for space weather perspective, occupy a quarter of the events ($\approx$24%).

(3) Within the LFG class, $\approx$25% events exhibiting end frequency below 50 kHz, are of special interest as this frequency corresponds to a heliocentric distance of $\approx$0.5 AU. Notably, cycle 24 drastically lacks such events in comparison to cycle 23 (12% versus 37%). In cycle 25, we could not find any events of this category within the observing period (January 2018–June2023).

(4) The LFG events show a clear inclination toward fastest CMEs and largest X-class flares.

In summary, the present study provides us insights about the statistical properties of DH type II radio bursts during cycles 23-25 but also widens our understanding about the conditions related to their origin in the near-Sun region and survival in the corona and interplanetary medium. Our results indicate that the three-element classification of radio loud CMEs – HFG, MFG, and LFG – has important implications. In particular, the HFG events imply propagation of the shock inside the complex magnetic environment and within faster solar wind condition of the low corona, whereas, the events of LFG category, representing the shock travelling from the corona to the interplanetary medium, are rather intriguing for understanding the space weather phenomena.

The research work at the Physical Research Laboratory (PRL) is funded by the Department of Space, Government of India. We gratefully acknowledge the WIND/WAVES type II burst catalog and LASCO CME catalog. We further acknowledge the SOHO, STEREO, GOES, and Wind missions for their open data policy. BJ thanks the organizers for providing the local hospitality.

References

Gopalswamy, N., Aguilar-Rodriguez, E., Yashiro, S., et al. 2005, Journal of Geophysical Research (Space Physics), 110, A12S07, doi: 10.1029/2005JA011158.

Gopalswamy, N., Mäkelä, P., & Yashiro, S. 2019, Sun and Geosphere, 14, 111, doi: 10.31401/SunGeo.2019.02.03.

Gopalswamy, N., Yashiro, S., Kaiser, M. L., Howard, R. A., & Bougeret, J. L. 2001, *JGR*, 106, 29219, doi: 10.1029/2001JA000234.

Gopalswamy, N., Yashiro, S., Xie, H., et al. 2008, *ApJ*, 674, 560, doi: 10.1086/524765.

Joshi, B., Ibrahim, M. S., Shanmugaraju, A., & Chakrabarty, D. 2018, *SoPh*, 293, 107, doi: 10.1007/s11207-018-1325-2.

Leblanc, Y., Dulk, G. A., & Bougeret, J.-L. 1998, *SoPh*, 183, 165, doi: 10.1023/A:1005049730506.

Michalek, G., Gopalswamy, N., & Xie, H. 2007, *SoPh*, 246, 409, doi: 10.1007/s11207-007-9062-y.

Patel, B. D., Joshi, B., Cho, K.-S., & Kim, R.-S. 2021, *SoPh*, 296, 142, doi: 10.1007/s11207-021-01890-6.

Patel, B. D., Joshi, B., Cho, K.-S., Kim, R.-S., & Moon, Y.-J. 2022, *SoPh*, 297, 139, doi: 10.1007/s11207-022-02073-7.

Press, W. H., Teukolsky, S. A., Vetterling, W. T., & Flannery, B. P. 1992, Numerical recipes in FORTRAN. The art of scientific computing.

Reiner, M. J., Kaiser, M. L., & Bougeret, J. L. 2007, *ApJ*, 663, 1369, doi: 10.1086/518683.

Syed Ibrahim, M., Joshi, B., Cho, K. S., Kim, R. S., & Moon, Y. J. 2019, *SoPh*, 294, 54, doi: 10.1007/s11207-019-1443-5.

Solar and Stellar Coronal Mass Ejections
Proceedings IAU Symposium No. 388, 2026
N. Gopalswamy, O. Malandraki, A. Vidotto & W. Manchester, eds.
doi:10.1017/S1743921324001534

Enhancing Triangulation of Interplanetary Type II Bursts through Wavevector Correction

Vratislav Krupar[1,2], Oksana Kruparova[1,2] and Adam Szabo[2]

[1]Goddard Planetary Heliophysics Institute, University of Maryland, Baltimore County, Baltimore, MD 21250, USA

[2]Heliospheric Physics Laboratory, Heliophysics Division, NASA Goddard Space Flight Center, Greenbelt, MD 20771, USA

Abstract. Type II radio bursts, characterized by their slow frequency drift, are associated with accelerated electron beams at coronal mass ejection (CME)-driven shock fronts. These bursts serve as crucial indicators of shock propagation and space weather phenomena. Recently, Krupar et al. (2024) introduced a wavevector correction technique, initially applied to type III bursts, to account for the scattering effects in the solar wind, thereby enhancing triangulation accuracy. In this study, we investigate the application of this correction to type II bursts, aiming to improve localization and understand their relation to CME dynamics. We compare the corrected type II source locations with white-light observations of CMEs and in situ measurements of interplanetary shocks. Our findings indicate a significant improvement in the correlation between radio triangulation results and both white-light and in situ data, underscoring the potential of this correction method to refine our understanding of CME propagation and its impact on space weather forecasting.

Keywords. Radio astronomy (1338); Radio bursts (1339)

1. Introduction

Solar transients are associated with intense radio emissions, particularly type II and type III radio bursts, across a broad frequency spectrum. These bursts occur when beams of suprathermal electrons interact with the surrounding plasma, producing Langmuir waves that transform into radio emissions at the plasma frequency (f_p) and its second harmonic ($2f_p$). As these electron beams propagate away from the Sun, the emissions occur at progressively lower frequencies, reflecting the decreasing density of the solar wind plasma (Wild 1950; Gopalswamy 2011). Type II radio bursts are generated by electron beams accelerated at shock fronts driven by high-speed coronal mass ejections (CMEs), while type III radio bursts originate from electrons impulsively released during solar flares and coronal jets.

Radio bursts are essential for understanding the link between energetic particles and their sources. However, several knowledge gaps limit their full potential. A significant challenge in solar radio astronomy is that all solar radio burst sources are embedded in magnetized turbulent plasma. Consequently, radio waves are affected by scattering and refraction in the solar corona and interplanetary medium, leading to an apparent source size that is larger and a source position that is altered (Steinberg et al. 1984).

Recently, Krupar et al. (2024) introduced a wavevector correction technique applied to type III bursts to account for the scattering effects in the solar wind, thereby enhancing

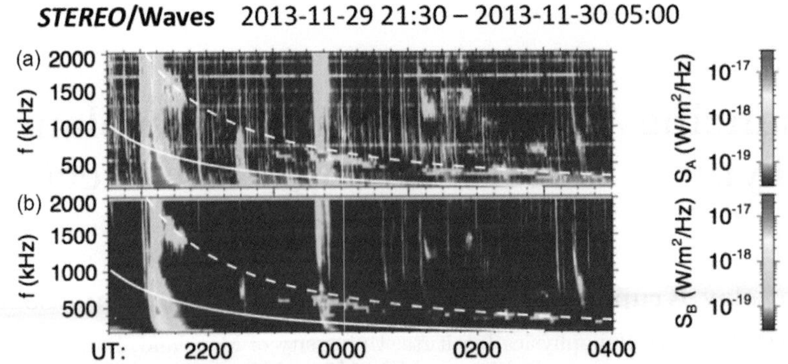

Figure 1. Radio measurements of type II and type III bursts. Radio flux density S observed by radio receivers onboard (a) STEREO-A and (b) STEREO-B. Solid and dashed lines denote results of the GCS model assuming emission at the leading edge and flanks, respectively. Adapted from Krupar et al. (2016).

triangulation accuracy. They found that the absolute values of the wavevector deviations from the Sun-spacecraft line are statistically 3.72 and 2.10 times larger than predicted by the density model, assuming fundamental and harmonic emission, respectively. This correction was successfully applied to type III bursts observed during the 2024 Mother's Day Solar storm (Kruparova et al. 2024). In this study, we compare the corrected type II source locations from 2013 November 29, with white-light observations of CMEs and in situ measurements of interplanetary shocks. We assess the radio source locations with and without the correction against these observations.

2. Observations and Analysis

On November 29, 2013, a CME was launched from the Sun at around 16:00 UT. This event was well-captured by three widely separated coronagraphs on board the STEREO-A, STEREO-B, and SOHO spacecraft. Utilizing these data, Krupar et al. (2016) employed the Graduated Cylindrical Shell (GCS; Thernisien et al. 2009) model to reconstruct the three-dimensional structure of the CME, tracking its evolution between 7 solar radii (R_S) and 16 R_S. Furthermore, the Self-Similar Expansion Fitting (SSEF; Davies et al. 2012) technique was applied to the time-elongation profile of the CME apex, as observed by the Heliospheric Imagers (HIs) of the Sun Earth Connection Coronal and Heliospheric Investigation (SECCHI) on board STEREO-A, extending the tracking range from 30 R_S to 140 R_S. The derived speeds and directions from both GCS and SSEF methods were found to be comparable, with the CME-driven interplanetary shock subsequently observed in situ by MESSENGER and STEREO-A at 0.4 au and 1 au from the Sun, respectively. The GCS and SSEF analyses provided reasonably accurate predictions for the arrival time of the CME-driven shock at these spacecraft.

The CME was accompanied by interplanetary type II and type III radio bursts, detected by both STEREO spacecraft on November 29-30, 2013 (Figure 1). Since the CME was launched from the far side of the Sun, ground-based high-frequency radio observations of the associated bursts were unavailable. Figures 2(a) and (b) present the radio flux density (S) measured by STEREO-A and STEREO-B, respectively. A type III burst was observed at around 22:00 UT on November 29, followed by intermittent type II bursts over the next seven hours.

The separation angle between the two STEREO spacecraft was 64°, allowing for precise triangulation of the radio sources of the type II bursts. Figure 2a shows the triangulated

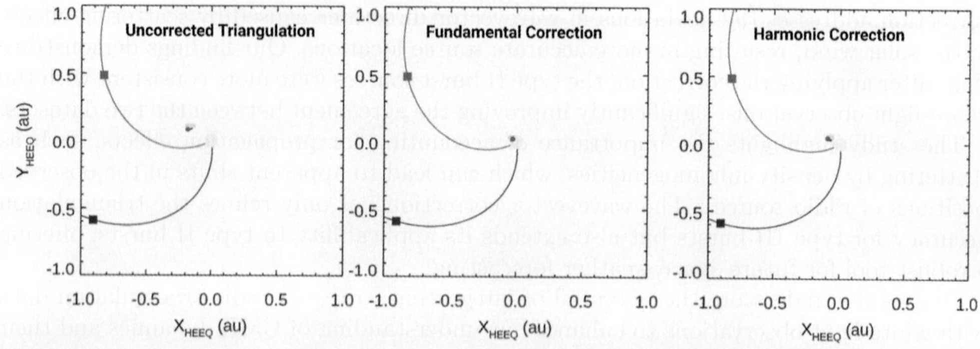

Figure 2. (a) Triangulated radio source locations between 1375 kHz (blue) and 1575 kHz (red) for the type II bursts observed on November 29-30, 2013, without applying wavevector correction. (b) Radio source locations with wavevector correction assuming fundamental emission. (c) Radio source locations with wavevector correction assuming harmonic emission.

radio sources between 1375 kHz (blue) and 1575 kHz (red) without applying the wavevector correction technique introduced by Krupar et al. (2024). These sources appeared at an average distance of 43.5 R_S from the Sun, significantly farther than the 10 R_S predicted by white-light observations at that time.

The wavevector correction technique accounts for the deviations in observed wavevector directions from those predicted by electron density models, which are attributed to scattering by density inhomogeneities in the solar wind. In their study, Krupar et al. (2024) found that the absolute values of wavevector deviations from the Sun-spacecraft line were statistically 3.72 and 2.10 times larger than predicted by density models, assuming fundamental and harmonic emissions, respectively. This discrepancy is believed to result from relative density fluctuations of approximately 0.40, as supported by the multiple flux-tube solar wind model (MULTI-VP).

Applying the wavevector correction assuming fundamental emission resulted in type II burst locations at approximately 9.9 R_S, closely aligning with the white-light observations of the CME front radial distance being around 10 R_S (Figure 2b). Conversely, when the correction was applied assuming harmonic emission, the type II bursts were positioned around 20.8 R_S, approximately double the distance suggested by white-light observations (Figure 2c). This correction effectively mitigates the propagation effects not only for type III bursts but also for type II bursts.

The implementation of wavevector correction techniques, as described by Krupar et al. (2024), provides a novel method for enhancing the accuracy of radio triangulation. This correction improves the reliability of tracking electron beams in the interplanetary medium and could be crucial for operational space weather forecasting. By integrating these corrected radio source locations with white-light CME observations, the accuracy of space weather predictions can be significantly improved, providing better insights into CME dynamics and their interactions with the solar wind. This advancement underscores the importance of precise measurement and analysis techniques in understanding solar phenomena and their broader implications on the heliosphere.

3. Conclusion

In this study, we investigated the CME event of November 29, 2013, and its associated interplanetary type II and type III radio bursts, using observations from the STEREO-A, STEREO-B, and SOHO spacecraft. A significant aspect of our research involved the application of wavevector correction techniques to the triangulated radio sources of type II bursts, following the methodology introduced by Krupar et al. (2024). This correction addresses the deviations in wavevector directions caused by scattering effects in the solar wind, resulting in more accurate source locations. Our findings demonstrate that, after applying the correction, the type II burst sources were more consistent with the white-light observations, significantly improving the agreement between the two datasets.

The study highlights the importance of accounting for propagation effects, such as scattering by density inhomogeneities, which can lead to apparent shifts in the observed positions of radio sources. The wavevector correction not only refines the triangulation accuracy for type III bursts but also extends its applicability to type II bursts, offering a robust tool for future space weather forecasting.

Our results underscore the potential of integrating corrected radio triangulation data with white-light observations to enhance our understanding of CME dynamics and their interaction with the solar wind. This comprehensive approach can lead to improved predictions of space weather events, which are crucial for mitigating their impacts on space-borne and ground-based technologies.

In conclusion, the wavevector correction technique represents a significant advancement in the field of solar and space physics, providing a more accurate framework for interpreting radio burst observations. Future research should focus on further refining these techniques and exploring their applications across a broader range of solar phenomena, thereby advancing our understanding of the complex interactions within the heliosphere.

References

Davies, J. A., Harrison, R. A., Perry, C. H., Möstl, C., Lugaz, N., Rollett, T., Davis, C. J., Crothers, S. R., Temmer, M., Eyles, C. J., & Savani, N. P. 2012, A Self-similar Expansion Model for Use in Solar Wind Transient Propagation Studies. *ApJ*, 750(1), 23.

Gopalswamy, N. Coronal Mass Ejections and Solar Radio Emissions. In *Planetary Radio Emissions VII, editors: Rucker, H. O., Kurth, W. S., Louarn, P., & Fischer, G., Proc. 7th Int. Workshop Planet. Sol. Heliospheric Radio Emissions (PRE VII), Austrian Acad. Sciences Press* 2011, pp. 325–342.

Krupar, V., Eastwood, J. P., Kruparova, O., Santolik, O., Soucek, J., Magdalenić, J., Vourlidas, A., Maksimovic, M., Bonnin, X., Bothmer, V., Mrotzek, N., Pluta, A., Barnes, D., Davies, J. A., Martínez Oliveros, J. C., & Bale, S. D. 2016, An Analysis of Interplanetary Solar Radio Emissions Associated with a Coronal Mass Ejection. *ApJL*, 823(1), L5.

Krupar, V., Kruparova, O., Szabo, A., Pinto, R. F., Maksimovic, M., & Martinez Oliveros, J. C. 2024, Enhancing Triangulation of Interplanetary Type III Bursts through Wavevector Correction. *ApJ*, 960(2), 101.

Kruparova, O., Krupar, V., Szabo, A., Lario, D., Nieves-Chinchilla, T., & Martinez Oliveros, J. C. 2024, Unveiling the Interplanetary Solar Radio Bursts of the 2024 Mother's Day Solar Storm. *ApJL*, 970(1), L13.

Steinberg, J. L., Hoang, S., Lecacheux, A., Aubier, M. G., & Dulk, G. A. 1984, Type III radio bursts in the interplanetary medium - The role of propagation. *A&A*, 140(1), 39–48.

Thernisien, A., Vourlidas, A., & Howard, R. A. 2009, Forward Modeling of Coronal Mass Ejections Using STEREO/SECCHI Data. *SoPh*, 256(1-2), 111–130.

Wild, J. P. 1950, Observations of the Spectrum of High-Intensity Solar Radiation at Metre Wavelengths. III. Isolated Bursts. *Austr. J. Sci. Res. A Phys. Sci.*, 3, 541.

Solar and Stellar Coronal Mass Ejections
Proceedings IAU Symposium No. 388, 2026
N. Gopalswamy, O. Malandraki, A. Vidotto & W. Manchester, eds.
doi:10.1017/S1743921325000481

Impact of Eruptive Events from Young Suns on Atmospheric Chemistry and Climates of Rocky Exoplanets

Vladimir S. Airapetian

NASA GSFC/SEEC and American University

Abstract. The search for habitable planets and signatures of extraterrestrial life is one of the most ambitious and consequential projects of our civilization. Observations taken with Kepler, Transiting Exoplanet Survey Satellite (TESS) and the Hubble Space Telescope have detected over 5600 exoplanets suggesting that majority of young rocky exoplanets in the habitable zones around magnetically active cool stars are exposed to extreme stellar coronal X-ray, Extreme UV (EUV) and particle fluxes. They have also revealed frequent (super) flares on young solar-like planet hosting stars, providing a mechanism by which host stars may have profound effects on the physical and chemical evolution of exoplanetary atmospheres. Recent data including our observational campaigns of EK Dra suggest that superflares from young solar analogs are accompanied by fast and energetic coronal mass ejections (CMEs). As CMEs propagate out from the stellar corona into interplanetary space at a few thousand km/s, they drive strong and extended shocks. Shocks are the sites of efficient acceleration of stellar energetic particles (StEPs) to tens of GeVs via diffusive shock acceleration. While CMEs interact with magnetospheres of rocky planets by disturbing them, StEPs can penetrate (exo)planetary atmospheres via open magnetic field and cause significant changes in atmospheric chemistry. Here, I will describe our recent results of observational data-constrained 3D MHD models of initiation and propagation of CMEs in coronae of young solar-like stars and acceleration of shock driven stellar energetic particles (StEP) to 40 GeV. I will then review our recent models of CME interaction with magnetospheres of rocky exoplanets that provide an efficient way for precipitation of StEPs into planetary lower atmospheres. I will also discuss how the flare/StEP modified chemistry may affect climates of rocky exoplanets, the critical factor of habitability. This model was recently validated with our laboratory experiments of proton irradiation of gas mixtures representative of a weakly reducing volcanicaly driven secondary atmospheres of young rocky exoplanets. I will discuss the implications of our recent theoretical simulations and laboratory experiments for the conditions required to find a prebiotic exoplanet around a solar-like stars and its pre-biosignatures.

Keywords. Solar analogs, superflare, CME, StEP, exoplanets, atmospheric chemistry, climate

1. Introduction

Recent discoveries of a great diversity of over 5600 exoplanetary systems around G, K and M dwarfs of various ages ignited a scientific question: How can we find a habitable planet in the solar neighborhood? These discoveries have ignited a search for potentially habitable exoplanets located within the habitable zone (HZ), the region around a planet hosting star within which a planet can sustain a standing body of liquid water on its surface (Kasting et al. 1993). Currently, astrophysicists have discovered over 200 rocky exoplanets within HZs mostly around M dwarfs, because the detections of small exoplanets are favored around the low-mass (0.1-0.5 solar mass), and thus low-luminosity stars (as low as 2000 times lower than the Sun's luminosity) with the radius of the inner boundary of the HZ around them varying between 0.04-0.16 times of the solar HZ. Specifically, the shorter orbital period and larger planet-to-star size ratio of a planet within the HZ of an M dwarf allow for easier transit detections. Thus, the proximity of HZs to these planet hosts suggests that exoplanets should receive up to 2000 times greater fluxes of quiescent (steady state chromospheric, transition region and coronal) form of ionizing radiation than that received by Earth from the Sun. This begs the important question: Can these fluxes be detrimental for the habitability of exoplanetary environments? The same question can be raised for the impact on habitability of an early Venus, Earth and Mars around young solar-like (G and K type) stars.

To assess the factors of habitability of rocky planets, we should thus gain the understanding of the detailed interaction between stars and exoplanets over geological timescales, the dynamical evolution of planetary systems, and atmospheric and internal dynamics. While this is a complex and multi-faceted problem, researchers made great strides in understanding the role of high-energy radiation, stellar winds, coronal mass ejections (CME) and stellar energetic particle (StPE) environments around planet hosting stars (of G, K and M dwarfs) on atmospheric chemistry, the rate of atmospheric escape from nitrogen rich secondary atmospheres and the impact on exoplanetary climates (Airapetian et al. 2016, 2017; Garcia-Sage et al. 2017; Dong et al. 2018). The outputs of magnetic activity specified as the perturbations traveling from stars to planets in the form of flares, winds, CMEs and StEP events are referred to as astrospheric space weather (see a comprehensive review in Airapetian et al. 2020).

The question of impact of stars on exoplanets is complex and requires an interdisciplinary approach, because to address it, we must start with the host star itself to determine its outputs and effects on the exoplanet environment, all the way from its magnetosphere to its surface. To understand whether an exoplanet is habitable at its surface, not only do we need to understand the changes in the chemistry of its atmosphere due to the penetration of energetic particles and their interaction with constituent molecules but also the loss of neutral and ionic species, and the addition of molecules due to outgassing from volcanic and tectonic activity. These effects will produce a net gain or loss to the surface pressure, and this will affect the surface temperature, as well as a net change in the molecular chemistry. Therefore, due to the complexity of the problem, we should study a set of interlinked research questions, all of which contribute pieces to the answer, with contributions from various disciplines involved in each topic.

The first line of studies should come from understanding the nature and forms of magnetic activity from planet hosts. Along with discoveries of exoplanets, Kepler space telescope also detected tens of thousands of flares from G, K and M dwarfs with solar-like energies to 10,000 times greater. The flares with energies greater that 10^{33} ergs are referred to as superflares (Maehara et al. 2012). These studies derived the statistics of these (super)flares and energy in the optical band and X-ray bands due to simultaneous observations with Chandra, XMMNewton and NICER (Getman & Feigelson 2021). Multi-wavelength campaigns of stellar flares have mostly been performed for clse by M dwarfs, while solar-like stars were not studied comprehensively.

In the meantime, because solar and stellar flares are the most powerful explosive phenomena in atmospheres of the Sun and cool dwarfs observed in the wide range of wavelengths, from radio to X-ray bands, the characterization of stellar flare physics can provide a critical impetus on the impact on (exo)planets on evoluion of (exo)planetary atmospheres and ist effect on habitability. Powerful solar flares are often accompanied by coronal mass ejections (CMEs) of hundreds of billions of tons of hot coronal material propagating into the interplanetary space at a few hundreds to a few thousands of km s^{-1}. Stellar flares are also found to be accompanied by ejection of prominence material known as filament or prominence eruptions, dimming of coronal EUV and X-ray flux, etc. (Namekata et al. 2024; Jin et al. 2022). As CMEs propagate out from the solar and stellar coronae into interplanetary space, they drive shocks. It is accepted that these shocks produce energetic protons and heavier ions at their fronts with energy up to a few tens of GeV via the diffusive shock acceleration (DSA), also known as the first-order Fermi acceleration (Zank et al. 2000; Hu et al. 2022). This process produces gradual SEP events that can last for a few days.

Recent MHD models and emerging observations of young solar-like stars suggest that the energetics and frequency of occurrence of stellar superflares are strongly coupled to the existence of strong (kiloGauss level) surface magnetic fields in large active region covering up 10% of the stellar surface (Airapetian et al. 2020; Namekata et al. 2024). They suggest that the standard models of magnetic reconnection driven superflares can be used to produce fast and energetic CME events that can induce extended shocks, the sites of particle acceleration (Hu et al. 2022).

The largest solar flare energy observed in modern times is about 10^{32} erg, however, recent studies have indirectly identified multiple possible extreme solar energetic events with energies up to 10^{34} ergs have occurred in the past millennia using cosmogenic radionuclides (^{14}C, ^{10}Be, ^{36}Cl) recorded in tree rings and ice cores (Usoskin et al. 2023). Furthermore, recent observations of solar-type stars have provided important insights into the rate of occurrence of superflares from the young solar-type stars with the energy up to 10^{36} ergs. Such energetic eruptive events can supply large energy fluxes to exoplanetary atmospheres in the form of extreme flare driven Extreme UV radiation, electromagnetic energy from CMEs, and particle energy fluxes from associated StEPs that can impact the erosion and chemistry of exoplanetary atmospheres. Hence, here we focus om discussing the recent efforts in observational and theoretical characterization of CMEs from young solar-like stars and their impact on chemistry and climates of terrestrial type exoplanets.

2. Coronal Mass Ejections on Active Stars: Observations

The evolution of the stellar magnetic field has dramatic effects on the frequency and energy of stellar superflares from active stars driven by the rapid release of free magnetic energy stored in the sheared and/or twisted strong fields typically associated with sunspots and active regions. The maximum energy of superflares from young G-K stars decreases with age from 10^{36} ergs (young stars with periods of 1-3 days) to 10^{34} ergs (periods greater than 20 days). Because large (white-light) solar flares are accompanied by CMEs with energy comparable (or exceeding) to solar flares, we can expect that stellar superflares from young solar-like stars should also be associated with energetic CMEs with energies up to 10^{36} ergs (Hu et al. 2022). The relationship between the eruptive flare and its resulting CME is well understood theoretically (Lynch et al. 2019). Because over 95% of powerful solar flares are accompanied by CMEs, and the physics of this correlation is understood, we can expect that active G-K stars with their frequent and powerful flares could also have correspondingly numerous energetic CMEs, another potentially important source of angular momentum loss (Vidotto 2021).

Stellar flares with the energy over 10^{33} erg or 10 times larger than the strongest solar flare (referred to as superflares) have never been reported in the modern solar observations. However, recent studies have identified multiple possible extreme solar energetic events with energies up to 10^{34} ergs have occurred in the past millennia using cosmogenic radionuclides (^{14}C, ^{10}Be, ^{36}Cl) recorded in tree rings and ice cores (Usoskin et al. 2023). Furthermore, recent observations of solar-type (G-type main-sequence) stars have provided important insights into the rate of occurrence of superflares from the young solar-type stars and the analogs of the present-day Sun. Vigorous observational studies of stellar flares over the last 30 yrs revealed that young rapidly-rotating solar-type stars show frequent superflares at a rate of 1 event per day, while mature slowly-rotating solar-type stars show superflares at much lower occurrence rates of about 1 event per a few thousand years (Maehara et al. 2012). These data suggest that superflares from the young Sun could have had significant impact on the magnetospheric and atmospheric environments of the young Venus, Earth, and Mars, and also imply a possibility that superflares can occur on the present-day, mature Sun. Such studies are critical for understanding the frequency and impact of these events on our civilization and its infrastructures. Thus, superflares on solar-type stars have been highlighted from the context of solar physics, planetary science , biochemistry and historical and geophysical studies (Airapetian et al. 2020; Kobayashi et al. 2023; Usoskin et al. 2023).

Observations of the nearby young solar-type stars in the first 300 Myr have provided insights into the time history of our Sun and its heliosphere, which have had a profound influence on the early Earth over the span of millions to billions of years. According to recent geophysical and fossil data, life on Earth is thought to have originated between 4.4 and 3.8 billion years ago (Westall et al. 2023). This time period is known as the "Hadean Eon", during which the Earth would have been subjected to intense XUV radiation and frequent solar flares and CMEs from the young Sun. The XUV radiation from the young Sun could have triggered photoionization processes and contributed to the atmospheric escape of the early Venus, Earth and Mars Airapetian et al. 2020. Moreover, the powerful energetic particles associated with flares/CMEs could have induced chemical reactions, resulting in the formation of greenhouse gases and prebiotic chemistry in the Earth's primitive atmosphere (Airapetian et al. 2016; Kobayashi et al. 2023). These processes may intimately contribute to the formation and disruption of habitable worlds, conditions crucial for the emergence and sustainment of life. As there is little direct evidence of the strong XUV flare radiation and energetic CMEs of the young Sun from the present-day Solar system, we need to infer them from observations of young solar-type stars at ages of a few tens to hundreds of millions of years characteristic of the Hadean period.

Thus, multi-wavelength campaign observations of young solar analogs in the first few Myrs can provide important constrains on space weather environments during the Hadean period on early Earth. They also provide critical constrains for developing sophisticated magnetohydrodynamic (MHD) models of CMEs and StEP events. Namekata et al. (2022) reported the detection of a blue wing absorption in H_α during a superflare on the young ($\sim$ 125 Myr) solar analogue EK Dra (G1.5V), simultaneously observed in the broad-band optical band (TESS) and H_α (SEIMEI) interpreted as an eruptive filament. Inoue et al. (2023) also presented the detection of a high velocity blue-wing emission feature as prominence eruption on the RS CVn system 1355 Ori. Recently, our team (PI Airapetian) has revisited EK Dra by performing a 12-day multi-wavelength campaign observations using TESS (optical), NICER (X-ray), and Seimei (H_α) telescopes. Our simultaneous optical and X-ray observations revealed three superflares with the bolometric energy of 1.5×10^{33} erg to 1.2×10^{34} erg. Two of these events show blueshifted H_α emission (see left panel of Figure 1) as a signature of two prominence eruptions at 690 km/s and 430 km/s and masses of 1.1×10^{19} g and 3.2×10^{17} g, respectively. The faster larger event also is a candidate of post-flare X-ray dimming with the amplitude of up to 10% as another signature of ejected material from the stellar corona (see right pane of Figure 1).

However, direct evidence of stellar CMEs, and associated high-energy particle events associated with stellar superflares on young (and matured) Sun-like stars have never been obtained so far. Radio observations that can provide transient signatures analogous to the solar Type II events associated with the plasma emission generated by the CME shock in the solar/stellar corona are required in addition to optical and X-ray signatures. This suggests that multiwavelength observational campaigns of young solar-like stars are critical in identifying elusive CMEs.

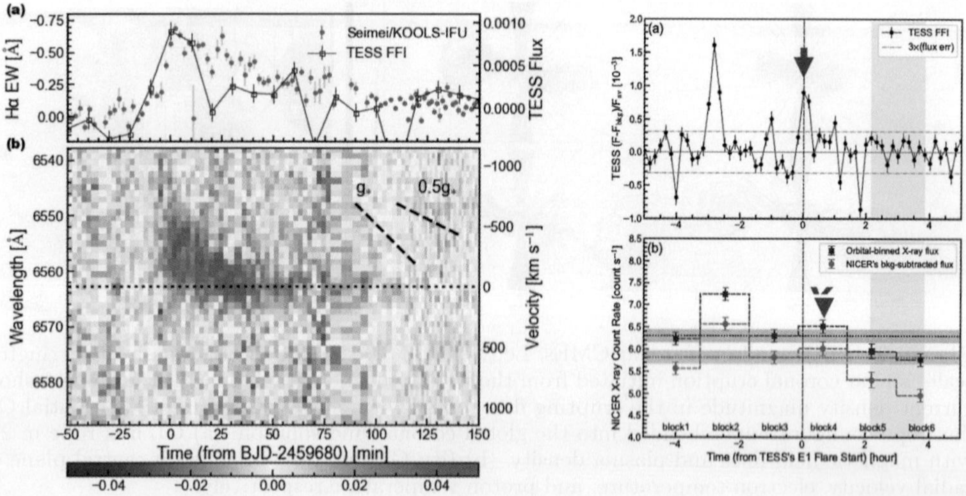

Figure 1. Time evolution of pre-flare subtracted H_α spectra (left panel) and X-ray emission (right panel) of the superflare observed on 2022 April 10. (a) Reference light curves in H_α (scaled on the left vertical axis) and white light (scaled on the right vertical axis). The color bar shows the non-dimensional intensity which is normalize by the continuum level.

3. Coronal Mass Ejections on Active Stars: Models

In order to develop an MHD model of a CME, we need first to model stellar coronal X-ray and EUV flux as an ingredient to simulate the stellar wind as another important factor affecting magnetospheres of rocky exoplanets. Recently, Alvarado-Gómez et al. 2018; do Nascimento et al. 2016; Boro Saikia et al. (2020) used Alfv/en Wave Solar Model (AWSoM) constrained with Zeeman Doppler Imaging (ZDI) magnetograms (Donati et al. 1997) of active stars as boundary for realistic three-dimensional magnetohydrodynamic models of the coronae and winds of active solar-like stars. Airapetian et al. (2021) have extended these models by using the fully resolved profiles of chromospheric lines derived from the Hubble Space Telescope observations to constrain the energy flux of Alfvén waves at the upper chromosphere that propagate upward and dissipate energy in the upper layers forming the stellar corona.

For stellar CMEs initiated from compact active regions, one effect associated with the enhanced magnetic field is the stronger magnetic confinement that could lead to more confined/failed stellar eruptions than in the solar case. Using a global 3D MHD model, it has been shown that a dipole field of 75 Gauss is capable of fully suppressing CMEs with energies up to 3×10^{32} erg (comparable to a large solar X-class flare/CME (Alvarado-Gómez et al. 2018). Using a potential field source surface model, a new study (Sun et al. 2021) estimates the torus-stable zone (TSZ) above a bipolar stellar active region embedded in a global dipole field and finds that in the active cool star conditions, the TSZ can extend significantly higher compared to solar conditions. This effect is believed to be one of the reasons causing so far sparse detection of CMEs on active stars (Drake 2013; Odert et al. 2017). Lynch et al. (2016) have shown that even in bipolar streamer distributions, the overlying, restraining closed flux is removed in a breakout-like way through an opening into the solar wind, thus enabling the eruption of the low-lying energized flux. Therefore, the evolution and interaction between the low-lying energized and overlying restraining fields are extremely important aspects of modeling eruption processes in solar and stellar coronae to correctly estimate the CME ejecta properties and energetics. Left panel of Figure 2 illustrates the recent ARMS 3D simulation results by Lynch et al. (2019) of a massive halo-type (width of 3600) energetic CME eruption based

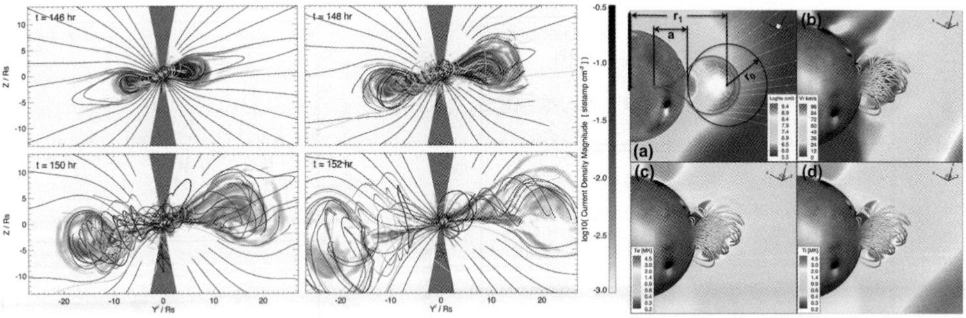

Figure 2. MHD models of stellar CMEs. Left: ARMS 3D simulation of a massive (Carrington scale) stellar coronal eruption initiated from the k1 Cet magnetogram. The contour planes show current density magnitude in the erupting flux rope (Lynch et al. 2019); Right: The initial GL flux rope configuration embedded into the global corona-wind solution. (a) GL flux rope in 2D with magnetic field lines and plasma density. (b)-(d): GL flux rope in 3D with central plane of radial velocity, electron temperature, and proton temperature,respectively.

on the observationally derived κ^1 Cet magnetogram. The entire stellar streamer-belt visible in this figure is energized via radial field-preserving shearing flows and the eruption releases $\sim 7 \times 10^{33}$ erg of magnetic free energy in 10 hours. Magnetic reconnection during the stellar flare creates the twisted flux rope structure of the ejecta and the 2000 km/s eruption creates a CME-driven strongly magnetized shock.

We have recently extended the corona-wind model of κ^1 Ceti Airapetian et al. (2021). by introducing an active regions using observational constrains from TESS rotationally modulated light curves. This model is consistent with the X-ray luminosity of the steay corona of the star, and thus provides realistic conditions of the steady inner corona to launch a CME. For this purpose, we have initialized of the Giblson-Low (GL) flux rope with the parameters describing its location, orientation, size, loaded mass, and free magnetic energy available in the eruption. The radius of the flux rope is constrained by the size of a stellar active region constrained by TESS light curves. The magnetic field strength parameter is then constrained by the free magnetic energy of the inserted active region. The loaded mass is constrained by the density of the stellar corona derived from X-ray line ratios [58] and the size of the active region.

4. Impact of CMEs and Associated StEPs of a Magnetized Rocky Planet on Atmospheric Chemistry

The simulation of the interaction of an energetic (Carrington-type) CME with the magnetosphere of a terrestrial-type planet is shown in Figure 1 of Airapetian et al. (2016). The distance of the subsolar magnetosphere from the planetary surface varies in response to changing dynamic pressure from the solar wind and a CME. Its boundary known as the standoff distance is determined from the balance between the magnetic pressure of the planetary magnetosphere and the wind dynamic pressure. The larger the dynamic pressure, the smaller the standoff distance. The left panel of Figure 1 of Airapetian et al. (2016) shows that the magnetospheric standoff distance moves from $10R_E$ (right panel) to $1.5R_E$ as the Carrington scale CME from the young Sun with the southward B_z component of the magnetic field as a result of the combined effect of magnetic reconnection between two fields and the CME dynamic pressure (Airapetian et al. 2016). The boundary of the magnetospheric open-closed field shifts to 36 degrees in latitude opening 70% of the Earth's magnetic field. The CME dynamic pressure drives large field aligned electric currents that dissipate in the ionosphere-thermosphere region (at about 110 km) of the Earth-like planet via ion-neutral frictional resistivity producing Joule heating rate reaching 4 W/m^2. This is over 20 times larger than that observed during the largest geomagnetic storms and will thus increase the upper atmospheric temperature and ignite enhanced ionospheric outflow. The models of associated atmospheric escape require the extension of sophisticated tools to consider multi-fluid effects for consistent treatment of neutral species in the ionosphere-thermosphere dynamics developed for the current and early Earth, Mars and Titan (Johnstone et al. 2018). The solar wind and a CME also compress the night-side magnetosphere and ignite magnetic reconnection at the night-side of the magnetosphere causing the magnetospheric storm as particles penetrate the polar regions of the planet. Also, the orientation of the magnetic field of the wind and CME as compared to that of the magnetospheric field controls the energy transfer from the wind to the planet, as it drives magnetic reconnection which in turn ignites particle acceleration and particle precipitation in the planetary ionosphere and thermosphere. This process is driven by electric fields that accelerate the ions against the neutrals resulting in current dissipation (Ohmic or Joule heating). In addition, the particle precipitation in the upper atmosphere impacts the local ionization and modifies Joule heating processes and atmospheric line excitation (i.e., auroral excitation, Schunk & Nagy 1980).

How can the environments of such terrestrial type planet within CHZ from their parent stars be hospitable to life? What would be the effects of SEPs accelerated in the CME shock on the atmospheres of early Earth, Mars and young rocky exoplanets? Could these sources have been beneficial in starting life on these (exo)planets? The answer is largerly depends on the atmospheric chemistry of a rocky planet and the flux and the frequency of an SEP event.

The initial atmospheric chemical composition is defined by the interplay between the atmospheric escape and volcanic outgassing of species including SO2, H2, CO2, H2S, and H2O, among other species, driven by the internal planetary dynamics (Airapetian et al. 2020). While XUV radiation mostly ionizes the upper atmosphere and contributes to the formation of the Earth's ionosphere and thermosphere (at up to 90 km from the ground), auroral electrons can reach the altitudes of $\sim$ 80 km, while relativistic electrons penetrate to the mesosphere at 40 km. Solar energetic protons with energies up to a few GeV can precipitate through the troposphere and reach the ground causing Ground Level Enhancement events as measured by ground monitors of neutrons, the byproducts of interactions of high-energy protons with atmospheric species. Impact electrons formed because of photo and collisional ionization can then interact with molecular nitrogen, carbon dioxide, methane and water vapor igniting a reactive chemistry in the Earth's lower atmosphere. Recent studies suggest that the early Earth's atmosphere was weakly reducing consisting of N_2, CO_2, CO and H_2O with only minor abundance of H_2, CH_4 and H_2S (Kasting et al. 1993; Lammer et al. 2018) or neutral.

The energetic particles from Galactic Cosmic Rays (GCR) and StEP events can provide a mechanism to deliver the particles directly to the lower atmosphere. The penetration depth depends on the frequency of collisions of protons with ambient molecules and for the Earth's atmosphere is scaled with the proton's energy (Jackman et al. 1980). This suggests that the protons with energy of 300 MeV can penetrate to the heights of 4.5 km above the ground. The collisions with molecules produce enhanced ionization of the atmosphere and form a broad energy distribution of secondary electrons at $\geq$ 35 eV. These electrons then get thermalized to lower energies in the atmosphere and as soon their energy reaches 10 eV, they become very efficient in breaking N2 into odd nitrogen with subsequent formation of NO_x. To study the pathways to complex organic molecules driven by energetic protons, Kobayashi et al. (2023) have performed laboratory experiments exposing gas mixtures of CO/CO_2, CO, N_2 and H_2O to 2.5 MeV protons. They reported production of amino acids precursors and nucleic acid bases as the results of secondary electron driven reactive chemistry. These experiments may have a direct

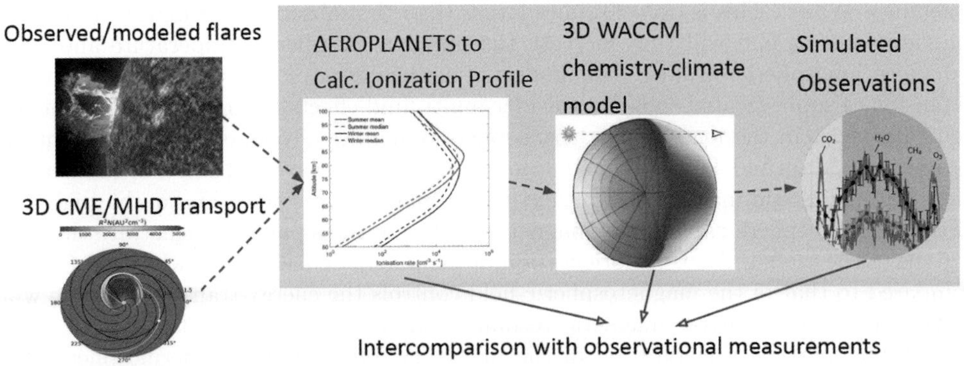

Figure 3. Schematic of the proposed workflow, illustrating the relationship between data and model components.

relevance to the atmospheric chemistry of young rocky exoplanets irradiated by flare and CME associated high energy protons.

To study the role of SEPs from the young Sun-like stars as the source of high energy protons in gas phase prebiotic chemistry in the atmosphere of early Earth, Airapetian et al. (2016, 2020) have applied a photo-collisional atmospheric chemistry model driven by the XUV flux and frequent stellar energetic particle events. Protons with energies of ≥ 0.3 GeV (at 0.5 bar atmosphere) precipitate into the middle and lower atmosphere (stratosphere and troposphere) and produce enhanced ionization, dissociation, dissociative ionization, and excitation of atmospheric species. The destruction of N_2 into ground state atomic nitrogen, N(4S), and the excited state of atomic nitrogen, N(2D), as the first key step toward production of biorelevant molecules. Reactions of these species with the products of subsequent dissociation of CO_2, CH_4, and H_2O produce nitrogen oxides, NO_x, CO and NH in the polar regions of the atmosphere. NOx then converts in the stratosphere into HNO_2, HNO_3 and its products including nitrates and ammonia.

The atmospheric model also predicts an efficient production of nitrous oxide, N2O, driven primarily through $N(4S) + NO_2 \rightarrow N_2O + O$; $NO + NH \rightarrow N_2O + H$. The recent model updated with the hard-energy spectrum of protons, vertical diffusion and Rayleigh scattering outputs the N2O production in the lower atmosphere by a factor ¿ 300 greater than the earlier model output of Airapetian et al. (2016). The particle acceleration via diffusive shock acceleration mechanisms on quasi-parallel strong shocks produces mostly SEPs with harder spectra (Hu et al. 2022). Laboratory experiments suggest that StEP driven chemistry predicts much more efficient production of HCN than that produced by lightning events for a weakly reducing ($N2 - CO/CO_2 - CH_4 - H_2O$) atmosphere on early Earth (Kobayashi et al. 1998, 2023). HCN is formed primarily due to neutral reactions including $NO + CH \rightarrow HCN + O$, $CH_2 + N(4S) \rightarrow HCN + H$, $CH_3 + N(4S) \rightarrow HCN + H + H$, and $CH + CN \rightarrow HCN + H$. As it forms in the stratosphere - troposphere region, HCN may subsequently rain out into surface reservoirs and initiate higher order chemistry producing more complex organics.

5. Impact of Extreme StEP Events on Exoplanetary Climates

To study the impact of a hard-spectra SEP event associated with a superflare of a young star, EK Dra (Model 3 in Hu et al. 2022), we used a self-consistent an SEP energy spectra as inout to Aeroplanets code to calculate the vetical profiles of ionization (ion pair production rate) and production of major species affecting climate including

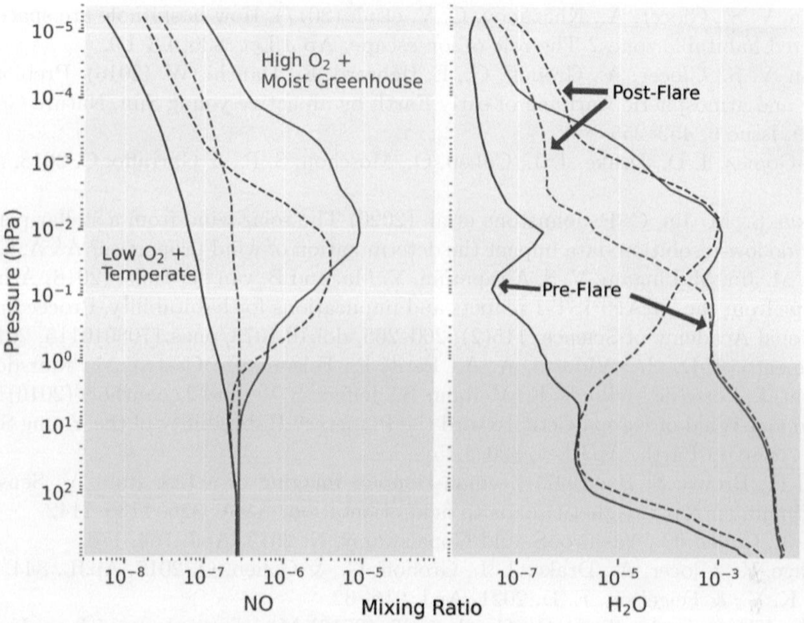

Figure 4. Pre-flare (solid curves) and post-flare (dashed-curves) global mean vertical profiles of HZ planet atmospheres with different assumptions in proton energy spectra and atmospheric condition using, for the first time, self-consistent flare energy and proton energy spectra relationships based on CME and MHD modeling. Lightly shaded regions show pressure ranges captured in conventional exoplanet GCMs.

NO_x species, N_2O and destruction of ozone. We use these results in conjunction with a stellar flare generator based on the MUSCLES campaign (Loyd et al. 2016) to calculate ionization profile to be used as inputs to Whole Atmosphere Community Climate Model (WACCM) (Marsh et al. 2013). We find enhanced column production of nitrogen oxides such as N_2O and NO_2, as demonstrated in Fig. 3. As incident protons are deposited into the lower atmosphere, and the general atmospheric response (e.g., enhancement of N2O) is increased by over three orders of magnitude than previous predictions leveraged from solar observations. Because of the elevated molecular density of the lower atmosphere and reduced exposure to photodissociation of species produced, N2O and NO2 persist for much longer timescales than what is chemically permitted in a non-perturbed oxic atmosphere.

Pre-flare (solid curves) and post-flare (dashed-curves) global mean vertical profiles of HZ planet atmospheres with different assumptions in proton energy spectra and atmospheric condition using, for the first time, self-consistent flare energy and proton energy spectra relationships based on CME and MHD modeling (refer to Fig 4). Lightly shaded regions show pressure ranges captured in conventional exoplanet GCMs.

Our preliminary result suggest that the enhancement of nitrous oxide, a potent greenhouse gas, can increase a global planetary temperature by 1K. This can have a large implications for climate evolution if the frequency of SEP events exceed the residence time of produced species. More results will be produced and reported in the coming months.

References

Airapetian, V. S., Jin, M. et al., (2021) One Year in the Life of Young Suns, ApJ, 916, 96.

Airapetian, V. S., et al., (2020) Impact of space weather on climate and habitability of terrestrial-type exoplanets, International Journal of Astrobiology, 19(2), 136.

Airapetian, V. S., Glocer, A., Khazanov, G. V. et al. (2017). How hospitable are space weather affected habitable zones? The role of ion escape. Ap J Let, 836, L3–L9.

Airapetian, V. S., Glocer, A., Gronoff, G., E. Hébrard, E., Danchi, W. (2016). Prebiotic chemistry and atmospheric warming of early Earth by an active young Sun. Nature Geoscience, Vol 9, Issue 6, 452–255.

Alvarado-Gómez, J. D., Drake, J. J., Cohen, O., Moschou, S. P., & Garraffo, C. 2018, ApJ, 862, 93.

Boro Saikia, S., M. Jin, C. P. Johnstone et al. (2020) The solar wind from a stellar perspective. How do low-resolution data impact the determination of wind properties? A&A, 635, A178.

Dong, C., M. Jin, M. Lingam, V. S. Airapetian, Y. Ma, and B. van der Holst (2018) Atmospheric escape from the TRAPPIST-1 planets and implications for habitability, Proceedings of the National Academy of Science, 115(2), 260–265, doi:10.1073/pnas.1708010115, 2018.

do Nascimento, J.-D., Jr.; Vidotto, A. A.; Petit, P.; Folsom, C.; Castro, M.; Marsden, S. C.; Morin, J.; Porto de Mello, G. F.; Meibom, S.; Jeffers, S. V.; and 2 coauthors (2016) Magnetic Field and Wind of Kappa Ceti: Toward the Planetary Habitability of the Young Sun When Life Arose on Earth. ApJ Let, 820, L15.

Donati, J-F., Brown, S. F. (1997) Zeeman-Doppler imaging of active stars. V. Sensitivity of maximum entropy magnetic maps to field orientation. A&A, 326, 1135–1142.

Drake, J. J., Cohen, O., Yashiro, S. and Gopalswamy, N. 2013, ApJ, 764, 170.

Garcia-Sage, K., Glocer, A., Drake, J. J., Gronoff, G., & Cohen, O. 2017, ApJL, 844, L13.

Getman, K. V., & Feigelson, E. D. 2021, ApJ, 916, 32.

Glocer, A., Kitamura, N., Toth, G., Gombosi, T. (2012) Modeling solar zenith angle effects on the polar wind. JGR, 117, A4, CiteID A04318.

Glocer, A., Kang, S-B., Airapetian, V. S. (2024) ApJ, to be submitted.

Güdel, M, Guinan, E. F., Skinner, S. L. (1997). The X-Ray Sun in Time: A Study of the Long-Term Evolution of Coronae of Solar-Type Stars. ApJ, 483, 947.

Hu, J., Airapetian, V. S., Li, G., Zank, G., Jin, M. (2022) Extreme energetic particle events by superflare-associated CMEs from solar-like stars, Science Advances, 8, eabi9743

Inoue, S., Maehara, H., Notsu, Y., et al. 2023, ApJ, 948, 9.

Jackman, C. H.; Frederick, J. E.; Stolarski, R. S. (1980). Production of odd nitrogen in the stratosphere and mesosphere - An intercomparison of source strengths. JGR, 85, 7495–7505.

Jin, M., M. C. M. Cheung, M. L. DeRosa, N. V. Nitta, and C. J. Schrijver (2022) Coronal Mass Ejections and Dimmings: A Comparative Study Using MHD Simulations and SDO Observations, ApJ, 928(2), 154.

Johnstone, C.P., Güdel, M., Lammer, H., & Kislyakova, K. G. (2018) Upper atmospheres of terrestrial planets: Carbon dioxide cooling and the Earth's thermospheric evolution, A&A, 617, A107, 36.

Kasting, J. F., D. P. Whitmire, and R. T. Reynolds (1993). Habitable Zones Around Main Sequence Stars. Icarus, 101, 108–128.

Kobayashi, K., Kaneko, T., Saito, Y., and Oshima, T. (1998) Amino acid production in gas mixtures by high energy particle irradiation. OLEB, 28, 155–165.

Kobayashi, K., Airapetian, V. S. et al. (2023) Formation of Amino Acids and Carboxylic Acidsin Weakly Reducing Planetary Atmospheres by Solar Energetic Particles from the Young Sun, Life, 13, 1103.

Lammer, H, Zerklem A. L., Gebauer, S. et al. 2018, Astron. Astrophys. Rev., 26(1), 72.

Leitzinger, M., Odert, P., & Heinzel, P. 2022, MNRAS, 513, 6058

Loyd, R. O. P., France, K., Youngblood, A. and 8 co-authors (2016). The MUSCLES Treasury Survey. III. X-Ray to Infrared Spectra of 11 M and K Stars Hosting Planets. ApJ, 824, 102.

Loyd, R. O. P., Mason, J. P., Jin, M., et al. 2022, ApJ, 936, 170

Lynch, B. J, Masson, S, Li, Y, Devore, C. R. et al. (2016) A model for stealth coronal mass ejections. J. Geophys. Res. 121, 10677.

Lynch, B. J., Airapetian, V. S., Kazachenko, M. D. et al. (2019). Modeling of a Carringtonscale Stellar Superflare and Coronal Mass Ejections from κ^1 Ceti, ApJ, 880, 97–109.

Maehara, H., et al. (2012). Astrophysics: Startling superflares. Nature 485, 478.

Marsh, D. et al. 2013, JGR, 118, 11442.

Miyakawa, S., Cleaves, H., Miller, S. L. (2002) The cold origin of life: A. Implications based on the hydrolytic stabilities of hydrogen cyanide and formamide, Origins of Life and Evolution of Biospheres 32(3):195–208.

Namekata, K., Airapetian, V. S., Petit, P. et al. (2024) Multi-wavelength Campaign Observations of a Young Solar-type Star, EK Draconis I. Discovery of Prominence Eruptions Associated with Superflares, ApJ, 961, 23.

Namekata, K., Maehara, H., Honda, S., et al. 2022, Nature Astronomy, 6, 241.

Odert, P., M. Leitzinger, A. Hanslmeier, and H. Lammer, Stellar coronal mass ejections I. - Estimating occurrence frequencies and mass-loss rates (2017), MNRAS, 472(1), 876.

Schunk, R. W. and Nahy, A. F. (1980) Ionospheres of the terrestrial planets, Reviews in Geophysics, 18, 813.

Sun, X., T. Török, and M. L. DeRosa, Torus-stable zone above starspots, MNRAS, 509(4), 5075.

Usoskin, I., Miyake, F., Baroni, M. 2023, Space Science Reviews, 219:73.

Veronig, A. M., P. Odert, M. Leitzinger, K. Dissauer, N. C. Fleck, and H. S. Hudson, Indications of stellar coronal mass ejections through coronal dimmings, Nature Astronomy, 5, 697.

Vidotto, A. A. 2021, Living Reviews in Solar Physics, 18, 1.

Westall, F. et al. 2023, Front. Astron. Space Sci., 04.

Zank, G. P., W. K. M. Rice, and C. C. Wu (2000) Particle acceleration and coronal mass ejection driven shocks: A theoretical model, Journal of Geophysical Research, 105(A11), 25,079–25,096.

Solar and Stellar Coronal Mass Ejections
Proceedings IAU Symposium No. 388, 2026
N. Gopalswamy, O. Malandraki, A. Vidotto & W. Manchester, eds.
doi:10.1017/S174392132400156X

Close-in exoplanets: nature's own Parker Solar Probe/Solar Orbiter

A. A. Vidotto[1], S. Bellotti[1], V. Bourrier[2], J. F. Donati[3], R. Fares[4], G. A. J. Hussain[5], J. Morin[6] and P. Petit[3]

[1]Leiden Observatory, Leiden University, PO Box 9513, 2300 RA, Leiden, the Netherlands
email: vidotto@strw.leidenuniv.nl

[2]Observatoire Astronomique de l'Université de Genéve, Chemin Pegasi 51b, 1290 Versoix, Switzerland

[3]Institut de Recherche en Astrophysique et Planétologie, Université de Toulouse, CNRS, IRAP/UMR 5277, 14 avenue Edouard Belin, F-31400, Toulouse, France

[4]Department of Physics, College of Science, United Arab Emirates University, P.O. Box No. 15551, Al Ain, UAE

[5]Science Division, Directorate of Science, European Space Research and Technology Centre (ESA/ESTEC), Keplerlaan 1, 2201 AZ, Noordwijk, The Netherlands

[6]Laboratoire Univers et Particules de Montpellier, Université de Montpellier, CNRS, F-34095, Montpellier, France

Abstract. Stellar outflows, in the form of winds and transient (bursty) mass ejections, drive the "exo-space weather" around exoplanets. The exo-space weather can be substantially different from that of solar system planets. In particular, close-in exoplanets experience much harsher stellar wind conditions than any planet in the solar system. These exoplanets are conveniently located in the region where the stellar wind is still accelerating, and they are thus nature's own "Parker Solar Probe" or "Solar Orbiter". Here, we present our recent efforts towards characterising the stellar wind properties of GJ 436. UV spectroscopic transits and optical spectropolarimetric observations of the host star, together with 3D MHD wind models, have allowed us to probe the wind of the planet hosting star GJ 436. This has implications for constraining exo-space weather and also for understanding the wind driving of cool dwarf stars.

Keywords. Stars: winds, outflows – Stars: individual: GJ 436 – planet-star interactions – planetary systems

1. Introduction

The M dwarf star GJ 436 hosts a warm-Neptune that is losing substantial amount of atmosphere (Kulow et al. 2014; Ehrenreich et al. 2015; Lavie et al. 2017; Dos Santos et al. 2019). As this atmosphere is lost to space, it interacts with the wind of its host star (Bourrier et al. 2016). This wind, formed by particles and magnetic fields, shape the exo-space weather around the exoplanet GJ 436 b. By modelling the observations of spectroscopic transits of GJ 436 b with atmospheric escape models that account for atmospheric interaction with the stellar wind, local properties of the stellar wind at the orbital distance of the planet, such as the wind speed and density, can be estimated; these can then be used to infer global properties of stellar winds, such as their mass-loss rates (Vidotto and Bourrier 2017; Vidotto 2021).

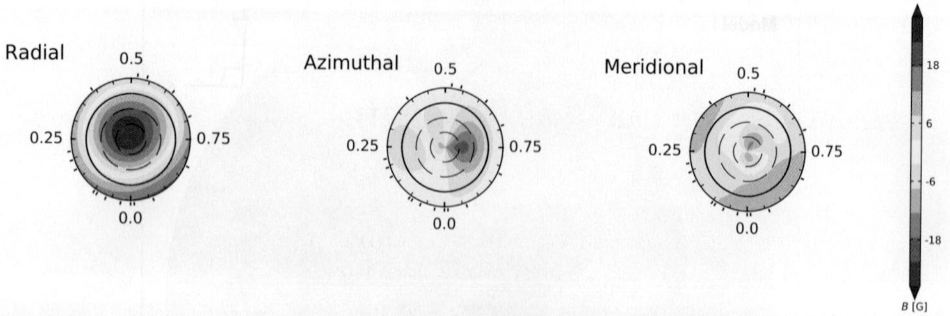

Figure 1. Observationally reconstructed magnetic field of GJ 436 using Zeeman Doppler Imaging. From left to right, we show the radial, azimuthal and meridional components of the stellar magnetic field in polar projection. Circles indicate different latitudes in the stellar surface, starting from 30 degrees (inner most one), until 120 degrees (outer-most one), with the equator shown as a solid circle. The radial component of the map is incorporated at the boundary of our stellar wind simulations. Figure adapted from Bellotti et al. (2023).

Here, we summarised the main results of our recent studies investigating the space weather around the gas giant planet GJ 436b. For full details of these studies, we refer the reader to Bellotti et al. (2023) and Vidotto et al. (2023).

2. The wind of GJ 436

The environment surrounding GJ 436 b consists of particles and magnetic fields that form the wind of the host star. In Vidotto et al. (2023), we characterised this environment by performing data-driven 3D MHD simulations of the stellar wind using the Alfvén wave driven wind model (Powell et al. 1999; Toth et al. 2012; van der Holst et al. 2014). At the boundary of our model, we included radial component of the reconstructed stellar surface magnetic field (Figure 1) derived by Bellotti et al. (2023).

In addition to the stellar surface magnetic map, other model inputs are the stellar properties (mass, radius and rotation period), the density and temperature at the transition region, and the Alfvén wave properties (boundary wave flux and the correlation length of the Alfvénic wave turbulence). Of these parameters, the least constrained ones are the wave properties. In our models, the energy density of the wave is assumed to be uniform at the inner boundary, which implies that the ratio between the base Alfven wave flux ($S_{A,0}$) and the stellar magnetic field ($B_\star$) is a constant: $S_{A,0}/B_\star$ =constant. Because this is poorly constrained for winds of cool dwarf stars, we assumed two values for $S_{A,0}/B_\star$: one that is similar to the canonical value used in the solar wind and one that is a factor of 10 smaller (referred to as Model I in Vidotto et al. (2023)). Figure 2 shows the 3D output of the latter model, after the simulation reached steady state.

In Figure 2, the grey streamlines are the magnetic field lines embedded in the wind. The orbital distance of the planet is shown by the translucent sphere at 14.56 $R_\star \simeq 0.028$ au. The contour shows the wind velocity and the Alfvén surface is shown by the translucent outer surface. This model predicts stellar wind mass-loss rates of 1.1×10^{-15} $M_\odot$ yr^{-1}. For comparison, the solar wind mass-loss rate is 2×10^{-14} $M_\odot$ yr^{-1}. We note most of the time the planet would be in the region of open magnetic field lines and that the orbital distance is nearly always within the Alfvén surface for our model, which can have implications for star-planet interactions (Folsom et al. 2020) and induced radio emission on the star (Kavanagh et al. 2021, 2022), for example.

The left panel of Figure 3 shows the trajectories of the planet as seen in the reference frame of the star during one rotational period of the star (about 44 days), during which

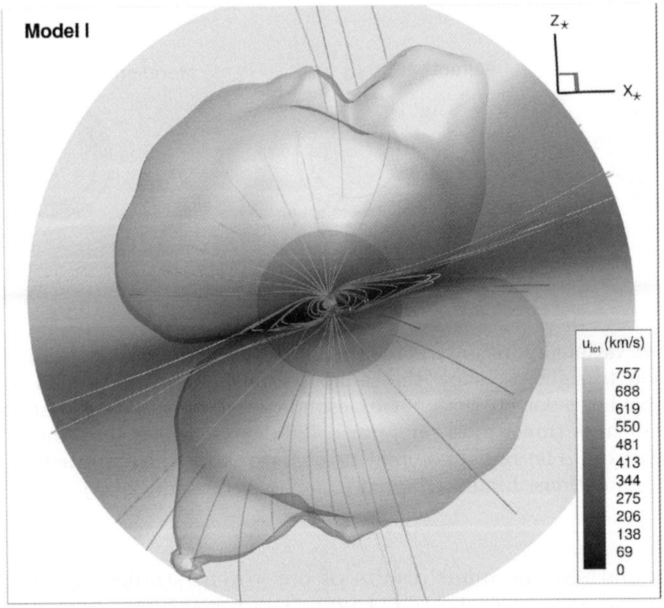

Figure 2. 3D view of the magnetised stellar wind for Model I presented in Vidotto et al. (2023).

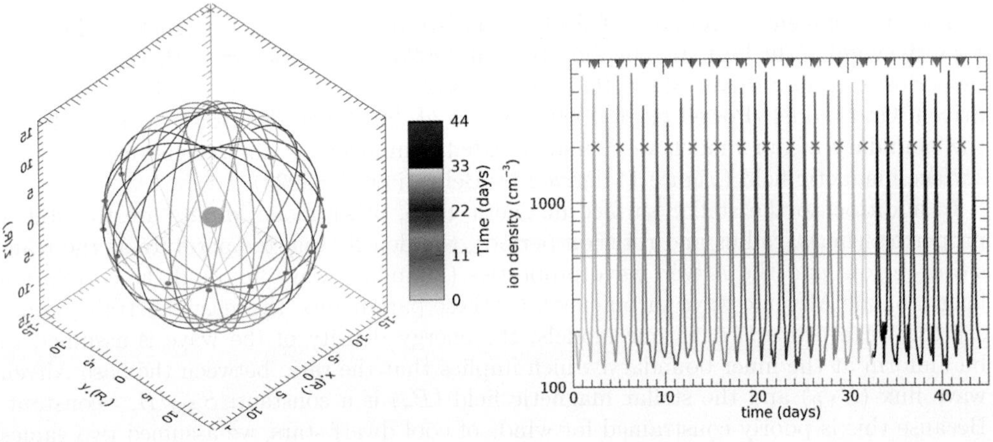

Figure 3. Left: Trajectories of the planet as seen in the reference frame of the star during one rotational period of the star (about 44 days, see colour bar), during which the planet completes about 16 orbital revolutions. The grey circles indicate the approximate times of mid-transit. Right: The stellar wind proton density at the orbit of GJ 436 b. The crosses indicate the wind values derived in the work of Bourrier et al. (2016, see text). Figures from/based on Vidotto et al. (2023)

the planet completes about 16 orbital revolutions. We extract the conditions of the stellar wind along these trajectories. The stellar wind proton density at the orbit of GJ 436 b, for example, is shown in the right panel of Figure 3. Colour represents the time evolution and follows the colour scheme as the left panel. The horizontal grey line is the time average of the stellar wind properties over one stellar rotation period (about 16 orbital periods of the planet). The grey arrows in the upper horizontal axis indicate approximate times of mid-transits. The crosses indicate the wind values derived in the work of Bourrier et al.

(2016) and they are plotted at mid-transit times. Note how they overlap really well with the values derived in our Model I, indicating it provides an excellent description of the wind of GJ 436. The density value of 2×10^3 cm^{-3} (grey crosses in the right panel of Figure 3) was derived from fitting the observations of Ly-α spectroscopic transits of GJ 436 b with atmospheric escape models that account for atmospheric interaction with the stellar wind. In addition to density, Bourrier et al. (2016) also provided estimates of the local wind speeds 85 km s^{-1} and our Model I also matches really well the velocity values (see our work Vidotto et al. 2023 for more details). Together, the wind speed and density are used to derive the mass-loss rate of GJ 436 (Vidotto and Bourrier 2017). All these quantities shape the space weather around the planet GJ 436 b.

3. Discussion and Conclusion

Recently, we investigated the space weather around the warm Neptune GJ 436b. We performed spectropolarimetric observations, which allowed us to reconstruct the stellar surface magnetic field (Bellotti et al. 2023). Using this information, we then performed 3D MHD simulations to model its 3D Alfvén-wave driven wind (Vidotto et al. 2023). By matching the local conditions of the stellar wind (e.g. ion density and speed) around GJ 436b derived in our simulations with previous transmission spectroscopic models (Bourrier et al. 2016), we suggest that the wind of GJ 436 is powered by a smaller flux of Alfvén waves than that powering the wind of the Sun. This suggests that the canonical flux of Alfvén waves assumed in solar wind models might not be applicable to the winds of old M dwarf stars, like GJ 436.

Additionally, compared to the solar wind, GJ 436's wind has a weaker acceleration and an extended sub-Alfvénic region. This is important because it places the orbit of GJ 436 b inside the region dominated by the stellar magnetic field (i.e., inside the Alfvén surface). Due to the sub-Alfvénic motion of the planet through the stellar wind, magnetohydrodynamic waves and particles released in reconnection events can travel along the magnetic field lines towards the star, which could power the anomalous ultraviolet flare distribution recently observed in the system (Loyd et al. 2023).

We refer the reader to Bellotti et al. (2023) and Vidotto et al. (2023) for more details of our studies, including the investigation into the power released by stellar wind-planet interactions in the system.

Acknowledgements

This project has received funding from the European Research Council (ERC) under the European Union's Horizon 2020 research and innovation programme (grant agreement No 817540, ASTROFLOW; grant agreement No 947634, SPICE DUNE; grant agreement No 740651, NewWorlds). RF acknowledges support from the United Arab Emirates University (UAEU) startup grant number G00003269. This work has been carried out within the framework of the NCCR PlanetS supported by the Swiss National Science Foundation under grants 51NF40_182901 and 51NF40_205606. We acknowledge funding from the French National Research Agency (ANR) under contract number ANR-18-CE31-0019 (SPlaSH). This work used the Dutch national e-infrastructure with the support of the SURF Cooperative using grant nos. EINF-2218 and EINF-5173. This work used the BATS-R-US tools developed at the University of Michigan Center for Space Environment Modeling and made available through the NASA Community Coordinated Modeling Center.

References

Bellotti, S., Fares, R., Vidotto, A. A., Morin, J., Petit, P., Hussain, G. A. J., Bourrier, V., Donati, J. F., Moutou, C., & Hébrard, É. M. 2023, The space weather around the exoplanet GJ 436b. I. The large-scale stellar magnetic field. *A&A*, 676, A139.

Bourrier, V., Lecavelier des Etangs, A., Ehrenreich, D., Tanaka, Y. A., & Vidotto, A. A. 2016, An evaporating planet in the wind: stellar wind interactions with the radiatively braked exosphere of GJ 436 b. *A&A*, 591, A121.

Dos Santos, L. A., Ehrenreich, D., Bourrier, V., Lecavelier des Etangs, A., López-Morales, M., Sing, D. K., Ballester, G., Ben-Jaffel, L., Buchhave, L. A., García Muñoz, A., Henry, G. W., Kataria, T., Lavie, B., Lavvas, P., Lewis, N. K., Mikal-Evans, T., Sanz-Forcada, J., & Wakeford, H. 2019, The Hubble PanCET program: an extensive search for metallic ions in the exosphere of GJ 436 b. *A&A*, 629, A47.

Ehrenreich, D., Bourrier, V., Wheatley, P. J., Des Etangs, A. L., Hebrard, G., Udry, S., Bonfils, X., Delfosse, X., Desert, J.-M., Sing, D. K., & Vidal-Madjar, A. 2015, A giant comet-like cloud of hydrogen escaping the warm Neptune-mass exoplanet GJ 436b. *Nature*, 522, 459–461.

Folsom, C. P., O Fionnagain, D., Fossati, L., Vidotto, A. A., Moutou, C., Petit, P., Dragomir, D., & Donati, J. F. 2020, Circumstellar environment of 55 cancri. the super-earth 55 cnc e as a primary target for star-planet interactions. *A&A*, 633, A48.

Kavanagh, R. D., Vidotto, A. A., Klein, B., Jardine, M. M., Donati, J.-F., & o Fionnagain, D. 2021, Planet-induced radio emission from the coronae of M dwarfs: the case of Prox Cen and AU Mic. *MNRAS*, 504(1), 1511–1518.

Kavanagh, R. D., Vidotto, A. A., Vedantham, H. K., Jardine, M. M., Callingham, J. R., & Morin, J. 2022, Radio masers on WX UMa: hints of a Neptune-sized planet, or magnetospheric reconnection? *MNRAS*, 514(1), 675–688.

Kulow, J. R., France, K., Linsky, J., & Loyd, R. O. P. 2014, Lyα Transit Spectroscopy and the Neutral Hydrogen Tail of the Hot Neptune GJ 436b. *ApJ*, 786, 132.

Lavie, B., Ehrenreich, D., Bourrier, V., Lecavelier des Etangs, A., Vidal-Madjar, A., Delfosse, X., Gracia Berna, A., Heng, K., Thomas, N., Udry, S., & Wheatley, P. J. 2017, The long egress of GJ 436b's giant exosphere. *A&A*, 605, L7.

Loyd, R. O. P., Schneider, P. C., Jackman, J. A. G., France, K., Shkolnik, E. L., Arulanantham, N., Cauley, P. W., Llama, J., & Schneider, A. C. 2023, Flares, Rotation, Activity Cycles, and a Magnetic Star-Planet Interaction Hypothesis for the Far-ultraviolet Emission of GJ 436. *AJ*, 165(4), 146.

Powell, K. G., Roe, P. L., Linde, T. J., Gombosi, T. I., & de Zeeuw, D. L. 1999, A Solution-Adaptive Upwind Scheme for Ideal Magnetohydrodynamics. *JCP*, 154, 284–309.

Toth, G., van der Holst, B., Sokolov, I. V., De Zeeuw, D. L., Gombosi, T. I., Fang, F., Manchester, W. B., Meng, X., Najib, D., Powell, K. G., Stout, Q. F., Glocer, A., Ma, Y.-J., & Opher, M. 2012, Adaptive numerical algorithms in space weather modeling. *Journal of Computational Physics*, 231, 870–903.

van der Holst, B., Sokolov, I. V., Meng, X., Jin, M., Manchester, IV, W. B., Toth, G., & Gombosi, T. I. 2014, Alfven Wave Solar Model (AWSoM): Coronal Heating. *ApJ*, 782, 81.

Vidotto, A. A. 2021, The evolution of the solar wind. *Living Reviews in Solar Physics*, 18(1), 3.

Vidotto, A. A. & Bourrier, V. 2017, Exoplanets as probes of the winds of host stars: the case of the M dwarf GJ 436. *MNRAS*, 470, 4026–4033.

Vidotto, A. A., Bourrier, V., Fares, R., Bellotti, S., Donati, J. F., Petit, P., Hussain, G. A. J., & Morin, J. 2023, The space weather around the exoplanet GJ 436b. II. Stellar wind-exoplanet interactions. *A&A*, 678, A152.

Solar and Stellar Coronal Mass Ejections
Proceedings IAU Symposium No. 388, 2026
N. Gopalswamy, O. Malandraki, A. Vidotto & W. Manchester, eds.
doi:10.1017/S1743921324001613

Preparing for Heliopolarimetry using New-generation Ground-based Radio Telescopes

Devojyoti Kansabanik[1]† and Angelos Vourlidas[2]

[1]Cooperative Programs for the Advancement of Earth System Science, University Corporation for Atmospheric Research, 3090 Center Green Dr, Boulder, CO, USA 80301

[2]The Johns Hopkins University Applied Physics Laboratory, 11101 Johns Hopkins Road, Laurel, MD 20723, USA

Abstract. Coronal mass ejections (CMEs) are large-scale ejections of magnetized plasma from the Sun and are associated with the most extreme space weather events. The southward component of CME magnetic fields (CME-B_z) is one of the crucial parameters determining its geoeffectiveness. Recent studies have shown that CMEs evolve significantly in the innermost part of the heliosphere ($\sim 20 - 90\ R_\odot$), and relying on extrapolations from low coronal heights can lead to wrong predictions of CME-B_z in the vicinity of Earth. Hence, measuring CME magnetic fields at these heights is important to improve CME-B_z prediction. A promising method to measure the CME-entrained magnetic field in the larger coronal heights and the innermost heliosphere is by measuring the changes in Faraday rotation (FR) of linearly polarized emission from background radio sources as their line-of-sight crosses the CME plasma. Here, we present the current preparation of new-generation ground-based radio telescopes for this purpose.

Keywords. Coronal Mass Ejections, Magnetic Field, Faraday Rotation, Heliopolarimetry

1. Introduction

Coronal mass ejections (CMEs) are large-scale expulsions of magnetized plasma from the Sun and the most important driver for space weather (SWX). Space weather refers to the environmental conditions in space as influenced by solar activities and their interactions with the Earth and other planetary bodies in the solar system. The geoeffectiveness of a CME is determined by its speed and the entrained magnetic field strength and topology, particularly the southward component (CME-B_z), the direction relative to the dipolar magnetic field of the Earth. Long periods of southward B_z lead to the opening of the magnetosphere and the injection of energetic particles into the terrestrial system which in turn disturbs the system in multiple ways. CMEs are routinely observed by white-light coronagraphs and heliospheric imagers at visible wavelengths, which capture the density structure of the CMEs. These decades-long observations have provided several geometrical and dynamical properties of the CMEs building our current understanding of CMEs (Howard et al. 2023). However, the white light observations cannot provide a direct measure of the CME-entrained vector magnetic fields.

Observations using modern radio telescopes provide some means to estimate the magnetic field of CMEs at coronal heights based on different types of radio emission (e.g. gyrosynchrotron and gyroresonance emission) (Carley et al. 2020; Vourlidas et al. 2020). However, beyond about $\sim$10 $R_\odot$ ($R_\odot$= solar radii), the optimal observing frequency for

† NASA Jack Eddy fellow hosted at the Johns Hopkins University Applied Physics Laboratory

these emissions falls below the ionospheric cutoff (<10 MHz). Since the strength of these emissions is generally too low for detection by space-based non-imaging radio instruments, this approach is only useful in the lower corona. However, interactions between different CMEs, ambient solar wind, and other heliospheric structures (like corotating or stream interaction regions) can lead to structural changes of CMEs in the larger coronal heights and a part of the inner heliosphere ($20 - 90\ R_\odot$) (Manchester et al. 2017; Braga et al. 2022). These deformations complicate the prediction of CME arrival times and CME-B_z at 1 AU from the information obtained when the CME was at lower coronal heights (Vourlidas et al. 2019). Hence, the measurement of CME magnetic fields at larger coronal heights and inner heliosphere is essential for improving SWX forecasting and predicting the geo-effectiveness of CMEs.

Measurement of changes in Faraday rotation (FR, Ferrière et al. 2021) of background linearly polarized radio sources due to coronal and heliospheric plasma (see Kooi et al. 2022, for a review) is known to be a promising remote-sensing tool for estimation the line-of-sight (LoS) magnetic field both in the corona and the inner heliosphere. We have coined the term, "Heliopolarimetry" to describe this observation technique. However, its application to SWX research remains limited due to observational and modeling challenges.

In this proceeding, we first review the current state of heliopolarimetry in Section 2. This will be followed by the capabilities of new-generation ground-based radio telescopes to perform heliopolarimetry at different solar elongations (Section 3) and the challenges involved (Section 4). We discuss the ongoing work for preparing these new-generation instruments for heliopolarimetry in Section 5.

2. Heliopolarimetry: Current Status

FR is the rotation of the plane of polarization of linearly polarized emission as it passes through magnetized plasma. FR is given by $\lambda^2\times$ RM, where the rotation measure (RM) (Brentjens, M. A. & de Bruyn, A. G. 2005) is the LoS integration of the product of electron density and LoS magnetic fields (RM $= \int n_e \vec{B}.d\vec{s}$) and λ is the wavelength of the radiation. White-light observations of Thomson scattered photospheric emission from the electrons in the CME plasma (Inhester 2015), observed using coronagraphs and heliospheric imagers, can provide the electron column density of a LoS, $N_e = \int n_e ds$. Combining both of these measurements can provide an estimate of the LoS averaged magnetic field ($<B>= \frac{RM}{N_e}$) of the CME.

Most of the previous FR experiments either used satellite beacons during short and infrequent periods of favorable observing geometry (e.g., Bird et al. 1985; Jensen et al. 2013; Wexler et al. 2019) or used high-frequency observations of a set of galactic/extragalactic linearly polarized radio sources (Kooi et al. 2017, 2021) mostly observing with the Jansky Very Large Array (JVLA, Kellermann et al. 2020). JVLA was primarily used due to its well-characterized polarization performance. These experiments were limited to coronal heights ($< 15\ R_\odot$) and could only provide sparse and temporally infrequent FR measurements due to the small field of view (FoV) of the instrument. Hence, the inversion of vector magnetic fields from the non-co-temporal measurements of $<B>$ invariably requires several assumptions and simplified models of the CME, which limits the benefit of heliopolarimetry for SWX research and application.

To extend the applicability of heliopolarimetry for SWX research, we need to expand the observation at higher coronal heights and into the inner heliosphere. As the CME propagates out in the heliosphere, both its electron density and magnetic fields decrease as well as its RM value. For a typical CME, the RM contribution by the CME drops from 10 rad/m^2 at $<2^\circ$ to 0.1 rad/m^2 at 90 $R_\odot$ (Oberoi & Lonsdale 2012). Since FR is proportional to λ^2, at low frequency, small changes in RM produce detectable changes in

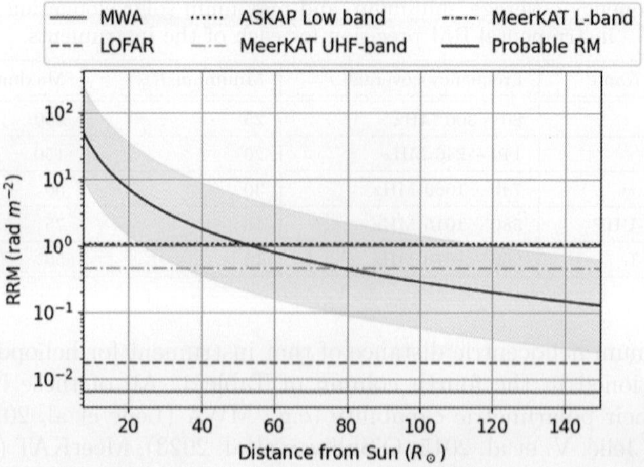

Figure 1. Probable variation of RRM of a CME with heliocentric distances is shown by the solid blue line, with an order of magnitude variance by light blue shade. Different horizontal lines represent the achievable precision in RRM with ~ 15 minutes of integration for different instruments in different observing bands.

FR. Hence, to achieve the required RM precision for heliopolarimetry at higher corona and inner heliosphere, one needs low-frequency ($\sim 80 - 1000$ MHz) observations. CME also expands in size as it propagates and covers a larger sky area. Wide FoV instruments are required to perform heliopolarimetry on a significant part of the CME.

3. Capabilities of New Generation Ground-based Radio Telescopes

Over the last decade, several new-generation radio interferometers have become operational, which provide some basic requirements of heliopolarimetry, wide FoV, wide frequency coverage from low to high frequency, polarization purity of the instrument, and good sensitivity over small temporal integration. These telescopes include – the Murchison Widefield Array (MWA, Tingay et al. 2013; Wayth et al. 2018), LOw Frequency ARray (LOFAR, van Haarlem et al. 2013), MeerKAT (Jonas & MeerKAT Team 2016) and Australian Square Kilometre Array Pathfinder (ASKAP, Hotan et al. 2021).

Each of these instruments operates at different frequency ranges with different observing bandwidths. To measure the RM contribution from the CME, we need precision in relative rotation measure (RRM = $\mathrm{RM_{obs}}$ - $\mathrm{RM_{source}}$) more than the RM of the CME. For a given instrument and frequency range, the maximum precision in the RRM is given as $\delta\mathrm{RRM} \sim \frac{\delta\phi_{\mathrm{FWHM}}}{\sqrt{2}\mathrm{SNR}}$. $\delta\phi_{\mathrm{FWHM}} = 2\sqrt{3}/\Delta\lambda^2$ is the half maximum of the full width of the rotation measure synthesis function (Brentjens, M. A. & de Bruyn, A. G. 2005), where $\Delta\lambda^2$ is the wavelength squared span across the observing bandwidth.

In Figure 1, the solid blue line shows the variation of probable RRM (Jensen et al. 2010; Oberoi & Lonsdale 2012) value of a CME with the distance from the Sun. Since, at higher corona and inner heliosphere, there is no measurement available, we have shown the region of an order of magnitude variation about the mean probable RRM value by the blue shade. Different horizontal lines represent the precision in RRM that can be achieved with ~ 15 minutes of integration of these instruments. We have taken the observing bandwidth mentioned in the second column of Table 1 for this calculation. We emphasize that these RRM estimates are only order-of-magnitude estimates and may be significantly different in a real scenario. The coronal and heliospheric region for which the value of probable RRM is higher than the achievable RRM precision of the instrument

Table 1. Frequency coverage, minimum, and maximum solar elongation allowed by the instrumental RM precision for each of the instruments.

Telescope Name	Frequency coverage	Minimum $R_\odot$	Maximum $R_\odot$
MWA	80 – 300 MHz	25	150
LOFAR	110 – 240 MHz	20	150
ASKAP-Low	740 – 1060 MHz	30	50
MeerKAT-UHF	580 – 1015 MHz	18	75
MeerKAT-L	900 – 1670 MHz	13	50

limits the maximum heliocentric distance of that instrument for heliopolarimetry. These values are mentioned in the fourth column of Table 1. All of these instruments have demonstrated their polarimetric capability (e.g., MWA (Lenc et al. 2017; Riseley et al. 2020), LOFAR (Jelić, V. et al. 2015; O'Sullivan et al. 2023), MeerKAT (Taylor & Legodi 2024), ASKAP (Thomson et al. 2023)) and conducted several polarization sky surveys. This survey data provides a reference RM of the sky for measuring changes in RM due to CMEs.

4. Preparing for Challenges with New Generation Instruments

While these new generation instruments are technically capable of providing high sensitivity and high precision polarimetry, the wide FoV poses some challenges when observing close to the Sun. To achieve similar sensitivity to the astronomical surveys, there are minimum solar elongations where these telescopes can be pointed. If they are pointed closer than this limit, the increase in system temperature of the instrument as well as radio emission from the Sun makes the calibration non-trivial and also reduces the sensitivity. The minimum solar elongations for these instruments at different observing bands are listed in the third column of Table 1. Considering both minimum and maximum solar elongations, these instruments can make heliopolarimetry possible over $\sim 15 - 150\ R_\odot$. This region has not yet been explored but holds important clues for the CME kinematics (e.g., Colaninno et al. 2013; Sachdeva et al. 2017). The suitable heliospheric region for each instrument are shown in Figure 2.

For observing as close as possible to the Sun, the main hurdle is to place the Sun within the first null of the primary beam. Given the chromaticity and wide observing bandwidth, it is impossible to exactly place the Sun at the null of the primary beam over the entire observing band. Hence, special care needs to be taken during calibration to account for the solar emission. Wideband low-frequency instruments, like the MWA and LOFAR and the upcoming Square Kilometre Array Observatory (SKAO)-low, can easily provide the required precision in RRM to measure the signature due to the CMEs up to $\sim 90\ R_\odot$. The achievable RRM precision by the MWA and LOFAR is ~ 0.01 rad m^{-2}, which is an order of magnitude better than the probable contribution by the CME. These precision limits are shown by the solid red and dashed green line in Figure 1.

However, the major hurdle for low-frequency observations targeting the larger solar elongations is the contribution of the ionospheric to the measured RM. Currently available and routinely used ionospheric RM correction tools, like ionFR (Sotomayor-Beltran, C. et al. 2013) and RMextract (de Gasperin et al. 2018), can routinely provide ionospheric RM at a precision of $\sim 0.1 - 0.3$ rad m^{-2}. This accuracy is sufficient to measure RM due to the CME up to $\sim$60 $R_\odot$. At much larger distances ($> 60\ R_\odot$), the RM contribution by the CME is comparable to the variability of ionospheric RM (Oberoi & Lonsdale 2012). This requires ionospheric RM estimation at a precision of ~ 0.01 rad m^{-2} with higher time resolution compared to the currently available measurements using GNSS satellites.

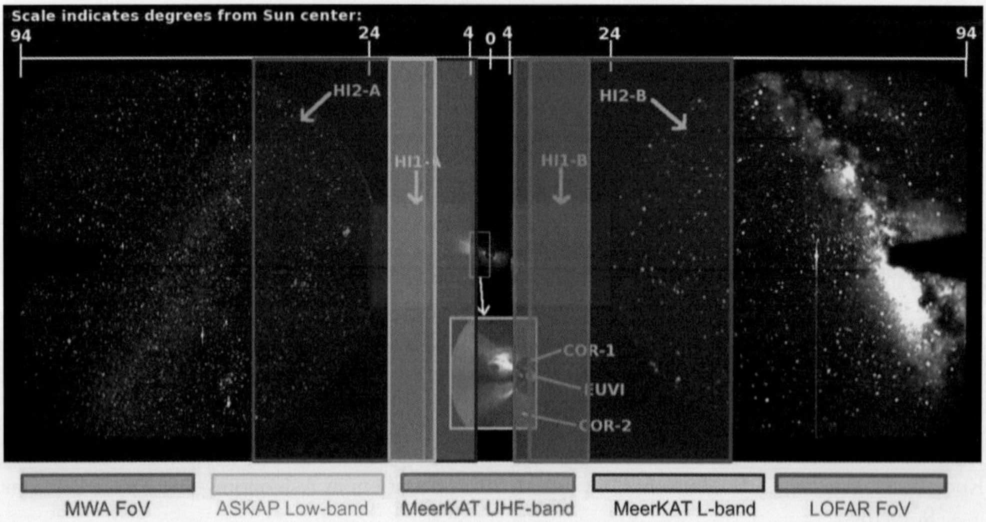

Figure 2. The ranges of heliocentric distances suitable for heliopolarimetry using the new-generation wide FoV instruments. The spans are only shown along the ecliptic plane, while the same is true for in any direction from the Sun. The background image is composed of images from coronagraph and heliospheric imagers of STEREO/SECIIII.

Efforts are underway to develop to meet the required RM-precision of ~ 0.01 rad m^{-2} using tri-band small satellites (Lanter et al. 2023).

5. Discussion

Although these new-generation wide FoV ground-based radio telescopes are capable of heliopolarimetry, no observations and studies have been done so far. Given the increasing numbers of CME, as we approach the peak of Solar Cycle 25, this is the ideal time to start heliopolarimetric observations with these instruments. We have recently initiated observations using some of these instruments and also look at suitable archival observations to solve different calibration challenges and demonstrate the measurement of $< B >$ along multiple LoS piercing through the CME. Once these measurements are available, they will be used for constraining CME flux rope models to understand how these new constraints improve the CME-B_z prediction at 1 AU.

Acknowledgements

This research was supported by the NASA Living with a Star Jack Eddy Postdoctoral Fellowship Program, administered by UCAR's Cooperative Programs for the Advancement of Earth System Science (CPAESS) under award #80NSSC22M0097. D.K. thanks the organizers of IAU Symposium 388 for providing travel support to attend the symposium and present this work. D. K. thanks John Morgan for coining the term Heliopolarimetry.

References

Bird, M. K., Volland, H., Howard, R. A., et al. 1985, Solar Physics, 98, 341.
Braga, C. R., Vourlidas, A., Liewer, P. C., et al. 2022, The Astrophysical Journal, 938, 13.
Brentjens, M. A., & de Bruyn, A. G. 2005, A&A, 441, 1217.
Carley, E. P., Vilmer, N., & Vourlidas, A. 2020, Frontiers in Astronomy and Space Sciences, 7.
Colaninno, R. C., Vourlidas, A., & Wu, C. C. 2013, Journal of Geophysical Research: Space Physics, 118, 6866–6879.

de Gasperin, F., Mevius, M., Rafferty, D. A., Intema, H. T., & Fallows, R. A. 2018, A&A, 615, A179.

Ferrière, K., West, J. L., & Jaffe, T. R. 2021, MNRAS, 507, 4968.

Hotan, A. W., Bunton, J. D., Chippendale, A. P., et al. 2021, PASA, 38, e009.

Howard, R. A., Vourlidas, A., & Stenborg, G. 2023, Frontiers in Astronomy and Space Sciences, 10

Inhester, B. 2015, arXiv e-prints, arXiv:1512.00651.

Jelić, V., de Bruyn, A. G., Pandey, V. N., et al. 2015, A&A, 583, A137.

Jensen, E. A., Bisi, M. M., Breen, A. R., et al. 2013, Solar Physics, 285, 83.

Jensen, E. A., Hick, P. P., Bisi, M. M., et al. 2010, Solar Physics, 265, 31.

Jonas, J., & MeerKAT Team. 2016, in MeerKAT Science: On the Pathway to the SKA, 1.

Kellermann, K. I., Bouton, E. N., & Brandt, S. S. 2020, The Very Large Array (Cham: Springer International Publishing), 319–390.

Kooi, J. E., Ascione, M. L., Reyes-Rosa, L. V., Rier, S. K., & Ashas, M. 2021, Solar Physics, 296, 11.

Kooi, J. E., Fischer, P. D., Buffo, J. J., & Spangler, S. R. 2017, Solar Physics, 292, 56.

Kooi, J. E., Wexler, D. B., Jensen, E. A., et al. 2022, Frontiers in Astronomy and Space Sciences, 9, 841866.

Lanter, F. P., Sutinjo, A., & Morgan, J. 2023, Advances in Space Research, 72, 5503, cOSPAR Space Weather Roadmap 2022-2024: Scientific Research and Applications.

Lenc, E., Anderson, C. S., Barry, N., et al. 2017, PASA, 34, e040.

Manchester, W., Kilpua, E. K. J., Liu, Y. D., et al. 2017, Space Science Reviews, 212, 1159.

Oberoi, D., & Lonsdale, C. J. 2012, Radio Science, 47.

O'Sullivan, S. P., Shimwell, T. W., Hardcastle, M. J., et al. 2023, MNRAS, 519, 5723.

Riseley, C. J., Galvin, T. J., Sobey, C., et al. 2020, PASA, 37, e029.

Sachdeva, N., Subramanian, P., Vourlidas, A., & Bothmer, V. 2017, Solar Physics, 292, 118.

Sotomayor-Beltran, C., Sobey, C., Hessels, J. W. T., et al. 2013, A&A, 552, A58.

Taylor, A. R., & Legodi, L. S. 2024, AJ, 167, 273.

Thomson, A. J. M., McConnell, D., Lenc, E., et al. 2023, PASA, 40, e040.

Tingay, S. J., Goeke, R., Bowman, J. D., et al. 2013, PASA, 30, e007.

van Haarlem, M. P., Wise, M. W., Gunst, A. W., et al. 2013, A&A, 556, A2.

Vourlidas, A., Carley, E. P., & Vilmer, N. 2020, Frontiers in Astronomy and Space Sciences, 7.

Vourlidas, A., Patsourakos, S., & Savani, N. P. 2019, Philos Trans A Math Phys Eng Sci, 377, 20180096.

Wayth, R. B., Tingay, S. J., Trott, C. M., et al. 2018, PASA, 35, e033.

Wexler, D. B., Hollweg, J. V., Efimov, A. I., et al. 2019, Journal of Geophysical Research: Space Physics, 124, 7761.

Solar and Stellar Coronal Mass Ejections
Proceedings IAU Symposium No. 388, 2026
N. Gopalswamy, O. Malandraki, A. Vidotto & W. Manchester, eds.
doi:10.1017/S1743921324002023

Application of the Global NM network for the Study of Solar Energetic Particles and their Space Weather Effects

Alexander Mishev and Nicholas Larsen

University of Oulu, Finland

Abstract. The study of strong solar energetic particles (SEPs) allows one to understand their acceleration and propagation in interplanetary space and gives the necessary basis to quantify the related cosmic-ray-induced terrestrial and space weather effects. During solar eruptive processes, such as solar flares and/or coronal mass ejections solar ions are accelerated to a high energy range. In most cases, the energy of the accelerated solar ions reaches several tens of MeV/nucleon, yet occasionally it exceeds 100 MeV/n or even reaches the GeV/n range. In the latter case, the energy of SEPs is enough, so that they induce a complicated particle shower in the earth's atmosphere, whose secondary particles reach the ground, eventually registered by ground-based detectors, such as neutron monitors (NMs). This particular class of events is known as ground-level enhancements (GLEs). Here we present a method for the analysis of GLEs, using neutron monitor data. Verification of the method with direct space-borne measurements is performed and good agreement is achieved. The derived SEP spectra give the necessary basis to model and quantify cosmic-ray-induced terrestrial and space weather effects following major solar eruptions. We present an example study, that is a full chain analysis of notable GLEs, namely their SEP energy/rigidity spectra derived using records from the global NM network and a model for computation of the induced space weather effects i.e. effective doses during those events.

Keywords. Cosmic Rays, Solar Energetic Particles, Neutron Monitor, Radiation dose

1. Introduction

Flux of high-energy particles, that is cosmic rays (CRs), consisting of protons (∼90%), Helium nuclei (∼9%), and some amount of heavier nuclei, constantly enter into the Earth's atmosphere. The bulk of CRs originate outside of the solar system, accelerated following supernova explosions, known as galactic cosmic rays (GCRs). GCRs reveal roughly a power-law energy spectrum, spanning from about 10^9 to 10^{21} eV/n (for details see e.g., Beatty et al. 2018, and references therein). High-precision measurements revealed the characteristics of CRs, e.g. AMS-02 (Alpha Magnetic Spectrometer – Aguilar et al. 2021). GCRs are slightly modulated in the Heliosphere by the solar wind (e.g., Potgieter 2013, and references therein).

When entering the atmosphere CRs initiate a nuclear-meson-electromagnetic shower of particles, resulting from consecutive interactions of the primary particle viz. proton and/or heavier nuclei with the atmospheric constituents. Those consecutive interactions produce large amounts of next-generation particles, which rain down to the ground if the energy of the primary is above about 0.5 GeV/n (e.g., Dorman 2004, and references therein). The secondary CR particles determine the complex radiation field in the atmosphere, particularly in the troposphere and stratosphere (e.g., Vainio et al. 2009). We emphasize that this complex atmospheric radiation field can be significantly enhanced by

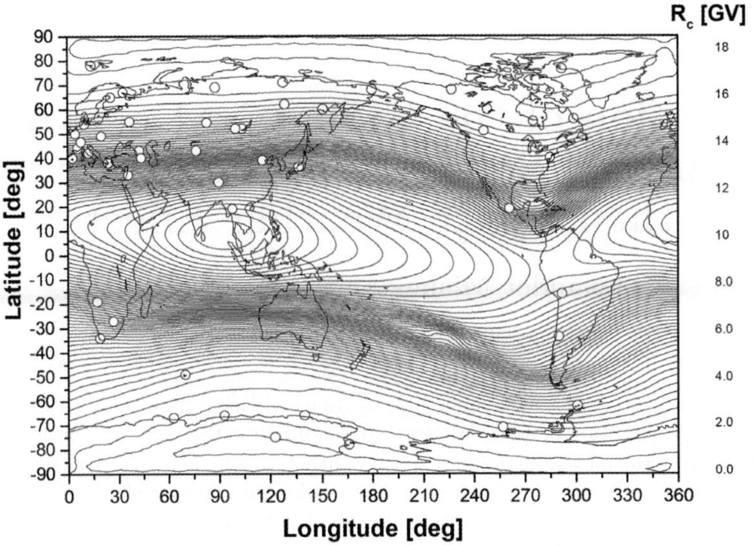

Figure 1. Map of the global NM network, stations given with open circles, within the lines of equal cutoff rigidity, the latter computed with IGRF model in quiet magnetospheric conditions.

sporadically accelerated particles, that is SEPs, produced following solar eruptions such as solar flares and/or coronal mass ejections (CMEs) (e.g., Klein and Dalla 2017, and references therein). In most cases, the SEP flux is greater than that of GCRs. Particular attention shall be paid when the energy of SEPs exceeds about 433 MeV/nucleon, producing secondaries observable at sea level (e.g., Miroshnichenko 2018; Mishev and Poluianov 2021). In this case, we speak about GLEs, events registered by ground-based NMs (for details see, Poluianov et al. 2017).

NMs have been used for continuous measurements of CR variations, and the study of various transients, and variations, including registration of GLEs (e.g., Simpson 2000), the latter specifically important for space weather purposes (e.g., Mavromichalaki et al. 2011, 2018). Herein, we present the application of the global NM network to provide specific space weather information, namely through a full-chain analysis to assess the radiation dose at flight altitudes, by the employment of verified methods and models (Mishev 2023). The full-chain analysis represents an unfolding of the SEP characteristics, particularly their spectra, computation of a global map of the cutoff rigidities and the final product is a 3-D dose rate in the atmosphere.

2. Global NM network and GLE analysis

The analysis of GLE using global NM records is performed using the geomagnetosphere as a giant spectrometer (e.g., Bütikofer 2018). NMs at different locations are sensitive to different energy/rigidity range of the primary CR spectrum and directions of arrival, the latter specifically important for the study of SEPs. An illustration of the global NM network with the stations at different cutoff rigidity is given in Fig. 1. Herein, we demonstrate the analysis of GLEs using NM records (for details see, Mishev et al. 2021), a method developed by Cramp et al. (1997), lately adopted by Bombardieri et al. (2006). The method was successfully used for the analysis of a plethora of GLEs (e.g., Mishev et al. 2022b; Papaioannou et al. 2022). We note, that the method was recently verified by direct space-borne measurements (Koldobskiy et al. 2019; Koldobskiy et al. 2021). The analysis is based on the modelling of the global NM response and subsequent unfolding of the SEP spectra. It includes the computation of cutoff rigidities and asymptotic

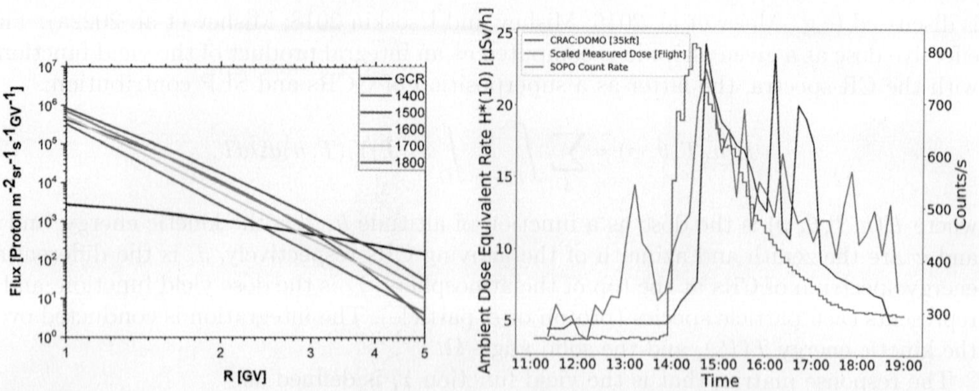

Figure 2. Left panel:SEP spectra during various stages of GLE #60 occurred on 15 April 2001. The black solid line denotes the average GCR flux. Right Panel: Count rate variation of SOPO NM and comparison between full chain model computations and measured ambient dose during GLE # 60.

cones of NMs, modeling their responses, and an optimisation of the difference between experimental and modelled NM responses.

Here, we present a notable example, that is GLE # 60 occurred on 15 April 2001, the so-called Easter event. This event was one of the strongest GLEs of solar cycle 23, e.g. the registered NM count rate increase was of about 225 % observed by the South Pole (SOPO) NM. The event followed increased solar activity over the end of March 2001 to mid-April 2001 period, leading to several M-class and X-class flares. The flare leading to the GLE # 60 was related to NOAA Active Region 9415, with coordinates S20 W85, therefore, well connected to the Earth.

The derived SEP spectra during GLE # 60 are depicted with a modified power-law rigidity spectrum, namely the proton flux $J_{\parallel}(P)$ with given rigidity P in GV, is given by:

$$J_{\parallel}(P) = J_0 P^{-(\gamma+\delta\gamma(P-1))}, \quad (1)$$

where the SEPs are arriving along the axis of symmetry, γ is the spectral index and $\delta\gamma$ represents the spectrum steepening. An illustration of several spectra during this event is shown in Fig. 2, whilst the table with all spectra is given in (Larsen and Mishev 2023). The derived GLE-causing SEP spectra were hard during the initial (14:00-14:30 UT) and main phase of the event (14:30–16:00 UT). Besides, a gradual enhancement of the particle flux was observed, whilst in the late stage of the event a moderate steady decrease of the particle flux is revealed. We would like to emphasize that after the main phase of the event, roughly corresponding to the peak SEP flux, the SEP spectra were soft.

3. Computation of radiation dose during GLEs

For the computation of the radiation dose, we employed the Cosmic Ray Atmospheric Cascade Dosimetric Model (CRAC:DOMO). It is a full target numerical model based on a response matrix formalism, that is yield functions, the latter derived via Monte Carlo simulations with PLANETOCOSMICS (Desorgher et al. 2005). The yield functions are the response as dose at given altitude (effective, ambient dose equivalent) versus monoenergetic particle flux (proton and/or alpha particle), precisely Monte Carlo obtained over a layered NRLMSISE-00 atmospheric model secondary particles, convoluted with fluence to dose coefficients, explicitly considering the radiation weighting factors, the latter derived using simulations with ICRP standard phantom. The ready to use tables are given in Larsen and Mishev (2023). A detailed description of the model and comparison is discussed (e.g., Meier et al. 2016; Mishev and Usoskin 2018; Mishev et al. 2022a). The effective dose at a given altitude is computed as an integral product of the yield functions with the CR spectra, the latter as a superposition of GCRs and SEP contribution:

$$E(h,T,\theta,\varphi) = \sum_i \int_{T(P_c)}^{\infty} \int_{\Omega} J_i(T) Y_i(T,h) d\Omega dT, \tag{2}$$

where $E(h,T,\theta,\varphi)$ is the dose as a function of altitude h, T is the kinetic energy, and θ and φ are the zenith and azimuth of the arriving CRs respectively, J_i is the differential energy spectrum of CRs at the top of the atmosphere, Y_i is the dose yield function, and i represents each particle species (proton or α-particle). The integration is conducted over the kinetic energy $T(P_c)$, and the solid angle Ω.

The response matrix, that is the yield function Y_i is defined as:

$$Y_i(T,h) = \sum_j \int_{T^*} F_{i,j}(h,T,T^*,\theta,\varphi) C_j(T^*) dT^* \tag{3}$$

where $C_j(T^*)$ is the fluence to effective/ambient dose conversion coefficient for a secondary particle of type j (neutron, proton, γ, e^-, e^+, μ^-, μ^+, π^-, π^+) with energy T^*, $F_{i,j}(h,T,T^*,\theta,\varphi)$ is the fluence of secondary particles of type j, produced by a primary CR particle of type i (proton or $\alpha-$particle) with a given primary energy T arriving at the top of the atmosphere from zenith angle θ and azimuth angle φ.

4. Examples of full chain analysis

Employing the derived spectra during GLE #60 by Larsen and Mishev (2023), presented in the left panel of Fig. 2, and Eq. 2, we computed the global map of the effective dose rate during this event at typical flight altitude, considering the SEP spectra evolution and compared with the actual flight profile PRG-JFK, when a dosimetric measurement was carried out (Spurny and Dachev 2001), (for details see their Figs. 6 and 7 in Larsen and Mishev 2023). A detailed comparison between our model computations of the ambient dose equivalent and Luilin measurements during PRG-JFK flight showed a good agreement as seen in the right panel of Fig. 2. We note, that herein the measurements of the Luilin are scaled in order to account for its limited sensitivity to the secondary hadronic component, specifically neutrons (Larsen and Mishev 2023). The model validation was performed against measurements by Liulin, that is a Si-semiconductor based dosemeter. Liulin measures the deposited energy and the amount of interacting particles within the active volume of the detector, i.e. giving the dose rate in silicon. We note, that Liulin type dosemeters are widely used for measurements of the absorbed dose in space and aviation altitudes (e.g. Dachev et al. 2023).

We also computed the ambient dose equivalent at FL350 during the strongest directly observed event, that is GLE # 5 on 23 February 1956, using recently reconstructed and revised spectra (details given in Hayakawa et al. 2024), and during the main phase of GLE #73 on 28 October 2021 as shown in Fig. 3. One can see that during GLE # 5, even considering the late phase of the event, that is at 07:00 UT, when the SEP spectra have been softer and with smaller flux compared to the initial phase of the event, the assessed dose is considerably greater than that during GLE # 60, and represents a notable threat, whilst during GLE # 73 the dose do not represented an issue. We emphasize that we considered an isotropic SEP flux in order to present a conservative approach for the dose (e.g., Copeland et al. 2008), whilst during the unfolding of SEP spectra the anisotropy is accounted. We note that the main uncertainty in this full chain analysis is due to SEP spectra reconstruction as discussed in (Miroshnichenko 2018), which eventually results to maximal range of errors of about 25 %.

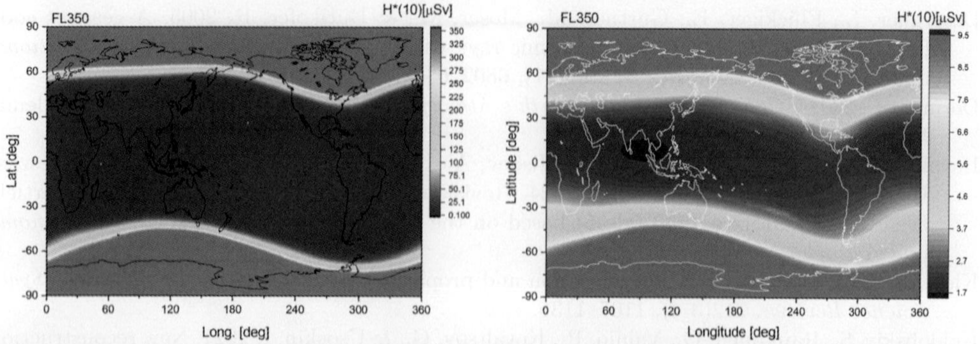

Figure 3. Global map of the 1h integrated ambient dose equivalent at FL350 during the late phase, 07:00 UT of GLE #5 on 23 February 1956 (left panel) and main phase, 17:30 UT, of GLE #73 on 28 October 2021 (right panel).

5. Conclusions

We presented a systematic study of the radiation dose at flight altitude during strong GLE events. We demonstrated the abilities of the global NM network to be used in a full-chain analysis of strong SEP events, and assessment of their space weather effect, that is exposure to radiation at flight altitude. The derived GLE-causing SEP spectra are used as input for corresponding radiation, allowing us to compute the exposure to radiation on a global scale explicitly accounting the SEP contribution.

Acknowledgments

This work was supported by the Academy of Finland (projects 330063 QUASARE, 24304650 GERACLIS, the HE projects ALBATROS and SPEARHEAD. We acknowledge the support of the ISSI for the Team No. 585 (REASSESS). This work was supported by the National Science Fund of Bulgaria under contract KP-06-H64/3.

References

Aguilar, M., Ali Cavasonza, L., Ambrosi, et al.,. 2021, The Alpha Magnetic Spectrometer (AMS) on the international space station: Part II Results from the first seven years. *Physics Reports*, 894, 1–116.

Beatty, J., Matthews, J., & Wakely, S. Cosmic rays. In *Review of Particle Physics* 2018, pp. 424–432. Physical Review D 98, 030001.

Bombardieri, D., Duldig, M., Michael, K., & Humble, J. 2006, Relativistic proton production during the 2000 July 14 solar event: The case for multiple source mechanisms. *Astrophysical Journal*, 644(1), 565–574.

Bütikofer, R. Ground-based measurements of energetic particles by neutron monitors. In *Solar Particle Radiation Storms Forecasting and Analysis, The HESPERIA HORIZON 2020 Project and Beyond* 2018, chapter 6, pp. 95–112. Springer Nature, Cham, Switzerland.

Copeland, K, Sauer, H., Duke, F., Friedberg, W., 2008, Cosmic radiation exposure of aircraft occupants on simulated high-latitude flights during solar proton events from 1 January 1986 through 1 January 2008, *Advances in Space Research*, 42(6), 1008–1029.

Cramp, J., Duldig, M., Flückiger, E., Humble, J., Shea, M., & Smart, D. 1997, The October 22, 1989,solar cosmic enhancement: ray an analysis the anisotropy spectral characteristics. *Journal of Geophysical Research*, 102(A11), 24 237–24 248.

Dachev, T., Litvak, M., Benton, E. et al., 2023. The neutron dose equivalent rate measurements by r3dr/r2 spectrometers on the international space station. *Life Sciences in Space Research*, 39, 43–51.

Desorgher, L., Flückiger, E., Gurtner, M., Moser, M., & Bütikofer, R. 2005, A geant 4 code for computing the interaction of cosmic rays with the earth's atmosphere. *International Journal of Modern Physics A*, 20(A11), 6802–6804.

Dorman, L. 2004,. *Cosmic Rays in the Earth's Atmosphere and Underground.* Kluwer Academic Publishers, Dordrecht.

Hayakawa, Hisashi, Koldobskiy, Sergey, Mishev, Alexander, Poluianov, Stepan, Gil, Agnieszka, Usoskina, Inna, & Usoskin, Ilya 2024, Revision of the strongest solar energetic particle event of 23 February 1956 (gle 5) based on the rediscovered original records. *Astronomy and Astrophysics*, 684, A46.

Klein, K.-L. & Dalla, S. 2017, Acceleration and propagation of solar energetic particles. *Space Science Reviews*, 212(3-4), 1107–1136.

Koldobskiy, S., Raukunen, O., Vainio, R., Kovaltsov, G., & Usoskin, I. 2021, New reconstruction of event-integrated spectra (spectral fluences) for major solar energetic particle events. *Astronomy and Astrophysics*, 647, A132.

Koldobskiy, S. A., Bindi, V., Corti, C., Kovaltsov, G. A., & Usoskin, I. G. 2019, Validation of the Neutron Monitor Yield Function Using Data from AMS-02 Experiment 2011 – 2017. *J. Geophys. Res. (Space Phys.)*, 124, 2367–2379.

Larsen, N. & Mishev, A. 2023, Analysis of the ground level enhancement gle 60 on 15 April 2001, and its space weather effects: Comparison with dosimetric measurements. *Space Weather*, 21(8).

Mavromichalaki, H., Gerontidou, M., Paschalis, P., Paouris, E., Tezari, A., Sgouropoulos, C., Crosby, N., & Dierckxsens, M. 2018, Real-time detection of the ground level enhancement on 10 September 2017 by a.ne.mo.s.: System report. *Space Weather*, 16(11), 1797–1805.

Mavromichalaki, H., Papaioannou, A., Plainaki, C. et al., 2011, Applications and usage of the real-time neutron monitor database. *Advances of Space Research*, 47, 2210–2222.

Meier, M., Trompier, F., Ambrozova, I., Kubancak, J., Matthiä, D., Ploc, O., Santen, N., & Wirtz, M. 2016, Concord: Comparison of cosmic radiation detectors in the radiation field at aviation altitudes. *Journal of Space Weather and Space Climate*, 6.

Miroshnichenko, L. 2018, Retrospective analysis of gles and estimates of radiation risks. *Journal of Space Weather and Space Climate*, 8, A52.

Mishev, A. 2023, Application of the global neutron monitor network for assessment of spectra and anisotropy and the related terrestrial effects of strong seps. *Journal of Atmospheric and Solar-Terrestrial Physics*, 243, 106021.

Mishev, A., Binios, A., Turunen, E., et al., 2022,a Measurements of natural radiation with an mdu liulin type device at ground and in the atmosphere at various conditions in the arctic region. *Radiation Measurements*, 154a, 106757.

Mishev, A., Kocharov, L., Koldobskiy, S., Larsen, N., Riihonen, E., Vainio, R., & Usoskin, I. 2022,b High-resolution spectral and anisotropy characteristics of solar protons during the gle n ° 73 on 28 October 2021 derived with neutron-monitor data analysis. *Solar Physics*, 297b(7), 88.

Mishev, A., Koldobskiy, S., Usoskin, I., Kocharov, L., & Kovaltsov, G. 2021, Application of the verified neutron monitor yield function for an extended analysis of the gle # 71 on 17 May 2012. *Space Weather*, 19(2), e2020SW002626.

Mishev, A. & Poluianov, S. 2021, About the altitude profile of the atmospheric cut-off of cosmic rays: New revised assessment. *Solar Physics*, 296(8).

Mishev, A. & Usoskin, I. 2018, Assessment of the radiation environment at commercial jet-flight altitudes during gle 72 on 10 September 2017 using neutron monitor data. *Space Weather*, 16(12), 1921–1929.

Papaioannou, A., Kouloumvakos, A., Mishev, A., Vainio, R., Usoskin, I., Herbst, K., Rouillard, A. P., Anastasiadis, A., Gieseler, J., Wimmer-Schweingruber, R., & Kühl, P. 2022, The first ground level enhancement of solar cycle 25 on 28 October 2021. *Astronomy and Astrophysics*, 660, L5.

Poluianov, S., Usoskin, I., Mishev, A., Shea, M., & Smart, D. 2017, Gle and sub-gle redefinition in the light of high-altitude polar neutron monitors. *Solar Physics*, 292(11), 176.

Potgieter, M. 2013, Solar modulation of cosmic rays. *Living Reviews in Solar Physics*, 10(3).

Simpson, J. 2000, The cosmic ray nucleonic component: The invention and scientific uses of the neutron monitor. *Space Science Reviews*, 93, 11–32.

Spurny, F. & Dachev, T. 2001, Measurements in an aircraft during an intense solar flare, ground level event 60, on April 15, 2001. *Radiation Protection Dosimetry*, 95(3), 273–275.

Vainio, R., Desorgher, L., Heynderickx, D., et al., 2009, Dynamics of the earth's particle radiation environment. *Space Science Reviews*, 147(3-4), 187–231.

Solar and Stellar Coronal Mass Ejections
Proceedings IAU Symposium No. 388, 2026
N. Gopalswamy, O. Malandraki, A. Vidotto & W. Manchester, eds.
doi:10.1017/S1743921324001157

Astrospheres of planet-hosting cool stars and their impact on exoplanetary habitability

Konstantin Herbst[1], Klaus Scherer[2], N. Eugene Engelbrecht[3], Juandre Light[3], Frederic Effenberger[2], Stefan E. S. Ferreira[3], Horst Fichtner[2], Jens Kleimann[2] and R. Du Toit Strauss[3]

[1]Institute for Experimental and Applied Physics, Kiel University, Germany
email: herbst@physik.uni-kiel.de

[2]Institute for Theoretical Physics IV, Ruhr-Universität Bochum, 44780 Bochum, Germany

[3]Centre for Space Research, North-West University, 2520, Potchefstroom, South Africa

Abstract. The impact of stellar energetic particles (StEPs) and galactic cosmic rays (GCRs) on exoplanetary atmospheres, and with that on transmission spectra and biosignatures like methane and ozone, recently became one of the foci of exoplanetary habitability studies. Deriving the exoplanetary radiation field, however, is an interdisciplinary task starting with astrospheric modeling. Although multi-wavelength observations of the Hubble Space Telescope, Kepler, and TESS helped to study the characteristics of planet-hosting cool stars. However, features like stellar wind speeds and mass-loss rates mandatory for astrospheric modeling can only be indirectly derived. Here, a brief review of the astrospheric modeling efforts of three M stars and the transport of cosmic rays within is presented. Using a unique suite of numerical models, the astrospheric plasma environment is modeled magnetohydrodynamically (3D), and the particle transport is modeled using an ab initio GCR modulation code. A recent model effort deriving flare-induced StEP intensities is also discussed.

Keywords. Astrospheres, Magnetohydrodynamic Modeling, Cosmic Rays, Stellar Energetic Particles, Modulation

1. Introduction

Due to their large number within the Galaxy, their small radii, and closer-in habitable zones (HZs), in particular K- and M-stars are prime targets to detect habitable rocky (Earth-like) exoplanets (see, e.g., Gillon et al. 2017; Dittmann et al. 2017). However, although planets in the HZ of their host stars could retain liquid water on their surfaces (see, e.g., Kasting et al. 1993; Noack et al. 2017), its presence within the HZ is not a direct indicator of habitability (see, e.g., Lammer et al. 2009). Possible reasons, for example, are the presence of strong stellar winds that can lead to the erosion of unprotected planetary atmospheres and extreme stellar particle events that can cause changes in the atmospheric chemistry, which may prevent the genesis and further evolution of life (see, e.g., Vidotto et al. 2013; do Nascimento et al. 2016; Herbst et al. 2019; Scheucher et al. 2020). The transport of GCRs through an astrosphere depends on its volume, structure, and turbulent state, which are governed by, for example, the stellar type, rotation rate, the stellar wind dynamics, the stellar activity, the magnetic field as inner boundary conditions and the local interstellar spectrum (LIS) as an outer boundary condition. In the case of the Sun, the latter has been recently measured by the two Voyager spacecraft (see, e.g., Stone et al. 2013; Richardson et al. 2021), so that earlier LIS models existing in the literature had to be updated (see, e.g., Herbst et al. 2017). By treating an astrosphere as a heliosphere-analog, the well-studied transport equation by Parker (1965) may be applied. Previous (semi-)analytical attempts, e.g., based on the simplified Force-Field approximation (see, e.g., Moraal 2013), have been used to derive the GCR flux within cool-star astrospheres by, e.g., Sadovski et al. (2018), Struminsky et al. (2017), Rodgers-Lee et al. (2021b), and Mesquita et al. (2022a). The results suggested that the flux of GCRs with energies below ~ 10 TeV can be neglected within all cool-star astrospheres.

That this might be a drastically simplified view of things and in itself might most likely lead to an erroneous view of things has first been reported in Herbst et al. (2020). Utilizing a 1D approach to solve Parker's transport equation, the authors showed that the opposite might be true.

It was further pointed out that updating the investigations to full 3D is mandatory to consider important physical processes such as particle drifts and turbulence to more reliably predict the GCR environment of exoplanets.

2. 3D Magnetohydrodynamic Astrospheric Modeling

The general set of the Euler equations, including the continuity-, momentum-, energy-, and induction equation for the magnetic field, is given as

$$\frac{\partial}{\partial t}\begin{bmatrix}\rho\\ \rho\,\mathbf{v}\\ e\\ \mathbf{B}\end{bmatrix}+\nabla\cdot\begin{bmatrix}\rho\mathbf{v}\\ \rho\,\mathbf{v}\otimes\mathbf{v}+\left(p+\frac{B^2}{8\pi}\right)\tilde{\mathbf{I}}-\frac{\mathbf{B}\otimes\mathbf{B}}{4\pi}\\ (e+p)\mathbf{v}-\frac{\mathbf{B}(\mathbf{B}\cdot\mathbf{v})}{4\pi}\\ \mathbf{v}\otimes\mathbf{B}-\mathbf{B}\otimes\mathbf{v}\end{bmatrix}=\begin{bmatrix}S^{c}\\ \mathbf{S}^{m}\\ S^{e}\\ \mathbf{0}\end{bmatrix}. \tag{1}$$

Here, ρ gives the mass density, $\mathbf{v}$ the velocity, P the thermal pressure, and $\mathbf{B}$ the magnetic field vector. Further, the terms on the right-hand side describe the source terms for charge exchange, electron impact, and photoionization (for details see Scherer et al. 2014), while

$$e=\frac{\rho u^2}{2}+\frac{p}{\gamma-1}+\frac{B^2}{8\pi} \tag{2}$$

represents the internal energy for an adiabatic index γ, which often is assumed as 5/3.

The above set of equations describes the (ionized) Magnetohydrodynamic (MHD) fluid while setting $\mathbf{B}=\vec{0}$ describes the neutral hydrogen fluid.

Single- and multi-fluid MHD results are often based on solving the 3D ideal MHD equations given by

$$\frac{\partial\rho}{\partial t}+\nabla\cdot\rho\mathbf{v}=0 \tag{3}$$

$$\frac{\partial\rho\mathbf{v}}{\partial t}+\nabla\cdot\left(\rho\mathbf{v}\otimes\mathbf{v}+\left(p+\frac{B^2}{2}\right)\tilde{I}-\mathbf{B}\otimes\mathbf{B}\right)=0 \tag{4}$$

$$\frac{\partial E}{\partial t}+\nabla\cdot\left[\left(\rho\mathbf{v}\otimes\mathbf{v}+\frac{\gamma}{\gamma-1}p+\frac{B^2}{2}\right)\mathbf{v}-(\mathbf{B}\cdot\mathbf{v})\mathbf{B}\right]=-R_L \tag{5}$$

$$\frac{\partial\mathbf{B}}{\partial t}+\nabla\cdot(\mathbf{v}\otimes\mathbf{B}-\mathbf{B}\otimes\mathbf{v})=0 \tag{6}$$

with ρ as the mass density, $\mathbf{v}$ as the velocity, p as the pressure, and $E=e+\frac{1}{2}\rho v^2+\frac{B^2}{2}$ as the total energy. $\tilde{I}$ represents the unit tensor while $\otimes$ referees to the dyadic/tensorial product. Cooling and heating functions ($-R_L$) are treated according to the discussion in Scherer et al. (2020).

Utilizing the CRONOS code (Kissmann et al. 2018), this set of equations is integrated over time on a fixed 3D grid until a steady state is approximately reached. The code is

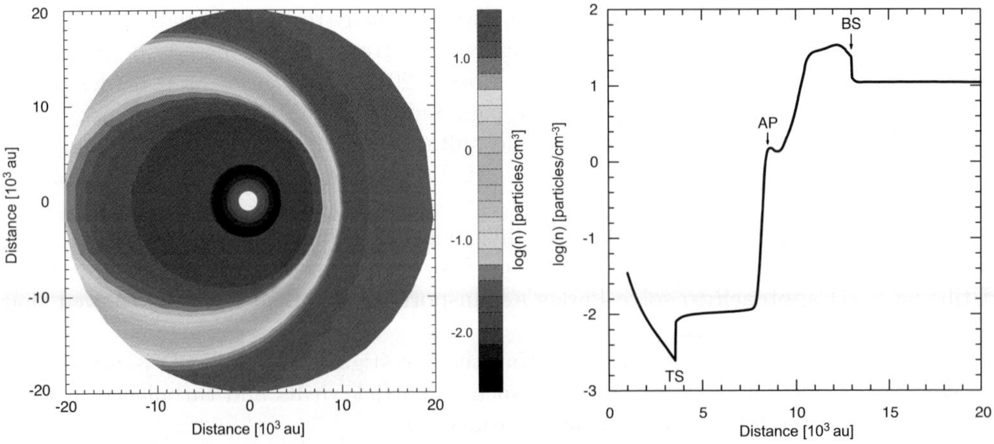

Figure 1. Left panel: Modeled astrosphere of V374 Peg. Right panel: Corresponding number density plot along the line of sight, with indications of the TS, AP, and BS distances. Taken from Herbst et al. (2020) ©AAS. Reproduced with permission.

based on a Riemann solver, and model runs are performed in a star-centered spherical grid with a resolution of $N_r \times N_\theta \times N_\phi$ cells (Scherer et al. 2020).

Important stellar input parameters to model astrospheric environments more reliably are the stellar wind speed ($v_{\rm sw}$) and wind density ($n_{\rm sw}$), the stellar mass loss rate ($\dot{M}_\star$), and the stellar magnetic field ($B_\star$). Also, information on the interstellar medium (ISM) is mandatory. Of particular interest is the ISM density ($n_{\rm ISM}$) and velocity ($v_{\rm ISM}$). Unfortunately, for most cool stars, only little information on these parameters can be found in the literature.

The results below are based on the single-fluid 3D MHD model efforts discussed in Herbst et al. (2020) and Herbst et al. (2022a). The first-ever multi-fluid 3D MHD model efforts for a cool star astrosphere are discussed in Scherer et al. (2024, under rev.).

2.1. *The astrosphere of V374 Pegasi*

V374 Peg, also known as GJ 4247, is an M3.5Ve† star with an effective temperature of 3440 K, known to be an ultrafast-rotating and fully convective star with a rotation period of only 0.44 days (Vida et al. 2016). In addition, V374 Peg most likely has an extraordinarily strong surface magnetic field of about $1.6 \cdot 10^8$ nT and a mass-loss rate of $4 \cdot 10^{-11}$ to $4 \cdot 10^{-10}$ $M_\odot$ yr^{-1} (Vidotto et al. 2011). Unfortunately, no information on the local ISM of V374 Peg is available in the literature, so sophisticated guesses based on heliospheric parameters had to be made. The authors choose a $T_{\rm ISM}$ of 9000 K, $v_{\rm ISM}$ of 30 km s^{-1}, a density of 11 cm^{-3}, and a magnetic field strength of $B_{\rm ISM} = 1$ nT.

Utilizing CRONOS, Herbst et al. (2020) modeled the astrosphere of V374 Peg on a star-centered spherical grid with a resolution of 1024 × 16 × 32 cells. The results are displayed in Fig. 1. As can be seen, the high stellar mass loss rate coupled with a fast stellar wind and the ultra-fast rotation period leads to a parsec-scale astrosphere (with the termination shock (TS) at $3.6 \cdot 10^3$ au, the astropause (AP) at $8.5 \cdot 10^3$ au, and a bow shock (BS) at $1.3 \cdot 10^4$ au), which is orders of magnitude larger than our heliosphere (TS at around 94 (84) au and HP at around 122 (119) au according to Voyager 1(2) data, e.g., Stone et al. 2019).

† data from http://simbad.u-strasbg.fr/simbad/

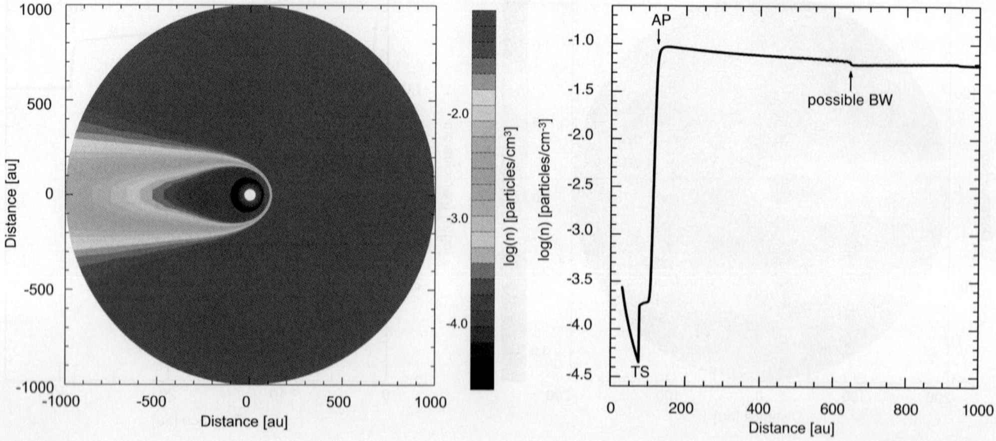

Figure 2. Left panel: Modeled astrosphere of Proxima Centauri. Right panel: Corresponding number density plot along the line of sight, with indications of the TS, AP, and BS distances. Taken from Herbst et al. (2020) ©AAS. Reproduced with permission.

2.2. *The astrosphere of Proxima Centauri*

Proxima Centauri, also known as GJ 551, is an M5.5Ve-type‡star and, with a distance from the Sun of only (1.3012 ± 0.0003) pc (Garraffo et al. 2016), our nearest neighbor. With that, the same ISM conditions as for the heliosphere can be assumed.

Proxima Centauri likely has a surface magnetic field of about 6 $\cdot 10^7$ nT and a mass-loss rate of approximately 2.84 $\cdot 10^{-15}$ $M_\odot$ yr^{-1} (according to Reiners and Basri 2008; Garraffo et al. 2016).

Utilizing CRONOS, Herbst et al. (2020) modeled the astrosphere of Proxima Centauri on a star-centered spherical grid with a resolution of 1024 × 64 × 32 cells. The results are shown in Fig. 2. As can be seen, with a TS at 53.7 au and an AP at 122 au, the astrosphere of Proxima Centauri is comparable in size and shape to the heliosphere. In addition, only a likely bow wave rather than a bow shock is seen, which is also in agreement with heliospheric numerical results (e.g., Zank et al. 2012, motivated by IBEX observations).

2.3. *The astrosphere of LHS 1140*

LHS 1140, also known as GJ 3053, is an M4.5-class star† located at a distance of 12.47 ± 0.42 pc. LHS 1140 is more than 5 Gyr old (Dittmann et al. 2017) and has a rather slow rotation period of ∼131 days (e.g., Lillo-Box et al. 2020).

At the time of the modeling efforts by Herbst et al. (2020), no information on stellar wind speeds, mass-loss rates, and stellar magnetic field strength was available in the literature. Because the star is a slow rotator and most likely inactive, by the time, the authors used lower limits reported for slow rotating stars. Thus, $B_\star$ was set to 0.33 μG (at 1 au), v_{sw} was assumed to be 250 km s^{-1}, and $\dot{M}_\star$ to be in the order of $5{\cdot}10^{-17}$ $M_\odot$ yr^{-1}.

The model results are shown in Fig. 4. Due to the presumably low mass-loss rate and the low stellar wind speed value, the modeled astrosphere – with a TS at 8 au, a AP at 11.3 au, and a BS at 28.9 au – is very small compared to the astrospheres of V347 Peg, the Sun, and Proxima Centauri.

‡ data from http://simbad.u-strasbg.fr/simbad/
† http://simbad.u-strasbg.fr/simbad/

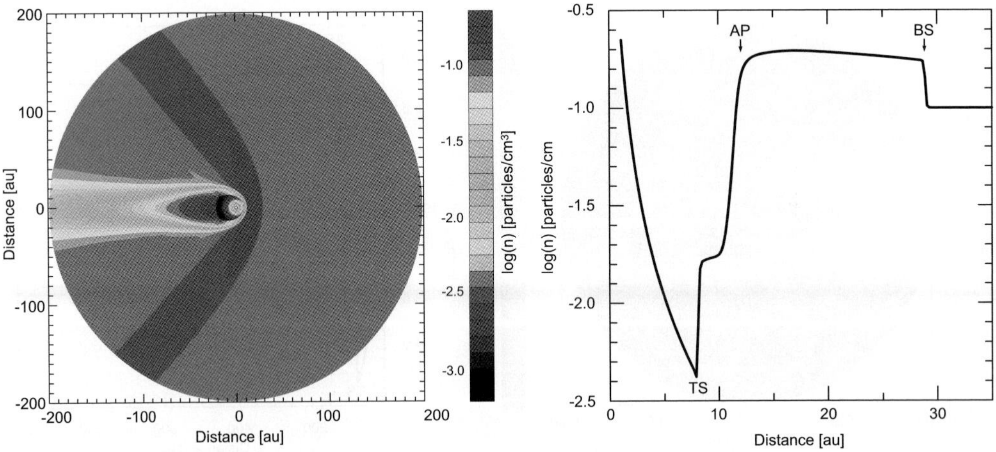

Figure 3. Left panel: Modeled astrosphere of LHS 1140. Right panel: Corresponding number density plot along the line of sight, with indications of the TS, AP, and BS distances. Taken from Herbst et al. (2020) ©AAS. Reproduced with permission.

The three examples show that the shapes and sizes of the astrospheres of cool stars can vary quite strongly and will directly impact the transport of GCRs therein (see Sec. 3), and thus, in some cases, could lead to serious radiation increases in the atmospheres of exoplanets within.

3. GCR Transport Modeling

Until 2020, it was commonly accepted that the modulation of GCRs within cool star astrospheres is so strong that they were to be neglected in exoplanetary habitability studies. However, Herbst et al. (2020) and Herbst et al. (2022a) could show that the opposite is the case and that analytic estimates and 1D approaches are not sufficient enough to predict the impact of GCRs on exoplanetary atmospheres properly. The authors proposed that 3D transport solutions are mandatory.

The transport of GCRs can be described by Parker's transport equation:

$$\frac{\partial f}{\partial t} = \nabla \cdot (\mathbf{K} \cdot \nabla f) - \mathbf{V}_{sw} \cdot \nabla f + \frac{1}{3} (\nabla \cdot \mathbf{V}_{sw}) \frac{\partial f}{\partial \ln p} + S(\mathbf{r}, p, t). \tag{7}$$

Here, $f = p^{-2} j_T$ represents the omnidirectional cosmic ray distribution function directly related to the cosmic-ray differential intensity j_T (Moraal 2013). While the diffusion and drift of the particles due to gradients and curvatures of the astrospheric magnetic field (AMF) are described by the 3D diffusion tensor $\mathbf{K}$, the outward convection with the stellar wind is given by $\mathbf{V}_{sw} \cdot \nabla f$ and adiabatic energy changes are represented by the term $1/3\,(\nabla \cdot \mathbf{V}_{sw})\,\partial f/\partial \ln p$, with $\mathbf{V}_{sw}$ as the stellar wind velocity (see, e.g., Potgieter et al. 2001; Engelbrecht et al. 2022). In addition, S($\mathbf{r}$,p,t) accounts for any potential cosmic ray sources or sinks in the astrosphere as a function of position $\mathbf{r}$, momentum p, and time t, which – for now – is neglected in the stellar context.

According to Burger et al. (2008), the 3D diffusion tensor $\mathbf{K}$ can be described in AMF aligned coordinates:

$$\mathbf{K}' = \begin{bmatrix} \kappa_{\perp,3} & \kappa_A & 0 \\ -\kappa_A & \kappa_{\perp,2} & 0 \\ 0 & 0 & \kappa_{\|} \end{bmatrix}, \tag{8}$$

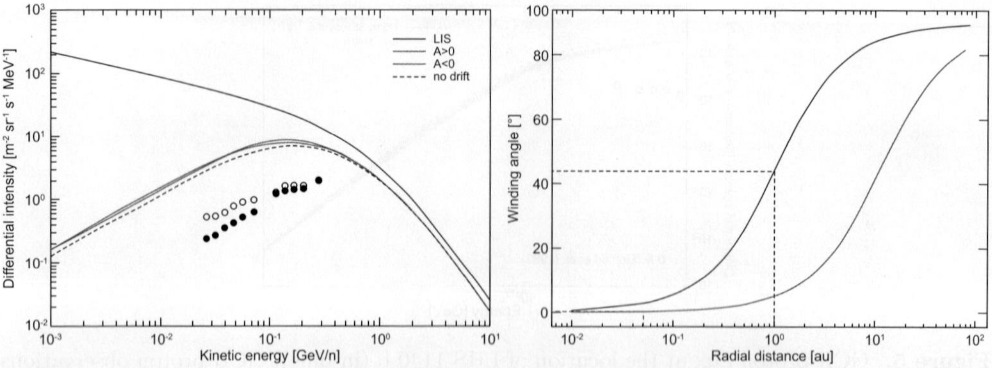

Figure 4. Left: Galactic cosmic ray proton intensities computed at the orbit of Proxima Centauri b for the different stellar rotation periods indicated in the legend. Also shown are the assumed local interstellar spectra (black line, Burger et al. 2008) and helium (Boschini et al. 2020). GCR intensities observed at Earth (McDonald et al. 1992; Marcelli et al. 2020) are shown as black and white circles and blue triangles, respectively, to guide the eye. Figure taken from Engelbrecht et al. (2024). Right: Astrospheric magnetic field winding angle of Proxima Centauri compared to the Sun. Here dashed lines indicate the winding angles at Proxima Centauri b (blue) and Earth (black).

with $\kappa_{\perp,\parallel} = v\lambda_{\perp,\parallel}/3$, where v represents the particle speed (Shalchi 2009). Utilizing the output of the 3D MHD astrospheric modeling, Eq. 7 is solved using a stochastic differential equation (SDE) approach as presented in Engelbrecht and Burger (2015). More detail on the SDE approach to solving the Parker transport equation can be found in Strauss and Effenberger (2017).

3.1. *GCR transport in the astrosphere of V374 Peg*

As discussed in Herbst et al. (2020), the differential intensities within the astrosphere of V374 Peg are strongly modulated because of the parsec-scale size astrosphere, and the impact of GCRs on exoplanetary habitability can be neglected. Note that, so far, no exoplanet has been found to orbit V374 Peg.

3.2. *GCR transport in the astrosphere of Proxima Centauri*

Proxima Centauri has two known planets. Of particular interest is the presumably rocky Earth-like planet Proxima Centauri b, which has a semimajor axis of (0.0485 ± 0.0041) au, a mass of $m_{\rm p} = 1.63^{+1.66}_{-0.72}\, m_{\rm Earth}$, a radius of $R_{\rm P} = 1.07^{+0.38}_{-0.31} R_{\rm Earth}$, and an orbital period of $11.186^{+0.001}_{-0.002}$ days (Bixel and Apai 2017).

Based on the input of the 3D MHD modeling and the 3D SDE cosmic ray modulation model discussed above, the derived GCR proton intensities calculated at Proxima Centauri b are shown in the left panel of Fig. 4 (see Engelbrecht et al. 2024, for further details). Here, the results during periods of positive magnetic polarity ($A > 0$, i.e., when positively charged particles drift inwards over the poles of the astrosphere, red line) and negative magnetic polarity ($A < 0$, where particles drift inward along the astrospheric current sheet, blue line), the assumed local interstellar spectrum (LIS, black line, by Burger et al. 2008), and observed GCR intensities at Earth (symbols, McDonald et al. 1992) are shown.

A direct comparison between the Earth-bound observations shows that the intensities calculated at the location of Proxima Centauri b are almost an order of magnitude higher. This is in direct contrast to the previously reported results using simpler GCR modulation

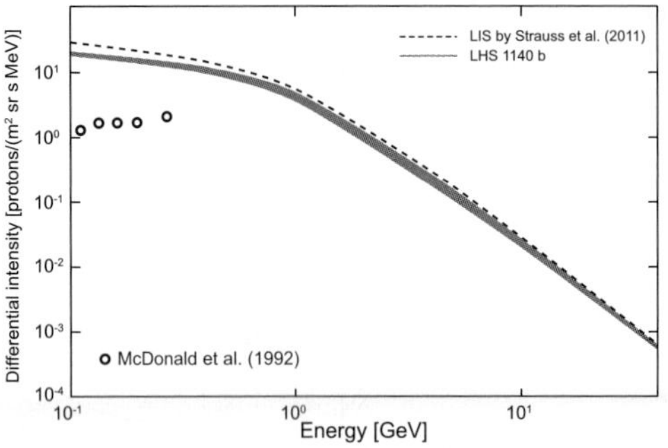

Figure 5. GCR proton flux at the location of LHS 1140 b (in blue). GCR proton observations at Earth (McDonald et al. 1992) are also displayed. Figure adapted from Herbst et al. (2020).

models (see, e.g., Sadovski et al. 2018; Mesquita et al. 2022a; Rodgers-Lee et al. 2021a,c; Mesquita et al. 2022b). In addition, the impact of 3D modeling neglecting drift effects is represented by the results highlighted as black dashed lines. It shows that compared to the no-drift solution, drift effects increase the intensities at lower energies by a factor of ~ 1.8. Further, note that intensities during $A > 0$ conditions are marginally smaller than those for $A < 0$, which contradicts what we observe at Earth. Engelbrecht et al. (2024) found the root of this odd behavior in the less wounded astrospheric magnetic field at the location of Proxima Centauri b, resulting from the slow stellar rotation rate (see right panel of Fig. 4).

3.3. *GCR transport in the astrosphere of LHS 1140*

The GCR modulation within a small astrosphere like that of LHS 1140 reported in Herbst et al. (2020) should be neglectable. Thus, the potential ocean world LHS 1140 b (Damiano et al. 2024), with a mass of $m_p = (6.65 \pm 1.82)m_{\text{Earth}}$ and a radius of $r_p = (1.43 \pm 0.10)r_{\text{Earth}}$, orbiting LHS 1140 at a distance of (0.0875 ± 0.0041) au (Dittmann et al. 2017), most likely is constantly bombarded by galactic cosmic rays of low and high energy.

This has been confirmed by Herbst et al. (2020) utilizing a 1D transport modeling. Because of its long rotation period of ~ 131 days (e.g. Lillo-Box et al. 2020) like in the case of Proxima Centauri, a strongly underwound magnetic field is present, which can facilitate the inward transport of GCRs (Engelbrecht et al. 2024). Most recently, Scherer et al. (2024, under rev.), utilizing the first 3D MHD multifluid astrospheric modeling results and the 3D solution of the above-mentioned GCR transport, showed that the modulation most likely is even less, suggesting that GCRs reach LHS 1140 b without any modulation, which, with that, might constantly be bombarded by the surrounding local ISM.

A new study by Light et al. (2024, in prep.) investigates the stellar rotational period as a simple-to-observe additional constraint on GCR modulation and, thus, on exoplanetary habitability. The study revisits Proxima Centauri and LHS 1140 and includes known stellar systems like TRAPPIST-1 and GJ 436.

4. 2D SEP Transport Modeling

Modeling the stellar energetic particle transport is in its infancy and - so far - has only been discussed by Hu et al. (2022) (CME-driven) and Engelbrecht et al. (2024) (flare driven). Since it is still under scientific debate whether coronal mass ejections (CMEs) do play an important role in the surroundings of cool stars, in Engelbrecht et al. (2024), we updated the solar numerical transport model discussed by Strauss and Fichtner (2015) (see also Strauss et al. 2020) to study the flare-driven particle transport around M stars.

Thereby, a 2D version of the focused transport equation given by

$$\frac{\partial f}{\partial t} + \nabla \cdot (\vec{v}_{gc} f) + \frac{\partial}{\partial \mu}\left(\frac{1-\mu^2}{2L} v f\right) = \frac{\partial}{\partial \mu}\left(D_{\mu\mu}\frac{\partial f}{\partial \mu}\right) + \nabla \cdot \left(\mathbf{D}_{\perp}^{(x)} \cdot \nabla f\right) \tag{9}$$

is solved. Here, $\vec{v}_{gc} = v_{||}\hat{b}$ gives the parallel streaming motion along the mean magnetic field, L is the magnetic focusing length with $L^{-1} = \nabla \cdot \hat{b}$, $D_{\mu\mu}$ represents the pitch-angle diffusion coefficient, and $D_{\perp}^{(x)}$ includes spatial diffusion perpendicular to the mean magnetic field (Strauss et al. 2017).

$D_{\mu\mu}$ is related to the parallel mean-free-path via (e.g. Bieber et al. 1994)

$$\lambda_{||} = \frac{3v}{8}\int_{-1}^{1} d\mu \frac{\left(1-\mu^2\right)^2}{D_{\mu\mu}(\mu)}, \tag{10}$$

where v is particle speed. Thereby, for $D_{\mu\mu}$ the following form (e.g., Dröge et al. 2010) is used

$$D_{\mu\mu} = D_0(1-\mu^2)\left[|\mu|^{1.67-1} + 0.05\right] \tag{11}$$

In addition, Eq. (10) is utilized to derive the D_0, and for the perpendicular diffusion coefficient the form

$$D_{\perp} = 2D_{\perp,0}|\mu|, \tag{12}$$

is assumed. Therewith, also the isotropic perpendicular diffusion coefficient can be calculated:

$$\kappa_{\perp} = \frac{1}{2}\int_{-1}^{+1} D_{\perp} d\mu, \tag{13}$$

which again is related to the perpendicular mean-free-path by $\lambda_{\perp} = 3\kappa_{\perp}/v$.

The inner boundary conditions close to the star are the most difficult to determine. In alignment with a proxy derived from solar remote-sensing observations, at the inner boundary of the model ($r = r_0$), the following boundary condition is set:

$$f(r = r_0, \phi, t) = CI_i \exp\left[-\frac{(\phi - \phi_i)^2}{2\sigma_i^2}\right]\delta(t - \tau_i), \tag{14}$$

with ϕ_i referring to the position of the flare region on the stellar disk and σ_i as the width/size of each event. Further, Eq. (14) contains a Gaussian distribution in azimuthal angle and a pulse-like temporal profile for each stellar energetic particle injection ($\delta(\ldots)$) and C is an elusive normalization constant that needs to be computed post facto (Strauss and le Roux 2019).

Finally, a time series of StEP events is injected based on actual observations by, e.g., *Kepler* or *TESS*.

First model results on the example of Proxima Centauri

In Engelbrecht et al. (2024), for the first time, we inject a time series of StEP events based on *TESS* measurements provided by Vida et al. (2019), and ended up with $i = 1, 2, 3, \ldots, 72$ StEP events occurring over the ~ 50 days of Proxima Centauri observations with TESS. However, *TESS* observations only indicate flaring on half of the stellar disk pointed to Earth. Thus, statistically speaking, the number of observed flares should be doubled for occulted stellar activity. Therefore, we mirrored the original *TESS* time series in time and added this second-time series to the original. The upper panel of Fig. 6 shows the assumed injections of stellar energetic particles into the numerical

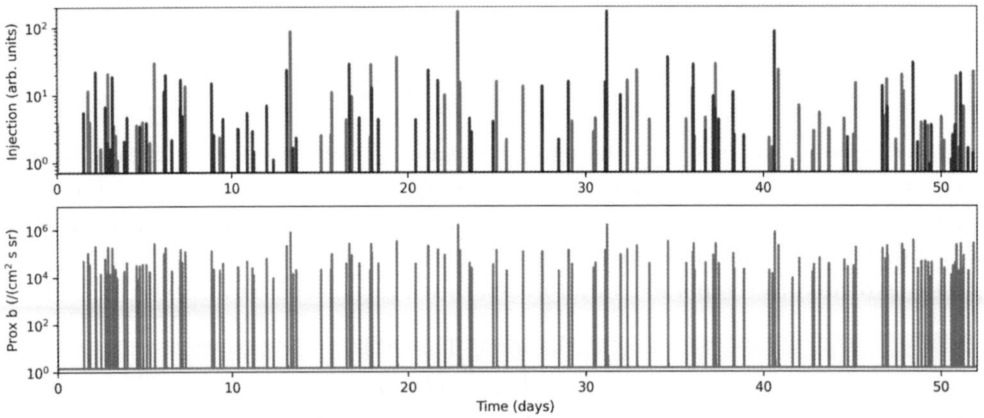

Figure 6. Upper panel: Assumed injections of StEPs into the numerical model as derived from TESS observations. Here, blue lines indicate actual TESS observations, while red represents the mirrored dataset. Lower panel: Simulated StEP intensities at Proxima Centauri b. Figures from Engelbrecht et al. (2024).

model derived from the *TESS* observations: blue lines represent the original data, while the mirrored dataset is highlighted in red.

Following the argumentation by Steyn et al. (2020), we assumed the observed flare energy to be a proxy for the StEP flux in the acceleration region. Thereby, the flare parameters determine the time of each flare, τ_i, and the flare energy for each event I_i, assuming that the StEP flux at the boundary is proportional to I_i. However, information on the position of the flare region on the stellar disk (ϕ_i) and the width/size of each event (σ_i) is lacking. Therefore, the parameters were generated as a walkaround by drawing a random number from a uniform distribution for ϕ_i and a Gaussian distribution for σ_i.

In the last step, we calculated the omnidirectional particle intensities according to

$$j_{\mathrm{SEP}} = \frac{\int_{-1}^{1} f d\mu}{\int_{-1}^{1} d\mu} = \frac{1}{2}\int_{-1}^{1} f d\mu, \qquad (15)$$

assuming a circumstellar source for the model runs and calculated the modeled j for a magnetically well-connected Proxima Centauri b.

However, because of the 2D nature and the stellar unknowns, the derived values - for now - must be normalized to known (solar) values. Therefore, we ran the model for the strongest *TESS* event with a bolometric energy of 1.74×10^{32} erg (Vida et al. 2019), which according to Benz (2017) corresponds to an SXR energy of 1.74×10^{-3} W/m^2 (X17.4). The latter is comparable with the flare that produced the strongest ground level enhancement (GLE) event ever observed, i.e., GLE05 that occurred on 23 February 1956 (e.g., Papaioannou et al. 2023). Assuming that the physical principles for such comparably low-energy flares are the same for all cool stars, known solar relations of upper solar limits can be utilized to derive the potential stellar particle intensities at Proxima Centauri b. According to Papaioannou et al. (2023), an X17.4 flare corresponds to an event-integrated $E > 10$ MeV flux of 3.85×10^3 pfu [#/(cm^2 s sr)]† at a distance of 1 au, which is higher than what has been derived for the GLE05 event ($\sim 10^4$ pfu).

To scale the event to the location of Proxima Centauri b (i.e., 0.04857 au), the value was scaled with $1/\mathrm{r}^2$, which resulted in an upper limit of the event-integrated E > 10 MeV proton flux of 1.632×10^6 pfu. Thus, for each of the 144 *TESS* events, the omnidirectional

† i.e., $2.81 \times 10^7 \cdot (1.74 \times 10^{-3}\ \mathrm{W/m^2})^{1.40} = 3.85\times10^3$ pfu (#/(cm^2 s sr))

particle intensities at 0.04857 au are scaled with

$$j_{\rm scaled} = \frac{j_{\rm SEP}}{\max(j) \cdot 1.632 \times 10^6 {\rm pfu}}, \tag{16}$$

with $\max(j) = 111.056$ (arb. units). The resulting event-dependent fluxes are shown in the lower panel of Fig. 6.

5. Discussion

Representing astrospheric modeling and GCR transport three-dimensionally is vital since the transport mechanisms that impact the modulation of charged energetic particles in any 3D astrosphere are of a 3D nature. Herbst et al. (2020), Engelbrecht et al. (2024), and Scherer et al. (2024, under rev.) showed that, despite the numerous uncertainties involved, a careful combination of extrapolations based on the heliospheric behavior of different plasma variables and MHD modeling may yield insightful results that are not achievable with modeling in less than 3D.

Our approach results in much higher GCR intensities at LHS 1140 b and Proxima Centauri b than at Earth. In contrast to previous results based on analytic and 1D methods (e.g., Sadovski et al. 2018; Mesquita et al. 2022a) we showed that GCRs must be considered when discussing the exoplanetary radiation environment and its impact on exoplanetary atmospheres, climate, chemistry - and with that biosignatures like methane and ozone - and exoplanetary habitability.

Moreover, the significance of GCR drifts and the unexpectedly strong impact of parallel diffusion strengthen the necessity of 3D transport modeling. For two of the three systems discussed, namely Proxima Centauri b and LHS 1140 b, GCR intensities are extremely high. Engelbrecht et al. (2024) suggested that a slow rotation rate coupled with relatively high stellar wind speed produces an underwound astrospheric magnetic field that significantly aids in the inward transit of GCRs. This suggests that stellar rotation periods could offer a comparatively easy-to-observe limitation on exoplanetary habitability. This hypothesis is currently under further investigation (Light et al. 2024, in prep.), with the first results suggesting that shorter rotation periods lead to a more wound astrospheric magnetic field and a decrease in the differential GCR intensities because the particles rely more on diffusion perpendicular to the astrospheric magnetic field as the field becomes more azimuthal.

The latter also provides a highway for stellar energetic particles around slowly rotating cool stars. Although it is still debated whether stellar energetic particles are accelerated to high energies at the shock fronts of CMEs, in the case of Proxima Centauri b, Engelbrecht et al. (2024) could show that even the stellar energetic particles accelerated by/in flares most likely strongly impact the exoplanetary atmosphere, and with that could lead to strong CR-induced atmospheric chemistry changes impacting the observation and interpretation of biosignatures like methane and ozone (see, e.g., Herbst et al. 2019, 2022b, 2024, for further information).

Acknowledgements. The authors gratefully acknowledge support from the German Research Foundation (Deutsche Forschungsgemeinschaft, DFG) under grant *HE 8392/2-1* to initialize this international collaboration, the Gauss Centre for Supercomputing e.V. (`www.gauss-centre.eu`) for funding this project by providing computing time through the John von Neumann Institute for Computing (NIC) on the GCS Supercomputer JUWELS at Jülich Supercomputing Centre (JSC), and the Center for High Performance Computations (CHPC) in Cape Town for partially funding it under the project *ASTRO1277*. KH acknowledges the support of the DFG priority program SPP 1992 "Exploring the Diversity of Extrasolar Planets (*HE 8392/1-1*)". KS acknowledges the support of the DFG grant *SCHE 334/16-1*. This work is based on the research supported partly by the National Research Foundation of

South Africa (NRF grant number 137793). Opinions expressed, and conclusions arrived at are those of the authors and are not necessarily to be attributed to the NRF.

References

Benz, A. O. 2017, Flare Observations. *Living Reviews in Solar Physics*, 14(1), 2.

Bieber, J. W., Matthaeus, W. H., Smith, C. W., Wanner, W., Kallenrode, M., & Wibberenz, G. 1994, Proton and electron mean free paths: The Palmer consensus revisited. *Astrophys. J.*, 420, 294–306.

Bixel, A. & Apai, D. 2017, Probabilistic constraints on the mass and composition of proxima b. *ApJ*, 836(2), L31.

Boschini, M. J., Della Torre, S., Gervasi, M., Grandi, D., Jóhannesson, G., La Vacca, G., Masi, N., Moskalenko, I. V., Pensotti, S., Porter, T. A., Quadrani, L., Rancoita, P. G., Rozza, D., & Tacconi, M. 2020, Inference of the Local Interstellar Spectra of Cosmic-Ray Nuclei $Z \leq 28$ with the GALPROP-HELMOD Framework. *Astrophys. J. Supp.*, 250(2), 27.

Burger, R. A., Krüger, T. P. J., Hitge, M., & Engelbrecht, N. E. 2008, A Fisk-Parker Hybrid Heliospheric Magnetic Field With a Solar-Cycle Dependence. *ApJ*, 674, 511–519.

Damiano, M., Bello-Arufe, A., Yang, J., & Hu, R. 2024, LHS 1140 b is a potentially habitable water world. *arXiv e-prints*, arXiv:2403.13265.

Dittmann, J. A., Irwin, J. M., Charbonneau, D., Bonfils, X., Astudillo-Defru, N., Haywood, R. D., Berta-Thompson, Z. K., Newton, E. R., Rodriguez, J. E., Winters, J. G., Tan, T.-G., Almenara, J.-M., Bouchy, F., Delfosse, X., Forveille, T., Lovis, C., Murgas, F., Pepe, F., Santos, N. C., Udry, S., Wünsche, A., Esquerdo, G. A., Latham, D. W., & Dressing, C. D. 2017, A temperate rocky super-Earth transiting a nearby cool star. *Nature*, 544(7650), 333–336.

do Nascimento, J. D., J., Vidotto, A. A., Petit, P., Folsom, C., Castro, M., Marsden, S. C., Morin, J., Porto de Mello, G. F., Meibom, S., Jeffers, S. V., Guinan, E., & Ribas, I. 2016, Magnetic Field and Wind of Kappa Ceti: Toward the Planetary Habitability of the Young Sun When Life Arose on Earth. *ApJ Lett.*, 820(1), L15.

Dröge, W., Kartavykh, Y. Y., Klecker, B., & Kovaltsov, G. A. 2010, Anisotropic Three-Dimensional Focused Transport of Solar Energetic Particles in the Inner Heliosphere. *ApJ*, 709(2), 912–919.

Engelbrecht, N. E. & Burger, R. A. 2015, Sensitivity of cosmic-ray proton spectra to the low-wavenumber behavior of the 2d turbulence power spectrum. *The Astrophysical Journal*, 814(2), 152.

Engelbrecht, N. E., Effenberger, F., Florinski, V., Potgieter, M. S., Ruffolo, D., Chhiber, R., Usmanov, A. V., Rankin, J. S., & Els, P. L. 2022, Theory of Cosmic Ray Transport in the Heliosphere, 218(4), 33.

Engelbrecht, N. E., Herbst, K., Strauss, R. D. T., Scherer, K., Light, J., & Moloto, K. D. 2024, On the comprehensive 3d modeling of the radiation environment of proxima centauri b: A new constraint on habitability? *The Astrophysical Journal*, 964(1), 89.

Garraffo, C., Drake, J. J., & Cohen, O. 2016, The Space Weather of Proxima Centauri b. *ApJ Lett.*, 833(1).

Gillon, M., Triaud, A. H. M. J., Demory, B.-O., Jehin, E., Agol, E., Deck, K. M., Lederer, S. M., de Wit, J., Burdanov, A., Ingalls, J. G., Bolmont, E., Leconte, J., Raymond, S. N., Selsis, F., Turbet, M., Barkaoui, K., Burgasser, A., Burleigh, M. R., Carey, S. J., Chaushev, A., Copperwheat, C. M., Delrez, L., Fernand es, C. S., Holdsworth, D. L., Kotze, E. J., Van Grootel, V., Almleaky, Y., Benkhaldoun, Z., Magain, P., & Queloz, D. 2017, Seven temperate terrestrial planets around the nearby ultracool dwarf star TRAPPIST-1. *Nature*, 542(7642), 456–460.

Herbst, K., Baalmann, L. R., Bykov, A., Engelbrecht, N. E., Ferreira, S. E. S., Izmodenov, V. V., Korolkov, S. D., Levenfish, K. P., Linsky, J. L., Meyer, D. M. A., Scherer, K., & Strauss, R. D. T. 2022, a Astrospheres of Planet-Hosting Cool Stars and Beyond. When Modeling Meets Observations, 218a(4), 29.

Herbst, K., Bartenschlager, A., Grenfell, J. L., Iro, N., Sinnhuber, M., Taysum, B., Wunderlich, F., Engelbrecht, N. E., Light, J., Moloto, K. D., Harre, J.-V., Rauer, H., & Schreier, F. 2024, Impact of cosmic rays on atmospheric ion chemistry and spectral transmission features of trappist-1e. *The Astrophysical Journal*, 961(2), 164.

Herbst, K., Grenfell, J. L., Miriam, S., & Wunderlich, F. 2022,b Increase: An updated model suite to study the influence of cosmic rays on exoplanetary atmospheres. *Astronomische Nachrichten*, 343b(4), e210072.

Herbst, K., Grenfell, J. L., Sinnhuber, M., Rauer, H., Heber, B., Banjac, S., Scheucher, M., Schmidt, V., Gebauer, S., Lehmann, R., & Schreier, F. 2019, A new model suite to determine the influence of cosmic rays on (exo)planetary atmospheric biosignatures. Validation based on modern Earth. *A&A*, 631.

Herbst, K., Muscheler, R., & Heber, B. 2017, The new local interstellar spectra and their influence on the production rates of the cosmogenic radionuclides ^{10}Be and ^{14}C. *JGR (Space Phys.)*, 122, 23–34.

Herbst, K., Scherer, K., Ferreira, S. E. S., Baalmann, L. R., Engelbrecht, N. E., Fichtner, H., Kleimann, J., Strauss, R. D. T., Moeketsi, D. M., & Mohamed, S. 2020, On the diversity of m-star astrospheres and the role of galactic cosmic rays within. *ApJ Lett.*, 897(2), L27.

Hu, J., Airapetian, V. S., Li, G., Zank, G., & Jin, M. 2022, Extreme energetic particle events by superflare-asssociated CMEs from solar-like stars. *Science Advances*, 8(12), eabi9743.

Kasting, J. F., Whitmire, D. P., & Reynolds, R. T. 1993, Habitable Zones around Main Sequence Stars. *ICARUS*, 101(1), 108–128.

Kissmann, R., Kleimann, J., Krebl, B., & Wiengarten, T. 2018, The CRONOS Code for Astrophysical Magnetohydrodynamics. *ApJ*, 236(2), 53.

Lammer, H., Bredehöft, J. H., Coustenis, A., Khodachenko, M. L., Kaltenegger, L., Grasset, O., Prieur, D., Raulin, F., Ehrenfreund, P., Yamauchi, M., Wahlund, J. E., Grießmeier, J. M., Stangl, G., Cockell, C. S., Kulikov, Y. N., Grenfell, J. L., & Rauer, H. 2009, What makes a planet habitable? *Astron. & Astrophys. Rev.*, 17(2), 181–249.

Light, J., Engelbrecht, N. E., Herbst, K., & Scherer, K. 2024, Constraining the Impact of Cosmic Rays on Exoplanetary Habitability. *Astrophys. J.*,. in prep.

Lillo-Box, J., Figueira, P., Leleu, A., Acuña, L., Faria, J. P., Hara, N., Santos, N. C., Correia, A. C. M., Robutel, P., Deleuil, M., Barrado, D., Sousa, S., Bonfils, X., Mousis, O., Almenara, J. M., Astudillo-Defru, N., Marcq, E., Udry, S., Lovis, C., & Pepe, F. 2020, Planetary system LHS 1140 revisited with ESPRESSO and TESS. *A&A*, 642, A121.

Marcelli, N., Boezio, M., Lenni, A., Menn, W., Munini, R., Aslam, O. P. M., Bisschoff, D., Ngobeni, M. D., Potgieter, M. S., Adriani, O., Barbarino, G. C., Bazilevskaya, G. A., Bellotti, R., Bogomolov, E. A., Bongi, M., Bonvicini, V., Bruno, A., Cafagna, F., Campana, D., Carlson, P., Casolino, M., Castellini, G., De Santis, C., Galper, A. M., Koldashov, S. V., Koldobskiy, S., Kvashnin, A. N., Leonov, A. A., Malakhov, V. V., Marcelli, L., Martucci, M., Mayorov, A. G., Mergè, M., Mocchiutti, E., Monaco, A., Mori, N., Mikhailov, V. V., Osteria, G., Panico, B., Papini, P., Pearce, M., Picozza, P., Ricci, M., Ricciarini, S. B., Simon, M., Sotgiu, A., Sparvoli, R., Spillantini, P., Stozhkov, Y. I., Vacchi, A., Vannuccini, E., Vasilyev, G. I., Voronov, S. A., Yurkin, Y. T., Zampa, G., & Zampa, N. 2020, Time Dependence of the Flux of Helium Nuclei in Cosmic Rays Measured by the PAMELA Experiment between 2006 July and 2009 December. *Astrophys. J.*, 893(2), 145.

McDonald, F. B., Moraal, H., Reinecke, J. P. L., Lal, N., & McGuire, R. E. 1992, The Cosmic Radiation in the Heliosphere at Successive Solar Minima. *J. Geophys. Res.*, 97(A2), 1557–1570.

Mesquita, A. L., Rodgers-Lee, D., Vidotto, A. A., Atri, D., & Wood, B. E. 2022, a Galactic cosmic ray propagation through M dwarf planetary systems. *MNRAS*, 509a(2), 2091–2101.

Mesquita, A. L., Rodgers-Lee, D., Vidotto, A. A., & Kavanagh, R. D. 2022,b The strong suppression of galactic cosmic rays reaching AU Mic b, c, and Prox Cen b. *MNRAS*, 515b(1), 1218–1227.

Moraal, H. 2013, Cosmic-Ray Modulation Equations. *Space Sci. Rev.*, 176, 299–319.

Noack, L., Snellen, I., & Rauer, H. 2017, Water in Extrasolar Planets and Implications for Habitability. *Space Sci. Rev.*, 212(1-2), 877–898.

Papaioannou, A., Herbst, K., Ramm, T., Cliver, E. W., Lario, D., & Veronig, A. M. 2023, Revisiting empirical solar energetic particle scaling relations. I. Solar flares. *A&A*, 671, A66.

Parker, E. N. 1965, The passage of energetic charged particles through interplanetary space. *Planet. Space Sci.*, 13(1), 9–49.

Potgieter, M. S., Burger, R. A., & Ferreira, S. E. S. 2001, Modulation of Cosmic Rays in the Heliosphere From Solar Minimum to Maximum: a Theoretical Perspective. *SSR*, 97, 295–307.

Reiners, A. & Basri, G. 2008, The moderate magnetic field of the flare star Proxima Centauri. *A&A Lett.*, 489(3), L45–L48.

Richardson, J. D., Cummings, A. C., Burlaga, L. F., Giacalone, J., Opher, M., & Stone, E. C. 2021, Using magnetic flux conservation to determine heliosheath speeds. *ApJ Lett.*, 919(2), L28.

Rodgers-Lee, D., Taylor, A. M., Vidotto, A. A., & Downes, T. P. 2021, a Stellar versus Galactic: the intensity of cosmic rays at the evolving Earth and young exoplanets around Sun-like stars. *MNRAS*, 504a(1), 1519–1530.

Rodgers-Lee, D., Vidotto, A. A., & Mesquita, A. L. 2021,b Charting nearby stellar systems: the intensity of Galactic cosmic rays for a sample of solar-type stars. *MNRAS*, 508b(4), 4696–4704.

Rodgers-Lee, D., Vidotto, A. A., & Mesquita, A. L. 2021,c Charting nearby stellar systems: the intensity of Galactic cosmic rays for a sample of solar-type stars. *MNRAS*, 508c(4), 4696–4704.

Sadovski, A. M., Struminsky, A. B., & Belov, A. 2018, Cosmic rays near proxima centauri b. *Astron. Lett.*, 44(5), 324–330.

Scherer, K., Baalmann, L. R., Fichtner, H., Kleimann, J., Bomans, D. J., Weis, K., Ferreira, S. E. S., & Herbst, K. 2020, MHD-shock structures of astrospheres: λ Cephei-like astrospheres. *MNRAS*, 493(3), 4172–4185.

Scherer, K., Fichtner, H., Fahr, H. J., Bzowski, M., & Ferreira, S. E. S. 2014, Ionization rates in the heliosheath and in astrosheaths. Spatial dependence and dynamical relevance. *A&A*, 563.

Scherer, K., Herbst, K., Engelbrecht, N. E., Ferreira, S. S., Kleimann, J., & Light, J. 2024, Modeling the astrosphere of LHS 1140. On the differences of 3D (M)HD single- and multi-fluid simulations and the consequences for exoplanetary habitability. *Astron. Astrophys.*,. Under review.

Scheucher, M., Herbst, K., Schmidt, V., Grenfell, J. L., Schreier, F., Banjac, S., Heber, B., Rauer, H., & Sinnhuber, M. 2020, Proxima centauri b: A strong case for including cosmic-ray-induced chemistry in atmospheric biosignature studies. *ApJ*, 893(1), 12.

Shalchi, A. 2009,. *Nonlinear Cosmic Ray Diffusion Theories*, volume 362 of *Astrophysics and Space Science Library*. Springer, Berlin.

Steyn, R., Strauss, D. T., Effenberger, F., & Pacheco, D. 2020, The soft X-ray Neupert effect as a proxy for solar energetic particle injection. A proof-of-concept physics-based forecasting model. *Journal of Space Weather and Space Climate*, 10, 64.

Stone, E. C., Cummings, A. C., Heikkila, B. C., & Lal, N. 2019, Cosmic ray measurements from Voyager 2 as it crossed into interstellar space. *Nature Astronomy*, 3, 1013–1018.

Stone, E. C., Cummings, A. C., McDonald, F. B., Heikkila, B. C., Lal, N., & Webber, W. R. 2013, Voyager 1 Observes Low-Energy Galactic Cosmic Rays in a Region Depleted of Heliospheric Ions. *Science*, 341, 150–153.

Strauss, R. D., Dresing, N., Kollhoff, A., & Brüdern, M. 2020, On the Shape of SEP Electron Spectra: The Role of Interplanetary Transport. *ApJ*, 897(1), 24.

Strauss, R. D. & Fichtner, H. 2015, On Aspects Pertaining to the Perpendicular Diffusion of Solar Energetic Particles. *ApJ*, 801(1), 29.

Strauss, R. D. & le Roux, J. A. 2019, Solar Energetic Particle Propagation in Wave Turbulence and the Possibility of Wave Generation. *ApJ*, 872(2), 125.

Strauss, R. D. T., Dresing, N., & Engelbrecht, N. E. 2017, Perpendicular Diffusion of Solar Energetic Particles: Model Results and Implications for Electrons. *ApJ*, 837(1), 43.

Strauss, R. D. T. & Effenberger, F. 2017, A Hitch-hiker's Guide to Stochastic Differential Equations. Solution Methods for Energetic Particle Transport in Space Physics and Astrophysics. *Space Sci. Rev.*, 212(1-2), 151–192.

Struminsky, A., Sadovski, A., & Belov, A. 2017, Cosmic Rays near Proxima Centauri b. *arXiv e-prints*, arXiv:1704.06168.

Vida, K., Kriskovics, L., Oláh, K., Leitzinger, M., Odert, P., Kővári, Z., Korhonen, H., Greimel, R., Robb, R., Csák, B., & Kovács, J. 2016, Investigating magnetic activity in very stable stellar magnetic fields. Long-term photometric and spectroscopic study of the fully convective M4 dwarf V374 Pegasi. *A&A*, 590.

Vida, K., Oláh, K., Kővári, Z., van Driel-Gesztelyi, L., Moór, A., & Pál, A. 2019, Flaring Activity of Proxima Centauri from TESS Observations: Quasiperiodic Oscillations during Flare Decay and Inferences on the Habitability of Proxima b. *ApJ*, 884(2), 160.

Vidotto, A. A., Jardine, M., & Helling, C. 2011, Prospects for detection of exoplanet magnetic fields through bow-shock observations during transits. *MNRAS*, 411, L46–L50.

Vidotto, A. A., Jardine, M., Morin, J., Donati, J. F., Lang, P., & Russell, A. J. B. 2013, Effects of M dwarf magnetic fields on potentially habitable planets. *A&A*, 557, A67.

Zank, G. P., Heerikhuisen, J., Wood, B. E., Pogorelov, N. V., Zirnstein, E., & McComas, D. J. 2012, Heliospheric structure: The bow wave and the hydrogen wall. *The Astrophysical Journal*, 763(1), 20.

Solar and Stellar Coronal Mass Ejections
Proceedings IAU Symposium No. 388, 2026
N. Gopalswamy, O. Malandraki, A. Vidotto & W. Manchester, eds.
doi:10.1017/S1743921324001832

CME-associated type-IV radio bursts: The solar paradigm and the unique case of AD Leo

Atul Mohan[1,2], Nat Gopalswamy[1], Surajit Mondal[3], Anshu Kumari[1] and G. Sindhuja[4]

[1]NASA Goddard Space Flight Center, 8800 Greenbelt Road Greenbelt, MD, 20771, USA

[2]The Catholic University of America, 620 Michigan Avenue, N.E. Washington, DC 20064, USA

[3]Center for Solar-Terrestrial Research, New Jersey Institute of Technology, NJ 07102-1982, USA

[4]Bangalore University, Mysore Rd, Jnana Bharathi, Bengaluru, Karnataka, IN

Abstract. The type-IV bursts, associated with coronal mass ejections (CMEs), occasionally extend to the decameter-hectrometric (DH) range. We present a comprehensive catalog of simultaneous multi-vantage point observations of DH type-IV bursts by Wind and STEREO spacecraft since 2006. 73% of the bursts are associated with fast ($>900\,\mathrm{km\,s^{-1}}$) and wide ($> 60^\circ$) CMEs, which are mostly geoeffective halo CMEs. Also, we find that the bursts are best observed by the spacecraft located within $|60^\circ|$ line of sight (LOS), highlighting the importance of LOS towards active latitudes while choosing target stars for a type-IV search campaign. In young active M dwarfs, CME-associated bursts have remained elusive despite many monitoring campaigns. We present the first detection of long-duration type-III, type-IV, and type-V bursts during an active event in AD Leo (M3.5V; $0.4\mathrm{M}_\odot$). The observed burst characteristics support a multipole model over a solar-like active region magnetic field profile on the star.

Keywords. Solar coronal mass ejections, Stellar coronal mass ejections, Radio bursts, Catalogs

1. Introduction

Coronal mass ejections (CMEs) often produce radio bursts associated with the propagating shock and the reconnection event (McLean and Labrum 1985). These include the type-II, complex long duration type-III and type-IV bursts, each of which have characteristic emission features in the dynamic spectrum (DS) (Gopalswamy 2011). The type-II bursts are associated with coronal and interplanetary (IP) shocks, while complex type-IIIs with a duration of the order of an hour are signatures of accelerated electrons propagating along open field lines (Gopalswamy et al. 2000). Meanwhile the type-IV burst which is the subject of this article, comes in two dynamic spectral variants: the moving type-IV which is a broadband emission with a drifting central frequency, and the stationary type-IV which is a broadband time-varying emission (Takakura & Kai 1961; Kundu & Spencer 1963). The moving type-IV burst is caused by flare-accelerated electrons trapped in a magnetised plasmoid moving with the CME (Smerd 1971; Schmahl 1972), while the latter variant is produced by supra-thermal electrons in the post-flare loops (Wild 1970). Due to its close association with various aspects of the CME, type-IV bursts are highly sought after in stellar CME search campaigns.

2. The solar paradigm

Like all radio burst types, the type-IV bursts were also identified in the sun and studied extensively over decades by contrasting with multi-waveband imaging and non-imaging data in the radio to the γ-ray band. Due to the availability of regular ground-based radio dynamic spectral observations since 1950s (Wild and McCready 1950), several statistical studies on meter waveband bursts exist. Using 15 years of data, Cane & Reames (1988) showed that metric type-IVs are mostly associated with strong long-duration soft X-ray (SXR) flares with a median flare strength of $\sim 3\times10^{-}5\,\mathrm{W\,m^{-2}}$ and a median duration of $\sim$1 h.

The decameter-hectometric (DH; 1 -14 MHz) type-IV bursts are rare continuations of the metric bursts, with 100% association to white-light CMEs (Hillaris et al. 2016; Mohan et al. 2024a). These events could be explored only after the advent of space-based radio observations with the launch of Wind and later the STEREO spacecraft. The CMEs and flares linked to DH type-IVs are much stronger than those associated with the type-IVs that terminate within the metric band. About 89% of the DH type-IVs are associated with M to X class (10^{-5} - $10^{-4}\,\mathrm{W\,m^{-2}}$) (Hillaris et al. 2016). Miteva et al. (2017) showed that of all the radio bursts, DH type-IV bursts show the highest association with solar energetic particle (SEP) events. All these properties make DH type-IV bursts highly relevant in space weather studies. Besides, the distribution of type-IV source longitudes are found to prefer $|60°|$ longitude (Cane & Reames 1988; Gopalswamy et al. 2016). There has been a debate on this being caused by either an inherent emission beaming (directivity) expected from the coherent plasma emission mechanism (Gopalswamy et al. 2016), or purely a propagation effect by which dense plasma structures along the line of sight (LOS) occult the emission (Pohjolainen & Talebpour Sheshvan 2020; Talebpour Sheshvan and Pohjolainen 2018). The effect of directivity can be crucial in stellar radio burst searches because choosing a candidate star with a high chance of type-IV detection depends on the LOS towards its active latitudes. However, the inferences on the LOS effect are based on either a few event studies with simultaneous multi-spacecraft data, or statistics on single spacecraft data. There have been no studies on a statistically significant sample of events with simultaneous multi-vantage point observations from multiple spacecraft, to systematically explore the LOS effect. Firstly, we need a comprehensive catalog of DH type-IV bursts observed the different spacecraft over the last couple of solar cycles.

2.1. *The multi-vantage point DH type-IV catalog*

The DH dynamic spectra from Wind (since 1996), STEREO-A and STEREO-B spacecraft (since 2006) form the data source. To search for type-IV events, three event lists were used; the LASCO CME catalog (Yashiro et al. 2004), a metric type-IV event list based on the daily reports from Space Weather Prediction Center (SWPC), and the DH type-II catalog (Gopalswamy et al. 2019). Since previous studies show a 100% association of the DH type-IV bursts with CMEs and since they are extensions of metric type-IVs, the union of the first two event lists should cover all DH type-IV events. The DH type-II list gives additional intervals to search in the DH band, in case some events were missed out due to data gaps. For each date-time period in the search list, the DS from each spacecraft was examined. In the case of a potential type-IV detection, the corresponding metric DS was searched for to confirm the signature. All dubious events were removed.

The final catalog has 139 DH type-IV bursts with a mean duration of $\sim$ 83 min and extending down to $\sim$7.8 MHz on average. The full catalog can be accessed online† All events had a white-light CME association in contrast to the type-IV bursts confined within the metric band (Kumari et al. 2021). 73% of the DH type-IVs are linked

† https://cdaw.gsfc.nasa.gov/CME_list/radio/type4/

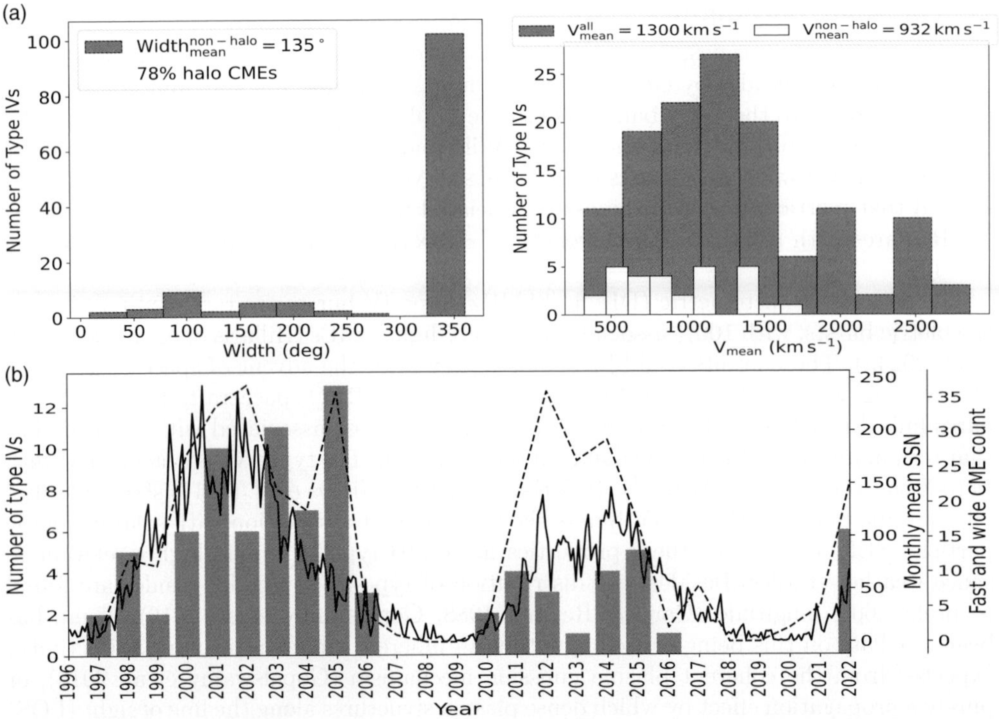

Figure 1. (a): Properties of CMEs associated with the DH type-IV bursts. (b): Solar cycle variation in the number of DH type-IV bursts (histogram), fast and wide CMEs (dotted), and mean sunspot number (solid).

to a fast (>900 km s^{-1}) and wide (> 60°) CME, and ~78% to halo CMEs (Howard et al. 1982; Gopalswamy et al. 2007) which are a class of relatively fast CMEs (mean speed, V_{mean}~1050 km s^{-1}) with nearly 360° angular width compared to regular CMEs. Figure 1a shows the properties of CMEs linked to DH type-IV bursts. With an average V_{mean}~1300 km s^{-1}, even the non-halo CMEs linked to DH type-IVs are fast CMEs. Figure 1b shows the strong correlation between the occurrence of DH type-IV bursts and fast-wide CMEs over two solar cycles.

To explore the effect of LOS on the observed type-IV DS, we closely examine the 48 bursts which had good quality observations from at least two spacecraft simultaneously. Figure 2 shows the different bursts numbered by IDs along with the observed source longitudes or equivalently AR LOS in various spacecraft. The best-view longitude of an event is the LOS that observed the brightest burst with the largest frequency-time extent, out of all recorded simultaneous multi-vantage point data. A clear concentration of best-view longitudes within |60°| is seen. Except for Event 31, in cases where the best-view longitude was > |60°|, there was no spacecraft observing the AR within the |60°| LOS. During Event 31, dense plasma structures of the neighboring AR caused significant LOS occultation to the spacecraft viewing from within < |60°| LOS. Hence, we note that in the absence of strongly occulting structures along the LOS, the spacecraft observing an AR within |60°| LOS or source longitude range always records the best-view of the burst. In general smaller the magnitude of the LOS angle, better the view as evidenced by the triangle (and square) markers that blanket the circles in Fig. 2. For more details, we refer the reader to Mohan et al. (2024a).

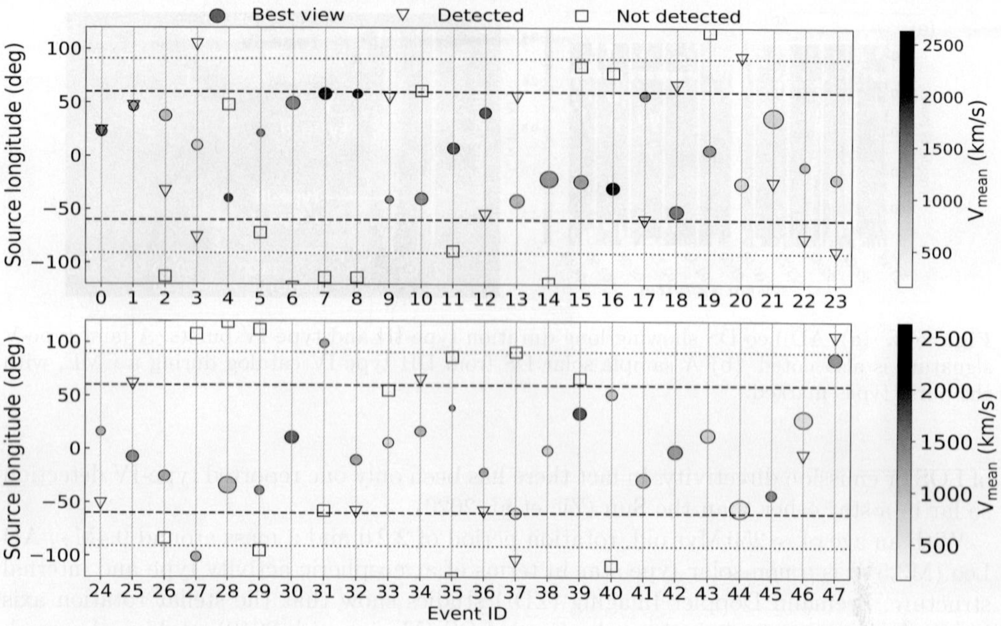

Figure 2. Events with good data from at least 2 spacecraft. Event ID simply labels each event. For each ID, the markers along the Y axis show the longitudes of the active region (AR) as seen by different spacecraft, providing the various lines of sight. Circle shows the LOS that gave the best view of the type-IV, its size denotes the burst duration and the color shows V_{mean}. Triangles mark the other detected longitudes and squares show non-detections. Horizontal lines mark $\pm 60^{\circ}$ and $\pm 90^{\circ}$ longitudes. Longitudes $> |90^{\circ}|$ pick out ARs that hosted high rising post-flare loops with type-IV emission visible beyond the limb.

3. The curious case of young active stars: Lessons from AD Leo

The cool main sequence stars, particularly the M dwarfs (dMs) within 0.6 - 0.1 $M_{\odot}$ mass range are known to show high flaring rates across the electromagnetic spectrum (e.g. Lacy et al. 1976; Bastian 1990; Dal 2020). The dM masses are spread across the $0.4 - 0.3 M_{\odot}$ band below which the stellar interior becomes fully convective, unlike a sun-like star that has an inner radiation zone (Chabrier & Baraffe 2000). In general, stellar atmospheres evolve with age as stars migrate from a fast-rotating active population ('C' branch) to a slow-rotating less active 'I' branch as they lose their angular momentum over time (Barnes 2003; Brown 2014). The stars above 1 Gyr, like our Sun, generally exist in the less active 'I' branch (Barnes 2003). Hence young (<1 Gyr) dMs with mass $\lesssim 0.4 M_{\odot}$ are expected to be very different from the sun in their internal structure and atmospheric activity type.

Young dMs demonstrate some of the highest flaring rates with a high ratio of flare-to-quiescent luminosity in X-ray and spectral lines (e.g., Noyes et al. 1984; Pizzolato et al. 2003; Vidotto et al. 2014; Newton et al. 2017; Yang & Liu 2019; Feinstein et al. 2020). But, despite the high activity, there has been no radio signatures of CMEs in the form of either a type-II or type-IV bursts in these stars, despite several hours to day long monitoring by various groups for over a few decades (Osten and Bastian 2008; Villadsen et al. 2014). The lack of type-II bursts is attributed to the strong magnetic field strengths in the corona which suppresses the shock formation (Alvarado-Gómez et al. 2022). However, the missing stationary type-IVs still remained a puzzle, as they do not need a shock but just a population of flare-accelerated electron beams within the post-flare loop. A possible reason for missing out on type-IV detection could be the effect

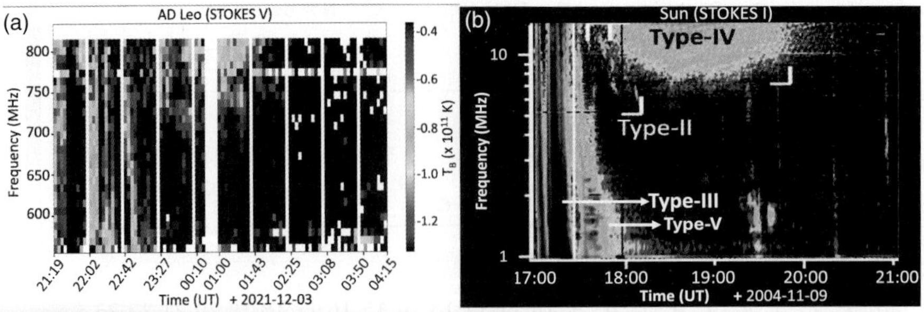

Figure 3. (a): AD Leo DS showing long duration type-III and type-IV bursts. A faint type-V signature is also noted. (b) A sample solar DS from DH type-IV catalog during a CME, with the burst types marked.

of LOS or emission directivity. In fact there has been only one reported type-IV detection so far in a star other than the Sun (Zic et al. 2020).

With an age of $\sim$ 250 Myr old, rotation period of 2.2 d and a mass around $0.4M_\odot$, AD Leo (M3.5V) is a non-solar type star in terms of atmospheric activity type and internal structure. Zeemann Doppler Imaging (ZDI) studies show that the stellar rotation axis has an inclination of $\sim$20° along the Earth LOS (Morin et al. 2010), making the south pole and the surrounding latitudes directly visible. In active young dMs, observations suggest that the flaring ARs are present in poleward latitudes than being focused within |45°| like in the Sun (Ilin et al. 2021). Besides, recent ZDI data show signs of a magnetic structure transition in AD Leo since 2020, whereby the axisymmetry of the large scale field dropped to around 60% from above 90% in the last decade (Bellotti et al. 2023). The reduction in the global poloidal field could present an opportunity for the massive flares to drive eruptions with a higher success rate as the strength of the Alfvén barrier proposed to effectively block eruptions or at least damp their energy so that a shock does not develop, will now be weaker. Using upgraded Giant Metrewave Radio Telescope (uGMRT), we observed AD Leo from 3 December 2021, 21:17 to 4 December 2021, 04:17 UT. We report the discovery of long-duration type-III bursts followed by a type-IV radio burst on a young dM. Besides being the first type-IV in a young dM, this is the first detection of a combination of CME-associated bursts in a star other than Sun, during an active event. Figure 3 highlights the correspondence of the AD Leo DS with a solar CME-associated DS.

The observed long-duration type-III, type-IV and type-V (type-III-like feature with a gradually varying duration as a function of frequency) signatures are likely powered by an electron cyclotron maser emission mechanism, as opposed to plasma emission processes (Ginzburg & Zhelezniakov 1958; Melrose 1970) seen in the solar case. This can be attributed to the stronger magnetic fields on AD Leo. Mohan et al. (2024b) compared different magnetic field profiles that are compatible with the observed burst emission properties and concluded that a solar-like AR magnetic profile cannot explain AD Leo bursts. Instead, a simple multipole expansion model with the relative strengths in different poles adopted from recent ZDI results (Bellotti et al. 2023), agrees better with the data.

4. Conclusion

The type-IV bursts are closely linked to solar CMEs, with the decameter-hectometric (DH) bursts being 100% associated with CMEs that cause interplanetary shocks. The stationary type-IV bursts are signatures of accelerated electron activity in the post-eruption arcades, while moving type-IV are produced by supra-thermal electron in moving plasmoids associated with the CME. We present a comprehensive multi-vantage point catalog of decameter-hectmetric (DH) type-IV bursts. They typically extend down to 7.8 MHz and have an average duration of 83 min. These bursts are associated mostly with fast (>900 km s^{-1}) and wide (>60°) CMEs, with 78% association to halo-CMEs making them highly geo-effective and space weather relevant. The flux and morphology of the observed type-IV bursts in the dynamic spectrum (DS) are significantly affected by the line of sight (LOS). It is found that, unless there are dense plasma structures along the LOS, a spacecraft that observes the flaring active region (AR) within |60°| always record the burst with the highest extent in the frequency-time plane, out of all recorded DS from various vantage points. Based on this, we propose that choosing stars that provide direct LOS to active latitudes can better the odds for type-IV detection.

We present the first detection of CME-associated long-duration type-III and type-IV bursts in a young (~250 Myr) active dM, AD Leo (M3.5V; $0.4M_{\odot}$). This is the fist case of detecting a combination of multiple CME-associated solar busts (long duration type-III, type-V and type-IV) producing a typical solar CME DS in a star other than the Sun. The stellar rotation axis inclination of ~ 20° along the LOS and the low axisymmetry fraction for the large-scale magnetic field over the recent years possibly contributed to the successful detection. The bursts are likely powered by an electron cyclotron maser (ECME) mechanism than the plasma emission processes that dominate in solar radio bursts. Besides, a multipole expansion model better fits the AR magnetic field profile than a solar-like profile given the type-IV characteristics.

References

Alvarado-Gómez, J. D., Drake, J. J., Cohen, O., et al. 2022, *AN*, 343, e10100.

Barnes, S. A. 2003, *ApJ*, 586, 464.

Bastian, T. S. 1990, *Solar Phys*, 130, 265.

Bellotti, S., Morin, J., Lehmann, L. T., et al. 2023, *A&A*, 676, A56.

Brown, T. M. 2014, *ApJ*, 789, 101.

Cane, H. V. & Reames, D. V. 1988, *ApJ*, 325, 895.

Chabrier, G. & Baraffe, I. 2000, *ARA&A*, 38, 337.

Dal, H. 2020, MNRAS, 495, 4529.

Feinstein, A. D., Montet, B. T., Ansdell, M., et al. 2020, *AJ*, 160, 219.

Ginzburg, V. L. & Zhelezniakov, V. V. 1958, *Soviet Astron*, 2, 653.

Gopalswamy, N. 2011, in *Planetary, Solar and Heliospheric Radio Emissions (PRE-VII)*, ed. H. O. Rucker, W. S. Kurth, P. Louarn, & G. Fischer, 325–342.

Gopalswamy, N., Akiyama, S., Mäkelä, P., Yashiro, S., & Cairns, I. H. 2016, arXiv e-prints, arXiv:1605.02223.

Gopalswamy, N., Kaiser, M. L., Thompson, B. J., et al. 2000, *GRL*, 27, 1427.

Gopalswamy, N., Mäkelä, P., & Yashiro, S. 2019, Sun and Geosphere, 14, 111.

Gopalswamy, N., Yashiro, S., & Akiyama, S. 2007, *JGR*, 112, A06112.

Hillaris, A., Bouratzis, C., & Nindos, A. 2016, *Solar Phys*, 291, 2049.

Howard, R. A., Michels, D. J., Sheeley, N. R., J., & Koomen, M. J. 1982, *ApJ*, 263, L101.

Ilin, E., Poppenhaeger, K., Schmidt, S. J., et al. 2021, *MNRAS*, 507, 1723.

Kumari, A., Morosan, D. E., & Kilpua, E. K. J. 2021, *ApJ*, 906, 79.

Kundu, M. R. & Spencer, C. L. 1963, *ApJ*, 137, 572.

Lacy, C. H., Moffett, T. J., & Evans, D. S. 1976, *ApJS*, 30, 85.

McLean, D. & Labrum, N. 1985, Solar Radio Astrophysics (Cambridge University Press).

Melrose, D. B. 1970, *AJP*, 23, 885.

Miteva, R., Samwel, S. W., & Krupar, V. 2017, *JSWSC*, 7, A37.

Mohan, A., Gopalswamy, N., Kumari, A., Akiyama, S., & G, S. 2024a, arXiv e-prints, arXiv:2406.00194.

Mohan, A., Mondal, S., Wedemeyer, S., & Gopalswamy, N. 2024b, arXiv e-prints, arXiv:2402.00185.
Morin, J., Donati, J. F., Petit, P., et al. 2010, *MNRAS*, 407, 2269.
Newton, E. R., Irwin, J., Charbonneau, D., et al. 2017, *ApJ*, 834, 85.
Noyes, R. W., Hartmann, L. W., Baliunas, S. L., Duncan, D. K., & Vaughan, A. H. 1984, *ApJ*, 279, 763.
Osten, R. A. & Bastian, T. S. 2008, *ApJ*, 674, 1078.
Pizzolato, N., Maggio, A., Micela, G., Sciortino, S., & Ventura, P. 2003, *A&A*, 397, 147.
Pohjolainen, S. & Talebpour Sheshvan, N. 2020, Advances in Space Research, 65, 1663.
Schmahl, E. J. 1972, *PASA*, 2, 95.
Smerd, S. F. 1971, *AJP*, 24, 229.
Takakura, T. & Kai, K. 1961, *PASJ*, 13, 94.
Talebpour Sheshvan, N. & Pohjolainen, S. 2018, *Solar Phys*, 293, 148.
Vidotto, A. A., Gregory, S. G., Jardine, M., et al. 2014, *MNRAS*, 441, 2361.
Villadsen, J., Hallinan, G., Bourke, S., Güdel, M., & Rupen, M. 2014, *ApJ*, 788, 112.
Wild, J. P. 1970, *PASA*, 1, 365.
Wild, J. P. & McCready, L. L. 1950, Australian Journal of Scientific Research A Physical Sciences, 3, 387.
Yang, H. & Liu, J. 2019, *ApJS*, 241, 29.
Yashiro, S., Gopalswamy, N., Michalek, G., et al. 2004, *JGR*, 109.
Zic, A., Murphy, T., Lynch, C., et al. 2020, *ApJ*, 905, 23.

Solar and Stellar Coronal Mass Ejections
Proceedings IAU Symposium No. 388, 2026
N. Gopalswamy, O. Malandraki, A. Vidotto & W. Manchester, eds.
doi:10.1017/S1743921324001558

A catalog of multi-vantage point observations of type-II bursts: Statistics and correlations

Atul Mohan[1,2], Nat Gopalswamy[1], Hemapriya Raju[3] and Sachiko Akiyama[1,2]

[1]NASA Goddard Space Flight Center, 8800 Greenbelt Road Greenbelt, MD, 20771, USA

[2]The Catholic University of America, 620 Michigan Avenue, N.E. Washington, DC 20064, USA

[3]Indian Institute of Technology, Indore, Simrol, Indore, 453552, India

Abstract. Coronal mass ejection (CME) often produces a soft X-ray (SXR) flare associated with the low-coronal reconnection and a type-II radio burst associated with an interplanetary (IP) CME-shock. SXR flares and type-II bursts outshine the background emission, making them sun-as-a-star observables. Though there exist SXR flare catalogs covering decades of observations, they do not provide the associated type-II luminosity. Besides, since radio burst emission could be beamed, the observed flux dynamic spectrum may vary with line of sight. Using long-term calibrated decameter-hectometric dynamic spectra from the Wind and STEREO spacecraft, we build a catalog of multi-vantage point observations of type-II bursts. Cross-matching with existing catalogs we compile the properties of the associated flare, reconnection, and the CME. Cross-correlation analysis was done between various parameters. Two novel metrics of flare and CME power show a strong correlation revealing a link between particle acceleration strengths in the low corona and IP space.

Keywords. Solar coronal mass ejections, Catalogs, Stellar coronal mass ejections

1. Introduction

Solar coronal mass ejections (CMEs) are a major driver of energetic space weather phenomena, like solar energetic particle events (SEPs), geomagnetic storms etc. CMEs are often associated with flares across the electromagnetic spectrum, from radio to gamma-rays (see, Howard et al. 2023, for an overview). The two major processes during a flare-CME event include the reconnection event that accelerates electrons leading to the SXR flare in the low corona, and the CME shock that accelerates particles across the corona and interplanetary space often leading to type-II radio bursts (e.g. Wild and McCready 1950; McLean and Labrum 1985; Gopalswamy 2011) in meter - kilometer (m-km) band. Gopalswamy et al. (2018b) showed that the low coronal flare-reconenction flux, $\phi_{\rm rec}$, correlates well with the CME kinetic energy revealing a link between the strengths of the low coronal and interplanetary impacts of the active phenomenon. However, despite decades of observations, we lack an understanding of the relationships, if any, between the observational proxies to the strengths of the particle acceleration in the low corona (peak SXR luminosity, $L_{\rm X}$) and interplanetary space (peak type-II luminosity, $L_{\rm R}$), and the physical properties of their drivers (i.e., flare and CME).

Meanwhile, when it comes to stellar CMEs, unlike the solar case, spatially resolved observations are impossible with current telescopes limiting the capability to directly infer crucial physical parameters, particularly $\phi_{\rm rec}$ and CME speed ($V_{\rm CME}$).

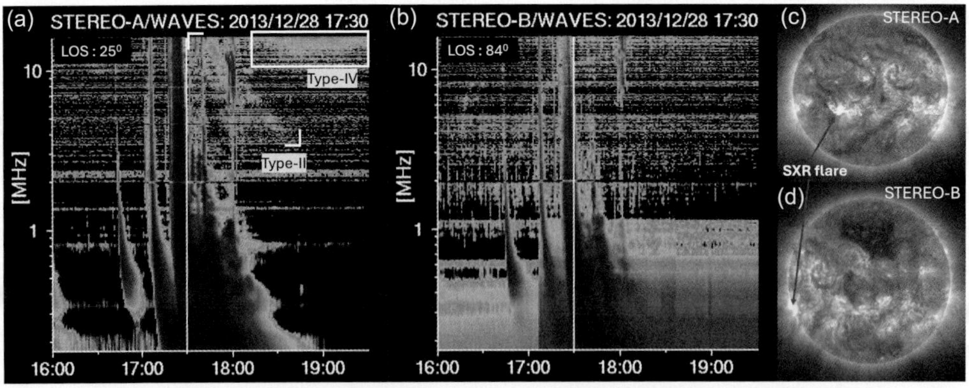

Figure 1. (a-b): DH DS from STEREO-A and B during a CME. The associated active region was at 25° and 84° lines of sight (LOS) in respective spacecraft. The type-II (marked by angular brackets) and type-IV (marked by box) seen in STEREO-A DS is absent in STEREO-B data. (c-d) The 195Å images from both spacecraft show the SXR flare, marked by arrows.

Hence, scaling laws linking sun-as-a-star L_X and L_R, to relevant spatially resolved physical properties are important to constrain stellar CME models and to understand the solar-stellar flare-CME phenomenon within a common framework. There exist SXR flare and CME catalogs, that report L_X, ϕ_{rec} (in some cases), V_{CME} etc. for decades of solar-CME observations, but they do not report simultaneous L_R for any associated type-II bursts. Besides, m - km burst emission is inherently beamed as demonstrated for different burst types (e.g. Dulk et al. 1980; Gopalswamy 2011; Saint-Hilaire et al. 2013; Mohan et al. 2024). Apart from inherent directivity, the line of sight (LOS) plasma and magnetic field structures can also impact the observed emission characteristics in the dynamic spectrum (DS) (e.g. Robinson and Cairns 1994; Arzner and Magun 1999; Talebpour Sheshvan and Pohjolainen 2018). Figure 1 shows an event observed by STEREO-A and STEREO-B spacecraft with the flaring active region at 25° and 84° lines of sight (LOS) respectively. The decameter-hectometric (DH) DS and the X-ray images from both spacecraft are compared. STEREO-A spacecraft observed the frequency drifting type-II burst and the long-duration stationary type-IV emission confined above 10 MHz (see, McLean and Labrum 1985, for an overview of the burst types). Clearly, STEREO-B does not observe both type-II and type-IV bursts, though both spacecraft see the SXR flare. Hence the LOS towards the active latitudes on the star can be crucial in choosing a target star with a high chance of detecting a radio burst. The first step to performing such detailed studies of the properties of the observed type-II bursts is to build a comprehensive catalog of multi-vantage point observations of type-II bursts listing their spectro-temporal properties and L_R, and the properties of the associated flares and CMEs. We present such a catalog and the initial results on the correlations between various paramters.

2. Data and event catalog

The calibrated DH band dynamic spectra recorded by Plasma Wave Investigation (WAVES) instruments onboard Wind, STEREO-A and STEREO-B spacecraft since Nov 2006 form a unique uniformly calibrated multi-vantage point dataset ideal for the goals of the study. This long-term database spans two solar cycles and the rising phase of the current cycle. The other aspects that make DH bursts interesting are the following. DH type-IIs are associated with interplanetary shocks driven by energetic CMEs. In the stellar CME studies, despite hours and days-long searches on stars with very high flaring rates in the meterwave band, not a single type-II burst has been found (e.g. Bastian 1990; Osten and Bastian 2008; Crosley and Osten 2018; Villadsen and Hallinan 2019). Numerical simulations suggest that due to the strong magnetic fields in the atmospheres of active stars, an eruptive event may not be able to generate shocks in the lower corona but probably at interplanetary heights, favoring DH band over the meterwave band (Odert et al. 2020; Alvarado-Gómez et al. 2022). Also, compared to their metric counterparts, the DH type-II bursts are associated with relatively stronger flares and faster CMEs (Gopalswamy 2011; Miteva et al. 2017). The DH type-II bursts are also strongly associated with SEPs and solar gamma-ray events making them more space weather relevant (Miteva et al. 2017; Gopalswamy et al. 2018c).

2.1. *The multi-vantage point flare- CME- type-II catalog*

The initial list of DH type-II bursts comes from Gopalswamy et al. (2019). For each event, we assign a 'data quality' metric based on the quality of the data from each spacecraft. The data quality metric is of the form WxAyBz, where $x,y,z \in 0,1,2$. W stands for Wind, A for STEREO-A and B for STEREO B, while the integer metric attached conveys the quality of the data stream. When 0 represents data gap, 1 represents a poor quality DS in which robust identification of emission features is difficult and 2 represents good data. All good dynamic spectra from different spacecraft are inspected and given an 'event quality' metric. This metric rates the quality of the event recorded. An event is rated along two axes: shape (S) and intensity (I). A type-II event with a well-discernible shape in the DS is given a rating of 3, while 1 represents an event whose shape is not very evident. This could be because it extends only within a narrow spectral band or because the event is in a crowded region of the DS where other burst types make the shape determination difficult. Similarly along the intensity axis, 3 represents a bright event while 1 denotes an event with a relatively low intensity in the region of the DS where it is detected. All dynamic spectra with a metric SxIy, where $x>1$ and $y>1$ are generally reliable. Of all the good dynamic spectra recorded for an event across spacecraft, the DS with the highest event quality metric identifies the observation line of sight that provided the best view of the event. The DS with the best event quality is used to estimate the L_R. The L_R was computed as the peak burst flux in the 3 - 7 MHz range. This frequency range was chosen because the type-III and type-II bursts are the major burst types in this range, and are relatively easily separable from one another due to their dynamic spectral morphology. Most long-duration broadband emission features that last for several hours to days, like the type-IV bursts are mostly confined above 7 MHz (Hillaris et al. 2016; Mohan et al. 2024). Hence in the higher frequency ranges often the type-II emission is mixed with type-IV bursts.

The aforementioned methodology led to a catalog of 163 DH type-II bursts with L_R estimates (hereafter 'full catalog'). As discussed earlier, since the DS can have multiple co-temporal burst emission features occurring around similar frequency ranges we bestow a 'reliability' metric on the L_R estimate. Reliability metric rates L_R estimate along two axes: Field crowdedness (F) and relative flux (R). F quantifies how crowded the frequency-time field of the DS is where the type-II is observed. If there are other co-temporal emission features in the 3–7 MHz band intervening with the type-II emission, then F is given a high value. A value of 2 is given when the crowdedness is so high that the L_R is unreliable, while F=1 implies that L_R is reliable but the possibility of contamination from other emission features is non-negligible. 0 denotes the best case where the type-II is not close to any other radio emission feature and hence L_R is robust. Meanwhile, the relative flux (R) metric rates the flux level of the burst in the 3–7 MHz range in contrast to the typical burst flux outside of this chosen band. A value of 1 denotes that the

DH type-II													Associated events									
Properties										Quality metrics			CME				Flare					SEP
Start_date	Start_time	End_date	End_time	Freq_start	Freq_end	Detection	Best Mission	Peak Flux	F_thresh	Data	Event	Reliability	CME_date	CME_time	Width	Speed	Src_loc	NOAA	GOES	Rec_flux	|B|	Flux (>10 MeV)
UT	UT	UT	UT	kHz	kHz			(log SFU)	log(SFU)				UT	UT	deg	km/s				Mx	G	$cm^{-2}\,s^{-1}\,sr^{-1}$
2011/02/15	02:10	2011/02/15	07:00	16000	400	Wind,STA,STB	Wind	6.1	3.7	W2A2B2	S3I3	F0R1	2011/02/15	02:24	360	669	S20W12	11158	X2.2	6.47e+21	286.0	nan
2011/06/02	08:00	2011/06/02	08:25	15000	4000	Wind,STB	STB	3.7	2.4	W2A2B2	S2I2	F0R1	2011/06/02	08:12	360	976	S19E25	11227	C3.7	9.59e+20	63.0	nan
2012/01/19	15:00	2012/01/20	02:45	16000	100	STA,STB,Wind	Wind	4.2	3.7	W2A2B2	S3I3	F0R1	2012/01/19	14:36	360	1120	N32E22	11402	M3.2	4.4e+21	443.0	nan
2012/03/10	17:55	2012/03/11	12:30	14000	300	STA,Wind	STA	4.25	2.6	W2A2B2	S2I2	F0R1	2012/03/10	18:00	360	1296	N17W24	11429	M8.4	7.86e+21	276.0	nan
2013/05/13	16:15	2013/05/13	19:10	8000	300	Wind,STA,STB	STB	5.7	2.4	W2A2B2	S3I3	F0R1	2013/05/13	16:07	360	1850	N11E85	11748	X2.8	6.42E+20	3	nan
2013/09/29	21:53	2013/09/30	21:00	14000	60	Wind,STA	Wind	4.8	3.7	W2A2B2	S2I3	F0R1	2013/09/29	22:12	360	1179	N17W29	EP	C1.3	3.48e+21	10.0	nan
2013/10/26	03:01	2013/10/26	04:00	14000	1800	Wind,STA	Wind	4.3	3.7	W2A2B2	S3I2	F0R1	2013/10/26	03:12	208	473	N09W44	11875	C4.5	3.93e+20	332.0	nan

Figure 2. A portion of List 2, that presents DH type-II events with reliable L_R,L_X,ϕ_{rec} and V_{CME} estimates. Refer text for details. All event lists and their description can be found at https://cdaw.gsfc.nasa.gov/CME_list/radio/multimission_type2/.

3–7 MHz flux is a good representative of the typical overall DH type-II burst flux level. A value of 0 would mean otherwise. So the F0R1 events have a very reliable L_R, which is also representative of the overall burst flux level.

For each event in the full catalog, L_X and V_{CME} were obtained by cross-matching the DH type-II event list with the catalogs maintained by the Coordinated Data Analysis Workshops (CDAW) group†. Events with good L_X and V_{CME} estimates were identified. Among those, the ones with an event quality SxIy, with x,y∈2,3 and reliability (F0R1) were chosen and ϕ_{rec} was estimated using the post-eruption arcade method (Gopalswamy et al. 2018a). Finally, a list of 25 'flare- CME- type-II' events (List 2) with reliable L_X,L_R,V_{CME}, and ϕ_{rec} estimates was produced that was used for further analysis. Figure 2 shows a portion of List 2. Peak_Flux provides radio flux density in 3 - 7 MHz band and F_thresh the mean background radio flux level. The time and frequency extents of the burst and the various quality metrics are also provided. The 'CME' section reports the maximum angular width and the speed of the interplanetary CME along with the time window of the event. Under the 'Flare' section, the location of the flaring active region as seen from the Earth (Src_loc), standard active region ID (NOAA), the peak SXR flux in the GOES scale‡, ϕ_{rec} in Mx unit and absolute magnetic filed strength at the flare region are provided. The link to the full catalog and event lists is in the caption of Fig. 2.

3. Results and discussion

Figure 3a shows the flux density distribution of the events in the full catalog and the sub-lists. A mean flux density of $\sim 3.5\times 10^4$ SFU is found for all lists. Besides L_X,L_R,V_{CME} and ϕ_{rec} we define two new metrics of flare and CME power combining these quantities. The 'flare power', $P_{flare} = \sqrt{L_X \phi_{rec}}$, combines the strength in the magnetic reconnection with that in flare emissivity. Meanwhile the 'CME power', $P_{CME} = \sqrt{L_R V_{CME}^2}$, estimates the combined power in the CME kinetic energy that drives the shock and the associated particle acceleration. Figure 3b shows the result from the Spearman correlation analysis performed between parameter tuples drawn from the set comprising of L_R, L_X, ϕ_{rec}, V_{CME}, P_{flare} and P_{CME}. Each cell in the matrix plot shows the correlation coefficient ($C_{R,C}$) between the row (R) and column (C) parameter. The cell color represents the p-value. The significant strong $C_{R,C}$ are highlighted by slanting text ($C_{R,C} \gtrsim 0.49$ & $p < 0.015$). Straight white text represent reliable weak correlations ($C_{R,C} < 0.49$ & $p<0.05$) and those in gray are insignificant ($p>0.05$; $C_{R,C} < 0.49$). The correlation coefficients between tuples that are related by definition are masked and will not be considered in the analysis. After the well-known strongly correlated tuple (ϕ_{rec},L_X) (Kazachenko

† https://cdaw.gsfc.nasa.gov/
‡ https://www.swpc.noaa.gov/products/goes-x-ray-flux

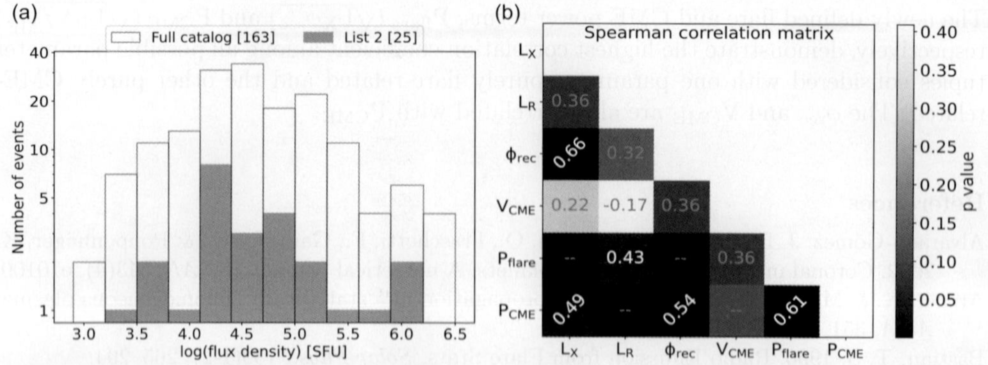

Figure 3. Statistical properties of the flare- CME- type-II events. (a): Flux density histogram. (b): Spearman correlation matrix for the various properties of the List 2 events. Numbers in each cell denote the correlation coefficient $C_{\mathrm{R,C}}$ corresponding to the row (R) and column (C) parameters. Rotated text highlights strong correlations, white normal text denotes reliable weak $C_{\mathrm{R,C}}$, and gray denotes the insignificant ones. Whenever R parameter is a function of C, $C_{\mathrm{R,C}}$ is not mentioned.

et al. 2017; Sindhuja and Gopalswamy 2020), (P_{CME},P_{flare}) is the strongly correlated pair with $C_{P_{\mathrm{CME}},P_{\mathrm{flare}}}=0.61$. The other strong correlations (green text) include (P_{CME},L_X) and (P_{CME},ϕ_{rec}). Of all the strongly correlated parameter tuples where one parameter is fully flare-related and the other fully CME-related, (P_{CME},P_{flare}) shows the highest $C_{\mathrm{R,C}}$ value. Note that the individual terms that makeup P_{flare} and P_{CME} are not significantly correlated with each other, making the P_{flare} - P_{CME} correlation robust. This correlation highlights the connection between the low-coronal and interplanetary particle acceleration events driven by the flare and the CME.

4. Conclusion

We present a catalog of decameter-hectometric (DH) type-II bursts with multi-vantage point observations from Wind and STEREO spacecraft. The data covers the interval from Nov 2006 to July 2023. Peak type-II flux in 3-7 MHz was found for all events along with the time-frequency extent of the burst. The full catalog has 163 events. For each type-II event, the catalog specifies the spacecraft that recorded the best view of the burst in terms of the time-frequency extent and morphology in the dynamic spectrum (DS). Also, the flare source location as seen from the Earth is mentioned. The DS of the best-viewed burst is rated along shape and intensity axes to help identify the events that demonstrated a well-characterizable shape with sufficient intensity. Along with the information on the source location and the relative position of the STEREO spacecraft, available from online resources† the catalog can be used to explore the line of sight dependency in the detection of radio bursts.

Cross-matching with the event catalogs compiled by the NASA-CDAW group, the properties of the associated soft X-ray (SXR) flare and interplanetary (IP) CME are compiled. The flare reconnection flux (ϕ_{rec}), peak SXR luminosity (L_X) and CME speed (V_{CME}) were either computed or compiled from existing catalogs, for all possible events. 25 events in the full catalog (List 2) had reliable estimates for L_X,L_R,ϕ_{rec} and V_{CME}. Using List 2 events, various correlations between flare, CME and type-II parameters were explored, to understand the interconnections, if any, between the particle acceleration processes in the low corona and IP space driven by the flare and the CME respectively.

† https://stereo-ssc.nascom.nasa.gov/cgi-bin/make_where_gif

The newly defined flare and CME power terms, P_{flare} ($\sqrt{L_X \phi_{rec}}$) and P_{CME} ($\sqrt{L_R V_{CME}^2}$) respectively, demonstrate the highest correlation coefficient among all possible parameter tuples considered with one parameter purely flare-related and the other purely CME-related. The ϕ_{rec} and V_{CME} are also correlated with P_{CME}.

References

Alvarado-Gómez, J. D., Drake, J. J., Cohen, O., Fraschetti, F., Garraffo, C., & Poppenhäger, K. 2022, Coronal mass ejections and exoplanets: A numerical perspective. *AN*, 343(4), e10100.

Arzner, K. & Magun, A. 1999, Radiowave propagation in a statistically inhomogeneous plasma. *A&A*, 351, 1165–1189.

Bastian, T. S. 1990, Radio Emission from Flare Stars. *Solar Phys.*, 130(1-2), 265–294.

Crosley, M. K. & Osten, R. A. 2018, Constraining Stellar Coronal Mass Ejections through Multi-wavelength Analysis of the Active M Dwarf EQ Peg. *ApJ*, 856(1), 39.

Dulk, G. A., Gary, D. E., & Suzuki, S. 1980, The position and polarization of Type V solar bursts. *A&A*, 88(1-2), 218–229.

Gopalswamy, N. Coronal Mass Ejections and Solar Radio Emissions. In Rucker, H. O., Kurth, W. S., Louarn, P., & Fischer, G., editors, *Planetary, Solar and Heliospheric Radio Emissions (PRE-VII)* 2011, pp. 325–342.

Gopalswamy, N., Akiyama, S., Yashiro, S., & Xie, H. A New Technique to Provide Realistic Input to CME Forecasting Models. In Foullon, C. & Malandraki, O. E., editors, *Proc. IAU Symposium No. 335* 2018,a, volume 335, pp. 258–262.

Gopalswamy, N., Akiyama, S., Yashiro, S., & Xie, H. 2018,b Coronal flux ropes and their interplanetary counterparts. *Journal of Atmospheric and Solar-Terrestrial Physics*, 180b, 35–45.

Gopalswamy, N., Mäkelä, P., & Yashiro, S. 2019, A Catalog of Type II radio bursts observed by Wind/WAVES and their Statistical Properties. *Sun and Geosphere*, 14, 111–121.

Gopalswamy, N., Mäkelä, P., Yashiro, S., Lara, A., Xie, H., Akiyama, S., & MacDowall, R. J. 2018,c Interplanetary Type II Radio Bursts from Wind/WAVES and Sustained Gamma-Ray Emission from Fermi/LAT: Evidence for Shock Source. *ApJ (Letters)*, 868c(2), L19.

Hillaris, A., Bouratzis, C., & Nindos, A. 2016, Interplanetary Type IV Bursts. *Solar Phys.*, 291(7), 2049–2069.

Howard, R. A., Vourlidas, A., & Stenborg, G. 2023, The evolution of our understanding of coronal mass ejections. *FrASS*, 10, 1264226.

Kazachenko, M. D., Lynch, B. J., Welsch, B. T., & Sun, X. 2017, A Database of Flare Ribbon Properties from the Solar Dynamics Observatory. I. Reconnection Flux. *ApJ*, 845(1), 49.

McLean, D. & Labrum, N. 1985,. *Solar Radio Astrophysics.* Cambridge University Press.

Miteva, R., Samwel, S. W., & Krupar, V. 2017, Solar energetic particles and radio burst emission. *Journal of Space Weather and Space Climate*, 7, A37.

Mohan, A., Gopalswamy, N., Kumari, A., Akiyama, S., & G, S. 2024, Inter-planetary type-IV solar radio bursts: A comprehensive catalog and statistical results. *arXiv e-prints*, arXiv:2406.00194.

Odert, P., Leitzinger, M., Guenther, E. W., & Heinzel, P. 2020, Stellar coronal mass ejections - II. Constraints from spectroscopic observations. *MNRAS*, 494(3), 3766–3783.

Osten, R. A. & Bastian, T. S. 2008, Ultrahigh Time Resolution Observations of Radio Bursts on AD Leonis. *ApJ*, 674(2), 1078–1085.

Robinson, P. A. & Cairns, I. H. 1994, Fundamental and harmonic radiation in type III solar radio bursts. *Solar Phys.*, 154, 335–360.

Saint-Hilaire, P., Vilmer, N., & Kerdraon, A. 2013, A Decade of Solar Type III Radio Bursts Observed by the Nançay Radioheliograph 1998-2008. *ApJ*, 762, 60.

Sindhuja, G. & Gopalswamy, N. 2020, A study of the observational properties of coronal mass ejection flux ropes near the sun*. *ApJ*, 889(2), 104.

Talebpour Sheshvan, N. & Pohjolainen, S. 2018, Visibility and Origin of Compact Interplanetary Radio Type IV Bursts. *Solar Phys.*, 293(11), 148.

Villadsen, J. & Hallinan, G. 2019, Ultra-wideband Detection of 22 Coherent Radio Bursts on M Dwarfs. *ApJ*, 871(2), 214.

Wild, J. P. & McCready, L. L. 1950, Observations of the Spectrum of High-Intensity Solar Radiation at Metre Wavelengths. I. The Apparatus and Spectral Types of Solar Burst Observed. *Australian Journal of Scientific Research A Physical Sciences*, 3, 387.

Solar and Stellar Coronal Mass Ejections
Proceedings IAU Symposium No. 388, 2026
N. Gopalswamy, O. Malandraki, A. Vidotto & W. Manchester, eds.
doi:10.1017/S1743921324001984

Historical extreme solar energetic particle events

Fusa Miyake

Institute for Space-Earth Environmental Research, Nagoya University, Nagoya, Japan

Abstract. Carbon-14 (^{14}C) analysis of tree rings has revealed multiple rapid increases in ^{14}C (^{14}C spikes) during the Holocene, e.g., in ca. 774 CE, 993 CE, 660 BCE, and 7176 BCE. These ^{14}C spikes have been replicated by several tree-ring ^{14}C records and other cosmogenic nuclide records, such as ^{10}Be and ^{36}Cl from ice cores. These events are considered to have been caused by extreme solar energetic particle (SEP) events, with the scale of the SEP events estimated to be several dozen times larger than the largest SEP event detected in modern observations. As such, studies on past SEP events using cosmogenic nuclides have advanced in recent years; however, several issues regarding such events remain unknown. This paper reviewed recent studies on the following topics: 1) further exploration of extreme SEP events over the long term, and 2) exploration of intermediate-sized SEP events between those detected by modern observations and the extreme SEP events detected by cosmogenic nuclides. Future research directions for these topics were also discussed.

Keywords. cosmogenic nuclide, ice core, tree ring, solar energetic particle, extreme SEP event

1. Introduction

Cosmic rays from outside the Earth produce cosmogenic nuclides such as ^{10}Be, ^{14}C, and ^{36}Cl, through a nuclear cascade shower in the atmosphere. Although cosmogenic nuclides are usually produced by galactic cosmic rays with an energy range from $\approx 10^7$ to $\approx 10^{21}$ eV, they are also produced by solar energetic particles (SEPs) suddenly released from solar eruptions, such as solar flares and coronal mass ejections (e.g., Miyake *et al.* 2019). Extreme SEP events can cause a sudden increase in the concentrations of these cosmogenic nuclides in the atmosphere above the baseline produced by galactic cosmic rays (e.g., Usoskin *et al.* 2023). After the production of cosmogenic nuclides, they are transported through the atmosphere and some are stored in natural archives, such as tree rings and ice sheets. Tree rings (^{14}C) and ice cores (^{10}Be and ^{36}Cl) are mainly used as samples to investigate extreme SEP events. Extreme SEP events are expected to be recorded as spikes in the concentrations of cosmogenic nuclides in these natural archives.

Previous studies have reported spikes in the concentrations of ^{14}C in tree rings and of ^{10}Be and ^{36}Cl in polar ice cores in ca. 774 CE, 993 CE, 660 BCE, and 7176 BCE (e.g., Miyake *et al.* 2012, 2013, 2015; Mekhaldi *et al.* 2015; Park *et al.* 2017; Brehm *et al.* 2022; Paleari *et al.* 2022). The origin of these spikes in the concentrations of cosmogenic nuclides has been attributed to not only extreme SEP events but also gamma-ray events, i.e., gamma-ray bursts and supernova explosions. However, the measured production ratios of ^{14}C, ^{10}Be, and ^{36}Cl could not be explained by a gamma-ray origin and were consistent with an SEP origin (e.g., Usoskin *et al.* 2023). The magnitude of these events is estimated to be several tens to a hundred times larger than the largest SEP event detected from modern observations (ground-level enhancement in February 1956: GLE #5) (Usoskin *et al.* 2023).

In addition to the four events mentioned above, several ^{14}C spikes in tree rings, namely, candidates for extreme SEP events, have been reported. Notably, the ^{14}C spikes in ca. 1279 CE, 1052 CE, 5259 BCE, 5410 BCE, and 12350 BCE have been confirmed by several tree samples (Brehm *et al.* 2021, 2022; Miyahara *et al.* 2022; Miyake *et al.* 2021; Bard *et al.* 2023). Since a gamma-ray origin cannot be ruled out with ^{14}C data alone, it will be essential to obtain ice core records in the future to identify the cause. However, since the variations in most of the ^{14}C spikes reported were similar to those of the four events identified with the SEP origin, the reported ^{14}C spikes may have the SEP origin. Since these ^{14}C spikes are recorded in trees worldwide, we can use them as time markers. The precise radiocarbon dating using ^{14}C spikes is called "spike matching," and it has been applied to fields of archeology, dendrochronology, stratigraphy, and natural disaster science (e.g., Wacker *et al.* 2014; Oppenheimer *et al.* 2017; Hakozaki *et al.* 2018; Sigl *et al.* 2015; Kuitems *et al.* 2021; Price 2023). The usefulness of ^{14}C spikes in a wide range of fields has motivated further exploration of spikes by measuring annual ^{14}C concentrations.

Multiple SEP events have been confirmed by acquiring cosmogenic nuclide data with a time resolution of $\approx$ 1 year over the past decade. However, cosmogenic nuclides data have not yet been collected for many years, even in the Holocene, and further exploration is required to investigate the characteristics of extreme SEP events. Also, a large-scale gap exists between SEP events detected by modern observations (past $\approx$ 70 years) and extreme SEP events detected by cosmogenic nuclides (e.g., Usoskin & Kovaltsov 2021), and an exploration of intermediate sized SEP events by cosmogenic nuclides is important. Chapters 2 and 3 discuss the current status and future research directions on these topics.

2. Long-term exploration of extreme SEP events

As introduced in Chapter 1, signatures of several extreme SEP events have been found in the analysis of cosmogenic nuclides. Since spikes in cosmogenic nuclide concentrations are characterized by a significant increase over one to several years and the following rapid decreases, it is crucial to conduct a cosmogenic nuclide analysis with a time resolution of one to a few years to capture such changes. Here, although the SEP event itself is on a time scale of several hours to several days, real-time cosmogenic variation is not recorded in natural archives due to several factors, such as the transport process of cosmogenic nuclides from their production in the atmosphere to their incorporation into archive samples and the seasonal dependence of tree and ice sheet formations. Therefore, most analyses of cosmogenic nuclides have been conducted with a time resolution of more than $\approx$ 1 year. So far, the exploration of extreme SEP events has mainly been performed using ^{14}C in tree rings due to sample availability and ease of analysis.

Annual ^{14}C data have been reported only for the past 2400-years in total, for the Holocene (past $\approx$ 12,000 years) according to Brehm *et al.* (2022). However, tree-ring samples whose ages were dendrochronologically dated are available for the Holocene. Several projects are on-going to measure annual ^{14}C data. For example, our research group is collecting annual ^{14}C data for the period of the past $\approx$ 8000 years using tree samples from Japan, Russia, and Finland. Therefore, it is expected that further continuous annual ^{14}C data during the Holocene will be available in the near future. It is considered that no extreme SEP events of magnitude much greater than the 774 CE event during the Holocene exist, according to the low-temporal resolution ^{14}C data (e.g., IntCal calibration curve Reimer *et al.* 2020) (e.g., Miyake *et al.* 2019). Therefore, tree-ring ^{14}C data currently being collected during the Holocene will elucidate the occurrence rate of extreme SEP events in the magnitude class of the 774 CE event.

On timescales beyond the Holocene, the availability of tree-ring samples is limited due to the small number of tree samples in the last glacial period. Therefore, in addition to excavating wood samples themselves and extending the dendrochronology of several tree-ring series, alternative archival samples, such as ice cores and lake sediments (a varve structure is found in Lake Suigetsu in Japan), should be considered for a continuous exploration of extreme SEP events. Since the use of ice cores is also limited and it is not apparent that lake sediments can be used to search for SEP events, further researches and efforts are necessary. However, such long-term exploration is essential to explore an upper limit on timescales beyond the Holocene and to improve event statistics. Therefore, the use of ice cores and the development of methods using alternative archives will be important in the future. In fact, the ca. 12350 BCE event reported recently by Bard *et al.* (2023) is the largest change in terms of the amount of ^{14}C increase, and it is not unexpected that even larger SEP events occurred before the Holocene.

3. Exploration of intermediate-sized SEP events

The ^{14}C spike in 5410 BCE is one of the smallest ^{14}C spikes ever reported, and its magnitude was estimated to be about 1/3 of that of the 774 CE event (Miyake *et al.* 2021; Brehm *et al.* 2022). Smaller ^{14}C spikes than this size cannot be detected by a normal ^{14}C analysis (a typical measurement error for a normal ^{14}C analysis is 2–3‰, which is sufficient to detect the largest ^{14}C spikes with ≈15–20‰increase, such as the 774 CE event, but events a fraction of that size are not significantly detectable). A large gap exists between SEP events detected by modern observations and extreme SEP events detected by cosmogenic nuclides due to a sensitivity problem in the analysis of cosmogenic nuclides (e.g., Usoskin & Kovaltsov 2021). However, let us suppose that the detection of intermediate-sized events in the gap is possible. In that case, it will be possible to discuss whether the occurrence rate of extreme SEP events detected by cosmogenic nuclides is an extension of the occurrence rate of events detected by modern observations (i.e., whether they are caused by a similar mechanism) or whether they follow a completely different distribution. Additionally, it is expected that intermediate-sized SEP events are caused by an extreme solar eruption that humankind has not experienced in the past decades, and its impact on space weather should be enormous. Therefore, understanding its occurrence rate and characteristics is very important for risk assessment in modern society.

To detect intermediate-sized events, Miyake *et al.* (2023) conducted ^{14}C measurements using tree samples from Alaska from 1844 to 1876 CE, including the Carrington event (1859) and the geomagnetic superstorm (1872). By measuring ^{14}C of the same year several times, the data errors were reduced to ≈ 0.8‰. However, no significant ^{14}C increase during the measurement period was detected, suggesting that no SEP event with a magnitude of ≈ 1/10 of the 774 CE event occurred. This gives the upper limit of the Carrington SEP event, assuming an energy spectrum of the SEP event (annual-integrated F_{200}: $<1 \times 10^9 \mathrm{cm}^{-2}$) (Miyake *et al.* 2023). Conducting similar ultrahigh-precision ^{14}C analyses in other periods may lead to the detection of intermediate-sized SEP events.

While Miyake *et al.* (2023) used tree samples from Alaska (the samples are considered to come from the southwest coastal area of Alaska, < 60°N) for the high-precision ^{14}C measurement, Uusitalo *et al.* (2024) used tree samples from a higher latitude area in Finland (>67°N), and they reported a significant ^{14}C increase after the Carrington event compared to the lower latitude data. Other studies have also reported some differences in ^{14}C variation based on latitude. For example, Miyake *et al.* (2022) reported that the ^{14}C increase during the 993 CE event occurred one year earlier in high-latitude tree samples (>60°N) than in low-latitude tree samples. Although it has not been confirmed that the significant ^{14}C enhancement in Finland data after the Carrington event is related to the solar events, such latitude differences of ^{14}C data may reflect the characteristics of SEPs, which have a large amount of nuclide production, particularly at high latitudes (>60°) due to geomagnetic cutoff effect and in the stratosphere (e.g., Poluianov *et al.* 2016;

Golubenko *et al.* 2022). In the future, it will be essential to develop a three-dimensional nuclide transport model to clarify the causes of such regional differences in ^{14}C. In addition, a correct understanding of such regional differences may lead to the detection of nuclide increases due to intermediate-sized SEP events (Uusitalo *et al.* 2024).

Another potential method for detecting intermediate-sized events is the ^{36}Cl analysis of ice cores. Compared to ^{14}C and ^{10}Be, ^{36}Cl has a production pathway from cosmic rays with lower energy, and it has a higher sensitivity for SEP events, particularly for SEP events with a soft energy spectrum (Mekhaldi *et al.* 2021). Given that the abundance of ^{36}Cl in ice sheets is less than that of ^{10}Be, a large number of ice core samples must be consumed to analyze ^{36}Cl with high sensitivity (4–5 times the amount of sample required for ^{10}Be). Therefore, in previous studies, ^{36}Cl data have often been obtained with a time resolution of 4–5 years. In contrast, ^{36}Cl data with a resolution of 1–2 years, which is suitable for detecting SEP events, was rarely obtained. In the future, such a high-temporal resolution ^{36}Cl analysis may enable the detection of intermediate-sized events.

4. Conclusion

In the past decade, understanding extreme SEP events based on cosmogenic nuclide data has deepened greatly. However, the number of extreme SEP events detected by cosmogenic nuclides is still small. In addition, the exploration of timescales beyond the Holocene has yet to progress. In addition to analyzing ^{14}C in tree rings, it will be necessary to explore methods that allow for the long-term investigation of extreme SEP events. Another research direction is to explore intermediate-sized SEP events that will bridge the gap between the small events in modern observations and extreme events. Their occurrence rates and underlying physical mechanisms are entirely unknown. To detect intermediate-sized events, it is crucial to conduct ultrahigh precision ^{14}C analysis of tree rings and ^{36}Cl (and ^{10}Be) analysis with high temporal resolution in ice cores.

Acknowledgement

F.M.'s fork was supported by JSPS Grant-in-Aids JP20H05643, JP23K17694, JP20H00035, JP23H00135, and JP22H00026. F.M. thank the ISSI international team activity (#510; SEESUP).

References

Bard, E., Miramont, C., Capano, M., *et al.* 2023, *Phil Trans R Soc A*, 381, 20220206.
Brehm, N., Bayliss, A., Christl, M., *et al.* 2021, *Nat Geosci*, 14, 10.
Brehm, N., Christl, M., Knowles, T.D.J., *et al.* 2022, *Nat Commun*, 13, 1196.
Golubenko, K., Rozanov, E., Kovaltsov, G., *et al.* 2022, *J Geophys Res Atmos*, 127, e2022JD036726.
Hakozaki, M., Miyake, F., Nakamura, T., *et al.* 2018, *Radiocarbon*, 60, 261.
Kuitems, M., Wallace, B.L., Lindsay, C., *et al.* 2021, *Nature*, 601, 388.
Miyake, F., Nagaya, K., Masuda, K., *et al.* 2012, *Nature*, 486, 240.
Miyake, F., Masuda, K., Nakamura, T. 2013, *Nat Commun*, 4, 1748.
Miyake, F., Suzuki, A., Masuda, K., *et al.* 2015, *Geophys Res Lett*, 42, 84.
Miyake, F., Usoskin, I., Poluianov, S. (eds.) 2019, IOP Publishing, Bristol.
Miyake, F., Panyushkina, I.P., Jull, A.J.T. *et al.* 2021, *Geophys Res Lett*, 48, e2021GL093419.
Miyake, F., Hakozaki, M., Kimura, K., *et al.* 2022, *Front Astron Space Sci*, 9, 886140.
Miyake, F., Hakozaki, M., Hayakawa, H., *et al.* 2023, *J. Space Weather Space Clim*, 13, 31.
Mekhaldi, F., Muscheler, R., Adolphi, F., *et al.* 2015, *Nat Commun*, 6, 8611.
Mekhaldi, F., Adolphi, F., Herbst, K., *et al.* 2021, *J Geophys Res Space Phys*, 126(8), e2021JA029351.

Miyahara, H., Tokanai, F., Moriya, T., *et al.* 2022, *Geophys Res Lett*, 49(5), e2021GL097201.
O'Hare, P., Mekhaldi, F., Adolphi, F., *et al.* 2019, *Proc Natl Acad Sci USA*, 116(13), 5961.
Oppenheimer, C., Wacker, L., Xu, J., *et al.* 2017, *Quaternary Science Reviews*, 158, 164.
Paleari, C.I., Mekhaldi, F., Adolphi, F., *et al.* 2022, *Nat Commun*, 13, 214.
Park, J., Southon, J., Fahrni, S., *et al.* 2021, *Radiocarbon*, 59(4), 1147.
Poluianov, S.V., Kovaltsov, G.A., Mishev, A.L. *et al.* 2016, *J Geophys Res Atmos*, 121, 8125.
Price, M. 2023, *Science*, 380, 6641.
Reimer, P.J., Austin, W., Bard, E., *et al.* 2020, *Radiocarbon*, 62(4), 725.
Sigl, M., Winstrup, M., McConnell, J.R., *et al.* 2015, *Natrue*, 523, 543.
Usoskin, I., Kovaltsov, G. 2021, *Geophys Res Lett*, 48(17), e2021GL094848.
Usoskin, I., Miyake, F., Baroni, M., *et al.* 2023, *Space Science Reviews*, 219, 73.
Uusitalo, J., Golubenko, K., Arppe, L., *et al.* 2024, *Geophys Res Lett*, 51, e2023GL106632.
Wacker, L., Guttler, D., Goll, J., *et al.* 2014, *Radiocarbon*, 56, 573.

Solar and Stellar Coronal Mass Ejections
Proceedings IAU Symposium No. 388, 2026
N. Gopalswamy, O. Malandraki, A. Vidotto & W. Manchester, eds.
doi:10.1017/S1743921324001285

All talk and no action: CMEs and prominences on M dwarfs

Moira Jardine

SUPA, School of Physics and Astronomy, North Haugh, St Andrews, Fife, UK
email: mmj@st-andrews.ac.uk

Abstract. One of the most important questions in the field of exoplanets is the impact that stellar CMEs (and their energetic particles) may have on exoplanetary atmospheres. This is of particular importance if the parent star is an M dwarf, since their powerful flares suggest (based on the solar analogy) that they will have commensurably powerful CMEs. Stellar CMEs have proved difficult to detect however, leading to the suggestion that the solar relationship between flares and CMEs does not extend down to the M dwarf regime. Two lines of evidence - both modelling of coronal evolution in response to flux emergence and surface transport and observations of the surface magnetic energy and helicity densities - point to the low surface differential rotation of M dwarfs as a possible culprit.

Keywords. stars: activity, stars: coronae, stars: magnetic fields

1. Introduction

M dwarfs have been popular targets for exoplanet searches for some time. Not only are they very numerous, but their low mass and luminosity make them good candidates for detection methods searching for exoplanets in the star's habitable zone, where exoplanets can sustain bodies of liquid water on their surfaces. The atmospheres of these planets are, however, particularly at risk from the damaging effects of energetic particles accelerated by stellar coronal mass ejections or CMEs (Airapetian 2020). The reason for this is straightforward - a star with a low mass is cool and so has a habitable zone that is close to the star. Exoplanets at these orbital radii therefore have high orbital frequencies. As a result, an exoplanet in the habitable zone would be continuously impacted by CMEs if the parent star had a rate of coronal mass ejections of only six times that of the present-day Sun at maximum (Khodachenko 2007).

When this calculation was first published, this seemed to be a very achievable rate, since M dwarfs produce frequent and powerful flares (Kowalski 2024). It was expected that a correspondingly high rate of CMEs would be detected. In the intervening years, however, clear detections of stellar CMEs have proved remarkably elusive, although there is growing evidence of prominence eruption (Drake 2013; Notsu 2013; Odert 2017; Moschou 2019; Leitzinger 2020; Maehara 2021; Kanodia 2022; Wu 2022; Namizaki 2023; Inoue 2023; see also Veronig 2021). It is not clear whether this represents some characteristic of the ejection process that makes it difficult to detect with current methods, or whether it represents an underlying low rate of ejection.

2. Coronal dynamics

In order to understand the processes that drive coronal dynamics we need to consider both the rate at which magnetic flux emerges through stellar surfaces, and the various surface processes that subsequently act on that flux, transporting it in meridional flows,

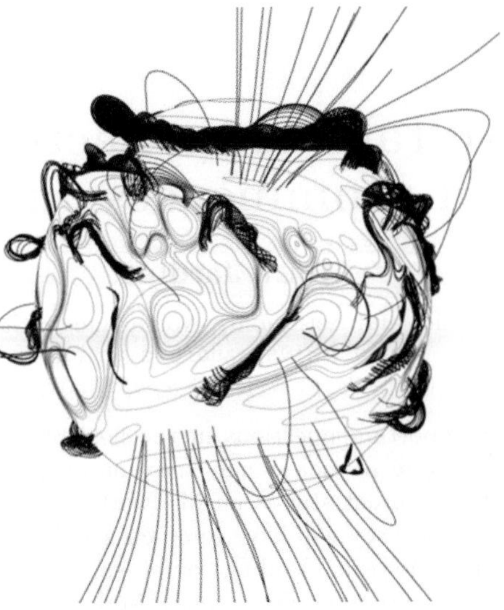

Figure 1. Twisted flux tubes formed in a stellar corona (taken from Gibb et al. 2016). Red and blue contours denote positive and negative surface magnetic field, respectively, and the black lines are field lines.

stretching and stressing it through surface differential rotation and causing it to diffuse. On the Sun, the fastest of these processes is the surface differential rotation. This has been measured for other stars (Marsden 2005; Marsden 2006; Barnes 2005; Reiners 2006; Jeffers 2011) where it is found to decrease with stellar mass, such that the lowest mass stars rotate almost as solid bodies. In order to assess the impact on coronal dynamics of this variation in shearing rates, Gibb (2014) undertook a pilot study considering the eruption and subsequent evolution of a single isolated flux tube. They found that the timescales for the formation and eruption of a twisted flux tube (the precursor of a CME) could be expressed as

$$\tau_{\rm formation} \propto \sqrt{\tau_{\rm lap}\tau_{\rm diffusion}} \tag{1}$$

and

$$\tau_{\rm erupt} \propto \tau_{\rm lap}, \tag{2}$$

where $\tau_{\rm formation}$ is the time to form a twisted flux tube, $\tau_{\rm lap}$ is the time for the equator to advance one full rotation ahead of the pole and $\tau_{\rm diffusion}$ is the diffusion timescale. Consequently, the lowest-mass stars, which have the largest values of $\tau_{\rm lap}$, have the longest timescales for the formation and eruption of these CME precursors. It seems that in the corona of an M dwarf, everything may just happen very slowly.

A subsequent full-star, long-term evolution study (see for example Fig. 1) considering a range of flux emergence and differential rotation rates, confirmed these trends (Gibb et al. 2016), but revealed a more complex interplay between the effects of flux emergence and differential rotation. The free magnetic energy (the maximum available for release) was found to depend principally on the flux emergence rate and hence ultimately on the stellar rotation rate. Changes in the differential rotation had little impact. This might suggest that M dwarfs which have high magnetic flux densities (and hence most likely high magnetic flux emergence rates) might be able to power frequent and powerful flares, even if they had little differential rotation. In contrast, the timescales for the formation and eruption of twisted flux tubes depended most sensitively on the differential rotation,

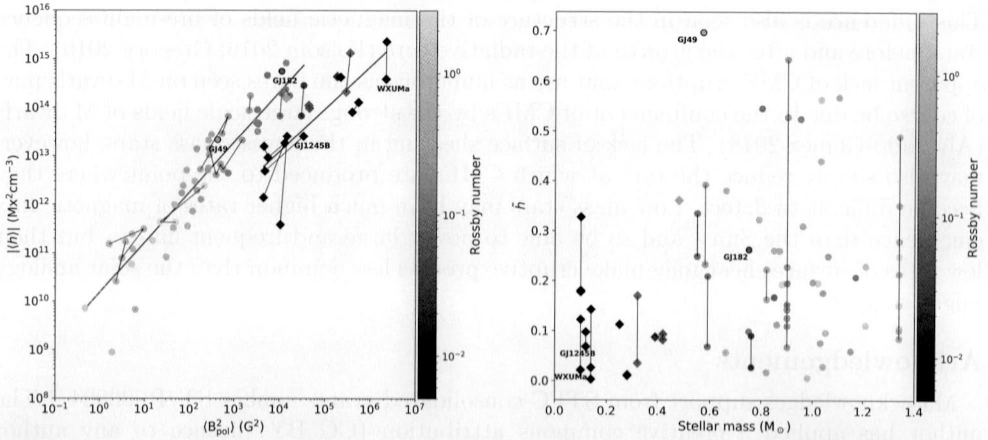

Figure 2. (Left) Absolute helicity density h for spherical harmonic degree $l \leq 4$ averaged across a single hemisphere versus the mean squared poloidal magnetic flux densities across the star. Diamonds represent stars with $M_\star < 0.5$ M$_\odot$. The dashed lines show the best fit of $|\langle h \rangle| \propto \langle B^2 \rangle^\alpha$. Best fit lines of stars with $M_\star < 0.5$M$_\odot$ and $M_\star > 0.5$M$_\odot$ are shown separately, resulting in $\alpha = 0.77 \pm 0.18$ and $\alpha = 1.11 \pm 0.07$ respectively. (Right) The efficiency of the production of helicity density h ($l \leq 4$), defined as $\tilde{h} \equiv |\langle h \rangle|/\langle R_\star B^2 \rangle$, versus stellar mass.

although high flux emergence rates could shorten these timescales. Thus the rate of CME eruption might be expected to vary more with differential rotation (and hence stellar mass) than flaring rates.

3. Mapping magnetic twist at stellar surfaces

It appears from the Gibb et al. (2016) study that in stellar coronae the two processes of flaring and eruption might show different trends with surface differential rotation. One way to test this would be to map the surface distribution of magnetic twist in stars with a range of rotation rates and masses (and hence surface differential rotation). This has already been done - although only for the large-scale field - using the spectropolarimetric technique of Zeeman-Doppler Imaging (Lund 2020). These authors mapped the large-scale surface "mean twist" or magnetic helicity density across the surfaces of over 100 stars. They found that the lowest mass stars, which lack a radiative core and therefore cannot power an interface dynamo, showed a relationship between the poloidal magnetic energy and mean twist that was different to that of stars that had a solar-like internal structure, with a convective outer zone and a radiative core (Fig. 2). At the same value of the poloidal magnetic energy, the lowest mass stars had significantly lower mean twist, since much of their (strong) poloidal fields did not link through the surface toroidal fields (Lund 2021). This can be seen in the ratio of the helicity density to the energy density, which is a measure of the efficiency with which stars can produce twist (Fig. 2). While stars of mass $M_\star > 0.5M_\odot$ have efficiencies approaching 1, the lower mass stars, with masses $M_\star < 0.5M_\odot$, have much lower values. This suggests that while they can generate very strong fields, the structure of these fields ensures that they do not have correspondingly large helicity densities.

4. Discussion

Both the observational and theoretical studies described above point to the fundamental difference between the magnetic fields produced by fully convective stars and those of stars like the Sun that have a tachocline and so can support an interface dynamo.

This difference is also seen in the structure of the magnetic fields of pre-main sequence stars, before and after the growth of the radiative core (Folsom 2016; Gregory 2016). The apparent lack of CME eruptions that are as numerous as the flares seen on M dwarfs may of course be due to the confinement of CMEs by the strong, large-scale fields of M dwarfs (Alvarado-Gómez 2018). The lack of surface shearing in the lowest-mass stars, however, may also simply reduce the rate at which CMEs are produced to the point where they become difficult to detect. Low mass stars may have much higher rates of magnetic flux emergence than the Sun - and so be able to power large and frequent flares - but their low rates of surface shear may make eruptive process less common than the solar analogy might suggest.

Acknowledgements

MJ acknowledges support from STFC consolidated grant number ST/R000824/1.The author has applied a creative commons attribution (CC BY) licence to any author accepted manuscript version arising.

References

Airapetian, V. S., Barnes, R., Cohen, O., et al. 2020, *International Journal of Astrobiology*, 19, 136
Alvarado-Gómez, J. D., Drake, J. J., Cohen, O., et al. 2018, *ApJ*, 862, 93
Barnes, J. R., Collier Cameron, A., Donati, J.-F., et al. 2005, *MNRAS*, 357, L1
Drake, J. J., Cohen, O., Yashiro, S., et al. 2013, *ApJ*, 764, 170
Folsom, C. P., Petit, P., Bouvier, J., et al. 2016, *MNRAS*, 457, 580
Gibb, G. P. S., Mackay, D. H., Jardine, M. M., et al. 2016, *MNRAS*, *MNRAS*, 456, 3624
Gibb, G. P. S., Jardine, M. M., & Mackay, D. H. 2014, *MNRAS*, *MNRAS*, 443, 3251
Gregory, S. G., Adams, F. C., & Davies, C. L. 2016, *MNRAS*, 457, 3836
Inoue, S., Maehara, H., Notsu, Y., et al. 2023, *ApJ*, 948, 9
Jeffers, S. V., Donati, J.-F., Alecian, E., et al. 2011, *MNRAS*, 411, 1301
Kanodia, S., Ramsey, L. W., Maney, M., et 2022, *ApJ*, 925, 155
Khodachenko, M. L., Ribas, I., Lammer, H., et al. 2007, *Astrobiology*, 7, 167
Kowalski, A. F.,2024, *Living Reviews in Solar Physics*, 21,
Leitzinger, M., Odert, P., Greimel, R., et al. 2020 *MNRAS*, 493, 4570
Lund, K., Jardine, M., Russell, A. J. B., et al. 2021, *MNRAS*, 502, 4903
Lund, K., Jardine, M., Lehmann, L. T., et al. 2020, *MNRAS*, 493, 1003
Maehara, H., Notsu, Y., Namekata, K., et al. 2021, *PASJ*, 73, 44.
Marsden, S. C., Waite, I. A., Carter, B. D., et al. 2005, *MNRAS*, 359, 711
Marsden, S. C., Donati, J.-F., Semel, M., et al. 2006, *MNRAS*, 370, 46
Moschou, S.-P., Drake, J. J., Cohen, O., et al. 2019, Drake, J. J., Cohen, O., Yashiro, S., et al. 2013, *ApJ*, 877, 105.
Namizaki, K., Namekata, K., Maehara, H., et al. 2023, *ApJ*, 945, 61
Notsu, Y., Shibayama, T., Maehara, H., et al. 2013, *ApJ*, 771, 127
Odert, P., Leitzinger, M., Hanslmeier, A., et al. 2017, *MNRAS*, 472, 876.
Reiners, A. 2006, *A&A*, 446, 267
Veronig, A. M., Odert, P., Leitzinger, M., et al. 2021, *Nature Astronomy*, 5, 697
Wu, Y., Chen, H., Tian, H., et al. 2022 *ApJ*, 928, 180

Solar and Stellar Coronal Mass Ejections
Proceedings IAU Symposium No. 388, 2026
N. Gopalswamy, O. Malandraki, A. Vidotto & W. Manchester, eds.
doi:10.1017/S1743921324001728

Observing Solar Coronal Mass Ejections from Space

Philippe L. Lamy

Laboratoire Atmosphères, Milieux et Observations Spatiales, CNRS & UVSQ, Guyancourt, France

Abstract. In this contribution to the panel discussion of the IAU Symposium N° 388 "Solar and Stellar Coronal Mass Ejections", I concentrate on white-light observations of solar coronal mass ejections (CMEs) from space and specifically address the following aspects: i) history of observations, ii) available catalogs of CMEs, iii) achievements of space observations of CMEs, iv) future of CME observation, and v) challenges and future directions.

Keywords. coronal mass ejections, solar space missions

1. Introduction

Following the discovery of CMEs, there has been a rapidly growing interest in their observation and their analysis motivated by their role in solar, coronal, interplanetary, and planetary physics as well as in space weather with in particular, far reaching consequences for human activities on Earth. Successive space solar missions have been implemented with among their objectives, the continuous observation of CME activity. As a consequence, the number of investigations of CMEs over the past four decades has exploded: several thousand refereed publications and at least 80 review articles, and we now see a rapidly growing field of star CMEs. It is obviously beyond the scope of this concise presentation of the space observations of CMEs to quote this extensive literature, but the interested reader is directed to recent reviews (e.g.,(Lamy et al. 2019)). This presentation is limited to space-based coronagraphic observations of CMEs and organized in five sections which briefly address the following aspects: i) history of observations, ii) available catalogs of CMEs, iii) achievements of space observations of CMEs, iv) future of CME observation, and v) challenges and future directions.

2. Brief History of CME Observations

After the first detection of a CME by the coronagraph onboard the seventh *Orbiting Solar Observatory* (OSO-7) in late 1971, the history of space observations of CMEs may be conveniently divided in three periods. The early years from 1971 to 1995 saw the first generation of space-borne coronagraphs: Skylab-ATM, P78-1 Solwind, and the Coronagraph/Polarimeter (C/P) on the *Solar Maximum Mission* (SMM). In January 1996, the Large Angle Spectrometric Coronagraph (LASCO) onboard the *Solar and Heliosphere Observatory* (SoHO) opened a golden age for CMEs, later complemented by the COR2 A and B coronagraphs onboard the STEREO twin spacecraft which offered two viewpoints until September 2014 when STEREO-B was lost and an enlarged field of view thanks to the inclusion of Heliospheric Imagers (HIs). The LASCO-C2 and -C3 remain the workhorses for CMEs investigations and for space weather forecasts with a record of quasi uninterrupted observations over 2.5 solar cycles (and hopefully three) in the range 2.5 – 30 $R_\odot$. The third period is characterized by the emergence of new players. The Wide Field Imager for Solar Probe (WISPR) onboard the *Parker Solar Probe* (PSP) offers for the first time views from the interior of a few CMEs when close to the Sun and is complemented by a similar instrument SoloHI onboard *Solar Orbiter*. The METIS ultraviolet+visible coronagraph onboard *Solar Orbiter* offers however limited observations due to restricted observational windows, thus precluding continuous monitoring. Consequently, only a handful of CMEs have been detected and analyzed so far.

3. Catalogs of CMEs

The very large number of CMES seen by LASCO soon imposed the necessity of their systematic detection and of a compilation of their properties. Following a first attempt performed by (St. Cyr et al. 2000) who reported the properties of 841 CMEs from January 1996 to June 1998, the Coordinated Data Analysis Workshop catalog (CDAW) was initiated based on the visual detection of CMEs by (Yashiro et al. 2004), see also (Gopalswamy et al. 2024) (this volume) for a second version. Biases inherent to visual detection were discussed since the 1980s, fostering the motivation for automated catalogs (Robbrecht et al. 2004). The resulting new generation comprises four catalogs that implement different detection techniques: CACTus (Robbrecht et al. 2004), SEEDS (Olmedo et al. 2005), ARTEMIS (Boursier et al. 2009) (Floyd et al. 2013), and CORIMP (Byrne et al. 2009) however limited to the [2000 – 2015] time interval. They all report the following basic properties: date of CMEs first appearance in C2, central position angle, angular width, speed (one or several). CDAW, SEEDS, and CORIMP include acceleration. These three catalogs, together with ARTEMIS, include mass and kinetic energy. Whereas automated methodologies are more objective since the detections are based on the use of thresholds fixed once for all (usually by test and trial), the listed properties are often inconsistent with each other and furthermore with the CDAW manual catalog. As an example, SEEDS reports about twice as many CMEs as CDAW, the other catalogs (except CORIMP) falling in-between. This is not a surprise in view of the intrinsic difficulty of identifying and characterizing CMEs which display a large diversity of apparent shapes. This situation clearly requires a complementary use of all catalogs when studying particular CMEs in order to ascertain their results. However, in their extensive comparison of the different catalogs, (Lamy et al. 2019) pointed out that there are also many points of convergence, particularly between SEEDS (once corrected for the image cadence) and ARTEMIS.

The CACTus and SEEDS catalogs have been adapted to incorporate the STEREO/COR2 data starting in 2017. In addition, the new MVC catalog was built from seven years of multi-viewpoint observations listing 3358 events simultaneously observed by the twin COR2 coronagraphs (Vourlidas et al. 2017). Finally, catalogs based on WISPR and SoloHI observations are in preparation.

4. Achievements of Space Observations of CMEs

Over the past decades, space-based observations of CMEs have led to numerous significant scientific achievements, revolutionizing our understanding of these complex and dynamic events. A synthetic list of the main scientific achievements is presented below.

- CMEs and solar cycle variability: Space-based observations have revealed that CME rates and properties vary significantly over the solar cycle, with implications for our understanding of solar cycle variability and its impact on the heliosphere. The CME occurrence rate closely tracks solar activity (Fig. 1 and 2) and particularly the solar radio flux at 10.7 cm, probably because it is a coronal index (although it is partly chromospheric) rather than a photospheric index.

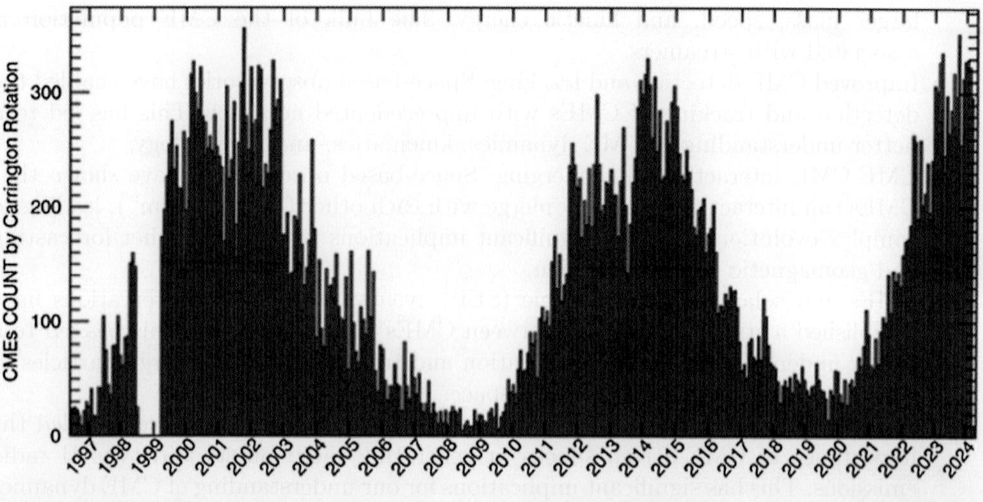

Figure 1. Monthly detection rate of CMEs reported in the ARTEMIS catalog (in blue) further corrected for the duty-cycle of LASCO (in red). Note the data gap in 1998 when SoHO went out of control.

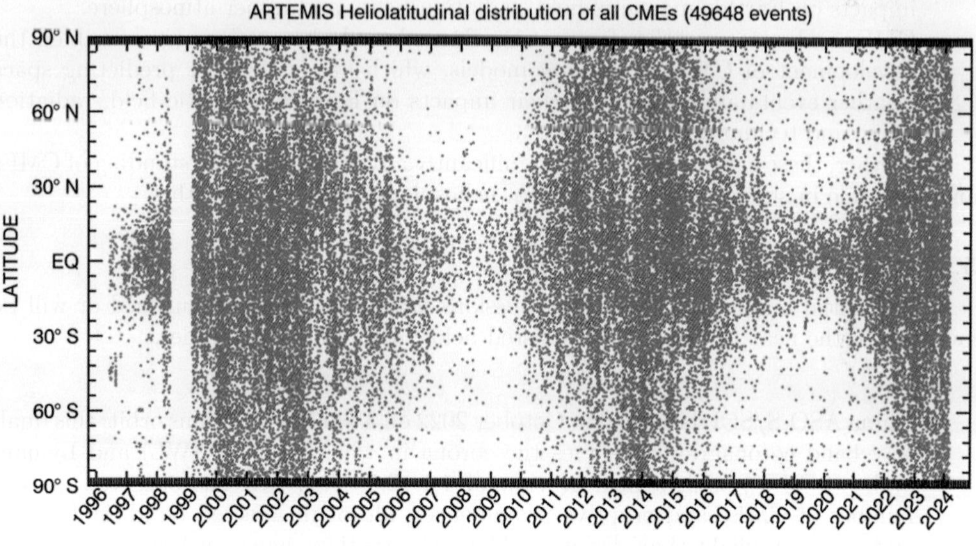

Figure 2. Heliolatitudinal apparent distribution of CMEs reported in the ARTEMIS catalog. Note the data gap in 1998 when SoHO went out of control.

- CMEs energy and helicity: Space-based observations have fostered our understanding of how magnetic energy is built up, stored, and released in magnetic flux systems and of the role of CMEs in removing large amounts of magnetic flux and helicity from the Sun, thus making room for the new solar cycle.
- CME initiation and acceleration mechanisms: Space-based observations have provided insights into the initiation and acceleration mechanisms of CMEs. For example, they have revealed the importance of magnetic reconnection and flux rope formation in CME initiation.
- CME associations: Space-based observations have revealed that only a small fraction (a few percent) of the overall population of CMEs is associated with C+M+X class flares and erupting prominences. These associated CMEs have significantly

larger mass, speed, and kinetic energy. The bulk of the CME population is associated with streamers.

- Improved CME detection and tracking: Space-based observatories have enabled the detection and tracking of CMEs with unprecedented accuracy. This has led to a better understanding of CME dynamics, kinematics, and morphology.
- CME-CME interactions and merging: Space-based observations have shown that CMEs can interact and possibly merge with each other ("cannibalism"), leading to complex evolution. This has significant implications for space weather forecasting and geomagnetic storm prediction.
- CMEs and solar energetic particle (SEP) events: Space-based observations have established a strong connection between CMEs and SEP events. This has led to a better understanding of the acceleration and transport of high-energy particles in the solar corona and interplanetary space.
- CME-driven shocks and radio bursts: Space-based observations have revealed the importance of CME-driven shocks in generating radio bursts and type II radio emissions. This has significant implications for our understanding of CME dynamics and their impact on the solar wind and magnetosphere.
- CMEs and geomagnetic storms: Space-based observations have demonstrated the critical role of CMEs in driving geomagnetic storms, which can have significant impacts on Earth's magnetic field, radiation belts, and upper atmosphere.
- CMEs and space weather forecasting: Space-based observations have enabled the development of CME forecasting models, which are critical for predicting space weather events and mitigating their impacts on Earth's magnetic field, radiation belts, and technological systems.

Altogether, these achievements have significantly advanced our understanding of CMEs and their role in shaping the solar corona, heliosphere, and space weather.

5. Future Observations of CMEs

A substantial number of solar space missions have been recently launched or will be launched in the near future. They are listed below in chronological order.

- China ASO-S/SCI: launched in October 2022 on a Sun-synchronous orbit, this dual-waveband coronagraph observes the corona in white-light (SCIWL) and Lyman-alpha (SCIUV) with a field of view of $1.1-2.5\ R_\odot$. SCI has observed quite a number of CMEs and eruptive prominences, although affected by ghost features and more straylight than designed. These observations have not been published so far.
- India Aditya-L1/VELC: launched in September 2023, it reached the L1 Lagrangian point on 6 January 2024. This emission line coronagraph has a field of view of $1.05-3\ R_\odot$. No observations have been reported so far. On the instrument website https://velc.iiap.res.in/quicklook, the tab "Quicklook Data" displays "TBD".
- NOAO GOES-U/CCOR-1: scheduled to launch in June 2024 on a geosynchronous orbit, this compact basic coronagraph without polarization capability has a field of view of $3.7-17\ R_\odot$. As a consequence of is orbit, the Earth will approach and or even transit its field of view every day.
- ESA Proba-3/ASPIICS : scheduled to launch in November 2024 on a highly elliptical Earth orbit, this will be the first externally occulted coronagraph comprising two distinct satellites (occulter and instrument) separated by 150 m in formation flight. It has a field of view of $1.1-3\ R_\odot$ and polarization capability and will operate typically six hours per day.

- NOAO SVFO/CCOR-2: scheduled to launch in mid 2025 at L1, this compact basic coronagraph is similar to CCOR-1 except for a field of view of $3.0-22\ R_\odot$.
- NASA PUNCH: scheduled to launch in April 2025 on a Sun-synchronous, low Earth orbit, it comprises four satellites, one with a coronagraph having a field of view of $6.0-32\ R_\odot$ and three with heliospheric imagers having a field of view of $18-180\ R_\odot$. These instruments have polarization capability.
- ESA Vigil/CCOR-3 (ex Lagrange): scheduled to launch in 2031 at L5, this satellite will embark a compact basic coronagraph similar to CCOR-2, that is a field of view of $3.0-22\ R_\odot$ and no polarization capability.

6. Challenges, Future Directions, and Conclusion

The difficulties of designing coronagraphs should not be underestimated and the absence of significant results from ASO-S/SCI and Aditya-L1/VELC indicates that mastering the question of straylight always remains a serious challenge. Whereas there is a substantial number of solar space missions that will target CMEs as seen in the above section, many of them implement simple coronagraphs at a single viewpoint to satisfy the basic need of space weather forecasting. To really make progress in the understanding of CMEs and addressing outstanding questions such as their initiation, their internal structure and their energetics, a comprehensive mission with global science objectives, multiple viewpoints, and continuous imaging over large fields of view is required. Several missions that combine observations from several of the Lagrangian points L1, L4, and L5 have been proposed, but only the combination of SVFO/CCOR-2 (at L1) and Vigil/CCOR-3 (at L5) has been approved so far.

The newly proposed MOST mission ("Multiview Observatory for Solar Terrestrial Science", Gopalswamy et al. 2024) is intended to move forward and to make substantial progress with respect to past missions. Briefly, it consists of four spacecraft at L4, L5, ahead of L4, and behind L5. The first two will have identical remote-sensing and in-situ instrument suites including coronagraphs and heliospheric imagers. Altogether, the four spacecraft will perform novel Faraday rotation measurements to measure the magnetic content of CMEs. Assuming an additional solar observatory at L1, accurate three-dimensional information will be obtained.

This is of utmost importance as detecting, tracking, and characterizing CMEs remain challenging as illustrated by the large uncertainties in the measured parameters, especially in the case of a single viewpoint, as pointed out in Section 2. Deep learning looks promising to make progress and several efforts are on-going. For instance, it has been implemented by Shan et al. (2024) to produce an automated "2D CME" catalog from one year of STEREO/COR1-A observations. Further applying the polarization technique, they upgrade it to a "3D CME" catalog that includes 3D propagation direction, speed, and 3D shape reconstruction. This clearly demonstrates the importance of polarization measurements for the 3D characterization of CMEs.

Outside the scope of this presentation, but worth mentioning, MHD modeling is fundamental to analyze and interpret the observations of CMEs. Several models are presently available, but improvements are needed to foster our understanding of their complexity.

References

Boursier, Y., et al. 2009, Solar Phys., 257, 125.
Byrne, J. P., et al. 2009, Astron. Astrophys., 495, 325.
Floyd, O., et al. 2013, Solar Phys., 288, 269.
Gopalswamy, N., et al. 2024, JASTP, 254, 106165.
Lamy, P., et al. 2019, Space Sci. Rev., 215, 39.
Olmedo, O., et al. 2005, Solar Phys., 248, 485.

Robbrecht, E., et al. 2004, Astron. Astrophys., 425, 1097.
Shan, J., et al. 2024, ApJ Sup., 272, 18.
St. Cyr, O. C., et al. 2000, J. Geophys. Res., 105, 18169.
Vourlidas, A., et al. 2017, ApJ, 838, 141.
Yashiro, S., et al. 2004, J. Geophys. Res., 109, A07105.

Author index